State of the Art and Progress in Production of Biohydrogen

Edited by

Nuri Azbar

Ege University
Turkey

David B. Levin

University of Manitoba
Canada

eBooks End User License Agreement

CONTENTS

FOREWORD

Fossil fuels (i.e., petroleum, natural gas and coal), which meet most of the world's energy demand today, are being depleted fast. Also, their utilization is causing global problems, such as the global warming, climate change, ozone layer depletion, acid rains, oxygen depletion and pollution, which are posing great danger for our environment and eventually for the life in the planet Earth. Many engineers and scientists agree that the solution to these global problems would be to replace the existing fossil fuel system by the Hydrogen Energy System. Hydrogen is the most efficient and the cleanest fuel. Its combustion will produce no greenhouse gases, no ozone layer depleting chemicals, little or no acid rain ingredients, no oxygen depletion and no pollution.

Of course, hydrogen is a synthetic fuel and it must be manufactured. There are various hydrogen manufacturing methods such as direct thermal, thermochemical, electrochemical, biological, etc. Among the hydrogen production methods, biological method has the potential of resulting in the most cost-effective hydrogen. Because of this, many research groups around the world are working on biological hydrogen production. In several cases, bench scale production systems have come up with encouraging results.

This eBook entitled "State of the Art and Progress in Production of Biohydrogen" covers the biological hydrogen production method authoritatively from A to Z, including mechanisms of hydrogen production, hydrogenase genes and enzymes, photosynthetic methods, photo fermentation, thermophillic fermentation, dark fermentation, metabolic engineering, electrohydrogenesis and immobilized cell systems.

I strongly recommend this excellent eBook to energy scientists, engineers and students who are interested in hydrogen production in general and biological hydrogen production in particular, as well as to industrial concerns who are looking for inexpensive hydrogen production technologies.

T. Nejat Veziroğlu

President, International Association for Hydrogen Energy

PREFACE

Three of the great challenges facing humanity in the 21st century are energy supply, climate change, and global food security. Although global energy demand is expected to continue to increase, the availability of low cost energy will continue to diminish. Coupled with increasing concerns about climate change due to the CO_2 release from the combustion of fossil fuels, there is now an urgent need to develop clean, renewable energy system. Hydrogen is a clean, zero carbon emission, and renewable energy carrier, with a high specific heat of combustion. Hydrogen can be used in internal combustion engines to generate mechanical power or in fuel cells to generate electricity. As hydrogen can be produced from many natural sources, it is expected to have a stable price in the future, independent of the fluctuation in price and availability of single sources. Hydrogen also allows flexibility in balancing centralized and decentralized power supply.

The use of biofuels for transport is becoming of increasing importance due to the environmental concerns relating to climate change, depleting fossil fuel reserves, and reducing reliance on imports. This is leading to international, national and regional focus on alternative energy sources. In the EU, transport is responsible for an estimated 21% of all greenhouse gas (GHG) emissions. A range of actions is being taken to reduce emissions from transport such as promoting the use of biofuels. Among alternative biofuels, hydrogen seems to be more advantageous due to the fact that it has a higher specific heat of combustion and does not contribute to the Greenhouse effect.

Among other alternative production methods such as pyrolysis, gasification, steam gasification, steam reforming, and electrolysis which are highly energy intensive processes, biological methods are promising both in terms of ecological and economical reasons. On the other hand, if biohydrogen systems are to become commercially competitive they must be able to synthesize H_2 at rates that are sufficient to power fuel cells of sufficient size to do practical work, and further research and development aimed at increasing rates of synthesis and final yields of H_2 are essential.

This eBook aims at contributing to this target. We hope that the chapters contained within this eBook will be useful and inspiring for many researchers and people who have interest in hydrogen.

Nuri Azbar
Ege University
Turkey

David B. Levin
University of Manitoba
Canada

List of Contributors

David B. Levin
Department of Biosystems Engineering, University of Manitoba, Winnipeg, Canada

Nuri Azbar
Ege University, Department of Bioengineering, Izmir, Turkey

Carrie Eckert
Biosciences Center, National Renewable Energy Laboratory, Golden, USA

Alexandra Dubini
Biosciences Center, National Renewable Energy Laboratory, Golden, USA

Jianping Yu
Biosciences Center, National Renewable Energy Laboratory, Golden, USA

Paul King
Biosciences Center, National Renewable Energy Laboratory, Golden, USA

Maria Ghirardi
Biosciences Center, National Renewable Energy Laboratory, Golden, USA

Michael Seibert
Biosciences Center, National Renewable Energy Laboratory, Golden, USA

Pin-C. Maness
Biosciences Center, National Renewable Energy Laboratory, Golden, USA

Taras K. Antal
Department of Biophysics, Faculty of Biology, Moscow State University, Moscow, Russia

Tatyana E. Krendeleva
Department of Biophysics, Faculty of Biology, Moscow State University, Moscow, Russia

Valdimir Z. Pashchenko
Department of Biophysics, Faculty of Biology, Moscow State University, Moscow, Russia

Andrew B. Rubin
Department of Biophysics, Faculty of Biology, Moscow State University, Moscow, Russia

Karin Stensjo
Department of Photochemistry and Molecular Science, The Ångström Laboratories, Uppsala University, Uppsala, Sweden

Esa Tyystjärvi
Department of Biochemistry and Food Chemistry, Plant Physiology and Molecular Biology, University of Turku, Turku, Finland

Seeram Ramakrishna
Nanoscience and Nanotechnology Initiative, National University of Singapore, 2, Singapore

Dmitry A. Los
Institute of Plant Physiology, Russian Academy of Sciences, Moscow, Russia

Robert Carpentier
Groupe de recherche en biologie végétale, Université du Québec à Trois-Rivières, Trois-Rivières, Canada: Robert

Hiroshi Nishihara
Department of Chemistry, School of Science, The University of Tokyo, Tokyo, Japan

Suleyman I. Allakhverdiev
Institute of Basic Biological Problems, Russian Academy of Sciences, Moscow, Russia

Basar Uyar
Department of Chemical Engineering, Kocaeli University, Kocaeli, Turkey

Gökhan Kars
Department of Biology, Science Faculty, Selçuk University, Konya, Turkey

Meral Yücel
Department of Biology, Middle East Technical University, Ankara, Turkey

Ufuk Gündüz
Department of Biology, Middle East Technical University, Ankara, Turkey

Inci Eroglu
Department of Chemical Engineering, Middle East Technical University, Ankara, Turkey

Anatrly A. Tsygankov
Institute of Basic Biological Problems RAS, Puschino, Russia

Daria N. Tekucheva
Institute of Basic Biological Problems RAS, Puschino, Russia

Patrick C. Hallenbeck
Département de microbiologie et immunologie, Université de Montréal, Montréal, Canada

Vikineswary Sabaratnam
Institute of Biological Science, Faculty of Science, University of Malaya, Kuala Lumpur, Malaysia

Mohammad A. Hassan
Faculty of Biotechnology and Biomolecular Sciences, Universiti Putra Malaysia, Serdang, Malaysia

Kaustubha Mohanty
Indian Institute of Technology, Guwahati, India

Debabrata Das
Indian Institute of Technology, Kharagpur, India

Ed W. J. van Niel
Lund University, Lund, Sweden

Karin Willquist
Lund University, Lund, Sweden

Ahmad A. Zeidan
Lund University, Lund, Sweden

Truus de Vrije
Food and Biobased Research, Wageningen, Netherlands

Astrid E. Mars
Food and Biobased Research, Wageningen, Netherlands

Pieternel A. M. Claassen
Food and Biobased Research, Wageningen, Netherlands

Richard Sparling
Department of Microbiology, University of Manitoba, Winnipeg, Canada

Carlo Carere
Department of Biosystems Engineering, University of Manitoba, Winnipeg, Canada

Thomas Rydzak
Department of Microbiology, University of Manitoba, Winnipeg, Canada

John Schellenberg
Department of Microbiology, University of Manitoba, Winnipeg, Canada

Nathan Wrana
Department of Biosystems Engineering, University of Manitoba, Winnipeg, Canada

Ilgi K. Kapdan
Department of Environmental Engineering, Dokuz Eylul University, Turkey

CHAPTER 1

Introduction: Biohydrogen in Perspective

David B. Levin[1*] and Nuri Azbar[2]

[1]University of Manitoba, Department of Biosystems Engineering, Canada and [2]Ege University, Department of Bioengineering, Turkey

Abstract: Hydrogen (H_2) is a versatile, clean burning, and renewable energy currency that can potentially displace the use of petroleum-based fuels in the transportation sector, which accounts for 74% of the total projected increase in liquid fuel consumption over the next 30 years. Demand for hydrogen is also expected to increase as it starts penetrating the transportation sector as a fuel: about 40 million tonnes of hydrogen per year would be required to fuel about 100 million fuel cell-powered cars after full market penetration. Hydrogen fuel can be produced from a diverse array of potential feedstocks including fossil fuels, water, and organic matter using various chemical and electrochemical methods. Biological hydrogen (biohydrogen) production, which employs the use of hydrogen producing microorganisms *via* light dependant or fermentative processes. Research on biohydrogen has increased dramatically in the past 6 years, with great emphasis on dark fermentation, but there are many scientific and engineering challenges that must be met by current and future biohydrogen researchers if these technologies are to be technically feasible and economically viable.

Keywords: Biohydrogen production, numbers of publications, future perspective.

INTRODUCTION

Hydrogen (H_2) is a versatile, clean burning, and renewable energy currency that can potentially displace the use of petroleum-based fuels in the transportation sector, which accounts for 80% of the total projected increase in liquid fuel consumption over the next 30 years [1]. It produces only heat and water upon combustion, releasing 286 kJ/mol of energy, and does not generate greenhouse gas emissions. It may be converted into electricity *via* proton exchange membrane fuel cells (PEMFC), a technology that is well understood. During this process, a platinum catalyst causes hydrogen to dissociate at the anode to electrons, used to generate a current, and protons, which permeate through a polymer electrolyte membrane and combine with oxygen to form water at the cathode.

About 38 Mt (5,000 petajoules) of hydrogen is produced worldwide annually, a market valued at about $60 billion [2]. Most (80 - 92%) is used in processing oil in refineries and producing chemicals such as ammonia and methanol. The rest is used in industrial processes, chemical manufacturing, and food preparation. The demand for hydrogen is expected to grow exponentially in the near future because hydrogen will be required to meet the demand for refining increasingly heavier, higher sulphur crude oils and oil sands and to meet more stringent regulations on the levels of sulphur in gasoline and diesel fuel. Demand for hydrogen is also expected to increase as it starts penetrating the transportation sector as a fuel: about 40 million tonnes of hydrogen per year would be required to fuel about 100 million fuel cell-powered cars after full market penetration. Canada is the largest per capita producer and user of hydrogen in the OECD: 3.4 million tonnes per year (yr) of hydrogen produced and 2.97 million tonnes per year consumed [2].

Despite its appeal in terms of energy content, environmental benefits, and utilization technology, hydrogen currently constitutes only about 3% of the world's total energy consumption [3]. Although it has the highest energy density of any known fuel by mass, three times that of gasoline (143 MJ/kg versus 46.9 MJ/kg), it has a very low volumetric energy density at ambient conditions due to its volatility. Consequently, gaseous hydrogen must be compressed into a liquid *via* energy intensive processes which leads to an economic and

***Address correspondence to David B. Levin:** University of Manitoba, Department of Biosystems Engineering, Canada; E-mail: levindb@cc.umanitoba.ca

Nuri Azbar and David B. Levin (Eds)

energetic drawback in regards to packaging, storage, and distribution. More importantly, current methods of hydrogen production are expensive, energetically intensive, and potentially environmentally harmful, and thus production of hydrogen may be one of the major technological barriers to overcome.

BIOHYDROGEN PRODUCTION

Hydrogen fuel can be produced from a diverse array of potential feedstocks including fossil fuels, water, and organic matter using various chemical and electrochemical methods [3-5]. Currently, over 80% of hydrogen production occurs *via* steam reformation of natural gas during which methane, the primary constituent of natural gas, is combined with high temperature steam (700-1000°C) in the presence of a catalyst, breaking it apart into H_2 and CO. The CO produced then reacts with water at high temperatures to further produce H_2 and CO_2 *via* a process known as the gas shift reaction. Although this process is well understood and less expensive than other methods, it is dependant on a potentially limited supply of natural gas, produces moderate levels of CO_2, and is very energy intensive. Alternatively, hydrogen may be produced *via* coal gasification, during which coal is converted into H_2 and CO_2 *via* a similar mechanism as natural gas reformation. Regardless of the abundance of coal, however, this process is more expensive and results in much higher CO_2 emissions than natural gas reformation since the production ratio of H_2 to CO_2 is roughly 1:1, when compared to the 4:1 ratio produced *via* natural gas reformation.

A number of alternative processes can be used for the production of hydrogen that do not result in a net increase in greenhouse gas emissions. The electrolysis of water, during which an electric current is passed through water, splitting it up into H_2 and O_2, is the second most commonly used method of hydrogen production. Although this process can be potentially emission free providing that clean, renewable sources such as wind, solar, hydro, or geothermal energy are used for the generation of electricity, it is the most energy intensive method of hydrogen production. Biomass gasification is an alternative method of hydrogen production that is potentially carbon neutral. In this process, biomass is broken down into H_2, CH_4, and CO, which can in turn be used for steam reformation and the gas shift reaction. Although these subsequent processes result in net CO_2 production, the overall process of biomass gasification is carbon neutral since the CO_2 produced is fixed *via* photosynthesis during the generation of new biomass. Furthermore, this process can utilize municipal and agricultural wastes, turning them into valuable commodities.

More recent emphasis has been placed on the development of biological hydrogen (biohydrogen) production, which employs the use of hydrogen producing microorganisms *via* light dependant or fermentative processes. A simple search of the Scopus database revealed that biohydrogen research has grown significantly over the past 15 years (Table **1**; Fig. **1**). Although this is not an exhaustive study the frequency of biohydrogen publications, it does illustrate that few biohydrogen papers were published in international peer reviewed journals between 1996 and 2003. The cumulative total of peer reviewed biohydrogen papers detected by Scopus between 1996 and 2003 was 29, while the cumulative total of peer reviewed biohydrogen papers detected by Scope between 2004 and 2010 was 574 (694 if we use the total of 180 papers projected for 2010). This is a 20 to 25-fold increase in biohydrogen publications over the past six years. The majority (> 48%) of biohydrogen publications are focused on dark fermentation (Fig. **2**).

Table 1: Numbers of Biohydrogen Papers Published in International Peer Reviewed Journals between 1996 and 2010.

Year	**Articles Found**
1996	1
2000	3
2001	4
2002	11
2003	10
2004	37
2005	40

2006	66
2007	72
2008	140
2009	159
2010 (Projected)	60 (180)
Total (excluding 2010 projection)	603
Total (including 2010 projection)	723

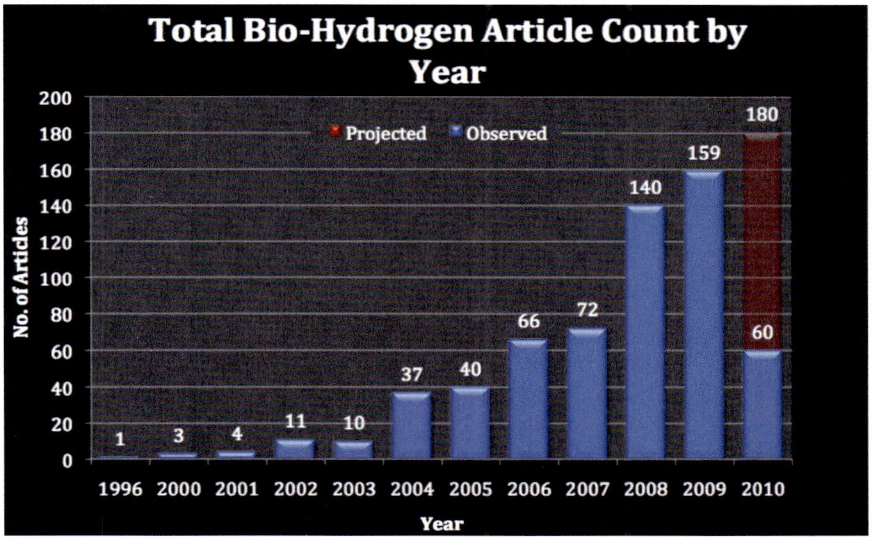

Figure 1: Numbers of Biohydrogen Papers Published between 1996 and 2010. The trend of increasing attention to biohydrogen production is shown by the increasing numbers of articles published in international peer reviewed journals. The 59 observed articles in 2010 were published in the first four months (January 1st to April 30th) of the year.

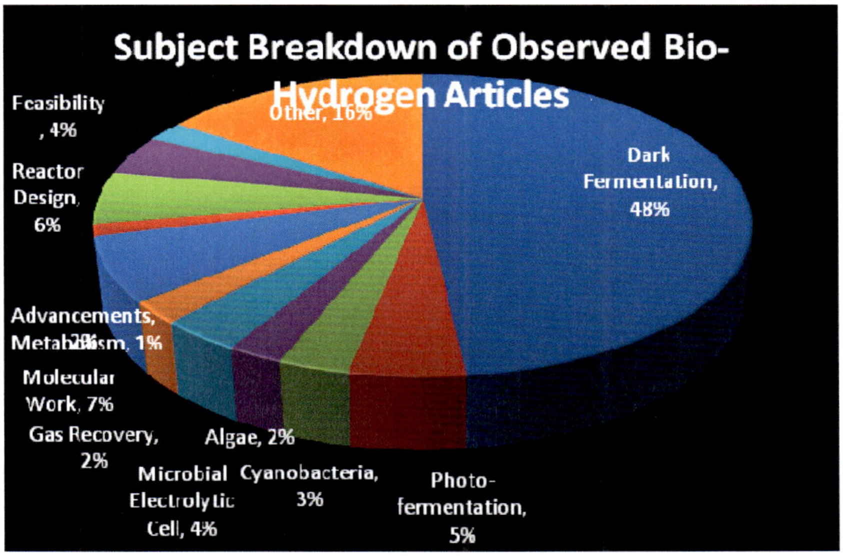

Figure 2: Subject Breakdown of Biohydrogen Articles. Frequency of biohydrogen research topics published between 1996-2010.

Biological hydrogen production is a potentially carbon neutral process that is carried out at lower temperatures and pressures, and is therefore less energy-intensive than thermochemical and electrochemical processes [6]. Furthermore, unlike chemical methods, which involve the conversion of nonrenewable fossil fuels into hydrogen, fermentative hydrogen production can utilize renewable carbohydrate-based substrates such as waste biomass from municipal, agricultural, and forestry sectors, while light driven processes utilize light energy, water, and/or CO_2 for the production of hydrogen [7].

A number of different light-dependent and fermentative biological processes may be employed in the production of hydrogen including direct biophotolysis (Green algae), indirect biophotolysis (Cyanobacteria), photo-fermentation (photosynthetic bacteria), and dark fermentation (fermentative bacteria) [4,7]. Each process has its advantages and disadvantages, but there is a general consensus that light-dependent processes for biological hydrogen production have limited potential for economically viable hydrogen production [5].

Light-Dependent Biohydrogen Production

There are severe restrictions on potential light-driven systems for biological H_2 production due to the diffuse nature of solar energy. Given that solar irradiation is as high as 6.6 GJ/m^2/year, Hallenbeck & Benemann [7] estimate that at 10% solar conversion efficiency, and a price of H_2 of \$15/GJ, light dependant processes translate to only \$10 worth of H_2/m^2/year. Note that this does not account for various operational costs and energy inputs, which include mixing, gas sparging, cooling, and gas separation. Furthermore, although solar conversion efficiencies are roughly 10% under low-light conditions, they fall to below 1% under high-light conditions since 90% of photons captured by photosynthetic pigments are decayed as heat or fluorescence due to the rate limiting step of electron transfer between photosystem II and photosystem I during dark reactions when compared to light capture during light reactions [8]. Thus, under high-light conditions the cost of light-dependant hydrogen production decreases to only \$1/m^2/year.

A number of methods of increasing photon use efficiency have been proposed. One solution involves the rapid mixing of dense cultures to create a 'flashing-light effect', during which each cell is exposed to millisecond flashes of high light intensity, followed by a 5 to 10-fold longer dark period in order to maximize the photon use efficiency of the entire culture by allowing the dark reactions to catch up with the light reactions in a particular cell at any given time [9]. The drawback of this method is that the high stirring rates required (25-35 cm/s) infringe impractical power inputs that make the process less economically feasible.

The use of optical fiber photoreactors, in which light harvested at a particular point is dispersed through the culture using fiber optics is another approach to increasing photon use efficiency, although this has been deemed technically and economically impractical [7]. An alternative approach is to develop photosynthetic cells with reduced pigment content in their light harvesting complexes that would allow fewer photons to be absorbed at high light intensities, and therefore less photons would be dissipated while waiting for the dark reactions to take place. Unfortunately, such strains are not selected for in nature since a larger light harvesting antenna size promotes survival over competitors that are less capable of harvesting light energy [10]. However, microalgae mutants in which light harvesting pigment content was reduced showed a 50% increase in productivity at high light intensities [11,12].

Light-Independent Processes: Dark Fermentations

Dark fermentation, in which carbohydrates are converted to H_2, CO_2, and organic acids, provides a promising alternative to light dependent processes, particularly when waste biomass is used as a feedstock for the generation of H_2. Since fermentation does not require a constant light supply, it can be run continuously using inexpensive and commercially used systems. Furthermore, hydrogen production rates are much higher using fermentations when compared to photosynthesis-based systems reducing bioreactor running costs [5]. A number of genera, including *Bacillus, Escherichia, Enterobacter, Ruminococcoi* and *Clostridium*, are capable of producing hydrogen *via* fermentation [4,5,13,14]. Some of these organisms not only operate at thermophillic (40-60°C) conditions, at which the solubility of H_2 is lower preventing product inhibition, but are also capable of degrading lignocellulose, a primary component of plant biomass [15,16].

BIOHYDROGEN: LOOKING FORWARD

Biohydrogen production offers potential to generate renewable H_2 from inexpensive "waste" feedstocks, but there are many scientific and engineering challenges that must be met by current and future biohydrogen researchers if these technologies are to be technically feasible and economically viable (see Chapter 14). It is no longer enough to simply demonstrate that one microorganism or another can produce H_2 from one or another substrate. Biohydrogen researchers must provide a deeper understanding of how H_2 is produced by different microorganisms under different circumstance, and provide testable hypotheses on how to increase H_2. The questions we need to ask and answer are not whether a microorganism synthesizes H_2, but why the microorganism does not produce more H_2? This book provides a state-of-the-art summary of the issues and challenges of biohydrogen production. It is dedicated to young bioydrogen scientists and engineers and will hopefully serve as a primer that will stimulate their imagination, and stir their passion, to create a better world through renewable hydrogen energy.

REFERENCES

[1] Independent Statistics and Analysis US Energy Information Administration [homepage on the Internet]. Washignton, DC: US Department of Energy; International energy outlook 2010. Available from: http://www.eia.gov/oiaf/ieo/liquid_fuels.html

[2] Natural Resources Canada. Canadian hydrogen survey – 2004/2005. Dalcor Consultants Ltd and Camford Information Services. June 2005.

[3] Das D, Veziroglu TN. Hydrogen production by biological processes: a survey of literature. Int J Hydrogen Energy 2001; 26: 13-28.

[4] Nath K, Das D. Improvement of fermentative hydrogen production: various approaches. Appl Microbiol Biotechnol 2004; 65: 520-9.

[5] Levin DB, Pitt L, Love M. Biohydrogen production: Prospects and limitations to practical application. Int J Hydrogen Energy 2004; 29: 173-85.

[6] Levin DB, Chahine R. Challenges for Renewable Hydrogen Production from Biomass. Int J Hydrogen Energy 2010; 35: 4962-9.

[7] Hallenbeck PC, Benemann JR. Biological hydrogen production: fundamentals and limiting processes. Int J Hydrogen Energy 2002; 27:1185-93.

[8] Kok B. Pilot-plant studies in the production of chlollera. In: Burlew JS, Ed. Algal culture: from laboratory to pilot plant. Washington DC: Carnegie Institute of Washington 1953; pp. 235-72.

[9] Kok, B. Photosynthesis. In: Gibbs M, Hollaender MA, Kok B, Krampitz LO, San Pietro A, Eds. Proceedings of the Workshop on Bio Solar Hydrogen Conversion; 1973 Sept 5-6; Bethesda, MD.

[10] Benemann JR. The Future of microalgae biotechnology. In: Cresswell RC, Rees TAV, Shah N, Eds. Algal biotechnology. London: Longman 1990; pp. 317-37.

[11] Nakajima Y, Ueda R. Improvement of microalgal photosynthetic productivity by reducing the content of light harvesting pigment. J Appl Phycol 1999; 11: 195-201.

[12] Nakajima Y, Ueda R. The effect of reducing light-harvesting pigment on marine microalgal productivity. J Appl Phycol 2000; 12: 285-90.

[13] Nandi R, Sengupta S. Microbial production of hydrogen: an overview. Crit Rev Microbiol 1998; 24: 61-84.

[14] Claassen PAM, van Lier JB, Lopez CAM, *et al.* Utilization of biomass for the supply of energy carriers. Appl Microbiol Biotechnol 1999; 52: 741-55.

[15] Levin DB, Carere CR, Cicek N, Sparling R. Challenges for biohydrogen production *via* direct lignocellulose fermentation. Int J Hydrogen Energy 2009; 34: 7390-403.

[16] Levin DB, Carere CR, Ramachandran U, Rydzak T, Saunders J. Fermentative biofuels: Prospects of practical application. In: Veziroglu TE, Muradov N, Eds. Carbon Neutral Fuels and Energy Carriers. Boca Raton Florida USA: CRC Press Taylor & Francis Group 2011; pp. 601-34.

CHAPTER 2

Hydrogenase Genes and Enzymes Involved in Solar Hydrogen Production

Carrie Eckert, Alexandra Dubini, Jianping Yu, Paul King, Maria Ghirardi, Michael Seibert and Pin-C. Maness*

Biosciences Center, National Renewable Energy Laboratory, Golden, CO, USA

Abstract: In both green algae and cyanobacteria, H_2 production is coupled to photosynthesis, which uses abundant water and sunlight as the source of electrons and energy, respectively. This process has immense potential for renewable H_2 production in future scaled-up processes. To generate H_2, green algae and cyanobacteria employ two different types of phylogenetically unrelated hydrogenases: [FeFe] enzymes in the former case and [NiFe] enzymes in the latter. Hydrogenases contain distinct metallo-clusters at their catalytic centers that require different suites of proteins for their assembly and maturation. Research in recent years has identified the genes encoding the hydrogenase proteins along with their maturation machineries, underlying regulatory controls, and enzyme catalytic mechanisms. One of the major challenges to be overcome is the deleterious effect of O_2 on hydrogenase structure and function. This chapter will examine the status of these topics in both classes of microbes, with the expectation that a more in-depth understanding will better guide the development of a robust system for sustained photobiological H_2 production.

Keywords: Hydrogen, hydrogenase, hydrogenase maturation, FeFe-hydrogenase, NiFe-hydrogenase, algae, *Chlamydomonas reinhardtii*, H-cluster, cyanobacteria, *Synechocystis*, *Rubrivivax gelatinosus*, HydA1, HydA2, Hox hydrogenase.

INTRODUCTION

Hydrogen can be produced thermochemically *via* pyrolysis and gasification, biologically *via* dark fermentation and photofermentation, and electrochemically *via* microbial electrolysis. All of these approaches utilize biomass as the feedstock, either directly or indirectly. Some of these processes have been categorized as first- (based on food crops) and second-generation (based on lignocellulose) technologies. Both entail the growth of biomass in the first step, and followed by its conversion to H_2 in the second step. Some of them represent nearer term technologies, and are in general less efficient from the perspective of land use. Longer term, third-generation processes (termed photoconversion, photobiological, or biophotolytic) harness sunlight to directly convert water to H_2 in a single step. Such a system would employ known photosynthetic processes in algae or cyanobacteria and with further development has the potential to produce large amounts of H_2 at solar-conversion efficiencies on the order of a hundred times greater than corn-grain ethanol, the current biofuel production standard, on an energy-conversion basis. To realize this potential, researchers have been studying hydrogenases, enzymes that catalytically combine protons and electrons to produce H_2 gas from the products of photosynthetic water oxidation. This article will highlight recent advances in our understanding of hydrogenase structure, catalytic function, maturation (enzyme assembly), gene structure and transcriptional regulation, and O_2 sensitivity. Due to a lack of crystal structure of the hydrogenases from either green algae or cyanobacteria, hydrogenase models developed in non-photosynthetic microbes will be used to gain more in-depth understanding. Basic knowledge in these areas will facilitate the development of applied systems to efficiently produce H_2 biologically from readily available solar and water resources.

GREEN ALGAL [FEFE]-HYDROGENASES

Enzyme Structure and Catalytic Mechanism

[FeFe]-hydrogenases are found in green algae and bacteria. Algal hydrogenases are soluble, bidirectional

***Address correspondence to Pin-C. Maness:** Biosciences Center, National Renewable Energy Laboratory, Golden, CO, USA; E-mail: PinChing.Maness@nrel.gov

Nuri Azbar and David B. Levin (Eds)

enzymes that belong to the class of [FeFe]-hydrogenases. They contain a metallo-cluster, termed the H-cluster, which is coordinated to the apoprotein by four cysteine residues contained within three conserved amino-acid motifs [1]. The H-cluster consists of a [4Fe-4S] cubane structure that is bridged by one of the cysteine residues to a [2Fe2S] subcluster [2,3]. The other three cysteines form ligands to the cubane. Interestingly, algal hydrogenases are the least complex [FeFe]-hydrogenases described so far, consisting of a single catalytic domain of about 48kDa [4]. They lack the additional [FeS] clusters found in most other bacterial [FeFe]-hydrogenases that act as electron-conducting pathways, or molecular wires. These hydrogenases function to mediate electron-transfer from the redox donor to the catalytic site. As a consequence, the natural electron donor to algal hydrogenases, ferredoxin, must dock very close to its catalytic site. Indeed, computational simulations suggest that the nature of the interactions between hydrogenases and ferredoxins is mostly electrostatic, involving specific surface residues on each enzyme [4-6]. Recent mutagenesis experiments demonstrated that genetic disruption of these residues, particularly of a set of lysines in the *Chlamydomonas reinhardtii* HYDA1 protein, does indeed eliminate electron transfer to ferredoxin [7].

One of the interesting aspects of all [FeFe]-hydrogenases characterized to date is the unique organometallic structure of the [2Fe2S] subcluster, which consists of CO and CN^- ligands coordinating the two Fe atoms (Fig. **1A**). The presumed function of these ligands is to act as electron-withdrawing moieties (CO) that stabilize the catalytic center during its redox cycle, and to possibly control the structure or basicity (CN^-) of the [2Fe2S] subcluster [8,9]. As discussed in Section 2.2, the biosynthesis of the H-cluster requires specific maturation enzymes to biosynthesize CO and CN^- ligands for the assembly and integration of the [2Fe2S] subcluster into the apoprotein. One of the only recently resolved questions regarding the [FeFe]-hydrogenase structure is the nature of the ligand bridging the [2Fe2S] moiety of the H-cluster. This ligand was originally modeled as a 1,3-propane dithiolate [10], subsequently proposed to be a dithiolmethylamine [11], and finally demonstrated by ^{14}N HYSCORE to be indeed the latter [12].

Fig. **2** describes the complete cycle of the various redox states that have been identified for [FeFe]-hydrogenases based on EPR, Mossbauer, and FTIR studies of the *Desulfovibrio* and *Clostridium* enzymes [13,14]. Only the *Desulfovibro* hydrogenase is known to form an inactive state (H_{inac}) when purified under aerobic conditions, transitioning to the catalytically active oxidized form (H_{ox}) through a three-electron reduction process involving an intermediate, the H_{trans} species. Reduction of H_{ox} occurs through a series of proton-coupled electron transfer steps, yielding a hydride anion intermediate. The hydride anion formed at the distal Fe (Fe_d) recombines with a nearby proton, yielding H_2 gas, which diffuses out of the catalytic site through gas channels or cavities [13,15], regenerating H_{ox}. The direct proton-donating moiety to Fe_d is thought to be the protonated nitrogen of the dithiolate bridge [13,16-18].

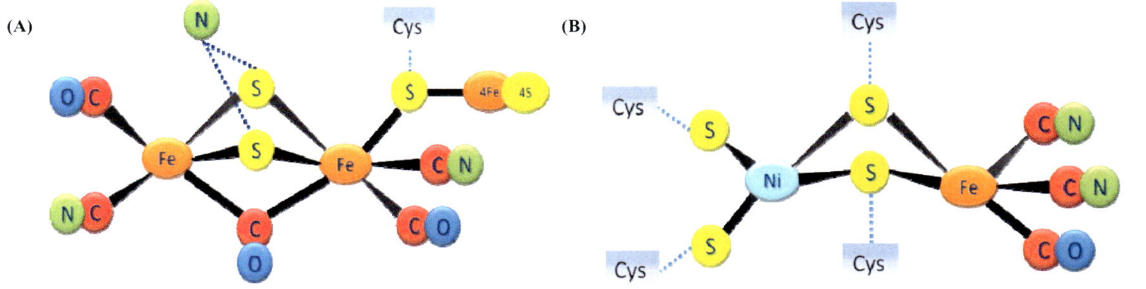

Figure 1: Chemical structure of the catalytic (A) H-cluster of [FeFe]-hydrogenase, and (B) [NiFe] cluster of the [NiFe]-hydrogenase. Fe_d: distal Fe atom.

Two other hydrogenase configurations have been electrochemically and spectroscopically characterized, namely the (i) $H_{ox\text{-}air}$, the irreversibly-oxidized form of the enzyme observed after exposure to air, and (ii) $H_{ox\text{-}CO}$, corresponding to a reversibly inactive, CO-bound form. The algal hydrogenases are particularly useful for EPR spectroscopy since they lack the additional [FeS] centers present in bacterial hydrogenases. As shown for the *Desulfovibrio* and *Clostridium* enzymes, EPR spectra of the as purified *C. reinhardtii* HYDA1 enzyme clearly have features that are consistent with the H_{ox} and $H_{ox\text{-}CO}$ states of the enzyme [19], while the reduced state H_{red} is EPR silent.

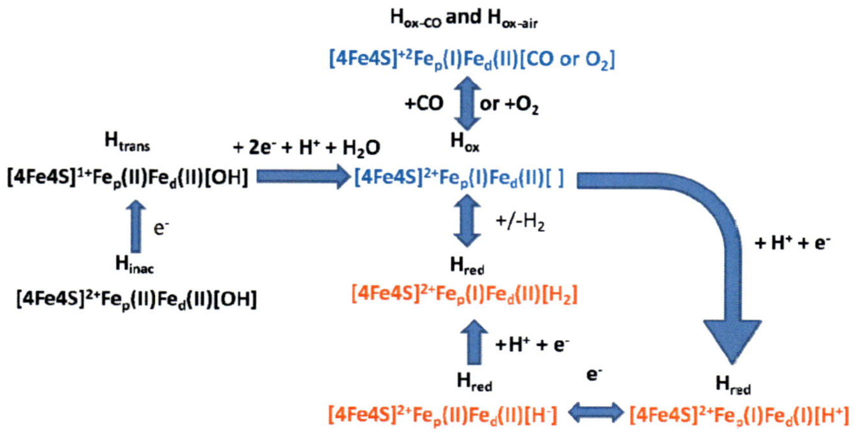

Figure 2: The various redox states of [FeFe]-hydrogenases. The redox state of the 2Fe subcluster Fe atoms and [4Fe4S] subcluster, and the chemical nature of bound ligands are shown for anaerobically inactive oxidized states (black), and the catalytically active oxidized (blue), and reduced (red) states. Catalytically active [FeFe]-hydrogenases are inactivated by CO (reversibly) and by O_2 (irreversibly). Fe_d: distal Fe; Fe_p: proximal Fe.

There are very limited reports on the proton transfer pathways in [FeFe]-hydrogenases. Protons are assumed to move through H-bonds between donor and acceptor residues, and involve coordinated water, as opposed to a tunneling mechanism, which is not viable at the long distances involved. Based on the crystal structure of the *Clostridium pasteurianum* [FeFe]-hydrogenase (CPI), Peters *et al.* [20] proposed the existence of several potential proton donor residues that are highly conserved in many [FeFe]-hydrogenases. These include a cysteine (C299 in CpI and C178 in *Desulfovibrio desulfuricans* hydrogenase (DdH)), a glutamate (E279 in CpI and E156 in DdH), a serine (S319 in CpI and S198 in DdH), and a second glutamate residue (E282 in CpI and E159 in DdH). Additional water molecules and perhaps the amine present in the dithiolate bridge may shuttle protons to the initial cysteine residue, while a second water molecule is proposed to connect the cysteine to the first glutamate [13].

Enzyme Maturation

The biosynthesis and maturation of [FeFe]-hydrogenase has not been completely elucidated yet. However, a set of proteins, HYDEF and HYDG, that are involved in the assembly of the H-cluster has been identified by screening *C. reinhardtii* insertional mutants that were unable to produce H_2 [21]. The HYDE domain of the HYDEF protein (HYDE and HYDF are found as separate proteins in other organisms) and HYDG protein belong to the Radical-S-adenosyl-methionine (SAM) superfamily, and coordinate the [FeS] clusters at the active site [21-25]. It is known that radical-SAM enzymes generally catalyze chemically difficult reactions such as C–H to C–S bond formation [26]. It is likely that HYDG and HYDE function synergistically to synthesize the dithiomethylamine bridging ligand as well as the CN^- and CO ligands of the [2Fe2S] H-cluster [20]. Biophysical characterization of purified, recombinant *Thermotoga maritima* HYDE and HYDG indicates that both proteins have [FeS] clusters chelated by a conserved cysteine motif found in all Radical-SAM proteins [24]. A C-terminal C-X7-C-X2-C motif that binds [FeS] centers is found in several HYDE homologs; as well as a second [FeS] motif that is only found in certain HYDE proteins [27]. The HYDG protein also contains a cysteine-containing C-terminal motif, C-X2-C-X22-C that binds [FeS] centers. Site-directed mutagenesis studies revealed that each of the cysteine motifs are required for a functional H-cluster assembly [28]. Recently, HydG activity was shown to increase in the presence of tyrosine, leading to a hypothesis that a tyrosine-derived dehydroglycine intermediate is the source for the H-cluster dithiol bridge [29,30]. Finally HydF has been identified as a GTPase based on its ability to hydrolyze GTP to GDP and the functional requirement of an intact GTPase domain for hydrogenase maturation [24,28]. The C-terminal domain also contains conserved cysteine and histidine residues arranged in a C-X-H-X(44-53)-HC-X2-C motif, suggesting [FeS] cluster coordination [21]. This observation was confirmed by the structural analysis of the reconstituted recombinant HYDF from *T. maritima* [24]. Recently HYDF was shown to act as a scaffold protein for H-cluster cofactor assembly and transfer to the hydrogenase apoprotein [24,29,31]. Spectroscopic analyses by EPR and FTIR techniques have shown that indeed the bi-nuclear part of the H-cluster is assembled in HYDF [32].

Fig. **3A** summarizes the sequential steps involved in [FeFe]-hydrogenases maturation. This complex maturation process requires the three maturases to work in concert. The maturases use common amino acids or metabolic products to synthesize a precursor to the [FeFe]-hydrogenase active site, which is then transferred to the structural enzyme [20]. HYDE and HYDG act on HYDF to assemble an active site precursor containing all the non-protein ligands of the binuclear Fe sub-cluster, which is subsequently transferred to the hydrogenase [20,32,33]. HYDE is probably responsible for the synthesis of CO and CN^- ligands and HydG for the dithiomethylamine ligand. The dithiomethylamine-bridged cluster is then transferred as a whole to HYDF to generate a complete, functional H-cluster for insertion to the [FeFe]-hydrogenase apoprotein [32]. A basis for this model was derived from X-ray structure of HYDE that displays three anion-binding sites, one of which binds SCN^- with high affinity [27].

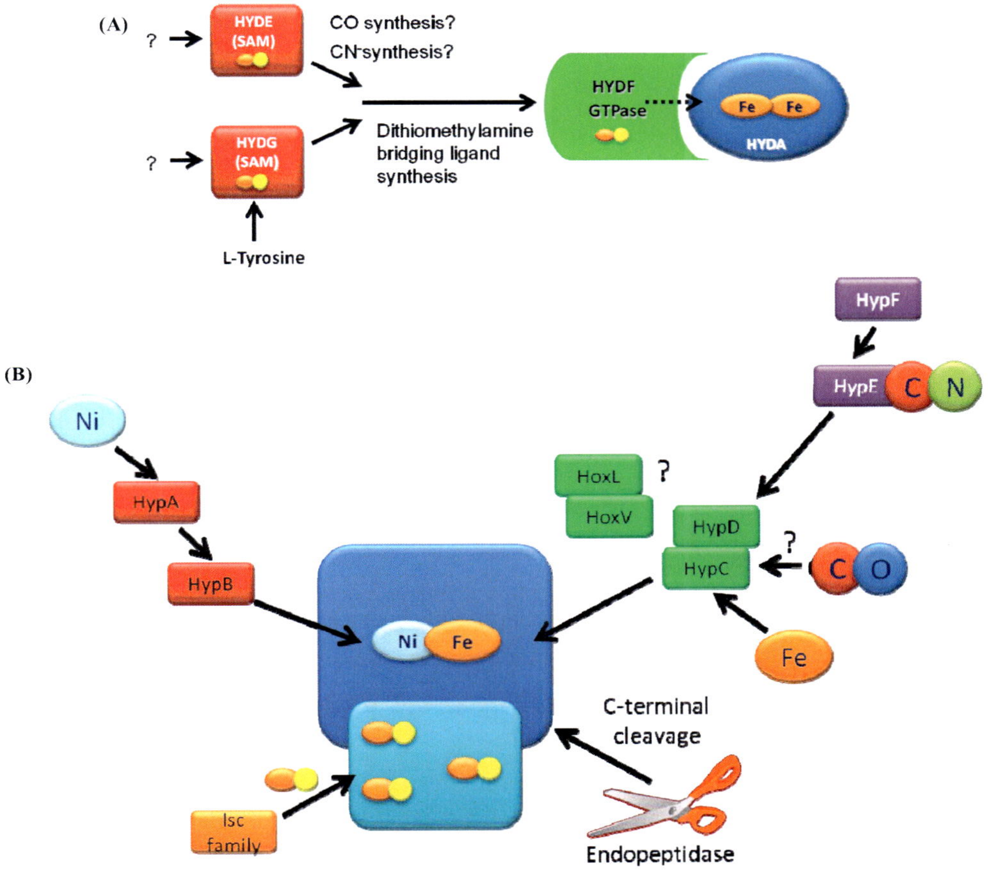

Figure 3: Maturation process of the (A) [FeFe]-hydrogenase, and (B) the [NiFe]-hydrogenase. SAM: S-adenosyl-methionine. Orange oval denotes Fe atom and yellow oval denotes sulfur atom of Fe-S cluster (Isc).

Gene Structure and Transcriptional Regulations

All the eukaryotic hydrogenase genes are nucleus-encoded, although their gene products are localized in the chloroplasts [4,34]. Algal [FeFe]-hydrogenases contain a transit peptide sequence at the N-terminal end of the protein that targets them to the specific organelle (Fig. **4A**). This signal peptide varies in length, depending on the organism. Differences in gene product length as well as in the number and localization of exons and introns have been reported as well. In particular, algal hydrogenases have two extra peptide insertions that are not observed in bacterial [FeFe]-hydrogenases (Fig. **4A**). These insertions correspond to extra loops found at the N- and C-terminal domains of the protein when compared to other [FeFe]-hydrogenases. In *C. reinhardtii*, transcription of the *HYDA1* and *HYDA2* genes, as well as the two [FeFe]-hydrogenase maturation genes, *HYDEF* and *HYDG* are induced under anaerobiosis [21,35-37]. Although the precise regulatory mechanisms are unknown, a number of metabolic mutants (*i.e., sta6, sta7)* with attenuated hydrogenase activity have also been isolated and shown to have decreased levels of hydrogenase transcripts [21,38]. This suggests that additional factors other than the lack of O_2 play a role in controlling *HYDA* gene transcription in this alga. Indeed in the starchless mutants, *sta6* [38,39] and *sta7* [21], hydrogenase *HYDA* gene transcription is significantly reduced relative to wild-type cultures under dark, anaerobic conditions [21]. These results suggest the importance of starch metabolism in inducing hydrogenase gene transcription under anaerobic conditions. The degradation of starch, which occurs under anaerobic conditions, can influence intracellular levels of NAD(P)H and/or the oxidation states of the plastoquinone pool, both of which are known to regulate gene transcription either directly or through signal transduction [40,41]. Experiments coupling varying lengths of the *HYDA1* promoter region to a promoterless arylsulfatase (*ARS*) reporter gene showed that the region between –128 and –21 relative to the *HYDA1* transcription start site is required for expression under anaerobic conditions [42].

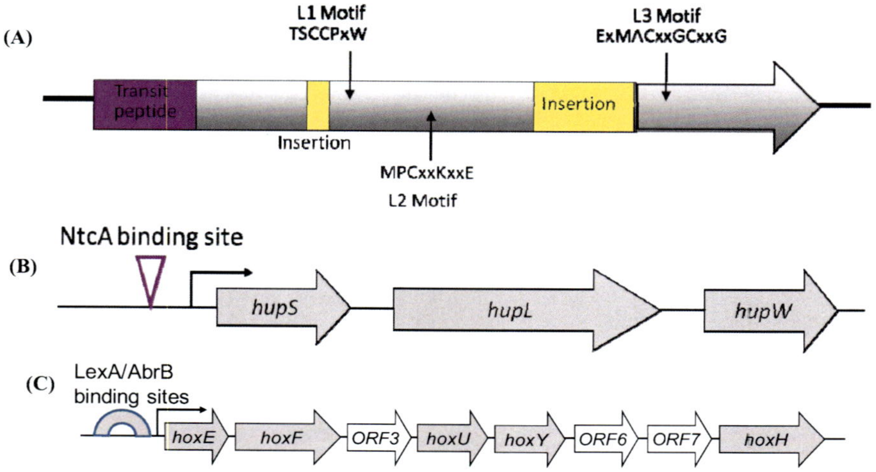

Figure 4: (A) [FeFe]-hydrogenase gene structure in green algae. The purple areas correspond to the transit peptide and the yellow areas to the insertions (see text for explanation). The *HYD* genes of algae also contain three motifs (L1, L2, L3), where the four cysteines that ligate the H-cluster are located. (B) The *hup* gene structure for the cyanobacterial uptake hydrogenase is from *Gloeothece* sp. ATCC 27152, whereas (C) the *hox* gene structure for the cyanobacterial bidirectional hydrogenase is from *Synechocystis* sp. PCC 6803.

Oxygen Sensitivity

The expression and activity of [FeFe]-hydrogenases in green algae are regulated by O_2 on at least three different levels: (i) transcription, (ii) maturation, and (iii) enzyme activity. The genes encoding HYDA1 and HYDA2 in green algae are up-regulated more than 100-fold under anaerobic conditions [36,37,43], in parallel with enzyme accumulation and hydrogenase activity [37,44,45], but the specific element involved in sensing O_2 and regulating hydrogenase expression are not known (see Section 2.3). Similarly, anaerobicity is required for transcription and activity of the hydrogenase maturation enzymes, and studies of maturation in heterologous and *in vitro* systems all show that formation of active enzymes occurs only in the absence of O_2.

Finally, the determination of sensitivity of [NiFe]- and [FeFe]-hydrogenase activities to O_2 is not an easy task, given that in many cases the resulting inactivity is irreversible, or in the case of [NiFe]-hydrogenases, O_2 generates a mixture hydrogenase redox states that reactivate at different rates [46]. An elegant electrochemical method was recently developed by Vincent *et al.* [47] and has been used to more accurately compare the O_2 sensitivity of bacterial hydrogenases under different conditions (oxidized, reduced, or CO-inhibited). The method is based on the competitive binding of CO and O_2 to the H-cluster, and the reversible *vs.* irreversible nature of the inhibition [3,47-51]. With respect to the algal [FeFe]-hydrogenase, the assay revealed that CO binding to the Fe_d of the H-cluster protects the [4Fe4S] cluster from O_2 damage [3,14,52]. The binding of CO (which reversibly inactivates the algal hydrogenase) was stronger than that of O_2 (which irreversibly inactivates the enzyme). Consequently, binding of CO prevented inactivation by O_2, allowing partial re-activation of the enzyme to occur once the two gases are flushed out.

Recent evidence demonstrated that [FeFe]-hydrogenase forms charge-transfer complexes with single wall carbon nanotubes (SWNTs) in solution, where the exchange of electrons between the enzyme and the SWNT results in changes in SWNT photoluminescence properties and Raman active modes. This was shown first for semiconducting-SWNTs (s-SWNTs), demonstrating that charge-transfer with [FeFe]-hydrogenase could be quenched by O_2 [53]. The latter was used to determine the inactivation rate constants of hydrogenases upon exposure to O_2 [54]. The results showed that the CaI [FeFe]-hydrogenase from *C. acetobutylicum*, prepared in the H_{red} state under 100% H_2, was inactivated 10-fold slower than if equilibrated in the H_{ox} state. Interestingly, it was also observed that when CaI was prepared in the H_{ox} state and complexed with metallic-SWNTs (m-SWNTs), a slower inactivation rate was observed that was similar to that of CaI in the H_{red} state, a phenomenon not observed with s-SWNT. The authors hypothesized that this could be explained by electron transfer from CaI $H_{ox,}$ (CaI oxidation) to the m-SWNT resulting in the formation of a H_{oxox} or possibly a H_{trans} state (refer to Fig. **2** for the various redox states).

CYANOBACTERIAL [NIFE]-HYDROGENASES

Enzyme Structure and Catalytic Mechanism

The [NiFe]-hydrogenases contain both Ni and Fe in their catalytic metallocluster. [NiFe]-hydrogenases are found throughout the domains of Archaea and Bacteria and can be divided into four subgroups, based on subunit composition, conserved cysteine motifs, and catalytic bias. These include (i) H_2 uptake hydrogenases (Hup) linking to respiration; (ii) cyanobacterial Hups linking to N_2 fixation and cytoplasmic H_2 sensor hydrogenases; (iii) bidirectional cytoplasmic hydrogenases (Hox); and (iv) membrane-associated, H_2-evolving, energy-converting hydrogenases (ECH) [13,55-57].

[NiFe]-hydrogenases are heterodimeric protein complexes in their simplest form. The heterodimer is composed of a large subunit (usually ~60 kDa in mass), containing the [NiFe] center, and a small subunit (usually ~30 kDa in mass), containing at least one accessory [FeS] cluster that functions in electron transfer between the catalytic site and its electron carrier. The [NiFe] and [FeS] moieties are coordinated by conserved cysteines present in either the large or the small subunits, respectively. The Fe atom of the [NiFe] site is ligated to one CO and two CN^- ligands, and conserved cysteine residues from the protein serve to coordinate the Ni atom and bridge the Ni and Fe atoms in the [NiFe] cluster (Fig. **1B**) [13,58]. The domain structure of [NiFe]-hydrogenases also contains hydrophobic channels linking the catalytic site and protein surface, allowing for gas diffusion [59,60]. The heterolytic splitting of H_2 is postulated to occur at the Ni atom of the active site, although this hypothesis is still under debate. The catalysis of H_2 production/oxidation involves multiple redox states of the active site, many involving the Ni moiety and the S-donor on one of the bridging cysteines. Studies involving the [NiFe]-hydrogenase from *D. vulgaris* (Miyazaki F) have elucidated the probable catalytic mechanism for these reactions [14,61-63]. Two oxidized states of the enzyme exist, Ni-A (unready) and Ni-B (ready). They differ in their rates of reactivation due to differences in the “O” species (postulated as –OH for Ni-B; -OOH for Ni-A) bound to the [NiFe] center; Ni-B reactivates within seconds while Ni-A requires a much longer time period. In both cases, an exchangeable proton is present in the active site, but it is not accessible in the Ni-A state, likely due to differences in bridging ligands. One-electron reduction of the Ni-A and Ni-B states results in EPR-silent, catalytically inactive Ni-SU (silent unready) and Ni-SI_r (silent ready) intermediates, respectively. This reduction is coupled to proton transfer within the active site. Reactivation of the enzyme in the Ni-SI_r state results in a Ni-SI_a (silent active) state. Further reduction of hydrogenases in the Ni-SI_a state leads to the paramagnetic Ni-C intermediate. The active site in this state is thought to bind a hydron or hydride. Further reduction of Ni-C results in the most reduced state, Ni-R. This EPR-silent state maintains a hydride and can exist in multiple protonated states, likely leading to the catalytic splitting of H_2 [14,62,64].

In this section, we focus on the encoding genes and biochemical information of [NiFe]-hydrogenases that are present in cyanobacteria, including the cyanobacterial Hup and the bidirectional Hox hydrogenase. The cyanobacterial Hup hydrogenases are found in N_2-fixing cyanobacteria linked to the nitrogenase enzyme, which produces H_2 in conjunction with N_2 fixation. The Hup hydrogenase functions to recycle the H_2 produced by the nitrogenase, conserving and restoring a source of reductant for the cell, providing ATP *via* the oxy-hydrogen reaction, and aiding in the removal of O_2, an inhibitor of both nitrogenase and hydrogenase [55,57]. The Hup hydrogenases are prototypic heterodimeric [NiFe] enzymes, consisting of a large subunit HupL and a small subunit HupS (Fig. **5A**). Studies of HupSL have demonstrated membrane localization, but how it associates with the membrane is still unknown since it lacks key targeting sequences/subunits found in conjunction with other Hup hydrogenases [55,65,66].

Bidirectional Hox hydrogenases are characterized by the association of large and small subunits with additional subunits that bind cofactors such as NAD(P)H (Fig. **5B**). These enzymes can catalyze both H_2 evolution and H_2 uptake depending on the redox state of the cell. With its low K_m for H_2, Hox hydrogenase may prefer the H_2-oxidation direction [67]. Although little data exists as to the cellular function of this hydrogenase, it has been proposed to play a role as an electron valve, balancing levels of available reductant in the cell [68]. Bidirectional hydrogenases are heteropentameric, and consist of large (HoxH) and small (HoxY) hydrogenase subunits similar to those of other [NiFe]-hydrogenases. In addition, the Hox hydrogenases contain three additional subunits (HoxEFU) that are collectively called the diaphorase. The diaphorase subunits contain additional [FeS] clusters and binding motifs for FMN and NAD(P) cofactors, and likely serves to donate/accept electrons from NAD(P)(H) for hydrogenase activity. The diaphorase components exhibit homology with subunits of the NDH-1 respiratory complex I, although no essential role in respiration has been discovered to date [56,69-71]. The Hox complex has been localized to the thylakoid membrane in some species of cyanobacteria, although this association, possibly *via* the HoxE subunit, is likely weak because purification of the complex reveals that it is largely in the soluble fraction [72-74].

Enzyme Maturation

The production of an active [NiFe]-hydrogenase is complex, species dependent, and involves multiple steps. Cyanobacteria contain homologues to the *hypABCDEF* genes characterized in *E. coli* and other organisms [75-80]. Most of what is known about the maturation of [NiFe]-hydrogenases has been deduced from studies of the maturation of the three hydrogenases in *E. coli* [13,25,81]. The first step involves synthesis and transfer of CO and CN^- ligands to the Fe atom in the active site. HypF transfers a carbamoyl phosphate group to a cysteine at the C-terminal of HypE where it is dehydrated, forming a HypE-SCN complex (Fig. **3B**). HypE then transfers the CN^- to the Fe of the active site through the concerted actions of HypC and HypD, which play a role in Fe coordination of the active site. Unlike CN^-, CO is thought to be synthesized from acetate or its precursor [82,83]. CO is also transferred to the active site, but the actual mechanism is not clear. HoxL, a HypC paralog, and HoxV are also involved in the maturation of the Hup hydrogenase of *Ralstonia eutropha*. Both function as a scaffold to deliver $Fe(CN^-)_2CO$ to the large subunit [84], although it is unknown whether these genes are present in cyanobacteria. The Ni incorporation into the active site is achieved by the concerted actions of HypB and HypA. Finally, the C-terminus of the large subunit is processed by an isoenzyme-specific endopeptidase to allow proper folding and maturation of

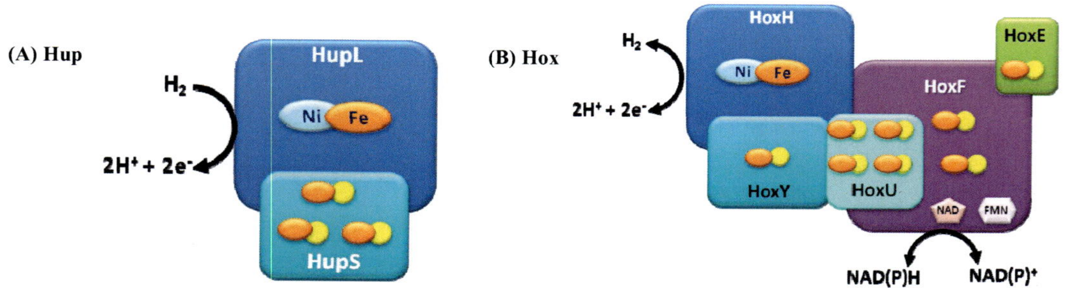

Figure 5: Cyanobacterial [NiFe]-hydrogenase subunit structure in (A) Hup uptake hydrogenase and (B) Hox bidirectional hydrogenase. Orange oval denotes Fe atom and yellow oval denotes sulfur atom of Fe-S cluster.

the active site. These endopeptidases seem to be highly specific to the large subunit it cleaves, with different endopeptidases present for different hydrogenases in a single organism [76,80]. In addition to the maturation of the active site, assembly of [FeS] clusters by the Isc protein family must also occur. In the [NiFe]-hydrogenase-containing bacteria *R. eutropha* and *R. leguminosarum*, gene products downstream of the *hup* and *hox* structural genes are present that seem to be required for proper processing of the small subunit but not the large subunit. However, it is not clear whether this is due to a lack of [FeS] insertion [85,86]. Large and small subunits can be processed independent of each other, although a lack of coordination of maturation may lead to improper targeting [25,87]. It also seems likely that protein-protein interactions between subunits may play a role in proper maturation since instability of the active site and [FeS] clusters in a *R. eutropha* [NiFe]-hydrogenase has been reported when a subunit of the complex is absent [88].

Gene Structure and Transcriptional Control

The first cyanobacterial *hupL* sequence was reported in 1995 in *Anabaena* sp. PCC 7120 (also named *Nostoc* PCC 7120) [89]. Subsequently, *hupS* and *hupL* genes were sequenced and characterized in numerous cyanobacteria, and more hydrogenase gene sequences have become available in recent years from genome sequencing projects. As is the case for most [NiFe]-hydrogenases, the structural genes *hupSL* are clustered in one operon, where *hupS* is upstream of *hupL* (Fig. **4B**). The first set of structural genes for the cyanobacterial Hox bidirectional hydrogenase (*hox*) was reported in 1995 from *Anabaena variabilis* ATCC 29413 [67] with more homologues reported later in other cyanobacteria. The physical organizations of the structural genes encoding the bidirectional hydrogenases are so far similar amongst different cyanobacterial species. The *hoxEFUYH* genes are usually clustered, with additional ORFs not related to hydrogenase function interspersed between some of the structural genes (Fig. **4C**).

In cyanobacteria, the *hyp* genes encoding hydrogenase maturation proteins are frequently clustered and located in the vicinity of the structural genes of one of the hydrogenases, with the notable exception in *Synechocystis* sp. PCC 6803, where the individual *hypABCDEF* genes are scattered throughout the genome [57]. The positions of the *hypW* and *hoxW* genes encoding specific endopeptidases vary within the cyanobacterial chromosome, and in some cases are found in the proximity of the corresponding hydrogenase structural genes [79]. Hoffmann demonstrated that *hypA1, B1, C, D, E* and *F* are required for an active bidirectional hydrogenase in *Synechocystis* sp. PCC 6803. The presence of a single copy of most of the *hyp* genes in cyanobacteria, regardless of the presence of only the uptake hydrogenase, the bidirectional hydrogenase, or both enzymes strongly suggests that they function in the maturation of both hydrogenases.

The transcription of the cyanobacterial *hupSL* genes is temporally regulated to coincide with heterocyst development in organisms that differentiate into these specialized, nitrogenase-containing cells. This was first reported when *Anabaena* PCC 7120 was transferred from non-N_2-fixing to N_2-fixing conditions [89]. The *hupSL* genes are co-transcribed, however, *hupW* is transcribed under both N_2-fixing and non-fixing conditions in this strain [76]. Analysis of *hup* promoter elements reveals putative binding sequences for the global cyanobacterial nitrogen regulator, NtcA, in several strains including *Gloeothece* sp. ATCC 27152 (Fig. **4B**), which was confirmed by gel-shift assays [57]. In the N_2-fixing, filamentous, non-heterocystous strain *Lyngbya majuscula* CCAP 1446/4, the gene encoding the putative uptake hydrogenase-specific endopeptidase, *hupW*, is located close to and can be co-transcribed with the *hupSL* structural genes. Alternatively it can be transcribed from its own promoter [90].

In *Synechocystis* sp. PCC 6803 as well as in *L. majuscula* CCAP 1446/4, the *hox* structural genes are co-transcribed [90]. In the latter strain the *hoxW* gene is in the same chromosomal region but transcribed from a different promoter. A bidirectional hydrogenase polypeptide is present in both anaerobic (induced) and aerobic (noninduced) cultures of *A. variabilis* despite being inactive under aerobic conditions. Hox subunits are also present in aerobic cultures of *Synechocystis* sp. PCC 6803. When cultured under limited nitrogen (1-2 mM sodium nitrate), enhanced rates of H_2 production under dark anaerobiosis were observed almost immediately in both *Synechocystis* sp. PCC6803 and *Gloeocapsa alipicola* CALU 743 [91,92]. Even though the hydrogenase activity in *G. alipicola* increased over time in darkness, the levels of *hoxY* and *hoxH* transcripts remained constant during the entire period. Sulfur limitation leads to increased H_2 production in *Synechocystis* sp. PCC6803, however, transcription of hydrogenase structural and maturation genes did not change under this condition [93]. Collectively, these observations suggest that the transcription and translation of the bidirectional hydrogenase genes is constitutive, at least with respect to O_2 and sulfur levels. However, Kiss *et al.* [94] found that *hox* transcript abundance in *Synechocystis* sp. PCC6803 decreases in the dark and recovers in the light. The transcript abundance is also modulated by the presence or absence of O_2, by inhibition of the Calvin cycle, and by inhibition of photosynthetic electron transport. It was suggested that unidentified O_2 and redox sensors may be involved in *hox* transcriptional regulation, in addition to the transcriptional regulators LexA and AbrB (see below).

Ferreira *et al.* [90] followed *hoxH, hupL, hoxW* and *hupW* transcription in *L. majuscula* cells grown under N_2-fixing and non-N_2-fixing conditions over a 12 h light/12 h dark cycle. The transcription of *hoxH, hoxW*, and *hupW* did not vary remarkably under the conditions tested, while the *hupL* transcript levels were significantly higher under N_2-fixing conditions with the peak occurring during the transition from the light to the dark, N_2-fixation phases. The latter is consistent with its role in N_2 fixation.

The genetic elements controlling the transcription of the *hox* genes have been studied the most in *Synechocystis* sp. PCC 6803. Transcription of the *hoxEFUYH* polycistronic unit in this strain is driven by a single promoter situated upstream of *hoxE* (Fig. **4C**), and the operon starts at -169 bp upstream of the *hoxE* start codon [95,96]. The transcription of *hoxE* was shown to be controlled by circadian rhythm, iron availability, and nitrogen starvation, while inorganic carbon limitation was shown to significantly repress the *hoxH* transcript [97]. The *hox* promoter has been studied using reporter gene analysis, and three regions of this promoter are known to be essential for its activity [95]. Two transcriptional regulators have been identified that bind at the *hox* promoter. A LexA-related protein binds to a region located between the nucleotides -198 and -338 bp, as well as -592 to -690 bp, respective to the translational start point (Fig. **4C**). Furthermore, a LexA-depleted mutant exhibited a decrease in bidirectional hydrogenase activity, reinforcing LexA's involvement in *hox* gene expression [95,96]. An AbrB-like protein also binds the *hox* promoter and positively regulates *hox* gene expression [98]. In addition, *hox* gene expression was significantly up regulated in *Sll0822* null mutant strain of PCC6803, suggesting that this protein may be a repressor of the *hox* operon [99].

Oxygen Sensitivity

Oxygen only reversibly inhibits the catalytic activity of most [NiFe]-hydrogenases, exerting no permanent damage to its structure [61-63,100]. Hydrogenases displaying this reversible property have been categorized in the literature as the "O_2-tolerant" hydrogenase, albeit little or no activity is detected in O_2. Moreover, O_2 does not affect transcription and translation (including maturation) events, at least for the Hox [NiFe]-hydrogenase in cyanobacteria (Section 3.3). Once anaerobic conditions are restored, the addition of a reducing agent such as sodium dithionite, H_2, or NAD(P)H to a [NiFe]-hydrogenase restores its activity fully. This accounts for a short burst of light-driven H_2 production when dark-adapted *Synechocystis* sp. PCC 6803 and *Oscillatoria chalybea* are exposed to light, followed immediately by a rapid H_2 uptake linked to CO_2 fixation [101,102]. The simultaneous accumulation of O_2 in the light immediately inhibits the hydrogenase activity. Once O_2 was consumed *via* respiration during the dark phase, H_2 production resumed from fermentation, suggesting a reactivation of the Hox bidirectional hydrogenase. Nevertheless, this observation cannot rule out any deleterious effect of O_2 on the energetics of electron transfer to the hydrogenase. In a *Synechocystis* M55 mutant defective in NAD(P)H dehydrogenase complex, more sustained H_2 production was observed, likely due to a combination of (i) a lower photosynthetic O_2 evolution rate and (ii) a negligible H_2-uptake rate [102,103]. Understanding the detailed mechanism of O_2 inactivation at the molecular level has important ramifications for applied H_2 production employing hydrogenases. Two different inactive and oxidized forms of [NiFe]-hydrogenase have been identified. The crystal structures of *D. fructosovorans* and *D. vulgaris* (Miyazaki F) reveal that exposure to O_2 under more reducing conditions generates a Ni-B ready state with a putative μ-hydroxo group bridging the [NiFe] center. On the other hand O_2 exposure under more oxidized environment yields a Ni-A unready state with a putative hydro-peroxide group bridging the [NiFe] diatomic ligands [61,63]. Ni-A and Ni-B do not interconvert; rather they must both proceed through the active state first, in which a hydride occupies the bridging position between the Ni and Fe atoms [61,100,104]. The O_2-tolerant membrane-bound, soluble, and sensor hydrogenases from the Knallgas bacterium *R. eutropha* form only the Ni-B ready state even under electron-deficient conditions, with no evidence for the presence of the Ni-A state [105]. This may represent an intrinsic requirement for being an O_2-tolerant hydrogenase.

The molecular basis governing O_2 tolerance has been studied in more detail in several [NiFe]-hydrogenases from both non-photosynthetic microbes and anoxygenic photosynthetic bacteria. This knowledge could serve as a model applicable to the [NiFe]-hydrogenases in oxygenic photosynthetic microbes. Two strategies have been recruited by Nature to surmount O_2 sensitivity amongst several [NiFe]-hydrogenases: (i) restricting O_2 gas access to the active site, (ii) utilizing unique [NiFe]-hydrogenase catalytic site architectures, and (iii) adjusting the redox poise for interconversion between active states and the inactive Ni-B and Ni-A states. Hydrogen gas diffusion into the [NiFe]-hydrogenase catalytic site is guided by hydrophobic gas cavities or channels [106,107], which presumably also permit O_2 access to the active site. Amino acid residues lining the hydrophobic gas channels are remarkably conserved among four O_2-sensitive [NiFe]-hydrogenases (*D. gigas, D. fructosovorans, D. desulfuricans*, and *D. vulgaris* (Miyazaki F) [107]. Two conserved residues closest to their active sites, *i.e.*, leucine and valine, are replaced by the more bulky phenylalanine and isoleucine, respectively, in the O_2-tolerant sensor hydrogenases from *R. eutropha* [108] and the photosynthetic bacterium *Rhodobacter capsulatus* [109]. However, when the two bulky amino acids in the wild-type sensor proteins were replaced by less bulky counterparts (Ile to Val; Phe to Leu) to enlarge the gas channel, they both become O_2 sensitive [110,111]. These experiments demonstrate the notion that limiting access of O_2 does provide O_2 tolerance. However, narrowing gas channel *via* the above strategy alone did not confer O_2 tolerance on the [NiFe]-hydrogenase from *D. fructosovrans.* [112]. Instead, replacing the Val and Leu with methionine residues dramatically improves tolerance to near ambient levels of O_2. The methionine substitutions not only slow down the inactivation kinetics but also accelerate its reactivation kinetics compared to wild type. The authors propose that methionine plays a dual role in (i) obstructing the gas channel for O_2 access and (ii) restoring the [NiFe]-hydrogenase catalytic site by a chemical reaction involving a sulfur atom with the peroxide species that inhibits normal activity.

The second strategy to cope with O_2 sensitivity has thus far only been reported in the O_2-tolerant NAD-reducing, soluble Hox hydrogenase from *R. eutropha.* Based on evidence from FTIR, chemical analyses, and N^{15}-enriched preparations, the *Ralstonia* Hox hydrogenase was found to contain four CN^- ligands, three of which bind to the Fe atom and one to the Ni atom of the active site. This unique configuration of $Ni(CN)Fe(CN)_3(CO)$ is different from the model depicted in Fig. **1B** of a "standard" [NiFe]-hydrogenase [113,114]. The maturation machinery of the Hox hydrogenase contains a unique *hypX* gene that is involved in the biosynthetic pathway and delivers the Ni-bound CN^- ligand. Deletion of *hypX* renders the Hox hydrogenase sensitive to O_2 [115]. The nickel-bound CN^- (along with the extra CN^- on the Fe) likely protects the catalytic center from binding O_2. The Hox bidirectional hydrogenase from *Synechocystis* sp. 6803 is a close relative of the Hox hydrogenase in *R. eutropha.* [102,116]. Yet the former was found to contain only two CN^- ligands chelating to Fe atom of the active site [117], similar to the "standard" [NiFe]-hydrogenase (Fig. **1B**). This is consistent with the reported transient production of H_2 under aerobic conditions catalyzed by the *Synechocystis* Hox hydrogenase [102].

CHALLENGES AND STRATEGIES FOR IMPROVING HYDROGENASE OXYGEN SENSITIVITY

Engineering O_2 tolerance in both the [FeFe]- and the [NiFe]-hydrogenases is an emergent area of research to support sustained photolytic H_2 production. It is equally crucial to develop a detailed understanding of both the inactivation mechanism associated with O_2-sensitive hydrogenase and how O_2-tolerant hydrogenases circumvent this challenge. Fundamental knowledge will inevitably guide the design of more O_2-tolerant hydrogenases. Three major approaches are being examined to address the O_2 sensitivity of [FeFe]-hydrogenases for fuel-production applications: (i) bioprospecting for enzymes from Nature with increased tolerance to O_2 inactivation [118], (ii) use of random mutagenesis followed by high-throughput screening to identify enzymes that are more tolerant to O_2 [119,120], and (iii) molecular engineering of hydrogenases to decrease O_2 access to their catalytic site [15,121,122]. Although the results are encouraging from a scientific perspective in understanding enzyme structure and function, none of these approaches has yielded biotechnologically significant organisms yet. A metagenomic approach from Sargasso Sea had yielded environmental DNA that encodes for a putative O_2-tolerant [NiFe]-hydrogenase, based on its similarity to the hydrogenase from *Thiocapsa roseopersicina* [123]. In a related approach, environmental sampling also led to the discovery of an O_2-tolerant [NiFe]-hydrogenase from *Rubrivivax gelatinosus* [124]. However, heterologous expression of [NiFe]-hydrogenases in other organisms is rather difficult. It entails co-expressing multiple maturation proteins since the maturation process is often hydrogenase- or species-dependent. Moreover, there is no guarantee that the foreign hydrogenase will produce H_2 *in vivo* by successfully linking to the photosynthetic pathways of the native hosts.

Recent kinetic calculations as well as electrochemical and EPR experiments suggest that one explanation for the variations in O_2 sensitivities of enzymes within the [FeFe]-hydrogenase family is that inactivation is governed by the rate of O_2 diffusion to their catalytic site, which differs for [FeFe]-hydrogenase from different organisms [46,49]. It is also likely that O_2 tolerance is conferred by multiple and dynamic mechanisms, such as a combination of gas channel narrowing along with substituting native amino acids near active sites with specific residues such as methionine. Obstructing the gas channel near the active site may also risk either lowering the catalytic activity of the enzyme or retarding the H_2 gas diffusion rate in and out of the enzyme. Indeed, the latter was demonstrated in *R. eutropha* where narrowing the gas channel decreased the affinity of its membrane-bound hydrogenase for H_2 while resulting in increasing O_2 sensitivity [50]. It remains to be seen if any of the current approaches to improving O_2 tolerance will provide an enzyme with properties sufficiently optimized (high turnover under aerobic conditions, a low sensitivity to O_2 inactivation, and long-term stability) to support future energy-production applications.

SUMMARY AND FUTURE DIRECTIONS

In this chapter, we have discussed advances in the basic understanding of [FeFe]- and [NiFe]-hydrogenases. These enzymes will be at the forefront of future technologies employing efficient, third-generation, photoconversion processes to generated H_2 from water using only sunlight as the driving force. Expansion of our knowledge about the structure, mechanisms, and regulation of [FeFe]-hydrogenases have been of particular note. The recognition of differences between algal and bacterial enzymes, the discovery of three maturation proteins and postulates about their function, and the discovery of various redox states of [FeFe]-hydrogenases analogous to those of [NiFe]-hydrogenases were unknown just 10 years ago. Similarly, great strides in the knowledge of [NiFe]-hydrogenases and their maturation (a more complex process compared to [FeFe]-hydrogenases) have been made as well as a more complete understanding of the relationships among the many redox states of these enzymes to O_2 sensitivity and catalysis. Although both types of hydrogenases can be heterologously expressed in non-native hosts, it has turned out that this is a less enzyme-specific process for [FeFe] when compared to the [NiFe] enzymes. Both types of hydrogenases are O_2 sensitive to different extents and progress is starting to being made on understanding the underlying mechanisms at the molecular level.

There are still a number of issues that will need to be examined and some might include (i) understanding the mechanisms involved in enzyme maturation processes, (ii) improving the maturation efficiency, especially in heterologous hosts, (iii) developing and refining a mechanistic understanding of the regulation of enzyme expression and the importance of hydrogenases and H_2 metabolism to the organism under various physiological conditions, (iv) understanding and devising strategies to optimize the network of biological pathways that involve hydrogenase, *in vivo*, (v) gaining a more in-depth understanding of the mechanisms of O_2 inactivation, (vi) significantly improving O_2 tolerance, and (vii) interfacing hydrogenases with light-capture and light-conversion molecules, devices, and electrodes for biomimetic applications in photolytic processes. Addressing and overcoming these technical challenges will improve the techno-economic feasibility of large-scale photosynthetic H_2 production.

ACKNOWLEDGEMENTS

The authors would like to acknowledge funding support from the US Department of Energy Office of Science and the EERE Fuel Cell Technologies Program.

REFERENCES

[1] Meyer J. [FeFe] hydrogenases and their evolution: a genomic perspective. Cell Mol Life Sci 2007; 64: 1063-84.

[2] Posewitz MC, Mulder DW, Peters JW. New frontiers in hydrogenase structure and biosynthesis. Curr Chem Biol 2008; 2: 178-99.

[3] Stripp ST, Goldet G, Brandmayr C, *et al.* How oxygen attacks [FeFe] hydrogenases from photosynthetic organisms. Proc Natl Acad Sci USA 2009; 106: 17331-6.

[4] Florin L, Tsokoglou A, Happe T. A novel type of iron hydrogenase in the green alga *Scenedesmus obliquus* is linked to the photosynthetic electron transport chain. J Biol Chem 2001; 276: 6125-32.

[5] Chang CH, King PW, Ghirardi ML, *et al.* Atomic resolution modeling of the ferredoxin: [FeFe] hydrogenase complex from *Chlamydomonas reinhardtii*. Biophys J 2007; 93: 3034-45.

[6] Long H, King PW, Ghirardi ML, *et al.* Hydrogenase/ferredoxin charge-transfer complexes: Effect of hydrogenase mutations on the complex association. J Phys Chem A 2009; 113: 4060-7.

[7] Winkler M, Kuhlgert S, Hippler M, *et al.* Characterization of the key step for light-driven hydrogen evolution in green algae. J Biol Chem 2009; 284: 36620-7.

[8] Song LC, Li CG, Ge JH, *et al.* Synthesis and structural characterization of the mono- and diphosphine-containing diiron propanedithiolate complexes related to [FeFe]-hydrogenases. Biomimetic H-2 evolution catalyzed by (mu-PDT)Fe-2(CO)(4)[(Ph2P)(2)N(n-Pr)]. J Inorg Biochem 2008; 102: 1973-9.

[9] Bruschi M, Greco C, Kaukonen M, *et al.* Influence of the [2Fe](H) subcluster environment on the properties of key intermediates in the catalytic cycle of [FeFe] hydrogenases: hints for the rational design of synthetic catalysts. Angew Chem Int Ed 2009; 48: 3503-6.

[10] Nicolet Y, Piras C, Legrand P, *et al. Desulfovibrio desulfuricans* iron hydrogenase: The structure shows unusual coordination to an active site Fe binuclear center. Structure 1999; 7: 13-23.

[11] Cao ZX, Hall MB. Modeling the active sites in metalloenzymes. 3. Density functional calculations on models for [Fe]-hydrogenase: Structures and vibrational frequencies of the observed redox forms and the reaction mechanism at the diiron active center. J Am Chem Soc 2001; 123: 3734-42.

[12] Silakov A, Wenk B, Reijerse E, *et al.* Spin distribution of the H-cluster in the H-ox-CO state of the [FeFe] hydrogenase from *Desulfovibrio desulfuricans*: HYSCORE and ENDOR study of N-14 and C-13 nuclear interactions. J Biol Inorg Chem 2009; 14: 301-13.

[13] Fontecilla-Camps JC, Volbeda A, Cavazza C, *et al.* Structure/function relationships of [NiFe]- and [FeFe]-hydrogenases. Chem Rev 2007; 107: 4273-303.

[14] Lubitz W, Reijerse E, van Gastel M. [NiFe] and [FeFe] hydrogenases studied by advanced magnetic resonance techniques. Chem Rev 2007; 107: 4331-65.

[15] Cohen J, Kim K, King PW, *et al.* Finding gas diffusion pathways in proteins: Application to O2 and H2 transport in Cpl [FeFe]-hydrogenase and the role of packing defects. Structure 2005; 13: 1321-9.

[16] Nicolet Y, de Lacey AL, Vernede X, *et al.* Cyrstallographic and FTIR spectroscopic evidence of changes in Fe coordination upon reduction of the active site of the Fe-only hydrogenase from *Desulfovibrio desulfuricans*. J Am Chem Soc 2001; 123: 1596-601.

[17] Fan HJ, Hall MB. A capable bridging ligand for Fe-only hydrogenase: Density functional calculations of a low-energy route for heterolytic cleavage and formation of dihydrogen. J Am Chem Soc 2001; 123: 3828-9.

[18] Tye JW, Lee J, Wang HW, *et al.* Dual electron uptake by simultaneous iron and ligand reduction in an N-heterocyclic carbene substituted [FeFe] hydrogenase model compound. Inorg Chem 2005; 44: 5550-2.

[19] Kamp C, Silakov A, Winkler M, *et al.* Isolation and first EPR characterization of the [FeFe]-hydrogenases from green algae. BBA-Bioenergetics 2008; 1777: 410-6.

[20] Peters JW, Szilagyi RK, Naumov A, *et al.* A radical solution for the biosynthesis of the H-cluster of hydrogenase. FEBS Lett 2006; 580: 363-7.

[21] Posewitz MC, King PW, Smolinski S, *et al.* Discovery of two novel radical S-Adenosylmethionine proteins required for the assembly of an active [Fe] hydrogenase. J Biol Chem 2004; 279: 25711-20.

[22] Casalot L, Rousset M. Maturation of the [NiFe] hydrogenases. Trends Microbiol 2001; 9: 228-37.

[23] Rubach JK, Brazzolotto X, Gaillard J, *et al.* Biochemical characterization of the HydE and HydG iron-only hydrogenase maturation enzymes from *Thermatoga maritima*. FEBS Lett 2005; 579: 5055-60.

[24] Brazzolotto X, Rubach JK, Gaillard J, *et al.* The [Fe-Fe]-hydrogenase maturation protein HydF from *Thermotoga maritima* is a GTPase with an iron-sulfur cluster. J Biol Chem 2006; 281: 769-74.

[25] Bock A, King PW, Blokesch M, *et al.* Maturation of hydrogenases. Adv Microb Physiol 2006; 51: 1-71.

[26] Wang SC, Frey PA. Binding energy in the one-electron reductive cleavage of S-Adenosylmethionine in lysine 2,3-aminomutase, a radical SAM enzyme. Biochemistry 2007; 46: 12889-95.

[27] Nicolet Y, Rubach JK, Posewitz MC, *et al.* X-ray structure of the [FeFe]-hydrogenase maturase HydE from *Thermotoga maritima*. J Biol Chem 2008; 283: 18861-72.

[28] King PW, Posewitz MC, Ghirardi ML, *et al.* Functional studies of [FeFe] hydrogenase maturation in an *Escherichia coli* biosynthetic system. J Bacteriol 2006; 188: 2163-72.

[29] Pilet E, Nicolet Y, Mathevon C, *et al.* The role of the maturase HydG in [FeFe]-hydrogenase active site synthesis and assembly. FEBS Lett 2009; 583: 506-11.

[30] Driesener RC, Challand MR, McGlynn SE, *et al.* [FeFe]-hydrogenase cyanide ligands derived from S-Adenosylmethionine-dependent cleavage of tyrosine. Angew Chem Int Ed 2010; 49: 1687-90.

[31] McGlynn SE, Shepard EM, Winslow MA, *et al.* HydF as a scaffold protein in [FeFe] hydrogenase H-cluster biosynthesis. FEBS Lett 2008; 582: 2183-7.

[32] Czech I, Silakov A, Lubitz W, *et al.* The [FeFe]-hydrogenase maturase HydF from *Clostridium acetobutylicum* contains a CO and CN-ligated iron cofactor. FEBS Lett 2010; 584: 638-42.

[33] Kuchenreuther JM, Stapleton JA, Swartz JR. Tyrosine, cysteine, and S-adenosyl methionine stimulate *In vitro* [FeFe] hydrogenase activation. PLoS One 2009; 4: e7565.

[34] Ueno Y, Kurano N, Miyachi S. Purification and characterization of hydrogenase from the marine green alga, *Chlorococcum littorale*. FEBS Lett 1999; 443: 144-8.

[35] Posewitz MC, King PW, Smolinski SL, *et al.* Identification of genes required for hydrogenase activity in *Chlamydomonas reinhardtii*. Biochem Soc Trans 2005; 33: 102-4.

[36] Mus F, Dubini A, Seibert M, *et al.* Anaerobic acclimation in *Chlamydomonas reinhardtii* - anoxic gene expression, hydrogenase induction, and metabolic pathways. J Biol Chem 2007; 282: 25475-86.

[37] Forestier M, King P, Zhang LP, *et al.* Expression of two [Fe]-hydrogenases in *Chlamydomonas reinhardtii* under anaerobic conditions. Eur J Biochem 2003; 270: 2750-8.

[38] Zabawinski C, Van den Koornhuyse N, D'Hulst C, *et al.* Starchless mutants of *Chlamydomonas reinhardtii* lack the small subunit of a heterotetrameric ADP-glucose pyrophosphorylase. J Bacteriol 2001; 183: 1069-77.

[39] Ball S. Regulation of starch biosynthesis. In: Rochaix JD, Goldschmidt-Clermont M, Merchant S, Eds. The molecular biology of chloroplasts and mitochondria in *Chlamydomonas*. Amsterdam, The Netherlands: Kluwer Academic 1998; pp. 549-67.

[40] Escoubas JM, Lomas M, LaRoche J, *et al.* Light intensity regulation of cab gene transcription is signaled by the redox state of the plastoquinone pool. Proc Natl Acad Sci USA 1995; 22: 10237-41.

[41] Pfannschmidt T, Schutze K, Fey V, *et al.* Chloroplast redox control of nuclear gene expression - A new class of plastid signals in interorganellar communication. Antioxid Redox Signal 2003; 5: 95-101.

[42] Stirnberg M, Happe T. Identification of a *cis*-acting element controlling anaerobic expression of hydA gene from *Chlamydomonas reinhardtii*. In: Miyake J, Igarashi Y, Rögner M, Eds. Biohydrogen III: renewable energy system by biological sola energy conversion. 1st ed. Oxford UK: Elsevier Science 2004; pp. 117-28.

[43] Dubini A, Mus F, Seibert M, *et al.* Flexibility in anaerobic metabolism as revealed in a mutant of *Chlamydomonas reinhardtii* lacking hydrogenase activity. J Biol Chem 2009; 284: 7201-13.

[44] Happe T, Mosler B, Naber JD. Induction, localization and metal content of hydrogenase in the green-alga *Chlamydomonas reinhardtii*. Eur J Biochem 1994; 222: 769-74.

[45] Happe T, Kaminski A. Differential regulation of the Fe-hydrogenase during anaerobic adaptation in the green alga *Chlamydomonas reinhardtii*. Eur J Biochem 2002; 269: 1022-32.

[46] Liebgott P, Leroux F, Burlat B, *et al.* Relating diffusion along the substrate tunnel and oxygen sensitivity in hydrogenase. Nat Chem Biol 2009; 6: 63-70.

[47] Vincent KA, Cracknell JA, Lenz O, *et al.* Electrocatalytic hydrogen oxidation by an enzyme at high carbon monoxide or oxygen levels. Proc Natl Acad Sci USA 2005; 102: 16951-4.

[48] Baffert C, Demuez M, Cournac L, *et al.* Hydrogen-activating enzymes: Activity does not correlate with oxygen sensitivity. Angew Chem Int Ed 2008; 121: 2082-4.

[49] Goldet G, Brandmayr C, Stripp ST, *et al.* Electrochemical kinetic investigations of the reactions of [FeFe]-hydrogenases with carbon monoxide and oxygen: Comparing the importance of gas tunnels and active-site electronic/redox effects. J Am Chem Soc 2009; 131: 14979-89.

[50] Ludwig M, Cracknell JA, Vincent KA, *et al.* Oxygen-tolerant H-2 oxidation by membrane-bound [NiFe] hydrogenases of *Ralstonia* species coping with low level H2 in air. J Biol Chem 2009; 284: 2159-68.

[51] Armstrong FA. Dynamic electrochemical experiments on hydrogenases. Photosynth Res 2009; 102: 541-50.

[52] Roseboom W, De Lacey AL, Fernadez VM, *et al.* The active site of the [FeFe]-hydrogenase from *Desulfovibrio deslufuricns*. II. Redox properties, light sensitivity and CO-ligand exchange as observed by infrared spectroscopy. J Biol Inorg Chem 2005; 10: 239-49.

[53] McDonald TJ, Svedruzic D, Kim YH, *et al.* Wiring-up hydrogenase with single-walled carbon nanotubes. Nano Lett 2007; 7: 3528-34.

[54] Blackburn JL, Svedruzic D, McDonald TJ, *et al.* Raman spectroscopy of charge transfer interactions between single wall carbon nanotubes and [FeFe] hydrogenase. Dalton Trans 2008; 40: 5454-61.

[55] Tamagnini P, Axelsson R, Lindberg P, *et al.* Hydrogenases and hydrogen metabolism of cyanobacteria. Microbiol Mol Biol Rev 2002; 66: 1-20.

[56] Vignais PM, Billoud B. Occurrence, classification, and biological function of hydrogenases: An overview. Chem Rev 2007; 107: 4206-72.

[57] Tamagnini P, Leitao E, Oliveira P, *et al.* Cyanobacterial hydrogenases: Diversity, regulation and applications. FEMS Microbiol Rev 2007; 31: 692-720.

[58] Heinekey DM. Hydrogenase enzymes: Recent structural studies and active site models. J Organomet Chem 2009; 694: 2671-80.

[59] Teixeira VH, Baptista AM, Soares CM. Pathways of H2 toward the active site of [NiFe]-hydrogenase. Biophys J 2006; 91: 2035-45.

[60] Galvan IF, Volbeda A, Fontecilla-Camps JC, *et al.* A QM/MM study of proton transport pathways in a [NiFe] hydrogenase. Proteins 2008; 73: 195-203.

[61] Ogata H, Hirota S, Nakahara A, *et al.* Activation process of [NiFe] hydrogenase elucidated by high-resolution X-ray analyses: Conversion of the ready to the unready state. Structure 2005; 13: 1635-42.

[62] Ogata H, Lubitz W, Higuchi Y. [NiFe] hydrogenases: Structural and spectroscopic studies of the reaction mechanism. Dalton Trans 2009; 37: 7577-87.

[63] Volbeda A, Martin L, Cavazza C, *et al.* Structural differences between the ready and unready oxidized states of [NiFe] hydrogenases. J Biol Inorg Chem 2005; 10: 239-49.

[64] Pandelia ME, Ogata H, Lubitz W. Intermediates in the catalytic cycle of [NiFe] hydrogenase: Functional spectroscopy of the active site. Chemphyschem 2010; 11: 1127-40.

[65] Tamagnini P, Leitao E, Oxelfelt F. Uptake hydrogenase in cyanobacteria: Novel input from non-heterocystous strains. Biochem Soc Trans 2005; 33: 67-9.

[66] Seabra R, Santos A, Pereira S, *et al.* Immunolocalization of the uptake hydrogenase in the marine cyanobacterium *Lyngbya majuscula* CCAP 1446/4 and two *Nostoc* strains. FEMS Microbiol Lett 2009; 292: 57-62.

[67] Schmitz O, Boison G, Hilscher R, *et al.* Molecular biological analysis of a bidirectional hydrogenase from cyanobacteria. Eur J Biochem 1995; 233: 266-76.

[68] Appel J, Phunpruch S, Steinmuller K, *et al.* The bidirectional hydrogenase of *Synechocystis* sp. PCC 6803 works as an electron valve during photosynthesis. Arch Microbiol 2000; 173: 333-8.

[69] Boison G, Schmitz O, Schmitz B, *et al.* Unusual gene arrangement of the bidirectional hydrogenase and functional analysis of its diaphorase subunit *HoxU* in respiration of the unicellular cyanobacterium *Anacystis nidulans*. Curr Microbiol 1998; 36: 253-8.

[70] Boison G, Bothe H, Hansel A, *et al.* Evidence against a common use of the diaphorase subunits by the bidirectional hydrogenase and by the respiratory complex I in cyanobacteria. FEMS Microbiol Lett 1999; 174: 159-65.

[71] Howitt CA, Vermaas W. Subunits of the NAD(P)-reducing nickel-containing hydrogenase do not act as part of the type-1 NAD(P)H-dehydrogenase in the cyanobacterium *Synechocystis* sp. PCC 6803. In: Peschek G, Löffelhardt W, Schmetterer G, Eds. The Phototrophic Prokaryotes. New York, USA: Kluwer Academic/ Plenum 1999; pp. 595-603.

[72] Kentemich T, Bahnweg M, Mayer F, *et al.* Localization of the reversible hydrogenase in cyanobacteria. Z Naturforsch C 1989; 44: 384-91.

[73] Serebriakova L, Zorin NA, Lindblad P. Reversible hydrogenase in *Anabaena variabilis* ATCC-29413-presence and localization in non-N-2-fixing cells. Arch Microbiol 1994; 161: 140-4.

[74] Schmitz O, Boison G, Salzmann H, *et al.* HoxE - a subunit specific for the pentameric bidirectional hydrogenase complex (HoxEFUYH) of cyanobacteria. BBA-Bioenergetics 2002; 1554: 66-74.

[75] Buhrke T, Bleijlevens B, Albracht SPJ, *et al.* Involvement of hyp gene products in maturation of the H-2-sensing [NiFe] hydrogenase of *Ralstonia eutropha.* J Bacteriol 2001; 183: 7087-93.

[76] Wunschiers R, Batur M, Lindblad P. Presence and expression of hydrogenase specific C-terminal endopeptidases in cyanobacteria. BMC Microbiol 2003; 3: 8.

[77] Oliveira P, Leitao E, Tamagnini P, *et al.* Characterization and transcriptional analysis of hupSLW in *Gloeothece* sp. ATCC 27152: An uptake hydrogenase from a unicellular cyanobacterium. Microbiology 2004; 150: 3647-55.

[78] Leitao E, Pereira S, Bondoso J, *et al.* Genes involved in the maturation of hydrogenase(s) in the nonheterocystous cyanobacterium *Lyngbya majuscula* CCAP 1446/4. Int J Hydrogen Energy 2006; 31: 1469-77.

[79] Hoffmann D, Gutekunst K, Klissenbauer M, *et al.* Mutagenesis of hydrogenase accessory genes of *Synechocystis* sp. PCC 6803 - additional homologues of *hypA* and *hypB* are not active in hydrogenase maturation. FEBS J 2006; 273: 4516-27.

[80] Devine E, Holmqvist M, Stensjo K, *et al.* Diversity and transcription of proteases involved in the maturation of hydrogenases in *Nostoc punctiforme* ATCC 29133 and *Nostoc* sp. strain PCC 7120. BMC Microbiol 2009; 9: 53.

[81] Forzi L, Sawers RG. Maturation of [NiFe]-hydrogenases in *Escherichia coli.* Biometals 2007; 20: 565-78.

[82] Roseboom W, Blokesch M, Bock A, *et al.* The biosynthetic routes for carbon monoxide and cyanide in the Ni-Fe active site of hydrogenases are different. FEBS Lett 2005; 579: 469-72.

[83] Lenz O, Zebger I, Hamann J, *et al.* Carbamoylphosphate serves as the source of CN-, but not of the intrinsic CO in the active site of the regulatory [NiFe]-hydrogenase from *Ralstonia eutropha.* FEBS Lett 2007; 581: 3322-6.

[84] Ludwig M, Schubert T, Zebger I, *et al.* Concerted action of two novel auxiliary proteins in assembly of the active site in a membrane-bound [NiFe] hydrogenase. J Biol Chem 2009; 284: 2159-68.

[85] Manyani H, Rey L, Palacios JM, *et al.* Gene products of the hupGHIJ Operon are involved in maturation of the iron-sulfur subunit of the [NiFe] hydrogenase from *Rhizobium leguminosarum* bv. viciae. J Bacteriol 2005; 187: 7018-26.

[86] Schubert T, Lenz O, Krause E, *et al.* Chaperones specific for the membrane-bound [NiFe]-hydrogenase interact with the Tat signal peptide of the small subunit precursor in *Ralstonia eutropha* H16. Mol Microbiol 2007; 66: 453-67.

[87] Dubini A, Sargent F. Assembly of tat-dependent [NiFe] hydrogenases: Identification of precursor-binding accessory proteins. FEBS Lett 2003; 549: 141-6.

[88] Loscher S, Gebler A, Stein M, *et al.* Protein-protein complex formation affects the Ni-Fe and Fe-S centers in the H(2)-sensing regulatory hydrogenase from *Ralstonia eutropha* H16. Chemphyschem 2010; 11: 1297-306.

[89] Carrasco CD, Buettner JA, Golden JW. Programmed DNA rearrangement of a cyanobacterial HUPL gene in heterocysts. Proc Natl Acad Sci USA 1995; 92: 791-5.

[90] Ferreira D, Pinto F, Moradas-Ferreira P, *et al.* Transcription profiles of hydrogenases related genes in the cyanobacterium *Lyngbya majuscula* CCAP 1446/4. BMC Microbiol 2009; 9: 67.

[91] Schutz K, Happe T, Troshina O, *et al.* Cyanobacterial H-2 production - a comparative analysis. Planta 2004; 218: 350-9.

[92] Sheremetieva ME, Troshina OY, Serebryakova LT, *et al.* Identification of hox genes and analysis of their transcription in the unicellular cyanobacterium *Gloeocapsa alpicola* CALU 743 growing under nitrate-limiting conditions. FEMS Microbiol Lett 2002; 214: 229-33.

[93] Zhang Z, Pendse ND, Phillips KN, *et al.* Gene expression patterns of sulfur starvation in *Synechocystis* sp PCC 6803. BMC Genomics 2008; 9: 344.

[94] Kiss E, Kos PB, Vass I. Transcriptional regulation of the bidirectional hydrogenase in the cyanobacterium *Synechocystis* 6803. J Biotechnol 2009; 142: 31-7.

[95] Gutekunst K, Phunpruch S, Schwarz C, *et al.* LexA regulates the bidirectional hydrogenase in the cyanobacterium *Synechocystis* sp. PCC 6803 as a transcription activator. Mol Microbiol 2005; 58: 810-23.

[96] Oliveira P, Lindblad P. LexA, a transcription regulator binding in the promoter region of the bidirectional hydrogenase in the cyanobacterium *Synechocystis* sp. PCC 6803. FEMS Microbiol Lett 2005; 251: 59-66.

[97] Oliveira P, Lindblad P. Transcriptional regulation of the cyanobacterial bidirectional *Hox*-hydrogenase. Dalton Trans 2009; 45: 9990-6.

[98] Oliveira P, Lindblad P. An AbrB-like protein regulates the expression of the bidirectional hydrogenase in *Synechocystis* sp. strain PCC 6803. J Bacteriol 2008; 190: 1011-9.

[99] Ishii A, Hihara Y. An AbrB-like transcriptional regulator, Sll0822, is essential for the activation of nitrogen-regulated genes in *Synechocystis* sp. PCC 6803. Plant Physiol 2008; 148: 660-70.

[100] Lamle SE, Albracht SPJ, Armstrong FA. The mechanism of activation of a [NiFe]-hydrogenase by electrons, hydrogen, and carbon monoxide. J Am Chem Soc 2005; 127: 6595-604.

[101] Abdel-Basset R, Bader KP. Physiological analyses of the hydrogen gas exchange in cyanobacteria. J Photoch Photobio B 1998; 43: 146-51.

[102] Cournac L, Guedeney G, Peltier G, *et al.* Sustained photoevolution of molecular hydrogen in a mutant of *Synechocystis* sp. strain PCC 6803 deficient in the type I NADPH-dehydrogenase complex. J Bacteriol 2004; 186: 1737-46.

[103] Gutthann F, Egert M, Marques A, *et al.* Inhibition of respiration and nitrate assimilation enhances photohydrogen evolution under low oxygen concentrations in *Synechocystis* sp. PCC 6803. BBA-Bioenergetics 2007; 1767: 161-9.

[104] Fernandez VM, Hatchikian CE, Cammack R. Properties and reactivation of two different deactivated forms of *Desulfovibrio gigas* hydrogenase. BBA-Protein Struct M 1985; 832: 69-79.

[105] Saggu M, Zebger I, Ludwig M, *et al.* Spectroscopic insights into the oxygen-tolerant membrane-associated [NiFe] hydrogenase of *Ralstonia eutropha* H16. J Biol Chem 2009; 284: 16264-76.

[106] Montet Y, Amara P, Volbeda A, *et al.* Gas access to the active site of Ni-Fe hydrogenases probed by X-ray crystallography and molecular dynamics. Nat Struct Biol 1997; 4: 523-6.

[107] Volbeda A, Montet Y, Vernede X, *et al.* High-resolution crystallographic analysis of *Desulfovibrio fructiosovorans* [NiFe] hydrogenase. Int J Hydrogen Energy 2002; 27: 1449-61.

[108] Kleihues L, Lenz O, Bernhard M, *et al.* The H2 sensor of *Ralstonia eutropha* is a member of the subclass of regulatory [NiFe] hydrogenase J Bacteriol 2000; 182: 2716-24.

[109] Vignais PM, Dimon B, Zorin NA, *et al.* HupUV proteins of *Rhodobacter capsulatus* can bind H-2: Evidence from the H-D exchange reaction. J Bacteriol 1997; 179: 290-2.

[110] Buhrke T, Lenz O, Krauss N, *et al.* Oxygen tolerance of the H-2-sensing [NiFe] hydrogenase from *Ralstonia eutropha* H16 is based on limited access of oxygen to the active site. J Biol Chem 2005; 280: 23791-6.

[111] Duche O, Elsen S, Cournac L, *et al.* Enlarging the gas access channel to the active site renders the regulatory hydrogenase *HupUV* of *Rhodobacter capsulatus* O-2 sensitive without affecting its transductory activity. FEBS J 2005; 272: 3899-908.

[112] Dementin S, Leroux F, Cournac L, *et al.* Introduction of methionines in the gas channel makes [NiFe] hydrogenase aero-tolerant. J Am Chem Soc 2009; 131: 10156-64.

[113] Happe RP, Roseboom W, Egert G, *et al.* Unusual FTIR and EPR properties of the H-2-activating site of the cytoplasmic NAD-reducing hydrogenase from *Ralstonia eutropha.* FEBS Lett 2000; 466: 259-63.

[114] Van der Linden E, Burgdorf T, Bernhard M, *et al.* The soluble [NiFe]-hydrogenase from *Ralstonia eutropha* contains four cyanides in its active site, one of which is responsible for the insensitivity towards oxygen. J Biol Inorg Chem 2004; 9: 616-26.

[115] Bleijlevens B, Buhrke T, van der Linden E, *et al.* The auxiliary protein HypX provides oxygen tolerance to the soluble [NiFe]-hydrogenase of *Ralstonia eutropha* H16 by way of a cyanide ligand to nickel. J Biol Chem 2004; 279: 46686-91.

[116] Appel J, Schulz R. Sequence analysis of an operon of a NAD(P)-reducing nickel hydrogenase from the cyanobacterium *Synechocystis* sp. PCC6803 gives additional evidence for direct coupling of the enzyme to NAD(P)H-dehydrogenase (complex I) BBA-Protein Struct M 1996; 1298: 141-7.

[117] Germer F, Zebger I, Saggu M, *et al.* Overexpression, isolation, and spectroscopic characterization of the bidirectional [NiFe] hydrogenase from *Synechocystis* sp. PCC 6803. J Biol Chem 2009; 284: 36462-72.

[118] Boyd ES, Spear JR, Peters JW. [FeFe] hydrogenase genetic diversity provides insight into molecular adaptation in a saline microbial mat community. Appl Environ Microbiol 2009; 75: 4620-3.

[119] Nagy LE, Meuser JE, Plummer S, *et al.* Application of gene-shuffling for the rapid generation of novel [FeFe]-hydrogenase libraries. Biotechnol Lett 2007; 29: 421-30.

[120] Boyer M, Stapleton J, Kuchenreuther J, *et al.* Cell-free synthesis and maturation of [FeFe] hydrogenases. Biotechnol Bioeng 2007; 99: 59-67.

[121] Ghirardi ML, King PW, Posewitz MC, *et al.* Approaches to developing biological H-2-photoproducing organisms and processes. Biochem Soc Trans 2005; 33: 70-2.

[122] Melis A, Seibert M, Ghirardi ML. Hydrogen fuel production by transgenic microalgae. Adv Exp Med Biol 2007; 616: 108-21.

[123] Maroti G, Tong Y, Yooseph S, *et al.* Discovery of [NiFe] hydrogenase genes in metagenomic DNA: Cloning and heterologous expression in *Thiocapsa roseopersicina*. Appl Environ Microbiol 2009; 75: 5821-30.

[124] Maness PC, Smolinski S, Dillon AC, *et al.* Characterization of the oxygen tolerance of a hydrogenase linked to a carbon monoxide oxidation pathway in *Rubrivivax gelatinosus*. Appl Environ Microbiol 2002; 68: 2633-6.

CHAPTER 3

Photosynthetic Hydrogen Production: Mechanisms and Approaches

T.K. Antal[1], T.E. Krendeleva[1], V.Z. Pashchenko[1], A.B. Rubin[1], K. Stensjo[2], E. Tyystjärvi[3], S. Ramakrishna[4], D.A. Los[5], R. Carpentier[6], H. Nishihara[7], S.I. Allakhverdiev[5,7,8]*

[1]Department of Biophysics, Faculty of Biology, Moscow State University, 119992, Moscow, Russia; [2]Department of Photochemistry and Molecular Science, The Ångström Laboratories, Uppsala University, Uppsala, Sweden; [3]Department of Biochemistry and Food Chemistry, Plant Physiology and Molecular Biology, University of Turku, Turku, Finland; [4]Nanoscience and Nanotechnology Initiative, National University of Singapore, Singapore, Singapore; [5]Institute of Plant Physiology, Russian Academy of Sciences, Moscow, Russia; [6]Groupe de Recherche en Biologie Végétale, Université du Québec à Trois-Rivières, Trois-Rivières, Québec, Canada; [7]Department of Chemistry, School of Science, The University of Tokyo, Tokyo, Japan and [8]Institute of Basic Biological Problems, Russian Academy of Sciences, Pushchino, Moscow, Russia

Abstract: The development of renewable fuels of the future is important for the replacement of depleting oil and natural gas reserves. Hydrogen is one of the most promising clean fuels, since its combustion yields only water. One of the visionary methods to obtain hydrogen at the expanse of solar energy is the use of photosynthetic microorganisms. Hydrogen production in phototrophs is coupled to the oxygenic and anoxygenic photosynthesis involving hydrogen-evolving enzymes, hydrogenases and nitrogenases. At the present time the efficiency of hydrogen photoproduction is not sufficiently high. Most hydrogen-evolving enzymes are inhibited by molecular oxygen, which creates a major barrier for the sustained hydrogen photoproduction in oxygenic phototrophs, such as green algae and cyanobacteria. However, several strategies have been applied to solve this problem, including spatial and temporal separation of water splitting and hydrogen evolution, and regulation of water splitting activity and respiration to maintain anoxic conditions. Anoxygenic photosynthesis can be used to drive hydrogen photoproduction in integrated systems including fermentative anaerobic organisms. In this review different mechanisms for hydrogen production in photosynthetic organisms and the latest advances in this area are discussed.

Keywords: Hydrogen production, hydrogenases, nitrogenases, ferredoxin, photosystem I, photosystem II, oxygen evolving complex, ATP, NADPH, *Nostoc*, *Synechocystis*, *Gloeocapsa*, *Synechococcus, Spirulina.*

INTRODUCTION

The Sun is the infinite source of energy for future ecologically safe sustainable production of fuels. The use of solar radiation for practical ends involves development of cost-effective systems for the conversion of light energy into energy of chemical bonds which can be stored as fuel. Hydrogen is one of the promising clean future fuels because its combustion yields only water [1-4]. At present, most of the world hydrogen is produced by reforming of fossil fuels, *e.g.* natural gas, which is accompanied by release of carbon into the environment. To achieve benefits of the process, hydrogen might be produced from non-fossil resources, such as water, using renewable energy sources. The energy for hydrogen production from water may come from irradiation of semiconductors, the heat accumulated in solar collectors, or electric voltage generated by renewable sources such as hydropower or wind [5,6]. The solar-driven hydrogen production by photosynthetic microorganisms [photohydrogen production] has been considered as a feasible alternative to photo-, thermo-, and electrolysis technologies [6-10].

Photosynthesis is the natural process allowing phototrophic organisms to use sunlight for the synthesis of

***Address correspondence to S.I. Allakhverdiev:** Institute of Basic Biological Problems, Russian Academy of Sciences, Pushchino, Moscow, Russia; E-mail: suleyman@issp.serpukhov.su

Nuri Azbar and David B. Levin (Eds)

carbohydrates from carbon dioxide. In photoautotrophic organisms the fixation of carbon dioxide includes two phases [11,12]. During the first phase light-harvesting pigments, such as chlorophylls or bacteriochlorophylls, collect the light energy. The absorbed energy then reaches reaction centers, where it is converted into chemical energy by a series of electron transfer steps that finally results in generation of a reducing agent, NADPH, and an endogenous fuel, ATP. During the second phase NADPH and ATP are consumed in the reactions of the Calvin-Benson cycle which catalyzes the synthesis of carbohydrates from carbon dioxide and water.

Photosynthesis proceeds in oxygenic (oxygen producing) or anoxygenic organisms. Oxygenic organisms such as higher plants, green, red, brown, and yellow algae, as well as cyanobacteria, use solar energy to extract electrons and protons from water; this results in oxygen evolution. Among oxygenic phototrophs, only green microalgae and cyanobacteria have been shown to sustain hydrogen production [13-15]. The simplest and most effective process would involve a direct transfer of electrons from water to hydrogen-evolving enzymes accompanied by oxygen and hydrogen evolution (so-called direct biophotolysis). However, oxygen is a strong inhibitor of hydrogen production that would therefore require oxygen-tolerant enzymes or adjustment of water splitting and respiration activity in order to attain anoxia. An alternative approach, indirect biophotolysis, involves separation between the water splitting reaction (aerobic stage) and hydrogen production (anaerobic stage). During the aerobic phase, photosynthesis enables carbon dioxide fixation and accumulation of carbohydrates that release reducing power for hydrogen production.

Anoxygenic photosynthesizers, such as green and purple bacteria, do not generate the necessary oxidizing potential to oxidize water and therefore take up electrons from other substrates such as hydrogen sulfide or exogenous organic compounds [16,17]. Some anoxygenic phototrophic organisms such as purple non-sulfur [PNS] bacteria are able to generate large amounts of hydrogen using light-induced extraction of electrons from small organic acids for hydrogen production [18]. The most economic method for hydrogen production using PNS bacteria involves sewage treatment or design of integrated biological systems in which different fermentation processes are combined [19,20].

The present review discusses fundamental aspects of relationships between hydrogen production, photosynthesis and other metabolic pathways in different groups of phototrophic microorganisms. The recent achievements in hydrogen production by green algae, non-nitrogen-fixing and nitrogen-fixing cyanobacteria, as well as PNS bacteria are reviewed.

PHOTOSYNTHESIS AND HYDROGEN BIOPRODUCTION

In green algae and cyanobacteria, photosynthetic pigments absorbing light energy are associated with two photosystems, photosystem I (PS I) and photosystem II (PS II) (Fig. **1**) [11,12]. The excitation energy is transferred from antenna pigments to chlorophyll reaction center molecules where the primary charge separation occurs accompanied by the formation of strong oxidants and reductants. The oxidation of water to molecular oxygen takes place in the oxygen evolving complex (OEC), a functional part of PS II [21,22]. The light-induced charge separation in the reaction center of PS II generates the strongly oxidizing cation radical $P_{680}^{+\bullet}$ which catalyses the water oxidation through a series of redox active components including the Mn_4O_4Ca cluster of the OEC [23]. Electrons extracted from water are transferred subsequently from PS II to PS I *via* the plastoquinone (PQ) pool, the cytochrome b_6f complex, and plastocyanin. PS I generates the electron flow that reduces the iron-sulfur protein ferredoxin (Fd). Further, electrons are transferred from the reduced Fd to $NADP^+$ with the formation of NADPH by a process catalyzed by the enzyme ferredoxin-NADP-reductase.

The OEC is located at the luminal side of the thylakoid membrane (Fig. **1**). Protons are released into the lumen by the OEC (water splitting) and during electron transport *via* the PQ pool that allows generation of a transmembrane electrochemical gradient. The latter drives the generation of ATP *via* the ATP-synthase (photophosphorylation) [24]. Free energy in the form of ATP, along with the reducing agent NADPH, is required to convert carbon dioxide into carbohydrates in the Calvin-Benson cycle [25].

Most non-oxygenic photosynthetic bacteria normally carry out photosynthesis under anaerobic conditions [16,17]. Unlike electron transport in oxygenic photosynthesis, the photosynthetic electron transport of purple bacteria is cyclic and involves two transmembrane proteins: the PS II - type photosynthetic reaction center and the cytochrome *bc₁* complex (Fig. **2**) [26]. Electron transfer between these redox complexes is mediated by ubiquinone and a *c*-type cytochrome. The reaction center catalyzes a light driven electron flow from cytochrome c_2 to ubiquinone while cytochrome bc_1 complex mediates re-oxidation of ubiquinone and reduction of cytochrome *c*. Reduction/oxidation of ubiquinone is linked to the generation of transmembrane proton gradient, which drives ATP synthesis. The photosynthetic electron transport is not a closed cycle as it communicates with different electron transport pathways predominately at the level of ubiquinone. A variety of exogenous substrates donate electrons into the electron transport pathways *via* dehydrogenases. Moreover, purple sulfur bacteria and green sulfur bacteria can use hydrogen sulfide as electron donor. In its turn, the ubiquinol can be re-oxidized with the formation of NADH or reduced Fd that requires ATP energy. The generated reductants and ATP are used for the fixation of dinitrogen or carbon dioxide.

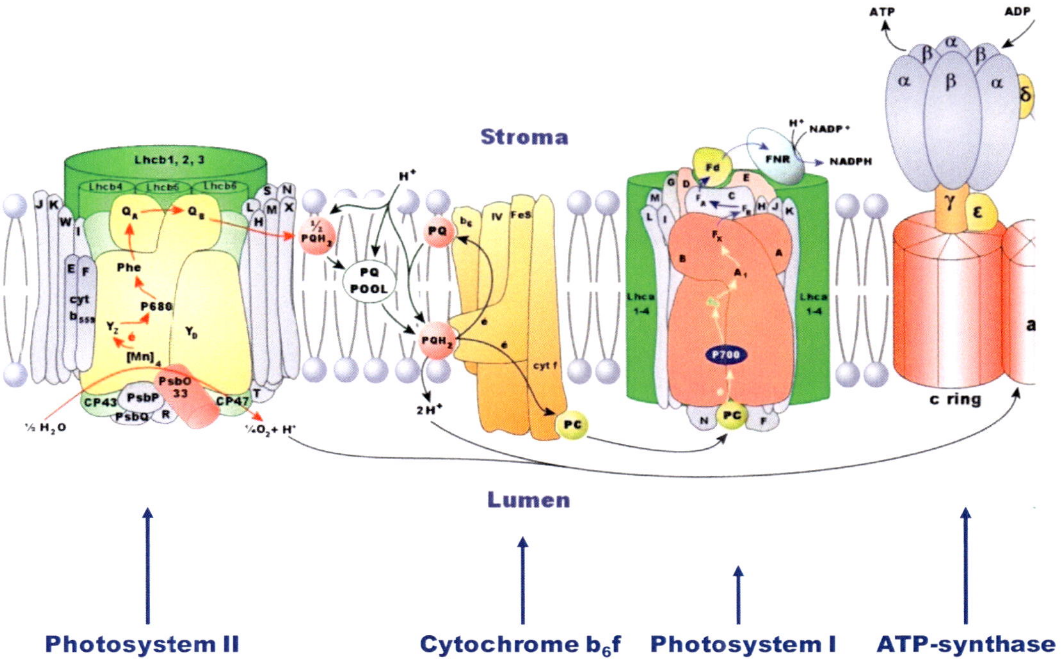

Figure 1: A scheme of the photosynthetic reactions in the chloroplast thylakoid membrane of plants and green algae. Light harvesting protein-pigment complexes bound to PS I and PS II capture light energy needed to drive water splitting and electron transport. Upon illumination, electrons derived from H_2O in PS II are transported along the photosynthetic electron transport chain *via* plastoquinones [PQ], cytochrome b_6f complex, plastocyanin [PC], photosystem I and ferredoxin [Fd] before being used for the production of NADPH by ferredoxin–$NADP^+$ oxidoreductase [FNR]. Simultaneously, protons are released into the thylakoid lumen by PS II and the PQ/PQH_2 cycle to generate a gradient which drives ATP production *via* ATP synthase. NADPH and ATP are used in Calvin-Benson reactions in the stroma for CO_2 fixation/carbohydrates synthesis [not indicated]. The similar reactions also take place in cyanobacterial thylakoids.

Under specific conditions molecular hydrogen can be produced directly in PS II, as it has been demonstrated in *Chlamydomonas reinhardtii* mutants lacking PS I and in preparations of PS II [27,28]. However, normally hydrogen production in phototrophic organisms is coupled to the photosynthetic electron transport chain *via* specific enzymes, hydrogenases (H_2ases) and nitrogenases (N_2ases), which may catalyze generation of molecular hydrogen under certain conditions using reduced Fd or NAD(P)H as an electron donor [reviewed in 29-32]. Reduced Fd is used as an electron donor for the [FeFe]-type H_2ases of green algae and all types of N_2ases. Fd has sufficiently low redox potential and normally requires light energy for its transition from the oxidized into the reduced state. For another hydrogen-evolving enzyme, the cyanobacterial bidirectional [NiFe]-H_2ase, the role of electron donor is played by NAD(P)H. NAD(P)H can be produced either by photosynthetic electron transport or by anaerobic catabolism of organic substrates.

H_2ases catalyze the simple reversible reaction: $2H^+ + 2X_{reduced} \leftrightarrow H_2 + 2X_{oxidized}$, where X is an electron donor, Fd or NAD(P)H. The direction of the reaction depends on the partial pressure of molecular hydrogen and amount of reducing equivalents in the cell. The [FeFe]-H_2ase functions predominately for hydrogen synthesis, while the reaction mediated by the bidirectional [NiFe]-H_2ase essentially depends on the molecular hydrogen content in the gas phase and can be inverted from the direct to reverse reaction at low partial pressure of hydrogen.

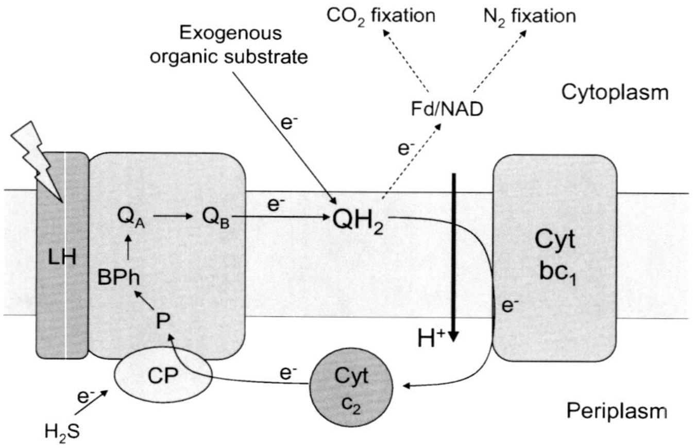

Figure 2: A scheme of the photosynthetic reactions in anoxygenic photosynthetic bacteria. Light is absorbed by bacteriochlorophylls in the light harvesting protein of PS and excitation is transferred to the pigment P of the reaction center, which absorbs photons between 870 nm and 960 nm depending on the species. Once P absorbs a photon, an electron is transferred to the bacteriopheophytin [BPh] *via* bacteriochlorophyll. Further reactions involve the subsequent transfer of an electron from BPh to the electron acceptor, Q_A and the exchangeable quinone molecule Q_B. This molecule is loosely associated with the protein and is fairly easy to detach in a fully reduced and protonated form. The reduced quinone QH_2 diffuses through the membrane to the cytochrome bc_1-complex where it is oxidised. In the process two protons are released to the periplasmic space. The electrons from the cytochrome bc_1-complex are then transferred through a soluble cytochrome intermediate, called cytochrome c_2, in the periplasm to the cytochrome subunit [CP] of PS. Exogenous substrates or H_2S are capable to donate electrons into the cyclic electron transport pathway, while the ubiquinol can be re-oxidized with the formation of NADH or reduced Fd. Dash arrows show reactions which require the energy of ATP.

The role of [FeFe]- and bidirectional [NiFe]-H_2ases in photosynthetic organisms is still a matter of discussions. Probably, hydrogen evolution can be an 'extra' electron sink under anaerobic conditions allowing the release of the excess reducing power in the cell when respiration is blocked [33-35]. Particularly, the burst of hydrogen observed transiently after exposure of dark adapted algal or cyanobacterial cells to the light [36] suggests that H_2ase utilizes photosynthetic electrons before Calvin-Benson reactions become re-activated by light.

N_2ases are complex proteins catalyzing the conversion of molecular nitrogen into ammonia in such photosynthetic organisms as nitrogen-fixing cyanobacteria and anoxygenic photosynthetic bacteria upon their transfer into nitrogen limiting conditions [37-39]. The reaction consumes ATP and has the general form:

$$N_2 + 8H^+ + 8e^- + 16ATP \rightarrow 2NH_3 + H_2 + 16ADP + 16Pi \quad \mathbf{(1)}$$

In this reaction, hydrogen is produced as a byproduct of dinitrogen fixation and normally can be re-cycled *via* active uptake by [NiFe]-H_2ase (uptake H_2ase) expressed simultaneously with N_2ase. Essential amount of hydrogen can be evolved in strains lacking the uptake H_2ase. In the absence of dinitrogen in the gas phase N_2ase catalyzes only hydrogen synthesis:

$$2H^+ + 2e^- + 4ATP \rightarrow H_2 + 4ADP + 4Pi \quad \mathbf{(2)}$$

Substantial energetic requirement for the N_2ase reaction decreases the efficiency of hydrogen production as compared to H_2ase which is independent of ATP availability. However, ATP hydrolysis forces N_2ase to operate exclusively in the direction of nitrogen fixation/hydrogen synthesis irrespectively of the hydrogen content in the gas phase.

Most of hydrogen-evolving enzymes are sensitive to molecular oxygen and can be promptly deactivated at low partial pressures of dioxygen [40,41]. The incompatibility between water splitting (oxygen evolution) and H_2ase/N_2ase-mediated reaction is a major barrier for sustained hydrogen photoproduction. However, different natural and artificial strategies have been invented for the protection of hydrogen-evolving enzymes from dioxygen, and they can be employed for large-scale sustained hydrogen production in the future [reviewed in 9,31,42].

Table **1** represents the main pathways for hydrogen production in phototrophic organisms and also includes general characteristics of the corresponding electron flows. Direct biophotolysis is an attractive mechanism, since hydrogen production is directly coupled to the highly efficient electron transport in the photosynthetic membranes. In practice, this process is transient due to the rapid inactivation of hydrogen-evolving enzymes by molecular oxygen produced in PS II. For the sustained hydrogen production, the process requires O_2-tolerant enzymes or specific conditions of cultivation to reduce oxygen content in the medium. A direct biophotolysis corresponding to a solar conversion efficiency of about 10% was demonstrated *in vivo* with the green microalgae *Chlamydomonas reinhardtii* under low light intensities and very low partial pressure of oxygen [43]. This approach avoids limitation by the light saturation effect and inhibition by photosynthetically produced oxygen. However, it involves the sparging of inert gas for permanent oxygen removal from the medium, that process is thus economically inefficient due to its high cost. The application of oxygen absorbers for the maintenance of anoxia in a cultivator [44] is also not practically justified. Another approach may involve the regulation of water splitting activity and respiration in order to maintain a low oxygen concentration in the cells. Melis *et al.* [14] proposed to use sulfur deficiency for hydrogen photoproduction by green algae. This stress induces a complex physiological response including starch accumulation and down-regulation of PS II activity that results in self-transition of cells into anaerobic conditions followed by the sustained hydrogen production under constant illumination during several days. The main portion of electrons needed in the process originates from the water splitting in PS II, while the minor portion comes from starch anaerobic catabolism linked to the H_2ase *via* the PS I-mediated electron flow (see Table **1**). Starch breakdown results in the formation of small organic acids/alcohols molecules and NAD(P)H. From NAD(P)H electrons are imported into the PQ pool by the NAD(P)H-PQ oxidoreductase and further can be transferred to the [FeFe]-H_2ase *via* cytochrome b_6f, PS I and Fd. Glycolysis has another important aspect in the described mechanism, *viz.* it supplies respiration with the reducing equivalents needed for the removal of oxygen produced in PS II and maintenance of virtually anaerobic conditions [45]. This requirement reduces the overall efficiency of hydrogen photoproduction since the process is limited by the amount of stored carbohydrates. The structure and function of [FeFe]-H_2ase and mechanisms of hydrogen photoproduction in sulfur deprived *C. reinhardtii* are reviewed in sections 3.1 and 4, respectively.

In indirect biophotolysis, the problem of the oxygen sensitivity of hydrogen-evolving enzymes is solved by separating water splitting and hydrogen evolution reactions in space or time (see Fig. **3**). In this case, solar energy and water are used to synthesize and store carbohydrates at the aerobic phase. During an anaerobic phase low potential electrons necessary for hydrogen production are released in the carbohydrates catabolism. Indirect biophotolysis is potentially less effective than the direct one due to the involvement of intermediate steps: synthesis, storage, and degradation of carbohydrates, establishment and maintenance of anoxic conditions. Since it does not require specific treatments or oxygen-tolerant enzymes to couple directly hydrogen evolution to water splitting, it has been successfully applied for the sustained hydrogen production in green algae and cyanobacteria [reviewed in 42,46,47]. In indirect biophotolysis, hydrogen production may require light energy for the coupling with carbohydrates catabolism *via* the PQ pool and PS I (see Table **1**). The less efficient pathway of hydrogen production involves the formation of reduced Fd and/or NAD(P)H upon glycolysis in the dark.

Table 1: The main pathways of hydrogen production in phototrophic organisms. Direct biophotolysis refers to hydrogen evolution coupled to water splitting at PS II *via* PS I. In indirect biophotolysis water splitting [aerobic phase] is separated from hydrogen evolution [anaerobic phase], and both processes are linked *via* subsequent synthesis, storage, and breakdown of carbohydrates. The index above reaction arrows indicate: *1* - photosynthetic electron transport involving subsequently PS II, PQ pool, cytochrome b_6f, plastocyanine, and PS I; *2* - anaerobic catabolism of organic substrate; *3* - photosynthetic electron transport involving PQ pool, cytochrome b_6f, plastocyanine, and PS I; *4* - reaction is mediated by ferredoxin–$NADP^+$ oxidoreductase.

Phototrophic organism/ conditions of cultivation	Type of the process	H_2-evolving enzyme	Electron transport pathways coupled to hydrogen production	Achievement of anoxic conditions
Green algae/ light/sulfur deficiency	Direct+indirect biophotolysis	[FeFe]-H_2ase	Water $\xrightarrow{1}$ Fd ~70%; Starch $\xrightarrow{2}$ NAD(P) $\xrightarrow{3}$ Fd ~30%	PS II down regulation
Nitrogen-fixing cyanobacteria/ light/nitrogen limitation	Indirect biophotolysis	N_2ase (requires ATP)	Carbohydrates $\xrightarrow{2}$ NAD(P) $\xrightarrow{3,4}$ Fd	Heterocyst formation (heterocystous strains), regulation of PS II activity and respiration
Green algae/ light-dark cycles/ nutrient limitation [optionally]	Indirect biophotolysis	[FeFe]-H_2ase	Starch $\xrightarrow{2}$ Fd	Dark respiration
Non-nitrogen-fixing cyanobacteria/ light-dark cycles/ nutrient limitation [optionally]	Indirect biophotolysis	Bidirectional [NiFe]-H_2ase	Glycogen $\xrightarrow{2}$ NAD(P)	Dark respiration
Anoxygenic phototrophs(PNS)/ light/nitrogen limitation	Photo-fermentation	N_2ase (requires ATP*)	Exogenous organic compounds $\xrightarrow{2}$ Fd (requires ATP*)	Anoxygenic photosynthesis

* - ATP is produced *via* non-oxygenic photosynthesis

Nitrogen-fixing cyanobacteria have developed natural mechanisms separating oxygen evolution from nitrogen fixation/hydrogen evolution reactions. The nitrogen-fixing heterocystous cyanobacteria protect N_2ase from the inactivation by dioxygen through the formation of specific cells, heterocysts, in which a micro-oxic environment is provided by thick cell walls to reduce gas flux and by down-regulation of PS II [48-51]. The carbohydrates produced in the light in vegetative cells are transferred into heterocysts where they are anaerobically catabolized with the formation of NAD(P)H. The latter is used to reduce Fd, electron donor for N_2ase, *via* photosynthetic electron transport including the PQ pool and PS I, or *via* the reverse action of ferredoxin–$NADP^+$ oxidoreductase (see Table **1**). Electron transport through the PQ pool is accompanied by the photo-phosphorylation process which provides N_2ase reaction with energy in the form of ATP. The N_2ase enzyme and hydrogen photoproduction in nitrogen-fixing heterocystous cyanobacteria are described in sections 3.2 and 6, respectively.

A large number of non-heterocystous cyanobacteria are also capable of nitrogen fixation [52,53]. These organisms do not produce any obviously differentiated cells and, generally, do not provide spatial separation between nitrogen fixation and photosynthetic production of oxygen, but rather achieve a temporal separation between these antagonistic processes. Mitsui and co-authors [54] showed, that the constantly illuminated synchronous culture of marine cyanobacteria *Synechococcus* spp. is capable of nitrogen fixation/hydrogen production and aerobic photosynthesis at different phases of the cell division cycle. Noteworthy, many nitrogen-fixing non-heterocystous cyanobacteria can grow and fix nitrogen under continuous illumination due to endogenous diurnal rhythms responsible for the regulation of PS II activity and respiration [55-57].

Some of non-heterocystous nitrogen-fixing cyanobacterial strains separate photosynthesis and nitrogen-fixation/hydrogen evolution by performing these functions during light and dark periods, respectively [58].

The similar strategy (light/dark cycles) is applicable for the non-nitrogen-fixing photosynthetic microorganisms (see Table **1**). Thus, the pioneering work of Gaffron and Rubin [59] demonstrated that the green alga *Scenedesmus* is capable of producing hydrogen gas fermentatively under dark anaerobic conditions using intracellular starch as a source of reducing equivalents. Based on these experiments, Miura and coauthors [60] proposed to use light and dark periods for accumulation of photosynthetic carbohydrates and fermentative hydrogen production, respectively. This approach is described in section 5.

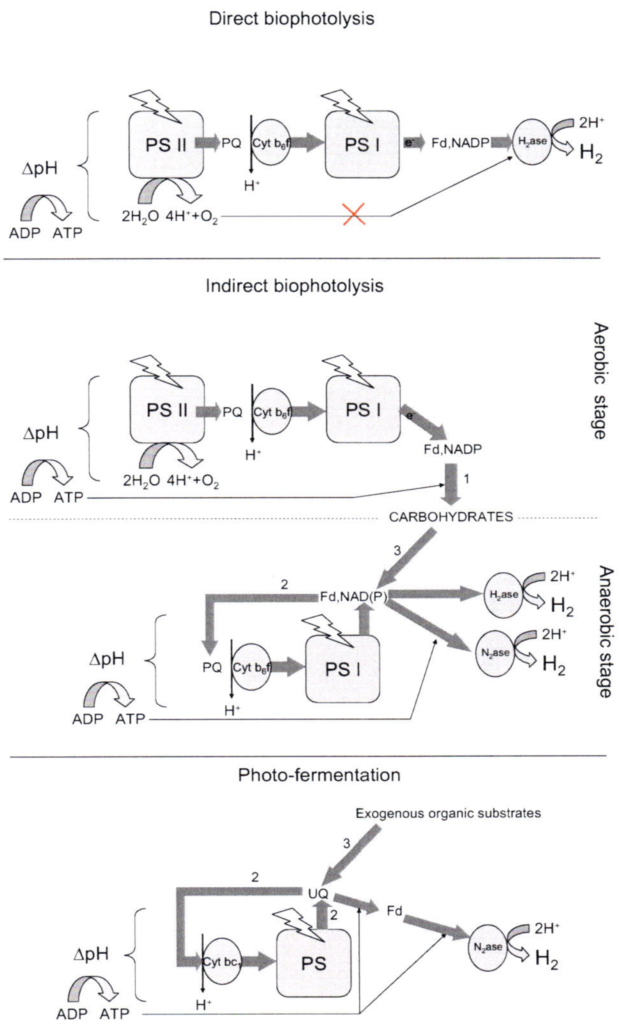

Figure 3: A scheme of the main electron transport pathways from substrates to the hydrogen-evolving enzymes *via* direct and indirect biophotolysis in green algae and cyanobacteria, and *via* photo-fermentation in PNS bacteria. 1 - Calvin-Benson reactions; 2 - cyclic electron flow around photosystem; 3 - anaerobic catabolism of carbohydrates. In indirect biophotolysis, reducing equivalents for hydrogen production can be generated either in the light [in heterocyst] or in the dark. Microorganisms producing hydrogen *via* indicated pathways are shown in Table **1**.

Unlike green algae and cyanobacteria, anoxygenic photosynthetic bacteria derive low potential electrons for nitrogen fixation/hydrogen production from exogenous inorganic or organic compounds (see Figs. **2**, **3**, Table **1**). In PNS bacteria which have been recognized as attractive organisms for hydrogen photoproduction, electrons are released from exogenous organic compounds and transferred into the photosynthetic chain at the level of ubiquinone cycling around the single bacterial photosystem upon illumination. The cyclic electron flow drives the photo-phosphorylation process and generates ATP for the dinitrogen fixation accompanied by hydrogen evolution [61,62]. Anoxygenic photosynthetic bacteria can use a wide variety of substrates for growth and hydrogen production. Due to the ability to use small-chain organic compounds for production of hydrogen, PNS bacteria are applicable for hydrogen production in combination with anaerobic heterotrophic bacteria processing biomass or food waste in dark fermentation [20,63,64]. The recent achievements in photo-fermentation process are reviewed in section 7.

HYDROGEN-EVOLVING ENZYMES

[NiFe]- and [FeFe] - H_2ases

Both [NiFe]- and [FeFe]- H_2ases use hydrogen as the only substrate or product of the reaction [35,65]. They catalyze both the uptake and evolution of hydrogen depending on the redox state and concentration of the components involved in the reaction which is important for the balancing of the redox potential in a cell. Thus, when the redox potential is low the excess of the reducing power is consumed in hydrogen synthesis, while the reverse reaction occurs when cell is deprived of reducing equivalents and molecular hydrogen is present in sufficient concentration. In general, the [NiFe]-H_2ases occur more often in conditions that favor the splitting of molecular hydrogen whereas the [FeFe]-H_2ases are mostly involved in hydrogen production.

The crystal structures of several [NiFe]-H_2ases and one [FeFe]-H_2ase from sulfate reducing bacteria, and another [FeFe]-H_2ase from carbohydrate-fermenting bacterium *Clostridium pasteurianum* have been previously reported [66-71]. Recently, the structure of the active site of [FeFe]-H_2ase from *C. reinhardtii* has been studied by X-ray absorption spectroscopy [72]. The H_2ases demonstrate the complex nature of the active sites containing two metal atoms and biologically unusual ligands, CN and CO [reviewed in 30,73-76]. Although the [NiFe]- and [FeFe]-H_2ases are phylogenetically unrelated, the fact that they share a unique property of the catalytic center, namely, CO and CN coordination to a chemically active Fe atom, indicates that such organization is central to biological hydrogen catalysis.

The active sites of [NiFe]- and [FeFe]-H_2ases are deeply buried within the protein globule. Therefore these enzymes require special pathways to transfer the substrates and products of the enzymatic reaction between the catalytic center and the molecular surface. The putative hydrophobic channel connects the protein surface with the active site to provide the exchange of hydrogen molecules and probably the access of molecular oxygen to the catalytic center [77,78]. Dioxygen is known to inactivate most of the known H_2ases, since hydrogen metabolism is restricted by anaerobic conditions. Oxygen may either reversibly or irreversibly oxidize the catalytic center as well as down regulate transcription of the H_2ase encoding genes. This high oxygen sensitivity constitutes a major obstacle in practical applications of phototrophic organisms for the production of hydrogen.

Among photosynthetic organisms cyanobacteria possess two types of [NiFe]-H_2ases: uptake and bidirectional, while in eukaryotic microorganisms including green algae only [FeFe]-type H_2ase was found.

The Algal [FeFe]-H_2ases

In 1942 Hans Gaffron and Jack Rubin discovered that the green alga *Scenedesmus obliquus* is capable of producing molecular hydrogen gas under anaerobic conditions in the absence of molecular hydrogen and carbon dioxide, whereas in their presence the reverse reaction occurs. Other green algae, such as *Chlorella* and *Chlamydomonas*, also produce hydrogen under strictly anoxic conditions.

An enzyme responsible for hydrogen evolution in green algae was purified from *C. reinhardtii* by Roessler and Lien [79] and assigned to the class of [FeFe]-H_2ases found also in some strict anaerobes, fungi, and protists [64,80]. Unlike other types of H_2ases, this enzyme is localized in chloroplasts or special organelles and rather involved in hydrogen production than splitting. Fd serves as the redox partner for the [FeFe]-H_2ase. In green algae the redox potential of Fd measured at pH 8.0 is sufficiently low (-390 to -425 mV) [81], and its transition from the oxidized to the reduced form is usually coupled to electron transport in PS I [82-84]. Most [FeFe]-H_2ases, including those found in green algae, are extremely active in generation of hydrogen, each molecule of the enzyme can produce up to 10^4 molecules of the gas per second at room temperature [85].

Algal [FeFe]-H_2ases are monomeric and consist of a catalytic protein only [30,64,86] (Fig. **4**). The H_2ase genes characterized in diverse green algal species encode a protein of 45-48 kDa with about a half sequence similarity related to the evolutionary conservative region [81,87]. The three dimensional structure of [FeFe]-H_2ase showed that the metallo-catalytic site, the so-called H-cluster, consists of a [4Fe4S] cubane linked through a protein cystein residue

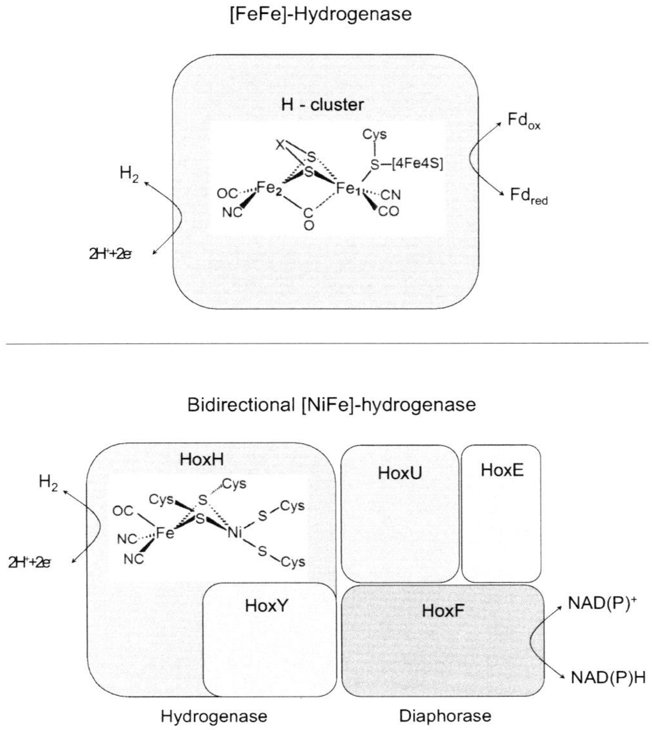

Figure 4: Schematic representation of algal [FeFe]- and cyanobacterial [NiFe]-H_2ases including the structure of the bimetallic catalytic centers.

to a bimetallic Fe-Fe subcluster [82,88,89]. Each Fe atom of the bimetallic subcluster is coordinated to CO and CN diatomic ligands, while two sulfur atoms and a small molecule (carbon monoxide or di(thiomethyl)amine) bridge Fe atoms. The Fe atom distal to the [4Fe4S] cluster (Fe_2 in Fig. **4**) has a vacant coordination site; it is therefore thought to be the position where dihydrogen or hydride ion binds during enzyme turnover.

In algal H_2ases electrons are delivered directly to the H-cluster by the reduced Fd [83,84]. The proposed mechanism of the catalytic reaction includes the double reduction of the Fe_2 in the bimetallic cluster and its stabilization in this form by the strong-field ligands CO and CN [89-91]. The next steps of the reaction involves the double reduction of a proton also bound to the Fe_2, followed by recombination between the resulting hydride anion and a second proton donated by either the di(thiomethyl)amine or by a nearby lysine or cysteine residue that results in the formation of molecular hydrogen.

Expression of the H_2ase genes is tightly regulated in green algae. In *C. reinhardtii,* for example, anaerobic conditions are prerequisite for the transcription of two H_2ase structural genes [87,92]. Additionally, two maturation genes are involved in the biosynthesis and assembly of the H-cluster of the enzyme [91]. Besides anaerobic conditions the additional factors such as sulfur deficiency and redox potential in the cell are involved in regulation of transcription [93,94]. However, the precise regulatory mechanisms are still unknown and more studies on this problem are required.

Due to the hydrophobic channel(s) in [FeFe]-H_2ase protein structure the active site located in the interior of the protein is accessible to oxygen, well-known inhibitor of the H_2ase reaction [72,78,95]. The algal H_2ases are extremely sensitive to oxygen which causes irreversible damage of the H-cluster of the enzyme with a half-time of a few seconds. The precise chemical nature of the species generated in the H-cluster after exposure to oxygen is not known. The catalytically active Fe_2 with an unoccupied coordination site was suggested to be a primary target for oxidation by molecular oxygen. [FeFe]-H_2ases are inactivated with a half-time of several hours even under strictly anaerobic conditions.

The Cyanobacterial Bidirectional [NiFe]-H_2ases

The [NiFe]-H_2ases constitute the most numerous and well studied class of bacterial H_2ases. Cyanobacteria possess two functionally different types of [NiFe]-H_2ases, an uptake and a bidirectional enzyme [29,30]. The former is responsible for taking up and recycling hydrogen produced as a side reaction from N_2ase enzyme and found predominately in nitrogen-fixing cyanobacteria. Unlike uptake H_2ases, bidirectional [NiFe]-H_2ases function in the directions of hydrogen uptake as well as hydrogen evolution depending on the physiological requirements of the cell [6,9,10]. An important intracellular redox intermediate NAD(P) is used as an electron donor/acceptor for the reaction. The redox potential of NAD(P) is about -350 mV at pH 8.0 that is higher than that of Fd and generation of NAD(P)H may involve either glycolysis or photosynthetic electron transport. Cyanobacterial bidirectional [NiFe]-H_2ase shows lower activity as compared to the algal [FeFe]-H_2ase; this difference can be explained by different properties of the catalytic center.

The bidirectional [NiFe]-H_2ase was firstly discovered by Houchins and Burris [96] in the cyanobacterium Anabaena 7120. Thereafter, this enzyme was found in nitrogen-fixing and non-nitrogen-fixing strains, including Nostoc, Synechocystis, Gloeocapsa, Synechococcus, Spirulina [97,98]. The bidirectional NAD(P)-linked H_2ases are found also in anoxygenic photosynthetic bacteria, e.g. Thiocapsa roseopersicina [99,100].

Cyanobacterial bidirectional H_2ases have more complex structure as compared to [FeFe]-H_2ase in green algae (Fig. **4**). The bidirectional H_2ase is a heteropentameric enzyme composed of the hydrogenase and diaphorase part. The hydrogenase part consists of the HoxH and HoxY subunits while the diaphorase part includes HoxU, HoxF and HoxE subunits [101-105]. The large subunit of the hydrogenase moiety, HoxH, contains the bimetallic catalytic center with Ni and Fe atoms. The small subunit, HoxY, has a putative [4Fe4S] cluster which likely participates in electron transfer to the large catalytic subunit. The diaphorase moiety is a flavoprotein homologous to subunits of respiratory complex I of the mitochondrial and bacterial respiratory chains. It contains NAD(P), FMN, and FeS binding sites and channels electrons between NAD(P) and the hydrogenase part in both directions. While the HoxFU is definitely involved in interaction with NAD(P) and electron transfer, the HoxE is likely a bridging subunit providing attachment of the enzyme to the thylakoid membrane. However, HoxE can be also involved in electron transport, since it contains a putative FeS cluster [102,105].The crystal structure of [NiFe]-H_2ases is available from several closely-related strains of sulfate reducing bacterium *Desulfovibrio* [reviewed in 73]. It revealed that the two metal atoms in the catalytic center, Ni and Fe, are held in close proximity by the sulfur atoms provided by two cystein residues of the protein (Fig. **4**). Like [FeFe]-H_2ase, the iron has CN and CO groups as ligands, whereas the nickel ion is coordinated by two additional cysteins [66].

The catalytic mechanism of bidirectional [NiFe]-H_2ase remains unclear. It is rather similar to that proposed for the bacterial uptake [NiFe]-H_2ase [40] but operating in reverse sequence for hydrogen production. The first step of hydrogen cleavage mediated by the uptake [NiFe]-H_2ase results in formation of a hydride anion and a proton in a bimetallic center. Then two electrons are extracted from the hidride and transferred to $NAD(P)^+$ *via* FeS clusters, while the released protons are delivered to the protein environment to complete the reaction [reviewed in 30,76].

Generally, the bidirectional [NiFe] - H_2ases are more tolerant to oxygen than [FeFe]-H_2ases [106,107]. Thus, the cyanobacterial bidirectional H_2ases remain active under microaerobic conditions and their inactivation by oxygen is reversible. The putative mechanism of inactivation of [NiFe] - H_2ases involves formation of an oxo- or a hydroxo-group in the catalytic center which bridges the Ni and Fe atoms and probably protects the [NiFe]-H_2ase from further oxidation [40]. In anaerobic conditions the oxo- or hydroxo-group is re-reduced in the presence of hydrogen or reducing agents, such as NAD(P)H, and the H_2ase activity is rapidly restored [36, 91].

Unlike [FeFe]-H_2ases in algae, the transcription of the bidirectional H_2ase genes in cyanobacteria (*hox*) is constitutive. However, the transcript levels of all *hox* genes are increased in anoxia [104,108]. Thus, the enzyme is likely present upon aerobic cultivation to account for the rapid appearance of activity and hydrogen production when anaerobic conditions are restored. Macronutrient deficiency positively regulates the transcription of *hox* genes [104,109]. This up-regulation is likely a part of the general response to stress regulated by the transcription repressor LexA [110]. As suggested in a recent study [111] abundance of *hox* transcripts in *Synechocystis* increases in the light and depends on the activity of the Calvin-Benson cycle.

The Role of H_2ase-Mediated Hydrogen Production in Photosynthetic Microorganisms

Establishment of anaerobic conditions in the dark leads to the transition of intracellular redox intermediates into the reduced state mainly due to the block of respiration and inactive state of Calvin-Benson cycle, both are the major electron sinks in a cell. Moreover, catabolism of the stored organic compounds in the dark would additionally increase the reducing power. The over-accumulation of the reducing equivalents may lead to a slowing down of the electron transport and the coupled phosphorylation process depleting the cell of ATP required for growth. The endogenous regulation of the redox potential in the cell would adjust the ratio of NAD(P)H to ATP with regard to the needs of an organism. It seems that the algal [FeFe]- and cyanobacterial bidirectional [NiFe]-H_2ases provide a regulatory mechanism for the regeneration of the oxidized state of intracellular redox intermediates under anaerobic conditions.

Transition of green algae or some cyanobacteria from dark to light conditions can be accompanied by a transient burst of hydrogen [36, 59]. This phenomenon is attributed to PS I-mediated short-term increase in the concentrations of reduced Fd and NADPH, their re-oxidation being slow because the enzymes of the Calvin-Benson cycle are inactive in dark adapted cells. The over-reduced state of the photosynthetic electron transport chain is known to facilitate generation of potentially harmful reactions with the formation of reactive oxygen species. One such reaction is electron flow to oxygen, the Mehler reaction [112], which results in the production of superoxide, O_2^-, at the acceptor side of PS I [113]. In this situation, H_2ases provide protection of photosynthetic apparatus from the damage caused during the short period between the onset of illumination and the light-induced activation of carbon dioxide fixation. Therefore, these enzymes provide a 'valve' for the removal of the excess of low potential electrons formed in the photosynthetic electron transport chain.

Nitrogenases

Dinitrogen fixation is utilized by numerous prokaryotes, including anoxygenic photosynthetic bacteria and several groups of cyanobacteria. N_2ases are metalloenzymes that catalyzes the key step in global nitrogen cycle: conversion of atmospheric dinitrogen into ammonia by coupled electron- and proton-transfer reactions [38]. The enzyme is expressed upon limitation of nitrogen sources under anaerobic or microaerobic conditions in photosynthetic bacteria and nitrogen-fixing cyanobacteria.

The reduction of one dinitrogen molecule by the N_2ase minimally requires 16 magnesium ATP (MgATP), eight protons, and eight electrons (see formulae 1) and involves the transient interaction of two protein components. Along with a dinitrogen fixation, one hydrogen molecule is synthesized in an obligatory side reaction consuming approximately two electrons, two protons, and four MgATP.

Three types of N_2ases have been characterized with regard to the metal composition in the active site [114,115]. Molybdenum N_2ase contains Mo and Fe atoms in the active site (FeMo-cofactor), vanadium N_2ase contains V and Fe atoms, whereas iron N_2ase does not contain any transition-metal atoms besides Fe. Studies on these three types of N_2ases have shown that the molybdenum N_2ase is the most efficient in nitrogen fixation providing higher yield for the process [114]. However, the V-N_2ase has about three times higher hydrogen production per one reduced dinitrogen than the Mo-nitrogenase [116].

The molybdenum containing enzyme is the most widely studied among N_2ases. It consists of the dinitrogenase *per se* (MoFe protein) and its redox partner nitrogenase reductase (Fe protein). Their X-ray crystallographic structures are available at various resolutions for *Clostridium pasteurianum* and

Azotobacter inelandii [reviewed in 32,117]. The MoFe protein is a $\alpha_2\beta_2$ heterotetramer with molecular weight of about 220 to 240 kDa encoded by *nifD* (α) and *nifK* (β) (Fig. **5**). Each α subunit harbors an active site, the FeMo-cofactor ($MoFe_7S_9N$·homocitrate), which binds a substrate and catalyzes reduction of dinitrogen bonds. A second [8Fe7S] metallocluster in MoFe protein, termed the P-cluster, is located between each α and β subunit and likely mediates electron transfer between the Fe-protein and the FeMo - cofactor [118]. Structural investigations of possible channels for substrate transport through the MoFe protein have revealed putative hydrophobic as well as hydrophilic channels [119]. However, the mechanism of substrate transfer to the active site remains unknown.

The Fe protein is a γ_2 homodimer of about 60 to 70 kDa encoded by *nifH.* It plays a specific role of mediating the transfer of electrons from an external electron donor like Fd or flavodoxin to the N_2ase [120]. The interaction with electron donors is provided by a [4Fe4S] cluster which bridges the γ subunits. Besides the [4Fe4S] cluster the Fe protein contains a MgATP binding sites. The MgATP hydrolysis is required to induce conformational changes of the Fe protein facilitating electron transport from the [4Fe4S] cluster of the Fe protein to the P-cluster of the MoFe protein.

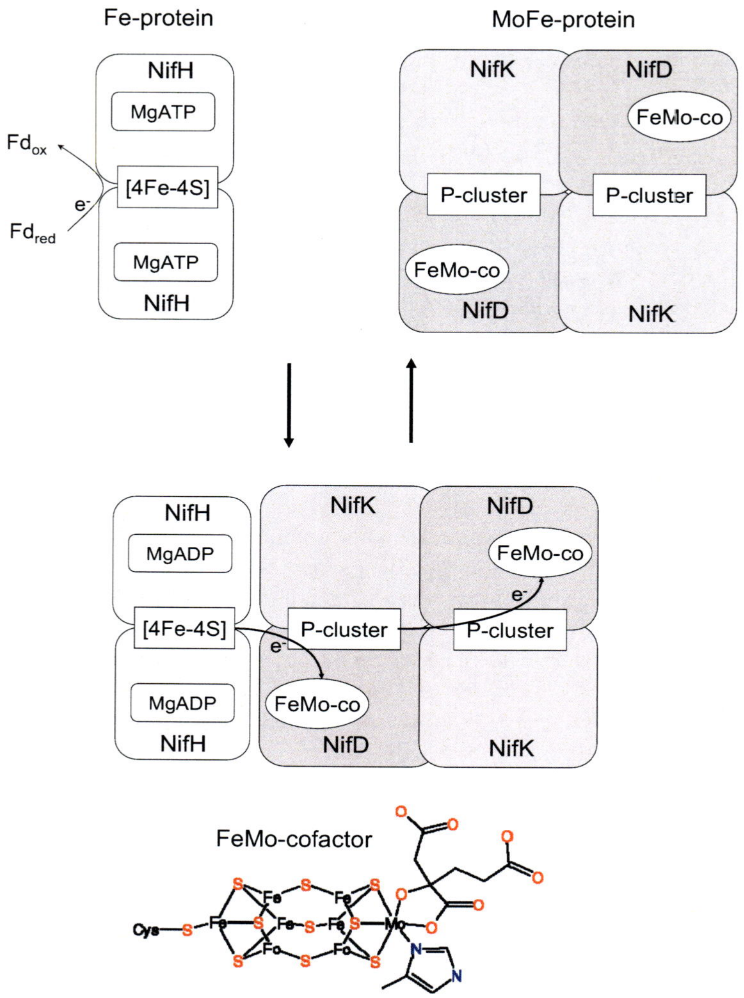

Figure 5: Schematic structure of the N_2ase including FeMo-cofactor and a suggested mechanism of electron transfer from the reduced Fd to the active site. The Fe-protein reduced by Fd binds to the MoFe-protein. Thereafter, two MgATP are hydrolyzed inducing conformational changes which enable electron transfer from the FeS cluster in the Fe-protein to the P-cluster and further to the FeMo-cofactor in the MoFe-protein. The complex dissociates again, MgADP is exchanged for MgATP, and the cycle starts over. When enough electrons are accumulated in catalytic centers, dinitrogen reduction to ammonia and protons reduction to hydrogen molecule occurs [not shown].

Proton transfer is also important for the action of cofactors and finally for dinitrogen fixation. Three putative proton-transfer routes have been suggested by structural and theoretical analysis of the N_2ase [39,119,121]: a water-filled channel running from the protein surface to the homocitrate ligand of FeMo-cofactor, and two hydrogen-bonded chains ended to the specific sulfur atoms in FeMo-co.

Low potential electrons for N_2ase reaction are generated *in vivo via* substrate oxidation, photosynthesis, or combination of these routes, depending on the organism. Electrons are transferred to flavodoxin or Fd, and further to the Fe protein of N_2ase which initiates a series of ATP-dependent redox cycles before dinitrogen is converted to ammonia (Fig. **5**). During each cycle the reduced Fe protein binds to the MoFe protein, and hydrolysis of MgATP occurs thereafter. After this the complex undergoes conformational changes facilitating the transfer of an electron from the Fe protein to the MoFe protein, *via* the P-cluster to the FeMo-cofactor. The complex dissociates again, MgADP is exchanged for MgATP, and the cycle can start again. Kinetic studies of the mechanism of dinitrogen fixation indicate that the rate-limiting step of the reaction is the slow dissociation of the two proteins, which determines the relatively low turnover rate of the process (6-7 s^{-1}) [122].

The maturation of the N_2ase is a complex process because of the structural complexity of the enzyme. It involves a multi-step process of transcription, and formation, insertion and activation of cofactors, and requires a complex mechanism for regulation [123,124]. N_2ase is encoded by a set of operons which includes regulatory genes (such as *nifLA*), structural genes (such as *nifHDK*) and other supplementary genes. A nif-regulatory gene product is involved in the mechanism of oxygen control of N_2ase synthesis.

Like H_2ases, the N_2ases are highly sensitive to molecular oxygen [65], which irreversibly oxidizes the FeS cofactors and down-regulate transcription of *nif* genes. This requires specific mechanisms for nitrogen fixers to avoid oxygen under aerobic conditions. One known exception, a N_2ase of *Streptomyces thermoautotrophicus*, is unaffected by the presence of molecular oxygen [125].

In the absence of dinitrogen in the gas phase the N_2ase is capable of catalyzing exclusively hydrogen production (see formula 2) that is the case *e.g.* in an argon atmosphere. This fact raises interest to this enzyme with regard to hydrogen production. Although the high energetic requirements and relatively low turnover rate would potentially lower the efficiency of the N_2ase-mediated hydrogen production as compared to that catalyzed by the H_2ase, the N_2ase enzyme is still attractive with respect to hydrogen production due to several advantages. Thus, while the H_2ase reaction is reversible and proceeds either in the direction of hydrogen synthesis or uptake depending on the redox state of electron donors and a partial pressure of hydrogen, the N_2ase operates strictly for hydrogen synthesis. Moreover, nitrogen fixing organisms have evolved several strategies for the protection of N_2ase from molecular oxygen under oxygenic environment that can be used also in artificial systems. In photosynthetic nitrogen-fixing organisms such as cyanobacteria and anoxygenic phototrophic bacteria the high energetic requirements of N_2ase are covered by the harvesting of solar light.

THE PHYSIOLOGY OF HYDROGEN PHOTOPRODUCTION

Hydrogen Photoproduction by Green Algae in Response to Sulfur Deficiency Stress

The physiological responses of green algae to macronutrient (nitrogen, sulfur, phosphorus) limitation include cessation of growth and cell division, as well as reduced photosynthesis, carbon dioxide assimilation, and accumulation of large amounts of starch [126-129]. Melis *et al.* [14] investigated the effect of sulfur deprivation with regard to hydrogen photoproduction by unicellular green algae *C. reinhardtii* supplied with exogenous acetate. They showed that after 1–2 days of adaptation to sulfur depletion, constantly illuminated cells produced volumetric amounts of hydrogen gas for a few days. The proposed protocol involves temporal separation between aerobic and anaerobic stages of incubation. During a first aerobic stage of sulfur deprivation, oxygenic photosynthesis drives accumulation of large amounts of starch, while during the second anaerobic stage, starch degrades upon glycolysis and the [FeFe]-H_2ase is induced, and sustained hydrogen production occurs. The transition into anaerobiosis is mainly due to a dramatic decrease in photosynthetic oxygen evolution [14,127,130]. Respiration stays at the same level or increases [14,131] leading to the consumption of oxygen from the medium.

The separation between aerobic (photosynthesis) and anaerobic (fermentation) phases suggests indirect biophotolysis occurs, as shown in Fig. **3**. However, 3-(3,4-dichlorophenyl)-1,1-dimethylurea (DCMU), an inhibitor of electron transport in PS II, significantly reduces hydrogen photoproduction by sulfur deprived *C. reinhardtii* [132,133] indicating that electrons for the H_2ase reaction are predominately extracted from water, *i.e.* in fact a direct biophotolysis is a major process in anaerobic phase. In general, hydrogen photoproduction in sulfur deprived green algae is coupled to both PS II-dependent and PS II-independent metabolic pathways (see Table **1**, Fig. **6**). The PS II-dependent pathway contributes usually about 70% of total hydrogen production and involves the residual water-splitting activity which does not exceed 20% of maximal observed under optimal conditions of cultivation. Electrons extracted from water are transported to the H_2ase *via* PS II and PS I.

The PS II-independent route involves the import of electrons from NAD(P)H produced upon starch catabolism into the photosynthetic chain at the level of the PQ pool and subsequent transfer to the Fd and H_2ase *via* PS I [36,134]. Recently, a type II NAD(P)H dehydrogenase has been evidenced in *C. reinhardtii* chloroplasts which is rather involved in non-photochemical reduction of PQs [135,136]. Sulfur deficiency essentially reduces the activity of the NAD(P)H dehydrogenase [137].

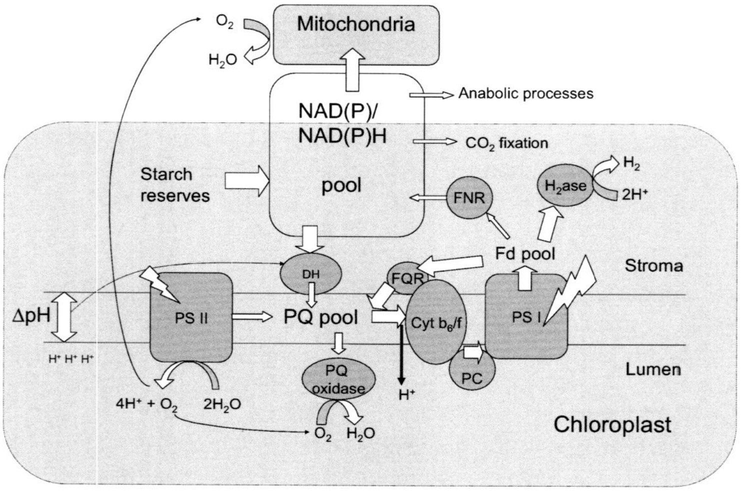

Figure 6: A tentative scheme of electron transport routes in sulfur deprived green algae during the hydrogen production phase [adapted from Antal *et al.* [137]**.** The electron flow pathways increased or reduced in sulfur deprived cells are indicated by thick or thin arrows, respectively. Besides residual activity of PS II, anaerobic catabolism of starch provides reducing equivalents for the H_2ase reaction. The PS II-independent pathway is mediated by the dehydrogenases [DH], PQs, and PS I. The activities of PS II and NAD(P)H dehydrogenase are significantly reduced under sulfur deprivation. The decrease in CO_2 fixation slows down linear electron transport from PS I to NADP, thus inducing cyclic electron transport around PS I which involves antimycin A - sensitive ferredoxin-plastoquinone reductase [FQR]. Cyclic electron flow reduces hydrogen production due to the competition for the reduced Fd. Moreover, cyclic electron flow maintains 'extra' ΔpH gradient which suppresses the activity of NAD(P)H dehydrogenase [dotted arrow]. The over-reduced state of the PQ pool induces transition of photosynthetic membranes from state 1 to state 2; the latter favors excitation of PS I and cyclic electron flow. The photosynthetically produced oxygen is removed predominately in mitorespiration.

PS II Regulation

When *C. reinhardtii* cells are transferred into sulfur-depleted medium, PS II undergoes the fastest degradation among components of the electron transport chain in the photosynthetic membranes [14,127,130]. The degradation of PS II and water splitting activity is attributed to down-regulation in the *de novo* biosynthesis of the D1 protein in PS II reaction center induced by sulfur deprivation [4,6,10,27,28,127]. This protein is known to suffer from frequently occurring oxidative damage under light illumination [138]. Inactivation of PS II finally leads to the establishment of anaerobic conditions in the cell suspension kept in a sealed bioreactor, followed by switching of the metabolism into the hydrogen production mode.

Besides the relatively slow process of PS II damage, a rapid (minutes) complete down-regulation of PS II activity occurs at the moment of the establishment of anaerobic conditions in sulfur deprived *C. reinhardtii* [139,140]. PS II photochemical activity under constant illumination was evaluated by measurement of the fluorescence parameter $\Delta F/F_M'=[F_M'-Ft]/F_M'$ with a pulse amplitude modulated fluorometer (PAM-2000). The rapid loss of the photochemical activity was due both to a reduction in maximal fluorescence level (F_M') and a rapid increase in steady state fluorescence (*Ft*). The latter has been attributed to an increase in the number of PS II centers with reduced primery electron acceptor quinone (Q_A) caused most likely by the rapid reduction of the PQ pool and other electron transport carriers due to lack of oxygen. Indeed, while slow inactivation of PS II requires sulfur to induce re-synthesis of core proteins, the rapid process is regulated by oxygen availability suggesting redox control of PS II activity [133,140]. At this stage of sulfur deprivation, respiration remains the only essential sink for the low potential electrons because the Calvin-Benson cycle and other anabolic processes are severely inhibited (Fig. **6**). PS II centers with reduced Q_A are unable to perform photochemical energy conversion and turn into inactive state until Q_A becomes re-oxidized again. The decrease in F_M' has been attributed to the decrease in PS II excitation owing to transition of photosynthetic membranes in state 1 (favors PS II excitation) to state 2 (favors PS I excitation). Indeed, state transition is regulated by the redox state of the PQ pool [141], which becomes completely reduced upon establishment of anaerobic conditions.

Shortly after the establishment of anaerobic conditions, gradual re-activation of PS II activity occurs [133,142]. This process is accompanied by induction and increase in hydrogen photoproduction suggesting H_2ase reaction enables water splitting *via* re-oxidation of the photosynthetic electron transport chain including the PQ pool and Q_A. The maximal re-activation of PS II during the anaerobic phase of sulfur deprivation is observed when *C. reinhardtii* cells are incubated autotrophically [142]. It seems that in the absence of exogenous acetate in the medium, the water splitting reaction is largely involved in hydrogen production as compared to hetero- or mixotrophic cultures.

The H_2ase-dependent regulation of the remaining water splitting activity in sulfur deprived *C. reinhardtii* under anaerobic conditions may represent a general mechanism for algal acclimation to the nutrient deficiency stress [143,45]. In starved cells under anaerobic conditions the H_2ase is the only sink for the excess of reducing power. When H_2ase is inactive, both linear and cyclic photosynthetic electron transport routes are inactive because their components are completely reduced. As consequence, the photo-phosphorylation process is completely blocked. For the same reason, substrate-level phosphorylation is slowed down. The oxidative phosphorylation in mitochondria is inhibited in the absence of oxygen. The inhibition of phosphorylation would deprive cells of energy supply in the form of ATP. The re-oxidation of photosynthetic electron carriers *via* release of molecular hydrogen permits photo-phosphorylation and re-activation of PS II; the latter is accompanied by oxygen evolution and, consequently, drives electron transport and oxidative phosphorylation in mitochondria. Moreover, the photosynthetic electron flow is coupled to starch catabolism *via* the activity of NAD(P)H dehydrogenase suggesting also a redox regulation of glycolysis *via* evolution of hydrogen. Therefore, the capacity of nutrient deficient green algae to evolve hydrogen in the light may represent a survival mechanism which adjusts cell metabolism to anaerobic conditions driving redox reactions and providing ATP for cell needs.

The Role of Starch

Sulfur deprivation of *C. reinhardtii* triggers a nearly ten-fold increase in cellular starch content and significant cease in cell division [130]. It seems that photosynthesis suffers in lesser extent than anabolic processes responsible for cell division, and this induces the storage of unconsumed energy in the form of starch. In heterotrophic cultures, exogenous acetate serves as a direct substrate for respiration during the aerobic phase of sulfur deprivation [14]. Consequently, starch formation becomes the only sink for electrons coming from water splitting activity.

The importance of starch catabolism in hydrogen production has been recognized from the earlier work [144]. As shown in [14] *C. reinhardtii* incubated heterotrophically does not uptake measurable amount of acetate during the anaerobic (hydrogen production) phase of sulfur deprivation. This indicates that sustained hydrogen production requires primarily the anaerobic catabolism of starch for the removal of oxygen produced due to the remaining water splitting activity in PS II and for the direct feeding of electrons to the process *via* non-photochemical reduction of the PQ pool [145,146]. Indeed, the observation that starchless *C. reinhardtii* mutants *sta6* and *sta7* have a reduced capacity to evolve hydrogen [147] testifies for the important role of starch as a key factor for the sustained hydrogen photoproduction in sulfur deprived green algae.

Cyclic Electron Transport around PS I and Hydrogen Photoproduction

Among other photosynthetic components, Rubisco is a primary target for sulfur deficiency stress. Reduced activity of the reactions of the Calvin-Benson cycle slows down the re-oxidation of the photosynthetic electron transport chain, thus leading to accumulation of reduced compounds. The over-reduced state of the PQ pool is known to induce transition of photosynthetic membranes from the state 1 to the state 2 in which excitation of PS I is favored [141]. Wykoff and co-authors [127] showed that sulfur deprivation induces a re-adjustment of the relative antenna cross-sections in favor of PS I in *C. reinhardtii*. The transition from state 1 to state 2 in *C. reinhardtii* might switch the electron transport from the linear to the cyclic mode [148] suggesting the latter is a major pathway in sulfur deprived cells.

During the anaerobic (hydrogen production) phase, cyclic electron transport and H_2ase reaction may compete with each other for reduced Fd (see Fig. **6**), as suggested by Kruse *et al.* [149]. In fact, as shown in Kruse *et al.* [149], *C. reinhardtii* strain *stm6* blocked in state 1 and thus incapable of cyclic electron flow produces more hydrogen than the wild type. On the other hand, the mutant also showed a modified respiratory metabolism which results in the accumulation of large amounts of starch, an important prerequisite for hydrogen production. This makes difficult to distinguish whether a decreased cyclic electron flow or an over-accumulation of starch is critical for the elevated hydrogen production in *stm6*. As shown in a recent study [137], cyclic electron flow around PS I mediated by the antimycin A -sensitive ferredoxin-quinone-reductase [FNR] slowed down hydrogen evolution in sulfur deprived *C. reinhardtii* by about two and half fold. This result has been assigned primarily to the competition between FNR and H_2ase for reduced Fd. Moreover, the authors suggested that cyclic electron transport around PS I promotes extra ΔpH gradient across the photosynthetic membrane which causes negative regulation of NAD(P)H dehydrogenase and, hence, reduces PS II -independent pathway of hydrogen production.

IMPROVEMENT OF HYDROGEN PHOTOPRODUCTION IN SULFUR DEPRIVED GREEN ALGAE

Different approaches have been proposed to increase hydrogen photoproduction in sulfur deprived *C. reinhardtii* [reviewed in 30,45,150,151,152]. The improvement has been achieved by optimization of chlorophyll concentration, medium composition, light and pH protocols in photobioreactors [145,153-155], synchronization of cell division [156], cell immobilization on a glass fiber matrix [157,158], continuous or semi-continuous regimes of cultivation [159,160].

Several mutants of *C. reinhardtii* have also been tested for the ability to produce hydrogen. As mentioned, Kruse and co-authors [149] designed a *C. reinhardtii* strain stm6 that accumulated large amounts of starch due to the modified respiratory metabolism and was blocked in state 1 and thus unable to maintain efficient cyclic electron flow around PS I. The mutant demonstrated a significant increase in the amount of evolved hydrogen under sulfur deficiency stress. Torzillo *et al.* [161] reported an increase in hydrogen production by one order of magnitude in sulfur deprived *C. reinhardtii* D1 protein mutant. This mutant is characterized by complex changes including low chlorophyll content, high photosynthetic capacity, and high respiratory rate, as compared to the wild type. The better performance of the *stm6* and *D1* mutants was mainly the result of a longer hydrogen production period. A sulfate permease is required to transport sulfate into the chloroplast. A mutant with altered sulfate permease activity was suggested as a candidate for hydrogen photoproduction without the need to deplete the culture medium of sulfate [162].

While starch accumulation in sulfur deprived *C. reinhardtii* is essential for hydrogen production, exogenous acetate is important for the establishment of anaerobic conditions in bioreactors [163]. However, the use of this organic acid for the large scale commercial hydrogen production would increase the cost of the process. As shown in the recent studies, photoautotrophic sulfur deprived *C. reinhardtii* is capable of establishing anaerobic conditions and producing hydrogen continuously when supplemented with carbon dioxide and incubated according to a special light protocol [142,164]. The light and pH regime for the photoautotrophic culture has been improved for cells grown under strong illumination [165].

INDIRECT BIOPHOTOLYSIS IN UNICELLULAR NON-NITROGEN FIXING CYANOBACTERIA

Indirect biophotolysis in green algae and non-nitrogen fixing cyanobacteria involves temporal separation between light-induced accumulation of carbohydrates and dark fermentative hydrogen production (see Table **1**). During the light phase, primary reactions of photosynthesis convert the energy of photons into chemical energy in the form of ATP and NADPH (Fig. **1**). Both NADPH and ATP are used for the fixation of carbon dioxide in the reactions of Calvin-Benson cycle where energy of NADPH and ATP is converted into the energy of carbohydrates. The carbohydrates can be stored in the form of starch in algae or glycogen in non-nitrogen-fixing cyanobacteria. Therefore, cells grown photoautotrophically accumulate carbohydrates that can be decomposed under anaerobic conditions *via* glycolysis providing cells with energy in the forms of ATP and reducing equivalents. Under dark anaerobic conditions, respiration and reactions of the Calvin-Benson cycle are inactive and activation of the [FeFe]- or bidirectional [NiFe]-H_2ase is an ultimate pathway for the utilization of the reducing equivalents released upon glycolysis. Although the efficiency of hydrogen production in the dark is lower than that observed in the light, the benefits come from the involvement of the natural light/dark cycles for the separation between aerobic and anaerobic phases [166].

Unicellular non-nitrogen-fixing cyanobacteria have been recognized with more potential for hydrogen production *via* indirect biophotolysis as compared with green algae [42]. Unlike [FeFe]-H_2ase in green algae, the cyanobacterial bidirectional [NiFe]-H_2ase is constitutively expressed under aerobic conditions, shows less sensitivity to inhibition by oxygen, and has NAD(P) as a redox partner (see section 3.1). Formation of NAD(P)H upon starch catabolism is an energetically more beneficial process than dark generation of the reduced form of Fd, an electron donor to [FeFe]-H_2ase. On the other hand, the operation of the bidirectional H_2ase in the direction of hydrogen evolution requires a large reducing power in the cells and low hydrogen content in the gas phase. Thus, the maintenance of maximal rate of hydrogen production involves the permanent removal of molecular hydrogen, *e.g.* by sparging with inert gas, that definitely decreases the net efficiency of the process.

The deficiency of macronutrients is the most frequently used approach for the enhancement of hydrogen production in non-nitrogen-fixing cyanobacteria. Thus, the lack of nitrogen in the medium is known to modify the metabolic processes in a way that significantly reduces the energetic requirements of the organism and results in over-accumulation of glycogen during the light phase which is essential for hydrogen production. For example nitrogen limited autotrophic cultures of *Spirulina platensis* and *Gloeocapsa alpicola* show a several fold increase in hydrogen production due to the over-accumulation of glycogen [167,168]. Moreover, limitation of nitrogen has a positive effect on the activity of [NiFe]-H_2ase and on the transcription of *hox* genes [104].

Unlike nitrogen deficiency, sulfur deprivation does not induce bleaching of photosynthetic pigments or degradation of important nitrogen-containing metabolites, such as pyridine nucleotides. However, sulfur deprivation modifies cell metabolism in the same manner as nitrogen limitation, causing over-accumulation of glycogen and an increase in the activity of bidirectional H_2ase at the catalytic level rather than at the transcriptional one [169,170]. Thus, a four-fold increase in evolution of hydrogen has been demonstrated in autotrophic cultures of unicellular non-nitrogen-fixing cyanobacteria *G. alpicola* and *Synechocystis* PCC6803 grown in sulfur depleted medium as compared to sulfur repleted culture [169]. The glycogen content in starved cells reached about 1g of glycogen per 1 g of protein after 7 days of deprivation and the maximum hydrogen production lasted for 8-12 hours, which is comparable to the duration of the night period. Hydrogen production achieved maximum level for both cyanobacterium at relatively low concentration of glycogen (0.4 g of glycogen per 1 g of protein) indicating that the process is rather limited by the rate of fermentation.

Essential improvement of hydrogen production by non-nitrogen-fixing cyanobacteria has been achieved through the optimization of cultivation conditions, such as nutrient supply and pH of the medium [171,172]. Thus, Burrows and co-authors [171,172] suggested that instead of completely starving cells of sulfur or another essential nutrient, the highest hydrogen production in *Synechocystis* sp. PCC 6803 occurs with a certain level of limitation of ammonia and sulfate and with an additional supply of cells with carbon in the form of bicarbonate anion. Such optimization of the medium composition resulted in a nearly 150-fold (!) greater hydrogen production as compared to cultures grown in a medium depleted of sulfur. This result has been attributed to a nearly 50-fold increase in glycogen content in the cells.

Immobilized cyanobacteria produce hydrogen at much higher volumetric rates than suspension cultures. The immobilization procedure alters cell metabolism to increase the reductant availability for the H_2ase enzyme and to prolong the duration of the hydrogen production period. Immobilized strains of nutrient limited *Microcystis aeruginosa* and *G. alpicola* showed the capacity to produce more hydrogen and during longer periods as compared to non-immobilized nutrient limited cultures [173,174] suggesting this method is applicable to improve the efficiency of hydrogen production by non-nitrogen-fixing cyanobacteria.

HYDROGEN PHOTOPRODUCTION IN NITROGEN-FIXING CYANOBACTERIA

Nitrogen-fixing cyanobacteria are considered as promising micro-organisms for photosynthetic hydrogen production [15,29,31,175,176]. Since N_2ase is an oxygen-sensitive enzyme, nitrogen-fixing cyanobacteria have evolved several strategies to combine nitrogen fixation with oxygen-evolving photosynthesis. In some of them, temporal separation between these processes takes place whereas in other cases only certain cells perform nitrogen fixation, *i.e.* a spatial separation occurs. The latter group involves heterocystous filamentous cyanobacteria which is being deprived of such nitrogen containing substances as ammonia, nitrate, or urea, differentiates non-randomly specific cells along the filament called heterocysts where the nitrogen fixation takes place (the scheme of metabolic pathways in heterocyst is shown in Fig. **7**). Heterocysts provide microaerobic conditions for the N_2ase reaction due to the: 1) absence of oxygen evolving activity of PS II, 2) high respiration rate, and 3) a thick cell envelope, made up of layers of polysaccharides and glycolipids that limits the influx of oxygen from the outside of the cell [48-51]. The ATP and reducing equivalents required for nitrogen fixation (formula 1) are produced in the heterocyst mainly by the cyclic electron flow around PS I and the oxidative pentose phosphate cycle, respectively (Fig. **7**). The reduced carbohydrates which is the substrate for the formation of NAD(P)H in heterocysts are transported from the neighboring vegetative cells.

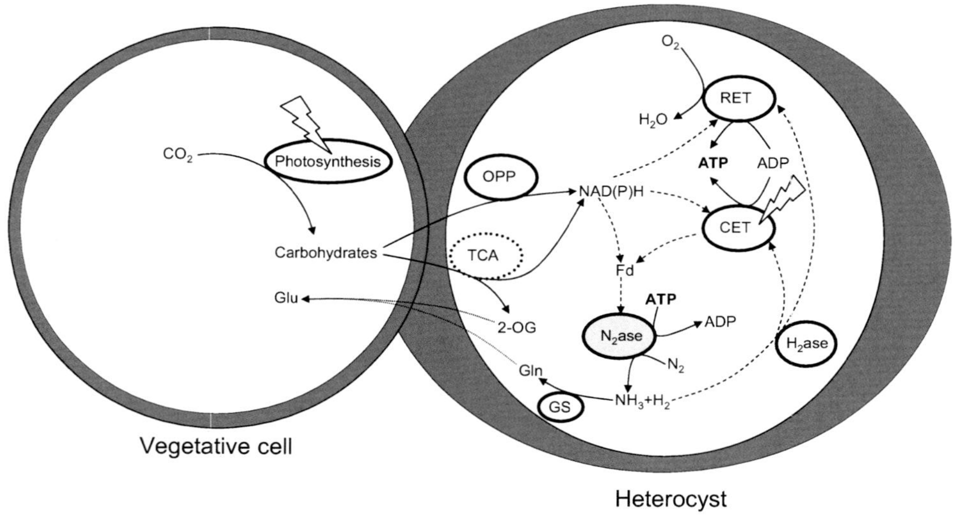

Figure 7: A scheme of heterocyst metabolism. Carbohydrates produced photosynthetically in vegetative cells are imported into heterocysts providing the dinitrogen fixing reaction with reducing equivalents *via* the oxidative pentose phosphate pathway [OPP] and the tricarboxylic acid cycle [TCA], which is incomplete in cyanobacteria [indicated by

dash]. Cylic electron transport around PS I [CET] and respiratory electron transport [RET] generate ATP for the N_2ase enzyme. Fixed nitrogen is converted into glutamine [Gln] by glutamine synthetase [GS]. 2-oxoglutarate [2-OG] produced by isocitrate dehydrogenase [not indicated] ís used as the carbon skeleton for further assimilation of glutamine into glutamate. The exact nature of the nitrogen compound exported from heterocysts is unidentified [dotted lines]. Hydrogen produced as a byproduct of dinitrogen fixation is decomposed by an uptake H_2ase and the released reducing equivalents are used in CET and RET. The redox reactions are indicated by dashed lines.

Non-heterocystous nitrogen-fixing cyanobacteria do not differentiate cells under nitrogen limitation and involve mostly a temporal separation between nitrogen fixation stage and oxygen-evolving photosynthesis [52,53]. In these cyanobacteria the establishment of microaerobic conditions in cells under illumination are achieved by down-regulation of PS II activity and an increase in respiration rate which are regulated by the endogenous circadian rhythms [55-57].

In nitrogen-fixing cyanobacteria, three enzymes, N_2ase (*Nif*), H_2-uptake H_2ase (*Hup*) and bidirectional H_2ase (*Hox*), are directly involved in hydrogen metabolism [29,30,52,98]. N_2ase produces hydrogen molecules as an obligatory side reaction along with the dinitrogen fixation (formula 1). Since this enzyme consumes substantial amounts of ATP, solar energy is required to drive photo-phosphorylation.

The bidirectional H_2ase (*Hox*) has been found in all non-nitrogen-fixing cyanobacteria and only in some nitrogen-fixing strains. It has a low *Km* for hydrogen and appears to function primarily in the uptake direction when molecular hydrogen is accumulated in gas phase. This would be expected to decrease N_2ase-mediated hydrogen production. Surprisingly, the *hoxH* mutant of *Anabaena* sp. PCC 7120 showed lower hydrogen production than the wild type strain [177] indicating the involvement of the bidirectional enzyme in hydrogen evolution rather than in hydrogen uptake.

The uptake H_2ase encoded by the *hupSL* genes functions to recycle hydrogen produced *via* the N_2ase-mediated reaction. Therefore, the net hydrogen evolution by N_2ase is negligible in most of nitrogen-fixing cyanobacteria. Since the uptake H_2ase is a major obstacle for the sustained hydrogen production, creation of *hup* mutants has proven to be an effective strategy to improve hydrogen production in nitrogen-fixing cyanobacteria. Among latter organisms the heterocystous *Nostoc* and *Anabaena* species and strains have been recognized as high hydrogen producers attracting an interest as a subject for genetic manipulations [reviewed in 31,98]. The uptake H_2ase mutants of *Anabaena* sp. strain PCC 7120, *Nostoc* ATCC 29133, and *Nostoc* sp. PCC 7422 showed a several fold increase in hydrogen production compared to the wild-type strain [177-179]. The highest energy conversion of light to hydrogen in the *hup* mutants of *Anabaena* and *Nostoc* strains attains about 1.0 % at moderate light intensity. The amount of hydrogen produced by the *hup* mutants of *Anabaena* and *Nostoc* under light aerobic conditions was comparable to that shown for sulfur deprived *C. reinhardtii* during anaerobic stage of incubation [14,178,179].

Two major limitations to the further increase in N_2ase-mediated hydrogen production concern the availability of ATP and reducing equivalents [180]. In order to identify processes that might compete with hydrogen production for energy and reducing power, the dynamics of the proteome of heterocystous cyanobacteria have been investigated by quantitative shot gun proteomics [181-183]. By comparing the proteomes of a *Nostoc punctiforme* uptake H_2ase mutant and the wild type, specific proteins and metabolic pathways have been identified as potential targets for further metabolic engineering improvements of N_2ase based photo-production of hydrogen [Dr. K. Stensjö, The Ångström Laboratories, Uppsala University; personal communication]. One such modification would be to introduce and oxygen scavenging module into the heterocyst to ensure a micro-oxic environment.

Besides *Nif1* expressed in heterocysts, the genome of *A. variabilis* encodes two functional N_2ases with cofactors that contain either molybdenum (*Nif2*) or vanadium (*Vnf*) at their active sites [reviewed in 184-186]. *Nif2* is induced under strictly anaerobic conditions and is localized in vegetative cells and heterocysts, whereas *Vnf* is expressed only in heterocysts and genes for this enzyme are expressed in the absence of molybdenum. Under certain conditions the alternative N_2ases can be involved in hydrogen production to increase the efficiency of the process. Thus, the vanadium N_2ase demonstrates higher ratio of hydrogen evolved per one reduced dinitrogen as compared with the molybdenum N_2ase [185]. The growing of *Anabaena hup* mutant with vanadate instead of molybdate in the medium forced the cells to use the vanadium N_2ase and, thus, to produce more hydrogen than does the molybdenum enzyme [116].

Unlike *Nif1* produced strictly in heterocysts, the alternative N_2ase *Nif2* is expressed in both vegetative cells and heterocysts [115,186] being involved in hydrogen production under anaerobic conditions in the presence of exogenous carbohydrates [187,188]. The uptake H_2ase does not interfere much with a *Nif2*-mediated hydrogen production because it is not expressed in vegetative cells [189]. Recently, the site-directed mutagenesis of *Nif2* subunit in *A. variabilis* was carried out to limit dinitrogen access to the catalytic center by blocking a putative gas channel [180]. The mutant grown with fructose under anaerobic conditions showed about four-fold higher hydrogen production in the presence of dinitrogen compared to the wild type, and hydrogen production was nearly as high as in an argon atmosphere when the N_2ase functions like a H_2ase (formula 2). The ability of nitrogen-fixing cyanobacteria to utilize organic compounds as substrate to produce hydrogen could be explored in the future in combination with wastewaters treatment.

Cyanobacteria constitute a large and heterogeneous group of photosynthetic micro-organisms widely spread in nature. One of the approaches to improve hydrogen photoproduction by these organisms can be the identification of the most promising hydrogen producers and optimization of the process *via* genetic manipulations [31,179]. Recently, the large-scale screening of 400 cyanobacterial strains isolated from different regions of the Baltic Sea and Finnish lakes for efficient hydrogen production has been carried out in a laboratory headed by E-M Aro in Turku University [190]. Ten strains have been shown to produce similar or more hydrogen than the most efficient producers of hydrogen, *Anabaena* PCC 7120 and *Nostoc punctiforme* ATCC 29133 uptake H_2ase mutants. All ten strains have been recognized as nitrogen-fixing heterocystous filamentous cyanobacteria thus suggesting a N_2ase-mediated hydrogen production occurs. Two of these strains, Calothrix 336/3 and XPORK 5E, have been selected for further genetic manipulations including *hup* genes knockout.

PHOTO-FERMENTATION IN PNS BACTERIA

Anoxygenic photosynthetic microbes, such as PNS bacteria, are capable of producing large amount of hydrogen in light in the presence of exogenous organic substrate [18,20,191]. They use solar energy to generate ATP, substantial amounts of which is needed to drive the N_2ase-mediated hydrogen production. The N_2ase-encoding genes are expressed in the light under anaerobic conditions and limitation of ammonium. Fermentation of exogenous compounds provides low potential electrons for the photosynthetic reactions and nitrogen fixation/hydrogen production (Table **1**, Fig. **3**). In terms of hydrogen production, the catabolism of organic substances in PNS bacteria has an advantage over fermentation by green algae and cyanobacteria as they have a higher substrate to hydrogen conversion efficiency. Thus, glucose is converted to carbon dioxide and hydrogen instead of organic acids or alcohols in fermentation. They are also able to use small-chain organic compounds, such as succinate, lactate, butyrate, malate, acetate, pyruvic acid, propionate, *etc.* as source of electrons for nitrogen fixation/hydrogen production. Rates of hydrogen photoproduction by anoxygenic phototrophs provided with small organic acids are typically high, reaching up to 20-30 mL of hydrogen per liter of culture per hour [192-195].

Different methods have been successfully applied to improve hydrogen photoproduction efficiency in PNS bacteria. They include, optimization of culture parameters and conditions of cultivation [19,195-197], the decrease in the light harvesting antenna cross-section that increases light capture efficiency [198], and knockout of the uptake H_2ase which splits molecular hydrogen and, thus, reduces hydrogen evolution [199, 200].

The N_2ase enzyme is expressed strictly under conditions of inorganic nitrogen limitation. To maximize hydrogen production activity in photosynthetic bacteria, it is important to diminish the suppression of N_2ases genes expression by inorganic nitrogen. Recently, the molecular engineering techniques have been applied to achieve constitutive expression of the N_2ase enzyme in the presence of ammonia [201]. Thus, mutants of PNS bacterium *Rhodopseudomonas palustris* were constructed which are capable of producing hydrogen in the presence of ammonium with a rate up to five times more than that in wild-type cells growing under nitrogen-fixing conditions. The increased rate of hydrogen production was due to the redirection of metabolism in such way that cells use the N_2ase as a hydrogen-producing enzyme rather than a catalyst for ammonia synthesis.

At present, the most profitable method from economical and ecological point of view implies an integrated system for hydrogen production which combines dark and photo-fermentation processes [9,19,20,63]. At the first step, large organic molecules are fermented to small compounds, typically organic acids, as well as to hydrogen by anaerobic heterotrophic bacteria. Biomass, wastes and by-products from the food and agricultural industries can be used as substrates to reduce the costs of the process. In the second step, small organic acids are converted to hydrogen and carbon dioxide through photo-fermentation by PNS bacteria. Such integration of dark fermentation with photo-fermentation can significantly increase the hydrogen yield due to the high efficiency for the conversion of chemical energy stored in the biomass to hydrogen. The recent studies on hydrogen production in PNS bacteria suggest a good potential for photo-fermentation within the integrated system [64,202].

In order to optimize such integrated systems, Melis and Melnicki [20] suggested co-cultivation of green algae and PNS bacteria in a photoreactor. Thus, anoxygenic photosynthetic bacteria are capable to absorb visible light [400 - 600 nm] and additionally near-infrared light [800 - 1010 nm] [203] that would result in better utilization of solar irradiance upon co-cultivation with green algae. Specific conditions of cultivation like nutrient deficiency can be applied to achieve anaerobic conditions in a photoreactor to increase hydrogen output by green algae. In this system, biomass and carbohydrates accumulated in the photoreactor are processed into organic acids and hydrogen by anaerobic heterotrophic bacteria in a fermentative process. The organic acids produced by fermentation are utilized in the photoreactor that is also accompanied by hydrogen evolution.

ACKNOWLEDGEMENTS

This work was supported, in part, by grants from the Russian Foundation for Basic Research (Nos: 09-04-01074, 09-04-91219-CT, 09-04-12303-ofi_m, 11-04-01389, 11-04-92690), and from the Molecular and Cell Biology Programs of the Russian Academy of Sciences, to DAL; by Japanes Society for the Promotion of Science (JSPS) Invitation Fellowship for Research in Japan, to SIA (ID No: S-09155); RC was supported by Engineering Research Council of Canada; ET was supported by Academy of Finland.

REFERENCES

[1] Turner JA. A realizable renewable energy future. Science 1999; 285: 687–9.

[2] Abraham S. Toward a more secure and cleaner energy future for America: national hydrogen energy roadmap; production, delivery, storage, conversion, applications, public education and outreach. U.S. Department of Energy. Washington, DC, 2002.

[3] Riis T, Hagen EF, Vie PJS, Ulleberg Ø. Hydrogen Production – Gaps and Priorities. IEA Hydrogen Implementing Agreement 2005.

[4] Allakhverdiev SI, Casal JJ, Nagata T. (Eds) Photosynthesis from molecular perspectives: towards future energy production. Photochem Photobiol Sci 2009; 8: 137–8.

[5] Crabtree GW, Dresselhaus MS, Buchanan MV. The Hydrogen Economy. Phys Today 2004; 57: 39–44.

[6] Allakhverdiev SI, Kreslavski VD, Thavasi V, *et al.* Photosynthetic energy conversion: hydrogen photoproduction by natural and biomimetic systems. In: Mukherjee A, Ed. Biomimetics, Learning from nature. Vienna, Austria: IN-TECH 2010; pp. 49-75.

[7] Rupprecht J, Hankamer B, Mussgnug JH, Ananyev G, Dismukes C, Kruse O. Perspectives and advances of biological H_2 production in microorganisms. Appl Microbiol Biotechnol 2006; 72: 442–9.

[8] Das D, Khanna N, Veziroğlu TN. Recent developments in biological hydrogen production processes. Chem Ind Chem Eng Q 2008; 14: 57–67.

[9] Ghirardi ML, Dubini A, Yu J, Maness PC. Photobiological hydrogen-producing systems. Chem Soc Rev 2009; 38: 52–61.

[10] Allakhverdiev SI, Kreslavski VD, Thavasi V, *et al.* Hydrogen photoproduction by use of photosynthetic organisms and biomimetic systems. Photochem Photobiol Sci 2009; 8: 148–56.
[11] Hall DO, Rao KK. Photosynthesis. 6th Edition. Cambridge, UK: Cambridge University Press 1999.
[12] Blankenship RE. Molecular mechanisms of photosynthesis. Oxford, UK: Blackwell Science 2002.
[13] Hall DO, Markov SA, Watanabe Y, Rao K. The potential applications of cyanobacterial photosynthesis for clean technologies. Photosynth Res 1995; 46: 159–67.
[14] Melis A, Zhang L, Forestier M, Ghirardi M, Seibert M. Sustained photobiological hydrogen gas production upon reversible inactivation of oxygen evolution in the green alga *Chlamydomonas reinhardtii*. Plant Physiol 2000; 122: 127–36.
[15] Sakurai H, Masukawa H, Dawar S, Yoshino F. Photobiological hydrogen production by cyanobacteria utilizing nitrogenase systems-present status and future development. In: Miyake J, Igarashi Y, Rögner M, Eds. Biohydrogen III. Renewable Energy System by Biological Energy Conversion. Amsterdam: Elsevier 2004; pp. 84–93.
[16] Olson JG, Ormerod JG, Amesz J, Stackebrandt E, Truper HG. Green photosynthetic bacteria. New York: Plenum Press 1988.
[17] Drews G, Imhoff JF. Purple phototrophic bacteria. In: Shively JM, Barton LL, Eds. Variations in autotrophic life. London: Academic Press 1991; pp. 51–97.
[18] Basak N, Das D. The prospect of purple non-sulfur (PNS) photosynthetic bacteria for hydrogen production: the present state of art. World J Microbiol Biotech 2007; 23: 31–42.
[19] Koku H, Eroglu I, Gunduz U, Yucel M, Turker L. Aspects of the metabolism of hydrogen production by *Rhodobacter sphaeroides*. Int J Hydrogen Energy 2002; 27: 1315–29.
[20] Melis A., Melnicki MR. Integrated biological hydrogen production. Int J Hydrogen Energy 2006; 31: 1563–73.
[21] Goussias C, Boussac A, Rutherford AW. Photosystem II and photosynthetic oxidation of water: an overview. Philos Trans R Soc Lond B Biol Sci 2002; 357: 1369–81.
[22] Kern J, Renger G. Photosystem II: Structure and mechanism of the water: plastoquinone oxidoreductase. Photosynth Res 2007; 94: 183–202.
[23] Barber, J. Crystal structure of the oxygen-evolving complex of photosystem II. Inorg Chem 2008; 47: 1700–10.
[24] Mitchell P, Moyle J. Chemiosmotic hypothesis of oxidative phosphorylation. Nature 1967; 213: 137–9.
[25] Bassham J, Benson A, Calvin M. The path of carbon in photosynthesis. J Biol Chem 1950; 185: 781–7.
[26] Jackson JB. Bacterial photosynthesis. In: Antony C, Ed. Bacterial energy transduction. London: Academic Press 1988; pp. 317–76.
[27] Boichenko VA, Allakhverdiev SI, Ladygin VG, Klimov VV. Functional conjunction of hydrogenase with photosystem II in whole cells of *Chlamydomonas reinhardtii* mutants. Dokl AN SSSR 1986; 290: 995–8.
[28] Maltsev SV, Allakhverdiev SI, Klimov VV, Krasnovsky AA. Hydrogen evolution by subchloroplast preparations of photosystem II from pea and spinach. FEBS Lett 1988; 240: 1–5.
[29] Tamagnini P, Leitão E, Oliveira P, *et al.* Cyanobacterial hydrogenases. Diversity, Regulation and Applications. FEMS Microbiol Rev 2007; 31: 692–720.
[30] Ghirardi ML, Posewitz MC, Maness PC, Dubini A, Yu J, Seibert M. Hydrogenases and hydrogen photoproduction in oxygenic photosynthetic organisms. Annu Rev Plant Biol 2007; 58: 71–91.
[31] Sakurai H, Masukawa H. Promoting R & D in photobiological hydrogen production utilizing mariculture-raised cyanobacteria. Mar Biotechnol 2007; 9: 128–45.
[32] Hoffman BM, Dean DR, Seefeldt LC. Climbing nitrogenase: toward a mechanism of enzymatic nitrogen fixation. Acc Chem Res 2009; 42: 609–19.
[33] Appel J, Phunpruch S, Steinmuller K, Schulz R. The bidirectional hydrogenase of *Synechocystis* sp. PCC 6803 works as an electron valve during photosynthesis. Arch Microbiol 2000; 173: 333–8.
[34] Appel, J, Schulz R. Hydrogen metabolism in organisms with oxygenic photosynthesis: hydrogenases as important regulatory devices for a proper redox poising? J Photoch Photobio B 1998; 47: 1–11.
[35] Vignais PM. Hydrogenases and H(+)-reduction in primary energy conservation. Results Probl Cell Differ 2008; 45: 223–52.
[36] Cournac L, Mus F, Bernard L, Guedeney G, Vignais P, Peltier G. Limiting steps of hydrogen production in *Chlamydomonas reinhardtii* and *Synechocystis* PCC 6803 as analysed by light-induced gas exchange transients. Int J Hydrogen Energy 2002; 27: 1229–37.
[37] Kranz RG, Cullen PJ. Regulation of nitrogen fixation genes. In: Blankenship RE, Madigan MT, Bauer CE, Eds. Anoxygenic photosynthetic bacteria. Netherlands: Kluwer Academic Press 1995; pp. 1191–208.

[38] Rees DC, Howard JB. Nitrogenase: standing at the crossroads. Curr Opin Chem Biol 2000; 4: 559–66.

[39] Igarashi RY, Seefeldt LC. Nitrogen fixation: the mechanism of the Mo-dependent nitrogenase. Crit Rev Biochem Mol Biol 2003; 38: 351–84.

[40] Volbeda A, Montet Y, Vernede X, Hatchikian EC, Fontecilla-Camps JC. High-resolution crystallographic analysis of *Desulfovibrio fructosovorans* [NiFe] hydrogenase. Int J Hydrogen Energy 2002; 27: 1449–61.

[41] Tosatto SCE, Giacometti GM, Valle G, Costantini P. Functional insights from the structural modelling of a small Fe-hydrogenase. Biochem Biophys Res Commun 2006; 339: 277–85.

[42] Yu J, Takahashi P. Biophotolysis-based hydrogen production by cyanobacteria and green microalgae. In: Mendes-Vilas A, Ed. Communicating current research and educational topics and trends in applied microbiology. Badajoz, Spain: Formatex 2007; pp. 79–89.

[43] Greenbaum E. Energetic efficiency of hydrogen photoevolution by algal water splitting. Biophys J 1988; 54: 365–8.

[44] Rosenkranz A, Krasna AJ. Stimulation of hydrogen photoproduction in algae by removal of oxygen by reagents that combine reversibly with oxygen. Biotechnol Bioeng 1984; 26: 1334–42.

[45] Melis A. Photosynthetic H_2 metabolism in Chlamydomonas *reinhardtii* (unicellular green algae). Planta 2007; 226: 1075–86.

[46] Benemann JR. Process Analysis and Economics of Biophotolysis of Water. IEA Hydrogen Implementing Agreement Technical Report, 1998.

[47] Hallenbeck PC, Benemann JR. Biological hydrogen production; fundamentals and limiting processes. Int J Hydrogen Energy 2002; 27: 1185–93.

[48] Fay P. Oxygen relations of nitrogen fixation in cyanobacteria. Microbiol Rev 1992; 56: 340–73.

[49] Campbell D. Interaction between the environment, intermediary metabolism and cellular differentiation in a filamentous cyanobacterium. B I Pasteur 1994; 92: 27–43.

[50] Haselkorn R, Schlictman D, Jones K, Buikema WJ. Heterocyst differentiation and nitrogen fixation in cyanobacteria. In: Elmerich C, Kondorosi A, Newton WE, Eds. Biological Nitrogen Fixation for the 21st Century. Dordrecht: Kluwer 1997; pp. 93–96.

[51] Wolk CP. Heterocyst formation. Annu Rev Genet 1996; 30: 59–78.

[52] Bergman B, Gallon JR, Rai AN, Stal LJ. N_2 fixation by non-heterocystous cyanobacteria. FEMS Microbiol Rev 1997; 19: 139–85.

[53] Gallon JR. N_2 fixation in phototrophs: adaptation to a specialized way of life. Plant Soil. 2001; 230: 39–48.

[54] Mitsui A, Kumazawa S, Takahashi A, Ikemoto H, Cao S, Arai T. Strategy by which nitrogen-fixing unicellular cyanobacteria grow photoautotrophically. Nature 1986; 323: 720 –2.

[55] Stal LJ, Krumbein WE. Aerobic nitrogen fixation in pure cultures of a benthic marine *Oscillatoria* (cyanobacteria). FEMS Microbiol Lett 1981; 11: 295–8.

[56] Huang T-C, Chow T-J. New type of N_2-fixing unicellular cyanobacterium. FEMS Microbiol Lett 1986; 50: 127–30.

[57] Gallon JR, Hashem MA, Chaplin AE. Nitrogen fixation by *Oscillatoria* spp. under autotrophic and photoheterotrophic conditions. J Gen Microbiol 1991; 137: 31–9.

[58] Mullineaux PM, Gallon JR, Chaplin AE. Acetylene reduction (nitrogen fixation) by cyanobacteria grown under alternating light-dark cycles. FEMS Microbiol Lett 1981; 10: 245–7.

[59] Gaffron H, Rubin J. Fermentative and photochemical production of hydrogen in algae. J Gen Physiol 1942; 26: 219–40.

[60] Miura Y, Ohta S, Mano M, Miyamoto K. Isolation and characterization of a unicellular marine green alga exhibiting high activities in dark hydrogen production. Agric Biol Chem 1986; 50: 2837–44.

[61] McEwan AG. Photosynthetic electron transport and anaerobic metabolism in purple non-sulfur photosynthetic bacteria. Antonie von Leeuwenhoek 1994; 66: 151–64.

[62] Klamt S, Grammel H, Straube R, Ghosh R, Gilles ED. Modeling the electron transport chain of purple non-sulfur bacteria. Mol Syst Biol 2008; 4: 156–62.

[63] Lee C, Chen P, Wang C, Tung Y. Photohydrogen production using purple nonsulfur bacteria with hydrogen fermentation reactor effluent. Int J Hydrogen Energy 2002; 27: 1309–13.

[64] Claassen PA, de Vrije T. Non-thermal production of pure hydrogen from biomass. Int J Hydrogen Energy 2006; 31: 1416–23.

[65] Vignais PM, Billoud B. Occurance, classification, and biological function of hydrogenases: an overview. Chem Rev 2007; 107: 4206–72.

[66] Volbeda A, Charon MH, Piras C, Hatchikian EC, Frey M, Fontecilla-Camps JC. Crystal structure of the nickel-iron hydrogenase from *Desulfovibrio gigas*. Nature 1995; 373: 580–7.

[67] Higuchi Y, Yagi T, Yasuoka N. Unusual ligand structure in Ni-Fe active center and an additional Mg site in hydrogenase revealed by high resolution X-ray structure analysis. Structure 1997; 5: 1671–82.

[68] Montet Y, Amara P, Volbeda A, *et al.* Gas access to the active site of Ni-Fe hydrogenases probed by X-ray crystallography and molecular dynamics. Nat Struct Biol 1997; 4: 523–9.

[69] Peters JW, Lanzilotta WN, Lemon BJ, Seefeldt L. C. X-ray crystal structure of the Fe-only hydrogenase (CpI) from *Clostridium pasteurianum* to 1.8 angstrom resolution. Science 1998; 282: 1853–8.

[70] Nicolet Y, Piras C, Legrand P, Hatchikian CE, Fontecilla-Camps JC. *Desulfovibrio desulfuricans* iron hydrogenase: the structure shows unusual coordination to an active site Fe binuclear center. Structure 1999; 7: 13–23.

[71] Matias PM, Soares CM, Saraiva LM, *et al.* [NiFe] hydrogenase from Desulfovibrio desulfuricans ATCC 27774: primary sequence determination, crystallographic refinement at 1.8 Å and modelling studies of its interaction with the tetra-haem cytochrome *c3*. J Biol Inorg Chem 2001; 6: 63–9.

[72] Stripp S, Sanganas O, Happe T, Haumann M. The structure of the active site H-cluster of [FeFe] hydrogenase from the green alga *Chlamydomonas reinhardtii* studied by X-ray absorption spectroscopy. Biochemistry 2009; 48: 5042–9.

[73] Lubitz W, Reijerse E, van Gastel M. [NiFe] and [FeFe] hydrogenases studied by advanced magnetic resonance techniques. Chem Rev 2007; 107: 4331–65.

[74] Fontecilla-Camps JC, Volbeda A, Cavazza C, Nicolet Y. Structure/function relationships of [NiFe]- and [FeFe]-hydrogenases. Chem Rev 2007; 107: 4273–303.

[75] Posewitz MC, Mulder DW, Peters JW. New frontiers in hydrogenase structure and biosynthesis. Curr Chem Biol 2008; 2: 178–99.

[76] Fontecilla-Camps JC, Amara P, Cavazza C, Nicolet Y, Volbeda A. Structure-function relationships of anaerobic gas-processing metalloenzymes. Nature 2009; 460: 814–22.

[77] Buhrke T, Lenz O, Krauss N, Friedrich B. Oxygen tolerance of the H_2-sensing [NiFe] hydrogenase from *Ralstonia eutropha* H16 is based on limited access of oxygen to the active site. J Biol Chem 2005; 280: 23791–6.

[78] Goldet G, Brandmayr C, Stripp ST, *et al.* Electrochemical kinetic investigations of the reactions of [FeFe]-hydrogenases with carbon monoxide and oxygen: comparing the importance of gas tunnels and active-site electronic/redox effects. J Am Chem Soc 2009; 131: 14979–89.

[79] Roessler PG, Lien S. Purification of hydrogenase from *Chlamydomonas reinhardtii.* Plant Physiol 1984; 75: 705–9.

[80] Vignais PM, Billoud B, Meyer J. Classification and phylogeny of hydrogenases. FEMS Microbiol Rev 2001; 25: 455–501.

[81] Cammack R, Rao KK, Bargeron CP, Hutson KG, Andrew PW, Rogers LJ. Midpoint redox potentials of plant and algal ferredoxins. Biochem J 1977; 168: 205–9.

[82] Florin L, Tsokoglou A, Happe T. A novel type of iron hydrogenase in the green algae *Scenodesmus obliguus* is linked to the photosynthetic electron transport chain. J Biol Chem 2001; 276: 6125–32.

[83] Stripp ST, Happe T. How algae produce hydrogen-news from the photosynthetic hydrogenase. Dalton Trans 2009; 45: 9960–9.

[84] Winkler M, Kuhlgert S, Hippler M, Happe T. Characterization of the key step for light-driven hydrogen evolution in green algae. J Biol Chem 2009; 284: 36620–7.

[85] Hatchikian EC, Forget N, Fernandez VM, Williams R, Cammack R. Further characterization of the [Fe] hydrogenase from *Desulfovibrio desulfuricans* ATCC 7757. Eur J Biochem 1992; 209: 357–65.

[86] Posewitz MC, Dubini A, Meuser J, Seibert M, Ghirardi ML. Hydrogenases, hydrogen production and anoxia in *Chlamydomonas reinhardtii*. In: Stern D, Harris E, Eds. The *Chlamydomonas* Source Book. Amsterdam: Elsevier 2008; pp. 217–55.

[87] Forestier M, King P, Zhang L, *et al.* Expression of two [Fe]-hydrogenases in *Chlamydomonas reinhardtii* under anaerobic conditions. Eur J Biochem 2003; 270: 2750–8.

[88] De Lacey AL, Sradler C, Cavazza C, Hatchikian EC, Fernandez VM. FTIR characterization of the active site of the Fe-hydrogenase from *Desulfovibrio desulfuricans.* J Am Chem Soc 2000; 122: 11232–3.

[89] Darensbourg MY, Lyon EJ, Zhao X, Georgakaki IP. The organometallic site of Fe-hydrogenase: models and entatic states. Proc Natl Acad Sci USA 2003; 100: 3683–8.

[90] Nicolet Y, de Lacey AL, Vernède X, Fernandez VM, Hatchikian EC, Fontecilla-Camps JC. Crystallographic and FTIR spectroscopic evidence of changes in Fe coordination upon reduction of the active site of the Fe-only hydrogenase from *Desulfovibrio desulfuricans*. J Am Chem Soc 2001; 123: 1596–601.

[91] Frey M. Hydrogenases: hydrogen-activating enzymes. Chembiochem 2002; 3: 153–60.

[92] Posewitz MC, King PW, Smolinksi SL, *et al.* Identification of genes required for hydrogenase activity in *Chlamydomonas reinhardtii*. Biochem Soc Trans 2005; 33: 102–4.

[93] Escoubas JM, Lomas M, LaRoche J, Falkowski PG. Light intensity regulation of cab gene transcription is signaled by the redox state of the plastoquinone pool. Proc Natl Acad Sci USA 1995; 92(22): 10237–41.

[94] Pfannschmidt T, Schutze K, Brost M, Oelmüller R. A novel mechanism of nuclear photosynthesis gene regulation by redox signals from the chloroplast during photosystem stoichiometry adjustment. J Biol Chem 2001; 276: 36125–30.

[95] Cohen J, Kim K, King P, Seibert M, Schulten K. Finding gas diffusion pathways in proteins: Application to O_2 and H_2 transport in CpI [FeFe]-hydrogenase and the role of packing defects. Structure 2005; 13: 1321–9.

[96] Houchins JP, Burris RH. Comparative characterization of two distinct hydrogenases from Anabaena sp. strain 7120. J Bacteriol 1981; 146: 215–21.

[97] Tamagnini P, Costa JL, Almeida L, Oliveira MJ, Salema R, Lindblad P. Diversity of cyanobacterial hydrogenases, a molecular approach. Curr Microbiol 2000; 40: 356–61.

[98] Tamagnini P, Axelsson R, Lindberg P, Oxelfelt F, Wünschiers R, Lindblad P. Hydrogenases and hydrogen metabolism of cyanobacteria. Microbiol Mol Biol Rev 2002; 66: 1092–2172.

[99] Rákhely G, Kovács ÁT, Maróti G, *et al.* Cyanobacterial type, heteropentameric, NAD^+ reducing [NiFe] hydrogenase in the purple sulfur photosynthetic bacterium, *Thiocapsa roseopersicina*. Appl Environ Microbiol 2004; 70: 722–8.

[100] Long M, Liu J, Chen Z, Bleijlevens B, RoseboomW, Albracht SP. Characterization of a HoxEFUYH type of [NiFe] hydrogenase from *Allochromatium vinosum* and some EPR and IR properties of the hydrogenase module. J Biol Inorg Chem 2007; 12: 62–78.

[101] Schmitz O, Boison G, Hilscher R, *et al.* Molecular biological analysis of a bidirectional hydrogenase from cyanobacteria. Eur J Biochem 1995; 233: 266–76.

[102] Appel J, Schulz R. Sequence analysis of an operon of NAD(P)-reducing nickel hydrogenase from the cyanobacterium Synechocystis sp. PCC 6803 gives additional evidence for direct coupling of the enzyme to NADP(H)-dehydrogenase (complex I). BBA-Protein Struct M 1996; 1298: 141–7.

[103] Boison G, Schmitz O, Mikheeva L, Shestakov S, Bothe H. Cloning, molecular analysis and insertional mutagenesis of the bidirectional hydrogenase genes from the cyanobacterium *Anacystis nidulans*. FEBS Lett 1996; 394: 153–8.

[104] Sheremetieva ME, Troshina OY, Serebryakova LT, Lindblad P. Identification of hox genes and analysis of their transcription in the unicellular cyanobacterium *Gloeocapsa alpicola* CALU 743 growing under nitrate-limiting conditions. FEMS Microbiol Lett 2002; 214: 229–33.

[105] Schmitz O, Boison G, Salzmann H, *et al.* HoxE – a subunit specific for the pentameric bidirectional hydrogenase complex (HoxEFUYH) of cyanobacteria. BBA-Bioenergetics 2002; 1554: 66–74.

[106] Kemner J, Zeikus JG. Purification and characterization of the membrane-bound hydrogenase from from *Methanosarcina barkeri* MS. Arch Microbiol 1994; 161: 47–54.

[107] Maness PC, Smolinski S, Dillon AC, Heben MJ, Weaver PF. Characterization of the oxygen tolerance of a hydrogenase linked to a carbon monoxide oxidation pathway in *Rubrivivax gelatinosus*. Appl Environ Microbiol 2002; 68: 2633–6.

[108] Axelsson R, Lindblad P. Transcriptional regulation of *Nostoc* hydrogenases: effects of oxygen, hydrogen, and nickel. Appl Environ Microbiol 2002; 68: 444–7.

[109] Antal TK, Oliveira P, Lindblad P. The bidirectional hydrogenase in the cyanobacterium *Synechocystis* sp. strain PCC 6803. Int J Hydrogen Energy 2006; 31: 1439–44.

[110] Oliveira P, Lindblad P. Transcriptional regulation of the cyanobacterial bidirectional Hox-hydrogenase. Dalton Trans 2009; 7: 9990–6.

[111] Kiss É, Kós PB, Vass I. Transcriptional regulation of the bidirectional hydrogenase in the cyanobacterium *Synechocystis* 6803. J Biotechnol 2009; 142: 31–7.

[112] Mehler AH. Studies on reactions of illuminated chloroplasts. I. Mechanism of the reduction of oxygen and other Hill reagents. Arch Biochem Biophys 1951; 33: 65–77.

[113] Asada K. Production and scavenging of reactive oxygen species in chloroplasts and their function. Plant Physiol 2006 141: 391–6.

[114] Eady RR. Structure-function relationships of alternative nitrogenases. Chem Rev 1996; 96: 3013–30.

[115] Thiel T. Nitrogen fixation in heterocyst-forming cyanobacteria. Chapter 5. In: Klipp W, Masepohl B, Gallon JR, Newton WE, Eds. Genetics and regulation of nitrogen fixing bacteria. Netherlands: Kluwer Academic Publishers 2004; pp. 73–110.

[116] Tsygankov AS, Serebryakova LT, Sveshnikov DA, Rao KK, Gogotov. Hydrogen photoproduction by three different nitrogenases in whole cells of *Anabaena variabilis* and the dependence on pH. Int J Hydrogen Energy 1997; 22: 859–67.

[117] Seefeldt LC, Hoffman BM, Dean DR. Mechanism of Mo-dependent nitrogenase. Annu Rev Biochem 2009; 78: 701–22.

[118] Ohki Y, Imada M, Murata A, *et al.* Synthesis, structures, and electronic properties of [8Fe-7S] cluster complexes modeling the nitrogenase P-cluster. J Am Chem Soc 2009; 131: 13168–78.

[119] Barney BM, Yurth MG, Dos Santos PC, *et al.* A substrate channel in the nitrogenase MoFe protein. J Biol Inorg Chem 2009; 14: 1015–22.

[120] Burgess BK, Lowe DJ. Mechanism of molybdenum nitrogenase. Chem Rev 1996; 96: 2983–3012.

[121] Durrant MC. Controlled protonation of iron-molybdenum cofactor by nitrogenase: a structural and theoretical analysis. Biochem J 2001; 355: 569–76.

[122] Lowe D, Thorneley R. The mechanism of *Klebsiella pneumoniae* nitrogenase action. Pre-steady-state kinetics of H_2 formation. Biochem J 1984; 224: 877–85.

[123] Rubio LM, Hernandez JA, Soboh B, *et al.* The role of Nif proteins in nitrogenase maturation. In: Dakora FD, Chimphango SBM, Valentine AJ, Elmerich C, Newton WE, Eds. Biological nitrogen fixation: towards poverty alleviation through sustainable agriculture. Cape Town, South Africa: Springer 2008; 325–8.

[124] Schwarz G, Mendel RR, Ribbe MW. Molybdenum cofactors, enzymes and pathways. Nature 2009; 460: 839–47.

[125] Ribbe M, Gadkari D, Meyer O. N_2 fixation by *Streptomyces thermoautotrophicus* involves a molybdenum-dinitrogenase and a manganese-superoxide oxidoreductase that couple N_2 reduction to the oxidation of superoxide produced from O_2 by a molybdenum-CO dehydrogenase. J Biol Chem 1997; 272: 26627–33.

[126] Kolber Z, Zehr J, Falkowski P. Effects of growth irradiance and nitrogen limitation on photosynthetic energy conversion in Photosystem II. Plant Physiol 1988; 88: 923–9.

[127] Wykoff DD, Davies JP, Melis A, Grossman A. The effects of phosphorus and sulfur deprivation on photosynthetic electron transport in *Chlamydomonas reinhardtii.* Plant Physiol 1998; 117: 129–39.

[128] Zhang Z, Shrager J, Jain M, Chang CW, Vallon O, Grossman A. Insights into the survival of *Chlamydomonas reinhardtii* during sulfur starvation based on microarray analysis of gene expression. Eukaryot Cell 2004; 3: 1331–48.

[129] Nguyen AV, Thomas-Hall SR, Malnoë A, *et al.* Transcriptome for photobiological hydrogen production induced by sulfur deprivation in the green alga *Chlamydomonas reinhardtii.* Eukaryot Cell 2008; 7: 1965–79.

[130] Zhang L, Melis A. Probing green algal hydrogen production. Philos Trans R Soc Lond B Biol Sci 2002; 357: 1499–509.

[131] Antal TK, Volgusheva AA, Kukarskih GP, Bulychev AA, Krendeleva TE, Rubin AB. Effects of sulfur limitation on the PS II functioning in *Chlamydomonas reinhardtii* as probed by the chlorophyll *a* fluorescence. Physiol Plantarum 2006; 128: 360–7.

[132] Ghirardi ML, Kosourov SN, Tsygankov AA, Seibert M. Two-phase photobiological algal H_2-production system. Proceedings of the 2000 DOE Hydrogen Program Review. NREL/CP-570-28890.

[133] Antal TK, Krendeleva TE, Laurinavichene TV, *et al.* The dependence of algal H_2 production on Photosystem II and O_2 consumption activities in sulfur deprived *Chlamydomonas reinhardtii* cells. BBA-Bioenergetics 2003; 1607: 153–60.

[134] Mus F, Cournac L, Cardettini V, Caruana A, Peltier G. Inhibitor studies on non-photochemical plastoquinone reduction and H_2 photoproduction in *Chlamydomonas reinhardtii.* BBA-Bioenergetics 2005; 1708: 322–32.

[135] Jans F, Mignolet E, Houyoux P-A, *et al.* A type II NAD(P)H dehydrogenase mediates light-independent plastoquinone reduction in the chloroplast of *Chlamydomonas.* Proc Natl Acad Sci USA 2008; 105: 20546–51.

[136] Desplats C, Mus F, Cuine S, Billon E, Cournac L, Peltier G. Characterization of Nda2, a plastoquinone-reducing type II NAD(P)H dehydrogenase in *Chlamydomonas* chloroplasts. J Biol Chem 2009; 284: 4148–57.

[137] Antal TK, Volgusheva AA, Kukarskih GP, Krendeleva TE, Rubin AB. Relationships between H_2 photoproduction and different electron transport pathways in sulfur-deprived *Chlamydomonas reinhardtii.* Int J Hydrogen Energy 2009; 34: 9087–94.

[138] Aro EM, Andersson B, Eds. Regulation of Photosynthesis. Dordrecht: Kluwer Academic Publishers, 2001.

[139] Antal TK, Krendeleva TE, Laurinavichene TV, *et al.* The relationship between photosystem 2 activity and hydrogen production in sulfur deprived *Chlamydomonas reinhardtii* cells. Dokl Biochem Biophys 2001; 381: 371–4.

[140] Antal TK, Krendeleva TE, Rubin AB. The photochemical activity of photosystem II in sulfur-deprived *Chlamydomonas reinhardtii* cells depends on the redox state of the quinone pool during the transition to anaerobiosis. Biofizika 2004; 49: 499–505.

[141] Allen JF, Bennet J, Steinback KE, Arntzen CJ. Chloroplast protein phosphorylation couples plastoquinone redox state to distribution of excitation energy between photosystems. Nature 1981; 291: 25–9.

[142] Kosourov S, Patrusheva E, Ghirardi ML, Seibert M, Tsygankov A. A comparison of hydrogen photoproduction by sulfur-deprived *Chlamydomonas reinhardtii* under different growth conditions. J Biotechnol 2007; 128: 776–87.

[143] Melis A, Happe T. Hydrogen Production. Green Algae as a Source of Energy. Plant Physiol 2001; 127: 740–8.

[144] Gibbs M, Gfeller RP, Chen C. Fermentative metabolism of *Chlamydomonas reinhardtii.* III. Photoassimilation of acetate. Plant Physiol 1986; 82: 160–6.

[145] Kosourov S, Seibert MS, Ghirardi ML. Effects of extracellular pH on the metabolic pathways in sulfur-deprived, H_2-producing *Chlamydomonas reinhardtii* cultures. Plant Cell Physiol 2003 44: 146–55.

[146] Hemschemeier A, Fouchard S, Cournac L, Peltier G, Happe T. Hydrogen production by *Chlamydomonas reinhardtii*: an elaborate interplay of electron sources and sinks. Planta 2008; 227: 397–407.

[147] Posewitz MC, Smolinski SL, Kanakagiri S, Melis A, Seibert M, Ghirardi ML. Hydrogen photoproduction is attenuated by disruption of an isoamylase gene in *Chlamydomonas reinhardtii.* Plant Cell 2004; 16: 2151–63.

[148] Finazzi G, Rappaport F, Furia A, Fleischmann M. Involvement of state transition in the switch between linear and cyclic electron flow in *Chlamydomonas reinhardtii*. EMBO Rep 2002; 3: 280–5.

[149] Kruse O, Rupprecht J, Bader K, *et al* . Improved photobiological H_2 production in engineered green algal cells. J Biol Chem 2005; 280: 34170–7.

[150] Hankamer B, Lehr F, Rupprecht J, Mussgnug JH, Posten C, Kruse O. Photosynthetic biomass and H_2 production by green algae: from bioengineering to bioreactor scale-up. Physiol Plantarum 2007 131: 10–21.

[151] Melis A, Seibert M, Ghirardi ML. Hydrogen fuel production by transgenic microalgae. Adv Exp Med Biol 2007; 616: 110–21.

[152] Rupprecht J. From systems biology to fuel – *Chlamydomonas reinhardtii* as a model for a systems biology approach to improve biohydrogen production. J Biotechnol 2009; 142: 10–20.

[153] Laurinavichene TV, Tolstygina IV, Tsygankov AA. The effect of light intensity on hydrogen production by sulfur-deprived *Chlamydomonas reinhardtii*. J Biotechnol 2004; 114: 143–51.

[154] Kim JP, Chang DK, Tai HP, Mi SK, Sang JS. Enhanced hydrogen production by controlling light intensity in sulfur-deprived *Chlamydomonas reinhardtii* culture. Int J Hydrogen Energy 2006; 31: 1585–90.

[155] Giannelli L, Scoma A, Torzillo G. Interplay between light intensity, chlorophyll concentration and culture mixing on the hydrogen production in sulfur-deprived *Chlamydomonas reinhardtii* cultures grown in laboratory photobioreactors. Biotechnol Bioeng 2009; 104: 76–90.

[156] Tsygankov A, Kosourov S, Seibert M, Ghirardi ML. Hydrogen photoproduction under continuous illumination by sulfur-deprived, synchronous *Chlamydomonas reinhardtii* cultures. Int J Hydrogen Energy 2002; 27: 1239–44.

[157] Laurinavichene TV, Kosourov SN, Ghirardi ML, Seibert M, Tsygankov AA. Prolongation of H_2 photoproduction by immobilized, sulfur-limited *Chlamydomonas reinhardtii* cultures. J Biotechnol 2008; 134: 275–7.

[158] Kosourov S, Seibert M. Hydrogen photoproduction by nutrient-deprived Chlamydomonas reinhardtii cells immobilized within thin alginate films under aerobic and anaerobic conditions. Biotechnol Bioeng 2008; 102: 50 – 8.

[159] Fedorov AS, Kosourov S, Ghirardi ML, Seibert M. Continuous hydrogen photoproduction by *Chlamydomonas reinhardtii* using a novel two-stage, sulfate-limited chemostat system. Appl Biochem Biotechnol 2005; 121: 403–12.

[160] Oncel S, Vardar-Sukan F. Photo–bioproduction of hydrogen by *Chlamydomonas reinhardtii* using a semi-continuous process regime. Int J Hydrogen Energy 2009; 34: 7592–602.

[161] Torzillo G, Scoma A, Faraloni C, Ena A, Johanningmeier U. Increased hydrogen photoproduction by means of a sulfur-deprived *Chlamydomonas reinhardtii* D1 protein mutant. Int J Hydrogen Energy 2009; 34: 4529–36.

[162] Chen HC, Newton AJ, Melis A. Role of SulP, a nuclear-encoded chloroplast sulfate permease, in sulfate transport and H_2 evolution in *Chlamydomonas reinhardtii*. Photosynth Res 2005; 84: 289–96.

[163] Fouchard S, Hemschemeier A, Caruana A, *et al.* Autotrophic and mixotrophic hydrogen photoproduction in sulfur-deprived *Chlamydomonas* cells. Appl Environ Microbiol 2005; 71: 6199–205.

[164] Tsygankov AA, Kosourov SN, Tolstygina IV, Ghirardi ML, Seibert M. Hydrogen production by sulfur-deprived *Chlamydomonas reinhardtii* under photoautotrophic conditions. Int J Hydrogen Energy 2006; 3: 11574–84.

[165] Tolstygina IV, Antal TK, Kosourov SN, Krendeleva TE, Rubin AB, Tsygankov AA. Hydrogen production by photoautotrophic sulfur-deprived *Chlamydomonas reinhardtii* pre-grown and incubated under high light. Biotechnol Bioeng 2009; 102: 1055–61.

[166] Miura Y, Akano T, Fukatsu K, *et al.* Stably sustained hydrogen production by biophotolysis in natural day/night cycle. Energ Convers Manage 1997; 38: S533–7.

[167] Aoyama K, Uemura I, Miyake J, Asada Y. Fermentative metabolism to produce hydrogen gas and organic compounds in a cyanobacterium, *Spirulina platensis.* J Ferment Bioeng 1987; 83: 17–26.

[168] Troshina O, Serebryakova L, Sheremetieva M, Lindberg P. Production of H_2 by the unicellular cyanobacterium *Gloeocapsa alpicola* CALU 743 during fermentation. Int J Hydrogen Energy 2002; 27: 1283–9.

[169] Antal TK, Lindblad P. Production of H_2 by sulphur-deprived cells of the unicellular cyanobacteria *Gloeocapsa alpicola* and *Synechocystis* sp. PCC 6803 during dark incubation with methane or at various extracellular pH. J Appl Microbiol 2005; 98: 114–20.

[170] Zhang Z, Pendse ND, Phillips KN, Cotner JB, Khodursky A. Gene expression patterns of sulfur starvation in *Synechocystis* sp. PCC 6803. BMC Genomics 2008; 9: 344–51.

[171] Burrows EH, Wong W-K, Fern X, Chaplen FWR, Ely RL. Optimization of pH and nitrogen for enhanced hydrogen production by *Synechocystis* sp. PCC 6803 *via* statistical and machine learning methods. Biotechnol Prog 2009; 25: 1009–17.

[172] Burrows EH, Chaplen FWR, Ely RL. Optimization of media nutrient composition for increased photofermentative hydrogen production by *Synechocystis* sp. PCC 6803. Int J Hydrogen Energy 2008; 33: 6092–9.

[173] Serebryakova LT, Tsygankov AA. Two-stage system for hydrogen production by immobilized cyanobacterium *Gloeocapsa alpicola* CALU 743. Biotechnol Prog 2007; 23: 1106–10.

[174] Rashid N, Song W, Park J, Jin H-F, Lee K. Characteristics of hydrogen production by immobilized cyanobacterium *Microcystis aeruginosa* through cycles of photosynthesis and anaerobic incubation. J Ind Eng Chem 2009; 15: 498–503.

[175] Shestakov SV, Mikheeva LE. Genetic control of hydrogen metabolism in cyanobacteria. Russ J Genet 2006; 42: 1272–84.

[176] Tsygankov A. Nitrogen-fixing cyanobacteria: A review. Appl Biochem Microbiol 2007; 43: 250–9.

[177] Masukawa H, Mochimaru M, Sakurai H. Disruption of the uptake hydrogenase gene, but not of the bidirectional hydrogenase gene, leads to enhanced photobiological hydrogen production by the nitrogen-fixing cyanobacterium *Anabaena* sp. Appl Microbiol Biotechnol 2002; 58: 618–24.

[178] Lindberg P, Schütz K, Happe T, Lindblad P. A hydrogen-producing, hydrogenase-free mutant strain of *Nostoc punctiforme* ATCC 29133. Int J Hydrogen Energy 2002; 27: 1291–6.

[179] Yoshino F, Ikeda H, Masukawa H, Sakurai H. High photobiological hydrogen production activity of a *Nostoc* sp. PCC 7422 uptake hydrogenase-deficient mutant with high nitrogenase activity. Mar Biotechnol 2007; 9: 101–12.

[180] Weyman PD, Pratte B, Thiel T. Hydrogen production in nitrogenase mutants in *Anabaena variabilis.* FEMS Microbiol Lett 2010; 304: 55–61.

[181] Stensjö K, Ow SY, Barrios–Llerena M, Lindblad P, Wright PC. An iTRAQ-based analysis to elaborate the proteomic response of *Nostoc* sp. PCC 7120 under N_2 fixing conditions. J Proteome Res 2007; 6: 621–35.

[182] Ow SY, Cardona T, Magnusson A, Lindblad P, Stensjö K, Wright PC. Quantitative shotgun proteomics of enriched heterocysts from *Nostoc* sp. PCC 7120 using 8-plex isobaric peptide tags. J Proteome Res 2008; 7: 1615–28.

[183] Ow SY, J Noirel, T Cardona, *et al.* Quantitative overview of N_2 fixation in *Nostoc punctiforme* ATCC 29133 through cellular enrichments and iTRAQ shotgun proteomics. J Proteome Res 2009; 8: 187–98.

[184] Thiel T. Characterization of genes for an alternative nitrogenase in the cyanobacterium *Anabaena variabilis.* J Bacteriol 1993; 175: 6276–86.

[185] Rehder D. Vanadium nitrogenase. J Inorg Biochem 2000; 80: 133–6.

[186] Thiel T, Lyons EM, Erker JC, Ernst A. A second nitrogenase in vegetative cells of a heterocyst-forming cyanobacterium. Proc Natl Acad Sci USA 1995; 92: 9358–62.

[187] Reddy PM, Spiller H, Albrecht SL, Shanmugam KT. Photodissimilation of fructose to H_2 and CO_2 by a dinitrogen-fixing cyanobacterium, *Anabaena variabilis.* Appl Environ Microbiol 1996; 62: 1220–6.

[188] Chen P-C, Fan S-H, Chiang C-L, Lee C-M. Effect of growth conditions on the hydrogen production with cyanobacterium *Anabaena* sp. strain CH3. Int J Hydrogen Energy 2008; 33: 1460–4.

[189] Weyman PD, Pratte B, Thiel T. Transcription of *hupSL* in *Anabaena variabilis* ATCC 29413 is regulated by NtcA and not by hydrogen. Appl Environ Microbiol 2008; 74: 2103–10.

[190] Allahverdiyeva Y, Leino H, Saari L, *et al.* Screening for biohydrogen production by cyanobacteria isolated from the Baltic Sea and Finnish lakes. Int J Hydrogen Energy 2010; 33: 1117–27.

[191] Rocha JS, Barbosa MJ, Wijfells RH. Hydrogen production by photosynthetic bacteria: culture media, yields, and efficiencies. In: Miyake J, Matsunaga T, San Pietro A, Eds. Biohydrogen II: an approach to environmentally acceptable technology. London: Pergamon 2001; pp. 3–32.

[192] Mao XX, Miyake J, Kawamura S. Screening photosynthetic bacteria for hydrogen production from organic acids. J Ferment Technol 1986; 64: 245–9.

[193] Nagamine Y, Kawasugi Y, Miyake M, Asada Y, Miyake J. Charaterization of photosynthetic bacterium *Rhodospirillum rubrum* RV for hydrogen production. J Mar Biotechnol 1996; 4: 34–7.

[194] Tsygankov AA, Fedorov AS, Laurinavichene TV, Gogotov IN, Rao KK, Hall DO. Actual and potential rates of hydrogen photoproduction by continuous culture of the purple non-sulphur bacterium *Rhodospirillum capsulatus*. Appl Microbiol Biotechnol 1998; 49: 102–7.

[195] Melnicki MR, Bianchi L, De Philippis R, Melis A. Hydrogen production during stationary phase in purple photosynthetic bacteria. Int J Hydrogen Energy 2008; 33: 6525–34.

[196] Tao Y, He Y, Wu Y, *et al.* Characteristics of a new photosynthetic bacterial strain for hydrogen production and its application in wastewater treatment. Int J Hydrogen Energy 2008; 33: 963–73.

[197] Laurinavichene TV, Tekucheva DN, Laurinavichius KS, Ghirardi ML, Seibert M, Tsygankov AA. Towards the integration of dark and photo fermentative waste treatment. 1. Hydrogen photoproduction by purple bacterium *Rhodobacter capsulatus* using potential products of starch fermentation. Int J Hydrogen Energy 2008; 33: 7020–6.

[198] Kondo T, Arakawa M, Hirai T, Wakayama T, Hara M, Miyaye J. Enhancement of hydrogen production by a photosynthetic bacterium mutant with reduced pigment. J Biosci Bioeng 2002; 93: 145–50.

[199] Franchi E, Tosi C, Scolla G, Della Penna G, Rodriguez F, Pedroni PM. Metabolically engineered *Rhodobacter sphaeroides* RV strains for improved biohydrogen photoproduction combined with disposal of food wastes. Mar Biotechnol 2004; 6: 552–65.

[200] Liang Y, Wu X, Gan L, Xu H, Hu Z, Long M. Increased biological hydrogen production by deletion of hydrogen-uptake system in photosynthetic bacteria. Microbiol Res 2009; 164: 674–9.

[201] Rey FE, Heiniger EK, Harwood CS. Redirection of metabolism for biological hydrogen production. Appl Environ Microbiol 2007; 73: 1665–71.

[202] Belokopytov BF, Laurinavichius KS, Laurinavichene TV, Ghirardi ML, Seibert M, Tsygankov AA. Towards the integration of dark- and photo-fermentative waste treatment. 2. Optimization of starch-dependent fermentative hydrogen production. Int J Hydrogen Energy 2009; 34: 3324–32.

[203] Nogi Y, Akiba T, Horikoshi K. Wavelength dependence of photoproduction of hydrogen by *Rhodopseudomonas rutila*. Agric Biol Chem 1985; 49: 35–8.

CHAPTER 4

Hydrogen Production *via* Photofermentation

Basar Uyar[1*], Gökhan Kars[2], Meral Yücel[3], Ufuk Gündüz[3] and Inci Eroglu[4]

[1]Department of Chemical Engineering, Kocaeli University, 41380, Kocaeli, Turkey; [2]Department of Biology, Science Faculty, Selçuk University, 42075, Konya, Turkey; [3]Department of Biology, Middle East Technical University, 06531, Ankara, Turkey and [4]Department of Chemical Engineering, Middle East Technical University, 06531, Ankara, Turkey

Abstract: Photofermentative hydrogen production is a bioprocess in which photosynthetic purple non-sulfur bacteria grow heterotrophically on organic acids like acetic acid, lactic acid and butyric acid and produce hydrogen using light energy under anaerobic conditions. Two enzymes are specifically involved in hydrogen production, namely nitrogenase and hydrogenase. While nitrogenases produce hydrogen under nitrogen-limited conditions acting as ATP-dependent hydrogenase, hydrogenases have the ability for both production and consumption of molecular hydrogen depending on the type of hydrogenase and physiological conditions. Photofermentation process can be achieved in a wide variety of conditions such as in batch or continuous mode, upon artificial or solar illumination, utilizing various carbon and nitrogen sources including food industry wastewater and dark fermentation effluents. Panel and tubular photobioreactors are the most applicable bioreactor types since they ensure simple design, reasonable material and production costs and high light energy utilization. Physiological parameters such as pH, temperature, medium composition and light intensity control the yield and hydrogen productivity of the bacteria. Hydrogen productivity and yield can also be increased by using genetically modified bacterial strains or immobilization of bacteria. Genetic studies focus on development of mutant strains by disrupting the uptake hydrogenase genes, altering pigmentation and blocking alternative by-product biosynthesis. Techno-economic evaluations show that photofermentative hydrogen production process is very near to the commercialization stage, however demo scale experience is necessary to solve some problems such as low rate of hydrogen production and the cost associated with photobioreactor scale-up. Furthermore, recent studies are trying to integrate photofermentation to dark fermentation to have an enhanced hydrogen production yield. Finally, the whole process could end up with a fuel cell application where the produced hydrogen is stored for future uses.

Keywords: Photofermentation, purple non-sulfur bacteria, nitrogenase, hydrogenase, photobioreactor, light energy, *R. sphaeroides*, *R. capsulatus*, organic acids, solar illumination.

INTRODUCTION

Photofermentation is a process in which organic compounds, like acetic acid, are converted into hydrogen and carbon dioxide with sunlight by photosynthetic bacteria. This process takes place under anaerobic conditions and can be combined to dark fermentation process to form a two-stage biohydrogen production process, where the organic substrate is completely converted into hydrogen and carbon dioxide [1].

Among photosynthetic bacteria, purple nonsulfur (PNS) bacteria comprise a very diverse group of microorganisms with respect to their carbon, nitrogen and energy metabolism. The members of this group of bacteria show a wide range of metabolic activities upon varying environmental conditions. Therefore, in many studies, these bacteria provided the opportunity to investigate basic metabolic events such as nitrogen fixation, carbon fixation, anoxygenic photosynthesis and membrane bioenergetics. PNS bacteria were initially told as "non-sulfur'' because they were thought not to use hydrogen sulfide as an electron donor while growing photoautotrophically. However, in addition to molecular hydrogen, PNS bacteria can use sulfide as an electron donor but not at high concentrations like purple sulfur (PS) bacteria. The well known

*Address correspondence to Basar Uyar: Department of Chemical Engineering, Kocaeli University, 41380, Kocaeli, Turkey; Tel: +90 2623033558; Fax: +90 2623591262; E-mail basar.uyar@kocaeli.edu.tr

Nuri Azbar and David B. Levin (Eds)

genera of PNS bacteria are *Rhodospirillum*, *Rhodopseudomonas* and *Rhodobacter* while *Thiocapsa* is the well known genera of PS bacteria among hydrogen producing microorganisms [2]. The biohydrogen production by PS and PNS bacteria confers several advantages over the other bioprocesses [3,4]. These are;

1. They can use a wide range of substrates for growth and hydrogen production, thus facilitating their potential to be used in waste treatment.

2. They have high substrate-to-hydrogen conversion efficiencies.

3. They show great metabolic versatility under different environmental conditions such as aerobic, anaerobic, with light or without light.

4. They can absorb energy at a wide range of the light spectrum and can withstand high light intensities.

5. Since they perform anoxygenic photosynthesis, there is no oxygen inhibition problem.

The representatives of photosynthetic bacteria commonly investigated for hydrogen production capabilities are *Rhodobacter sphaeroides*, *Rhodobacter capsulatus*, *Rhodopseudomonas palustris*, *Rhodospirillum rubrum*, and *Thiocapsa roseopersicina* [5]. In the following sections the aspects of hydrogen production in photosynthetic purple non-sulfur bacteria will be reviewed. These include the hydrogen production metabolism, the development of low-cost photobioreactors for efficient hydrogen production, optimization and possible improvements of bacterial strains and finally the potential applications of photofermentative hydrogen production processes.

Metabolism

Hydrogen production by PNS bacteria occurs under illumination in the presence of an inert, anaerobic atmosphere (such as argon), from the breakdown of organic substrates such as malate and lactate. There should be a nitrogen limitation (*i.e.* a high C/N ratio) in the culture medium, to direct the excess energy and reducing power towards hydrogen production. Several individual components make up the overall production system and these may conveniently be grouped as: i) the enzyme systems, ii) the carbon flow – specifically the TCA cycle and iii) the photosynthetic membrane apparatus. These groups are interconnected within the hydrogen production scheme by means of the exchange of electrons, protons and ATP (Fig. **1**) [4].

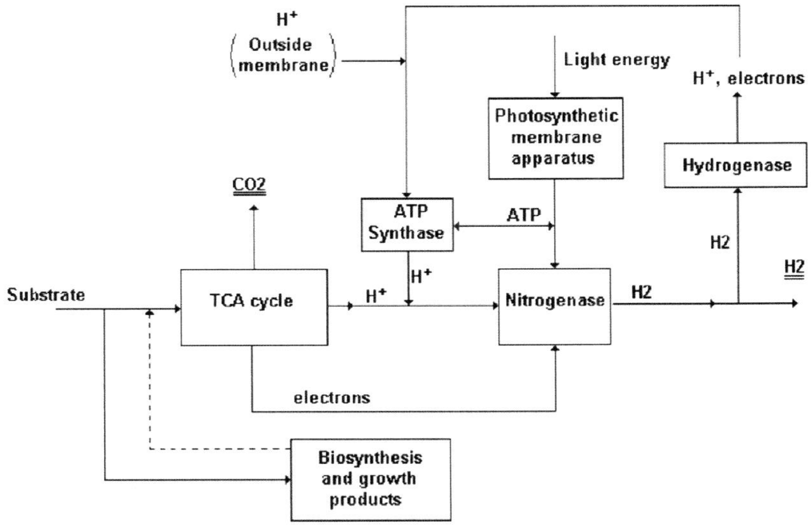

Figure 1: The overall scheme of hydrogen production by purple non-sulfur bacteria [4].

Enzymes

Hydrogen production and consumption in photosynthetic bacteria are mediated by nitrogenases and hydrogenases. Both enzymes are metalloproteins containing Fe/S clusters. Much progress has been made in the elucidation of gene expression, structure and regulation of these key enzymes. There exist three genetically distinct types of nitrogenase systems (nif, vnf, anf) according to the metal ion that they contain [3]. All three nitrogenase systems consist of two dissociable component metalloproteins, component 1, dinitrogenase, molybdenum-iron (MoFe) protein, vanadium-iron (VFe) protein, and iron-iron (FeFe) protein; and component 2, dinitrogenase reductase, only iron (Fe) protein). The [MoFe] nitrogenaseis the most widespread and best studied one among others. The hydrogenases are more diverse and recently three major groups of hydrogenases are distinguished according to their metal atoms. Certain common features of nitrogenases and hydrogenases are compared in Table **1**.

Table 1: Comparative properties of nitrogenases and hydrogenases [4].

Property	**Nitrogenase**	**Hydrogenase**
Substrates	ATP, H^+, N_2 and electrons	H_2
Products	H_2 and/or NH_4^+	H^+ and electrons
Number of subunits	Two (Mo-Fe and Fe)	One or more
Metal components	Mo, Fe	Ni, Fe
Optimum temperature	30°C (*A. vinelandii*)	55°C (*Rs. rubrum*) 70°C (*Rb. capsulatus*)
Optimum pH	7.1-7.3 (*A. vinelandii*)	6.5-7.5 (*Rb. sulfidophilus*)
Inhibitors, repressors	NH_4^+, O_2, CO	CO, EDTA, O_2
Stimulators	Light	H_2 (*Rb. sphaeroides*)

Nitrogenase

The basic function of the nitrogenase is to fix the molecular nitrogen into ammonia which is the useable form of nitrogen for the organisms. Hydrogen evolution by PNS bacteria is attributed mainly to this enzyme.

The nitrogenases isolated from a wide range of nitrogen-fixing microorganisms, including photosynthetic bacteria are very similar to each other in terms of their size, amino acid composition and reactivity. It consist of two proteins, a molybdenum - iron containing protein (also known as MoFe protein or dinitrogenase) and an iron-containing protein (Fe protein, or dinitrogenase reductase) which is the obligate electron donor to MoFe protein. The MoFe protein is a 200-250 kDa $\alpha_2\beta_2$ tetramer of the *nifD* and *nifK* gene products, containing molybdenum (2), iron (30) and sulfur (30) atoms. The Fe protein is a 60 kDa homodimer and encoded by the *nifH* gene [6,7].

It is an intrinsic property of nitrogenase to reduce protons along with nitrogen and to produce hydrogen as an obligatory side product [8,9]. One mole of hydrogen and two moles of ammonia are formed by Mo-nitrogenase catalyzed reaction at the expense of 16 moles of ATP:

$$N_2 + 8H^+ + 8e^- + 16ATP \rightarrow 2NH_3 + H_2 + 16ADP + 16P_i$$

The reaction is irreversible and can provide continuous production of hydrogen even under an atmosphere of 100% hydrogen, but it consumes large amounts of energy in the form of ATP molecules. ATP is regenerated by photophosphorylation during nitrogen fixation by photosynthetic bacteria at the expense of light energy [8]. However, in the absence of molecular nitrogen, total electron flux is directed to the protons and all the electrons are utilized for hydrogen production resulting in the production of one mol of H_2 at the expense of four ATP using two electrons [10]:

$$2H^+ + 2e^- + 4ATP \rightarrow H_2 + 4ADP + 4P_i$$

Since efficient operation of nitrogenase requires considerable amounts of ATP and reducing power, synthesis and activity of this enzyme in response to environmental stimuli are subject to strict regulatory controls at transcriptional and post-translational levels [4]. In photosynthetic bacteria, the physiological regulation of nitrogenase activity involves inactivation by covalent modification of the Fe protein of nitrogenase. This inactivation can be reversed by an activating factor (or activating enzyme) which is an extrinsic membrane protein [10].

The primary inhibitor/repressor of nitrogenase is oxygen which interferes with biological nitrogen fixation at different levels. At genetic level, oxygen represses the nitrogenase synthesis by down regulation of the expression of *nif* genes [7]. Furthermore, due to its oxidative properties oxygen causes irreversible damage to the enzyme by degrading the Fe-S cofactors. Dinitrogenase reductase (Fe protein) as being the site for covalent modification is more sensitive to inactivation by O_2 than dinitrogenase (MoFe protein) [11]. The other suggested reason for the inhibition of nitrogenase activity by O_2 is that oxygen and nitrogenase compete for a limited supply of electrons leading to decreased nitrogenase activity [12]. The nitrogenase activity of the cells might be restored partially or completely by reestablishing anaerobic conditions after respiring the traces of oxygen present in the culture. Furthermore, once the anaerobic conditions are set, new nitrogenase enzymes will be synthesized.

Ammonium is another inhibitor of the nitrogenase. It represses both the synthesis and activity of nitrogenase enzyme. NtrB/C and GlnB/K are the two-component systems which sense the fixed nitrogen status of the cell and accordingly regulate the *nif* operon and *nifA* gene. The gene product of *nifA* which is a member of enhancer-binding proteins (EBPs) is a transcriptional activator of *nif* structural genes in bacteria and thereby regulates the activity of *nif* genes [3]. The presence of ammonium ion in the culture medium causes a decrease in the expression levels of *nif* operon and *nifA* gene and results in a decrease in hydrogen production.

At the post-translational level, the nitrogenase enzyme activity is reversibly inactivated in a way that an ADP-ribose group from NAD^+ is attached to an arginine residue in one subunit of the homodimeric NifH protein resulting in NifH inactivation (switch-off). This process is catalysed by dinitrogenase reductase ADP-ribosyltransferase (DraT). However, this inactivation is reversible such that when the added ammonium is exhausted by cellular metabolism, the ADP-ribose group is removed by dinitrogenase reductase activating glycohydrolase (DraG) leading to NifH activation (switch-on).

Availability of certain metal ions such as molybdenum and iron which are found in the structure of the nitrogenases is critical for the hydrogen production. Cells grown in media without molybdenum and iron could not produce significant amounts of hydrogen. Moreover, the expressions of structural nitrogenase genes, *nifK* and *nifD*, were shown to be reduced significantly in the molybdenum and iron starved cells. On the contrary, increasing the concentrations of the molybdenum and iron to certain levels stimulates hydrogen production and gene expressions [9]. However, if the bacteria are capable of synthesizing 'alternative' nitrogenases (*e.g.* those containing iron instead of molybdenum), molybdenum deficiency may be tolerated [4]. In addition,the presence of Mo was shown to repress such alternative nitrogenases in *Rb. capsulatus* [13].

As stated earlier molecular nitrogen and air inhibit or decrease the hydrogen production of nitrogenase [14,15], however, nitrogenase biosynthesis is strongly stimulated by light. It was observed that nitrogenase represents up to 40% of the cytoplasmic proteins in *Rb. capsulatus* under high illumination [15,16]. It is also worth noting that a diurnal pattern of illumination (one involving alternating periods of light and dark) rather than continuous light results in a more stable nitrogenase activity [16].

Hydrogenase

The other key enzyme in the H_2 metabolism is the hydrogenase. Hydrogenases are structurally and functionally diverse group of metalloproteins and they catalyze a simple chemical reaction:

$$H_2 \leftrightarrow 2H^+ + 2e^-$$

They have been roughly classified according to their preferences for H_2 production or consumption as "uptake", "bidirectional" or "H_2 evolving" hydrogenases. Actually, the direction of the hydrogenase catalyzed reaction may change depending on the redox potential of the components which are able to interact with the enzyme. In the presence of hydrogen and an electron acceptor, it will act as a hydrogen uptake enzyme; in the presence of an electron donor of low potential, it may use the protons from water as electron acceptors and release hydrogen [17]. However, recently three major groups of hydrogenases are distinguished according to their metal atoms at their active sites: nickel-iron [NiFe]-hydrogenases, iron-iron [FeFe]-hydrogenases, no iron sulphur cluster and no nickel containing enzymes (Hmd enzyme, metal free hydrogenase or [Fe]-Hydrogenases) [3].

[NiFe]-hydrogenases are the most numerous and best studied class of hydrogenases of bacteria and consists of the large subunit (α-subunit, ~60 kDa) containing the bimetallic active site and the small subunit (β-subunit, ~30 kDa) containing the Fe-S clusters [17]. Membrane bound uptake hydrogenases (i.e; hupSL and hynSL), hydrogen sensors (i.e; hupUV), NADP-reducing (i.e; HydDA), bidirectional NADP/NAD-reducing (i.e; hoxYH) and energy converting membrane associated H_2 evolving hydrogenases are the subgroups of [NiFe] hydrogenases [3, 17]. Some phototrophic bacteria including *Rb. capsulatus* and *Rb. sphaeroides* have only membrane bound uptake hydrogenase and sensor hydrogenases. The function of uptake hydrogenase in these bacteria is to split hydrogen into protons and electrons thereby decreasing the efficiency of hydrogen production. Since hydrogen production is attributed mainly to nitrogenase, hydrogen producing activity of hydrogenase is negligible (if any) [18]. Uptake hydrogenase has been inactivated to enhance the hydrogen production efficiency in many studies [3,19,20] which were discussed in "Strain Improvement" section.

SUBSTRATES AND CARBON FLOW

PNS bacteria are capable of utilizing a wide variety of substrates as carbon and nitrogen sources for growth, but only a portion of these would support high rates and yields of hydrogen production. There are different substrate utilization pathways and they may differ among species or among different strains of the same species.

Two criteria are frequently used to evaluate the hydrogen production performance on a specific substrate. The first is the hydrogen production rate, which is the volume of hydrogen produced on a basis of the amount of culture in a unit time (mL H_2/L.h). For outdoor operations hydrogen production rate was reported based on ground area as well (L H_2/day. m^2 ground area). The second useful parameter for characterizing microbial hydrogen production is the substrate conversion efficiency, which is a measure of how much of the substrate has been utilized for hydrogen production rather than growth or alternative biosynthesis. It was determined as the ratio of moles of hydrogen that have actually been produced *versus* per moles of hydrogen expected theoretically through the stoichiometric conversion of the substrate according to the following hypothetical reaction:

$$C_xH_yO_z + 2(x-z)H_2O \rightarrow (y/2+ 2x-z)H_2 + xCO_2$$

When substrate to hydrogen conversion efficiency and hydrogen production rate are considered together, the organic acids lactic acid, malic acid, succinic acid and pyruvic acid show the highest performance. The lowest rates have been obtained when the substrate was propionic acid and butyric acid. Significant rates could be obtained from sugars, glucose and fructose but the conversion efficiencies with these substrates are considerably lower than those for organic acids [21].

The difference in the amounts and rates of hydrogen evolution is possibly due to the differences among the substrates in terms of the oxidation states, metabolic patterns of utilisation and the availability of pathways alternative to hydrogen production for the substrate, such as polysaccharide or poly-β-hydroxybutyrate (PHB) formation. Hydrogen production essentially takes place through the action of an anaerobic light-dependent citric acid cycle (TCA cycle). Inhibition of the cycle by any means inhibits the hydrogen production [22]. A simplified schematic of the carbon flow in purple non-sulfur bacteria is given in Fig. **2**.

As seen in Fig. **2**, sugars and structurally similar components like glycerol are utilised through the Embden-Meyerhoff and Entner-Doudoroff pathways. Lactate, acetate and butyrate are converted to pyruvate and acetyl-CoA before entering the TCA cycle. Malate, succinate and fumarate are the intermediates of TCA cycle and should be replenished by additional pathways [3,22]. The substrate assimilation pathways for hydrogen production were discussed below in more detail.

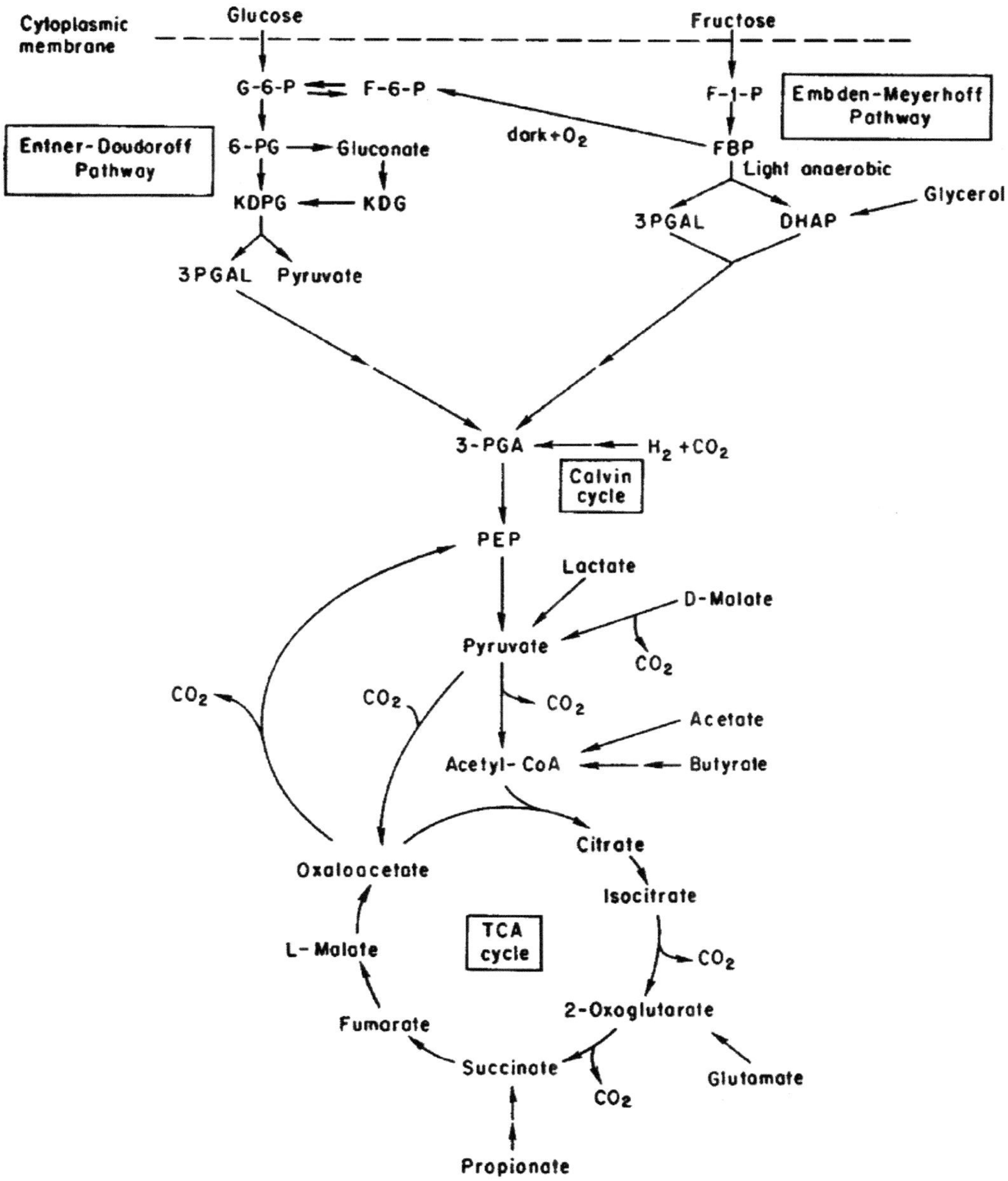

Figure 2: A simplified overall scheme of the carbon metabolism in purple non-sulfur bacteria [23]

Malate

Malate follows two different routes, depending on the optical isomer. L-malate is metabolised by the TCA cycle while D-malate enters the carbon flow after being converted to pyruvate. Possibly due to this difference, the amounts of hydrogen produced from these two isomers are different. *Rs. rubrum* produces more hydrogen using L-malate than D-malate which favours the accumulation of PHB, rather than hydrogen [24].

Acetate

The utilization of acetate or other compounds, which enter the central metabolism at the level of acetyl-coenzyme A (CoA), poses a unique problem: the conversion of acetyl-CoA to all other cell components. Acetyl-CoA catabolism generally proceeds *via* the TCA cycle. However, growth is impossible if C_4 acids drained from TCA cycle for biosynthetic reactions are not replenished. Most facultative anaerobic and aerobic bacteria use the glyoxylate cycle for acetate assimilation. The key enzymes for this pathway are isocitrate lyase and malate synthase. However, some of the purple nonsulfur bacteria *Rs. rubrum*, *Rb. capsulatus* and *Rb. sphaeroides*, do not possess isocitrate lyase but can grow on acetate, using it as the sole source of carbon and energy. The assimilation of acetate in these bacteria occurs through other alternative pathway(s).

Several studies using isotopically labelled acetate or studies at the level of enzyme activity revealed the presence of citramalate cycle for the assimilation of acetate in *Rs. rubrum* and *Rb. capsulatus*. The key steps of this cycle are the condensation of acetyl-CoA and pyruvate to citramalate, which is subsequently converted to propionyl-CoA and glyoxylate. Propionyl-CoA is carboxylated and converted to succinate and glyoxylate condenses with another molecule of acetyl-CoA to malate from which pyruvate is formed [25].

A completely novel acetate assimilation pathway fundamentally different from glyoxylate cycle was proposed for *Rb. sphaeroides*. In this pathway, two molecules of acetyl-CoA and one molecule CO_2 are converted *via* acetoacetyl-CoA and mesaconyl-CoA to glyoxylate and propionyl-CoA. Condensation of glyoxylate with acetyl-CoA yields malate, while the established reactions of propionate metabolism result in carboxylation and conversion of propionyl-CoA to succinate [26].

Pyruvate

Pyruvate is utilised practically by all strains of purple non-sulfur bacteria. In *Rb. sphaeroides* or *Rb. capsulatus*, pyruvate assimilation under photoheterotrophic conditions starts by conversion of pyruvate into acetyl-CoA or oxaloacetate [22]. When nitrogenase is active, pyruvate is also a good substrate for hydrogen production [27]. Significance of pyruvate is that, under appropriate conditions, it becomes the starting point for fermentative metabolism. Seven different species of purple non-sulfur bacteria including *Rb. sphaeroides* were able to ferment pyruvate [28].

Butyrate

Butyrate is easily converted to the reserve material PHB since it is the monomer of that material. Therefore, when butyrate is present as a substrate, the carbon flow is directed towards the synthesis of PHB rather than hydrogen production. Possibly because of this ease of conversion, butyrate is among the least efficient substrates for hydrogen production [21, 29].

The comparative kinetic analysis of the biomass growth, substrate consumption and hydrogen production of photosynthetic *Rb. sphaeroides* O.U 001 in media containing different organic acids as carbon source (malate, acetate, propionate, lactate, butyrate) was investigated for photofermentative hydrogen production. The highest hydrogen production rate was obtained in malate (24 mL H_2/L.h) and the highest biomass concentration was obtained in acetate containing media (1.65 g/L). The substrate conversion efficiencies for different organic acids were found to vary between 14 – 50% [29].

Sugars

For growth, glucose and fructose are good carbon sources, though not as efficient as organic acids. Growth on sugars generally results in acidification of the medium due to the excretion of acidic end-products by the bacteria [30]. The end product is usually gluconate.

In *Rb. sphaeroides*, fructose was catabolised predominantly *via* the Embden-Meyerhoff pathway under phototrophic conditions while it was catabolised mainly by the Entner-Doudoroff pathway under anaerobic conditions. Glucose, on the other hand, was utilized *via* the Entner-Doudoroff pathway under both phototrophic and aerobic conditions [31]. In *Rb. capsulatus*, both the glucose and the fructose were assimilated *via* the Entner-Doudoroff pathway. Sucrose was hydrolysed by the action of a constitutive cytoplasmic sucrase [32].

THE PHOTOSYNTHETIC APPARATUS AND LIGHT CONVERSION EFFICIENCY

The Photosynthetic Apparatus

The photosynthetic apparatus of the PNS bacteria absorb light energy (photons) and converts it into chemical bond energy (ATP) which in turn used for growth and hydrogen production.The photosynthetic apparatus is composed of transmembrane protein complexes and an ATPase complex that allows ATP synthesis at the expense of a proton gradient. Light is collected by the light harvesting components and channeled into the reaction center, which initiates a cyclic electron flow involving several carriers, resulting in the generation of a proton gradient. Finally, ATPase utilizes this gradient to generate ATP [4]. The photosynthetic device of purple bacteria consists of only one photosystem which is fixed in the intracellular membrane. The photosystem itself does not generate an oxidant that is powerful enough to split water.

Under photoheterotrophic conditions, the electrons that are liberated from the organic carbon sources are pumped through a large number of electron carriers. During the electron transport, protons are pumped through the membrane, a proton gradient is developed (high and low H^+), and this is dissipated by the ATP synthase enzyme to generate ATP. The extra energy in the form of ATP can be used to elevate the electrons further *via* reverse electron transport to the electron acceptor ferredoxin. Under nitrogen-fixing circumstances, these electrons can be used, using extra "ATP energy", by the nitrogenase enzyme to reduce molecular nitrogen into ammonium and when molecular nitrogen is not present, this enzyme can, again with the help of extra energy in the form of ATP, reduce protons into hydrogen gas, with the electrons derived from the ferredoxin [33].

Oxygen is the primary regulator of the synthesis of the photosynthetic apparatus, its presence immediately repressing the synthesis of bacteriochlorophyll molecules. The effect is reversible and synthesis resumes once the oxygen tension is eliminated.

Light also controls synthesis but its effect is not as strong as oxygen. Under an anaerobic atmosphere, the number of photosynthetic membrane vesicles produced varies inversely with the incident light intensity. Synthesis does not stop even under total darkness, but repeated subculturing in this manner eventually results in the total cessation of the bacteriochlorophyll synthesis.

From the standpoint of hydrogen production, it is essential to provide and maintain an ATP flux to the cell that exceeds the growth requirement, in order to meet the ATP requirements of the nitrogenase. The light intensity necessary to achieve maximum nitrogenase activity may be considerably higher than that necessary to maintain growth [4].

Light Conversion Efficiency

A performance evaluation parameter that has gained widespread acceptance to evaluate the photobiological hydrogen production process is the solar (or light) conversion efficiency. This parameter is the ratio of the total energy (heat of combustion) value of the hydrogen that has been obtained to the absorbed light energy [4,34].

It is calculated by

$$\eta = [(33.61 \cdot \rho_{H2} \cdot V_{H2})/(I \cdot A \cdot t)] \cdot 100$$

where V_{H2} is the volume of produced hydrogen in L, ρ_{H2} is the density of the produced hydrogen gas in g/L, I is the light intensity in W/m^2, A is the irradiated area in m^2 and t is the duration of hydrogen production in hours.

The light conversion efficiencies obtained for photosynthetic purple non-sulfur bacteria ranges between 1–5% on the average and there are several factors that affect the light conversion efficiency of photofermentation processes [33]. High efficiencies were generally obtained at low light intensities while high light intensities such as mid-day sunlight result in lower light conversion efficiencies. As an example, light conversion efficiency of *Rb. sphaeroides* decreased from 1.11% to 0.25% as light intensity increased from 88 to 405 W/m^2 [35].

Alterations of light intensity and light spectrum upon passage of the light through the bioreactor affect light energy conversion efficiency to hydrogen. As the depth of the photobioreactor increases, the light energy decreases but the efficiency of the conversion of light to hydrogen increases [35,36]; reducing the bioreactor depth from 4 to 2 cm resulted in an approximately four-fold increase in the energy conversion efficiency [37].

Photobioreactor design also affects the efficiency. Light shade bands were set on the surface of a photobioreactor for spatial dispersion of light under the excessive light conditions. In such a photobioreactor, light conversion efficiency was increased from 1.1 to 1.4% in outdoor (illuminated by sunlight) and from 1.6 to 2.1% in indoor (artifical illumination) conditions [38]. Several approaches were made to increase the light conversion efficiency. It is suggested that light energy conversion by coupling to suitable pigments and proteins may increase light utilization efficiency. In such an example, cultures of *Rb. sphaeroides* were combined with packed cells of *Halobacterium salinarum* that contains a retinal transmembrane protein bacteriorhodopsin which acts as a light-driven proton pump and the light energy conversion efficiency was increased from 0.6% to 2.3% [39]. Bacteria with altered light harvesting systems also show increased light conversion capability, which were discussed in "Strain Improvement" section.

Photobioreactors

A bioreactor is a device or system that supports a biologically active environment. A photobioreactor can be defined as a culture system in which light has to pass through the transparent reactor's wall to reach the cultivated cells that carry out a light dependent biological process [40].

The construction and successful operation of the photobioreactors is of critical importance for photofermentative hydrogen production. A successful photobioreactor concept must ensure simple design, low material and production costs and high utilization of light. A reactor for photofermentative hydrogen production has to meet several conditions. Since the hydrogen gas has to be collected, a prerequisite of the photobioreactor is that it must be impermeable to hydrogen to be an enclosed system. It should be possible to maintain a monoculture for an extended time (it must be practical to sterilize the reactor). Preferably sunlight is the energy source. The productivity of photobioreactor is light limited, and a high surface- to-volume ratio is required for an efficient photobioreactor [33].

Bioreactor Types

Various photobioreactor types constructed from different materials can be employed for photofermentative hydrogen production processes; including channel plate bioreactors made of acrylic sheet [41], pneumatically agitated flat-panel photobioreactors made of a stainless-steel frame and polycarbonate panels [42], glass bottles with different sizes [34,43,44]. Apart from those custom-made photobioreactors, commercially available fermenters can also be used after modification for biohydrogen production [45]. Nevertheless, flat panel and tubular photobioreactors are commonly preferred because they show the highest efficiencies most probably due to their high illumination area [33,46,47].

Flat panel reactors consist of a rectangular transparent box with a depth of only 1–5 cm. The height and width can be varied to some extent, but in practice only panels with a height and width both smaller than 1 m have been studied. Usually the panels are illuminated from one side by direct sunlight and the panels are placed vertically, or inclined *versus* the sun. A disadvantage of these systems is that the power consumption of aeration (or mixing with another gas) is high, although mixing is usually necessary in any reactor.

Many scaled-up versions of photobioreactors consist of repeating many of the smaller photobioreactor units, with its practical implications. Since the scaled-up reactor consists of only one unit, it is still practical to sterilize it. A series of panel photobioreactors is shown in Fig. **3**.

Tubular photobioreactors consist of long transparent tubes with diameters ranging from 3 to 6 cm, and lengths ranging from 10 to 100 m. The culture liquid is pumped through these tubes by means of mechanical or gas-lift pumps. The tubes can be positioned in different ways. The shape of the light gradient in the tubes is similar in most designs. Also with respect to liquid mixing, *i.e.* light/day cycling, the circumstances in most designs are similar. The length of the tubes is limited because of accumulation of gas. One way to scale-up is to connect a number of tubes *via* manifolds [33]

A pilot tubular bioreactor used for hydrogen production by *Rb. capsulatus* is shown in Fig. **4**. The photobioreactor consisted of 9 tubes and total illuminated surface area was 2 m^2.

Figure 3: Panel photobioreactors (4 L each) with internal cooling coils. The bioreactors were operated using *Rhodobacter capsulatus* in outdoor conditions (picture taken in summer 2009, in Ankara, Turkey).

Figure 4: Outdoor operated pilot tubular photobioreactor (80 L volume) with internal cooling coils (picture taken in summer 2009, in Ankara, Turkey).

INDOOR AND OUTDOOR OPERATIONS

Process Technology for Indoor operations

In indoor conditions, photobioreactors are illuminated by artificial light. Indoor operation of photobioreactors has several advantages over the outdoor operations such that the light input can be customized by manipulating light related parameters like the wavelength, intensity of the light and the illumination pattern. Continuous illumination is possible to obtain the highest productivity. Any possible light inhibition can be prevented easily. Since light source locations can be chosen, photobioreactors can be stacked in a more compact manner to obtain more productivity per unit ground area. In indoor systems, due to the controlled light input, temperature control of the photobioreactor is also easier. On the other hand, the requirement of electricity to provide illumination is a drawback of indoor operation which creates a significant operating cost and contradicts with the renewable concept of the biohydrogen technology.

Most laboratory research on photofermentative hydrogen production was carried out in indoor conditions. Recent indoor studies showed that the average hydrogen production rate and substrate conversion efficiency by *Rb. sphaeroides* O.U.001 from malic acid in a 1 L annular photobioreactor operating in indoor conditions was about 6.5 mL H_2/L.h and 75%, respectively [48]. In another research where a photobioreactor was operated in fed-batch mode, *Rp. palustris* produced hydrogen at average rates of 9.0 and 7.3 mL H_2/L.h in the presence of malic acid and acetic acid as carbon sources, respectively. Here, irradiating the photobioreactor from two opposite sides increased the yield by 14% compared to irradiating the photobioreactor from only one side [49]. Other notable findings of indoor researches were discussed in "Physiological Parameters" section.

Process Technology for Outdoor Applications

The biological production of hydrogen at a competitive cost in industrial scales can only be possible in photobioreactors which are illuminated by sunlight, operating in outdoor conditions. However, growth and hydrogen productivity of photosynthetic bacteria in an outdoor bioreactor system is strongly affected by fluctuations in temperature and light intensity due to the day-night cycle and due to seasonal, geographic and climatic conditions. Therefore, the process must be adapted to the outdoor conditions by examining these uncontrolled parameters and the response of the system to the changes in these parameters.

Depending on the location and season, the number of hours per day when the light intensity is high enough to support photosynthetic cell growth can be very short. In the absence of light energy or some other metabolizable organic carbon source in the medium, cells begin to metabolize the intracellular components to maintain the cell growth, thus leading to a decrease in cell weight. In order to overcome the problems of diurnal variations in solar light intensity and prolonged periods of bad weather, an illumination system with integrated solar and artificial light sources can be used. In such a system, solar light is used for illumination during the day but when the solar light intensity decreases below a set value (during cloudy days and at night) the system switches automatically to artificial light source, thus ensuring continuous light supply to the reactor [50].

The orientation (position) of the outdoor photobioreactor panel is a configurable parameter that affects the sunlight received by the bioreactor plate. When the front surface of the flat panel reactor was inclined towards to the south with inclinations of 30° and 60° for summer and winter, respectively, it receives the major thrust of solar irradiance. The back panel as well as the side walls are illuminated by diffuse and reflected light at rather low photon flux densities which is very effective for photosynthesis [51,52]. On the other hand, when the illuminated surface was placed in an east-west facing orientation, the productivity is higher than that in a north-south facing orientation, because more solar energy was received in the former case than in the latter [53].

A major problem in outdoor operations is the increase of temperature to lethal levels due to the absorbed sunlight by photosynthetic bacteria, thus cooling methods (such as shading and water spraying) are needed to be adapted to the system [54].

Solar energy conversion efficiency is high (7%) at low intensities of sunlight (below 100 W/m^2), but drops down to 2% at high intensities (1000 W/m^2) in agar immobilized *Rb. sphaeroides* 8703 cells [55]. Hydrogen production in plane type photobioreactor with different depths was studied for 12 days and a light conversion efficiency of 1.1 % was reported [56]. Hydrogen production was thought to be enhanced by using laser dye filter transforming the absorbed wavelengths into longer ones, which are more effective in photosynthesis. A 1.4 L column photobioreactor was made of two glass flasks with transparent jackets filled with laser dye filter. In trials carried out at Sahara desert, *Rb. sphaeroides* DSM 9483 which was obtained by directed evolution, produced hydrogen at a rate of 85 mL H_2/L.day [57].

Rs. rubrum grown in an outdoor reactor having a total area of 8m^2 using lactate produced hydrogen constantly with exchange of the half of the medium every fifth day for almost two months, indicating that there was no significant contamination of the reactor although no sterile methods or substrates were applied. Approximately 40 L H_2/day.m^{2}ground area was obtained. The photobioreactor utilized was in the form of hollow channel plates made of acrylic glass placed at an east-west position [41].

Hydrogen production under illumination simulating the daily sunlight irradiation pattern was compared to hydrogen production in outdoor conditions in a study using *Rb. sphaeroides* RV as the photosynthetic bacterium. In the outdoor conditions, hydrogen production by *Rb. sphaeroides* RV varied depending on the sunlight intensity received. The maximum hydrogen production rate was 63 mL H_2/L.h under a 4.5cm light path and the average light energy conversion efficiency was 1.1%. In the indoor conditions, under the simulated illumination, the maximum hydrogen production rate was 75 mL H_2/L.h with a light energy conversion efficiency of 1.0% [34].

Rb. sphaeroides O.U.001 used different organic acids (malate, lactate, and acetate) and olive mill waste water as carbon sources and accumulated byproducts such as PHB and carotenoid in addition to hydrogen production, in an eight liter flat plate solar bioreactor that operated in outdoor conditions [47].

Batch and Continuous Operating Modes

In terms of mode of operation, reactors can be divided into two broad categories, batch reactors and continuous reactors. When working in batch mode, reactants are loaded into a suitable reactor, the system is closed and reaction is allowed to take place as a function of time. The products are discharged at the end (or during the course) of the operation. When working in continuous mode, reactants are continuously fed into the reactor and emerge as continuous stream of product.

The choice between continuous and batch operation is often related to the production capacity. The principal advantage of continuous systems is their high capacity, while the principal advantage of batch reactor is its flexibility and adaptability. In practice, the selection criteria is the economics and it is based on capital investment and operating costs [58]. On the other hand, it should also be noted that long-term continuous operation can cause contamination problems in biological reactors [41].

Photobioreactors can be operated in batch or continuous modes for biohydrogen production, however almost all of the academic research was carried out in batch reactors so far and continuous reactor operations are scarce. Notable findings obtained in batch mode were discussed in "Physiological Parameters" section.

In continuous photobioreactors, the hydraulic retention time (HRT) which is the average length of time that a soluble compound remains in the reactor, is an important factor influencing the cell dry weight (cdw), hydrogen percentage, hydrogen production rate, light conversion efficiency and substrate conversion efficiency. An HRT of 72 h was reported for *Rb. capsulatus* in a 1.5 L continuous photo-bioreactor, using acetate, butyrate and propionate as substrates [59]. In a comparable case, using butyric and acetic acid from dark fermentation as substrate, *Rp. palustris* WP3-5 produced hydrogen in a continuous bioreactor with 96 h HRT. Hydrogen production rate and the cell concentration were fairly stable with a mean value of 26 mL H_2/L.h and 8.21 g/L for a 10 day operation, respectively [60].

A long term stable operation of a continuous panel photobioreactor by *Rb. capsulatus* on acetate and glutamate containing feed has been achieved with an average hydrogen production rate of 2.5 mL H_2/L.h during a 3-month operation with 384 h HRT, cell concentration was stable in the range of 0.6-0.7 g/L. Although the rate and yield values were lower compared to the batch reactors, the continuous operation had the advantage of long term run [61].

Another long term stable continuous operation lasted for three months using *Rp. palustris* R-1 and acetate, propionate, butyrate and ethanol as substrates. An HRT of 180 h resulted in hydrogen production rate of 4-12 mL H_2/L.h during the early stage and 12-21 mL H_2/L.h in the later stage [62].

In fed batch cultures, each repetitive substrate addition in the absence of cellular growth diminished rates and yields of hydrogen production and it was suggested that some bacterial growth (about 10% of the batch bioreactor) was necessary for cellular enzymatic maintenance and repair processes, which were needed to sustain high rates and yields of hydrogen production [63]. This finding suggests that in continuous systems, HRT should be long enough to minimize cell growth rate allowing a greater proportion of the cellular metabolic activity to be directed toward hydrogen production.

Scale-Up

Obviously, the ultimate goal of biohydrogen research with the photosynthetic bacteria is the operation of large-scale bioreactors under natural sunlight. There are several problems related to the operation of large-scale photobioreactors even when conditions optimized in smaller scale reactors and vigorous species are employed. One of the major problems in scale-up is the diffuculty in providing sufficient light intensity for the cells. The self-shading effect of the culture increases with increasing reactor thickness and increasing cell concentration. This effect seriously limits the abundance and homogeneous distribution of light inside the photobioreactor [27]. Hydrogen production is proportional to the intensity of supplied light and moreover lack of light could lead to the shifting of the organisms into alternative modes of growth.

The problem of self-shading can be alleviated by the use of a high surface to volume ratio for the photobioreactor. It was also observed that hydrogen bubbles tend to form next to the surfaces, so increasing the surface area may promote desorption of hydrogen from media [27]. This solution introduces the additional problems of large surface (land) requirement and difficulties in handling and manipulation of the reactor contents (*e.g.* sterilization and mixing). A more elaborate solution to self-shading is internal illumination by a bundle of optical fibers inside the reactor, which brings the possibility of more compact photobioreactors, much like the conventional cylindrical fermenters used [50], albeit at an increasing cost. Finally, genetic tailoring of the photosynthetic apparatus may partially solve the self-shading problem.

The total absence of light within the diurnal cycle is obviously a problem, but not detrimental, since hydrogen production recovers more or less to the same rate once illumination is resumed. In order to improve hydrogen production, an artificial light source connected to an intensity sensor can automatically switch on at night and on cloudy days and off under solar light in sunny days [50].

The application of mixing is not mandatory but beneficial for hydrogen production because it facilitates the removal of hydrogen gas, and distributes the cells within the culture, aiding the homogeneous exposure of the microorganisms to light and substrate. In case of lack of mixing, the light, substrate and cell distribution inside the bioreactor panel becomes more non-uniform as the panel size increases [27].

Mechanical mixing in large-scale systems is difficult due to the usually large ratio of the surface to volume in reactors. Therefore mixing is usually carried out by sparging an inert gas or recycled headspace gas through the culture. However the former results in the dilution of the produced gas and the latter requires additional power input.

The conceptual floating-type photobioreactor is intended to make use of the natural movement of the waves so as to provide an alternative method of mixing [62]. Apart from the depth of the photobioreactor, the height of the photobioreactor may also be a limiting parameter for scale up. The photobioreactor height means the maximum hydrogen bubble travel distance through culture, that is, a formed hydrogen bubble at the bottom of the photobioreactor has to travel all the way up through the medium. The hydrogen gas may be absorbed and used by the bacteria *via* the uptake hydrogenase enzyme which can result in decreased hydrogen productivity. Moreover, increasing the height of photobioreactor increases the pressure at the bottom and the bacteria may be sensitive to high pressure [27].

For large scales, another concern is the difficulty of the sterilization of reactor and medium. It is suggested that for a large scale application the bioreactor can only be operated under non-sterile conditions [41]. There are very few successful examples of scale-up photobiological hydrogen production using photosynthetic bacteria. Successfully tested photobioreactors at large scales include: 6 L acrylic plate reactor with *Rp. palustris* R-1 [62], 8 L flat plate acrylic bioreactor operating in outdoor conditions by *Rb. sphaeroides* O.U.001 [47], 11 L polyacrylate flat plate reactor with *Rb. sphaeroides* RV [64], 25L flat plate acrylic bioreactor by *Rb. capsulatus* [27], 65 and 80 L tubular photobioreactors made up of low-density polyethylene with *Rb. capsulatus* [65,66].

OPTIMIZATION AND POSSIBLE IMPROVEMENTS

If biohydrogen is to be produced by whole microorganisms, it must be taken into account that hydrogen output represents an energy loss for the cell and that the microbial metabolic network has evolved for the rational use of energy and optimization of specific growth rate. Therefore, regulatory networks control the biosynthesis and activity of the involved enzymes; it is necessary to understand how they function before being able to maximize the production of hydrogen.

There are several parameters determining the overall efficiency of hydrogen production and these parameters needed to be optimized or improved for enhanced hydrogen production. Optimization of physiological parameters, immobilization of microorganisms and strain improvements will be discussed in detail in the next sections. Physiological parameters include pH and temperature of the culture, C and N sources in the media, light effect and minor nutrient composition in the media. These aspects of the process significantly influence the hydrogen production. The usage of immobilized cells has also paramount effect on hydrogen production. Strain improvement *via* metabolic engineering is a promising approach for the improvement of biological hydrogen production by known microorganisms, particularly as concerns the redirection and optimization of the flow of reducing equivalents to the hydrogen-producing enzymes (nitrogenase or hydrogenase). Its application, however, will require a detailed understanding of the metabolic pathways and regulatory circuits involved, many of which are extremely complex. It is also not known at what extent metabolic engineering could improve hydrogen production [8].

A combination of systematic experimental design, gene knockouts, nutritional studies, heterologous expression, and artificial environments will be needed in order to achieve biohydrogen production levels that can compete with existing non-renewable production methods [67].

Physiological Parameters

Physiological parameters such as pH, temperature, medium composition and light intensity affect the growth and hydrogen production of purple non-sulfur bacteria. The bacteria can grow in a limited range of these parameters (*i.e.* non-acidic pH, mesophilic temperatures) and produce hydrogen under even more limited conditions (*i.e.* high C/N ratio, neutral pH, under illumination). It is of primary importance to determine the ranges and optimum values of physiological parameters to obtain high and stable hydrogen production rate and yield.

pH

Growth and photoproduction of hydrogen in *Rb. sphaeroides* O.U.001 was regulated by pH. There is no appreciable biomass growth at pH below 5.5. Biomass yield is high at a pH range of 6.0–6.8 and it decreases as the pH increases. Hydrogen is not produced below pH 5.5. At pH 6.5-7.0 hydrogen production is observed at early stationary phase and above 7.5 hydrogen production is delayed until mid to late stationary phase. Highest hydrogen production rate is obtained between pH 6.5-7.5 and the rate decreases as the pH increases. Hydrogen production does not occur at pH values above 9.0. The decrease in hydrogen production at a pH away from the optimum is much more pronounced at acidic than at alkaline pH. The final pH of the medium usually changes to higher values during hydrogen assays, which was explained as being due to consumption of substrate (an organic acid) [43,68,69].

Temperature

Hydrogen production from purple non-sulfur bacteria is an enzymatic process and thus it has an optimum temperature generally between 30°C and 35°C (reviewed in [69]). Most purple non-sulfur bacteria are unable to grow at temperatures above 39°C [27,70]. From 226 isolates of anoxygenic phototrophic purple non-sulfur bacteria, only 13 strains could grow under light-anaerobic conditions at 40°C and only one of these strains showed photoproduction of molecular hydrogen [71]. The best candidate for high-temperature operation seems to be *Rb. centenum*, which is reported to grow optimally at 40-42°C and up to 45°C [70], however it can not grow at high substrate concentrations which limits its application for hydrogen production operations. It was also observed that daily temperature fluctuations between 15-40°C result in a decrease in hydrogen production by 50% when compared to cultures grown at a constant temperature (30°C) [54].

N Sources

Photosynthetic purple non-sulfur bacteria can be adapted well to different types of N sources however glutamate has been found to be the most suitable nitrogen source for hydrogen production [4,27,72]. In *Rb.*

sphaeroides O.U.001, increasing concentration of glutamate results in an increase in biomass yield. On the contrary, hydrogen production is high at lower concentrations of glutamate (1-2 mM) and decreases if the glutamate concentration is higher. Increasing glutamate concentration from 2 to 10 mM decreases hydrogen production by 55-80%. Hydrogen production completely stops at a glutamate concentration of 17 mM [43,72]. Urea can be used as a substitute to glutamate although some hydrogen yield loss occurs: *Rb. sphaeroides* O.U.001 produced 1.24 L H_2/L in media containing 0.5 mM of urea compared to 1.45 L H_2/L in media containing 2 mM of glutamate [27].

C Sources

Hydrogen production depends on the type and concentration of carbon source fed to bacteria. Though a wide variety of substrates can be used by photosynthetic bacteria for growth, only a portion of these is suitable for hydrogen production. The preferred substrates are anions of organic acids such as acetic, butyric, propionic, lactic and malic acid while sugars such as glucose and sucrose do not seem to be efficient. Industrial effluents may also be used. On the basis of available literature, malate and lactate are the most favorable substrates giving the highest hydrogen production rates and the highest substrate to hydrogen conversion efficiencies. Butyrate was not that efficient for hydrogen production due to the competitive production of PHB [4,29]. *Rb. sphaeroides* and *Rb. capsulatus* are also capable of using mixtures of acetate, butyrate and propionate efficiently for hydrogen production [29,59]. Hydrogen production of *Rb. sphaeroides* O.U.001 was found to be increasing with increasing malate concentrations; 0.5, 0.9 and 1.3 L H_2/L were obtained at 7.5, 15 and 30 mM malate concentrations, respectively, when N source was 1 mM glutamate [72].

Hydrogen productivity of three other strains of *Rb. sphaeroides* tested in media containing 21, 42, 84, 168 mM of acetate and highest rates were obtained in 42 mM concentration, increasing acetate concentration further reduced hydrogen production rate [73]. *Rb. sphaeroides* SCJ, was examined for its potential in hydrogen production at 10–100 mM butyrate concentrations. As the amount of butyrate increased up to 25 mM, hydrogen production rised, and then decreased and ceased at 100mM. Additionally, lag phase increased with butyrate concentration, possibly indicating substrate inhibition [74].

Minor Nutrients

Iron is an essential metal for photosynthetic bacteria; it is in the cofactor of the hydrogenase and nitrogenase enzymes in photosynthetic bacteria. It is also present in ferredoxin. Therefore, changes in Fe concentration in the media may affect hydrogenase activity and electron transport chain of the bacteria. Indeed, the effect of ferrous ion (0–57 μM) on photoheterotrophic hydrogen production was studied and the results showed that hydrogen production by *Rb. sphaeroides* was significantly suppressed when Fe^{2+} was limited. Hydrogen production increased linearly with an increase in Fe^{2+} concentration in the range of 0-29 μM; reaching a maximum at 43 μM [75]. In a similar study, it was reported that there was a growth delay and almost no hydrogen production in media which is not containing iron indicating the vitality of iron for cellular functions and for the hydrogen evolution. It was also reported that high iron concentration (1 mM) caused a decrease in the hydrogen production which might be due to a toxic effect [9].

Molybdenum is another metal that dramatically affects hydrogen production since it is in the cofactor of the nitrogenase enzyme. There was almost no hydrogen production in medium where molybdenum is absent and little hydrogen production in 0.0165 μM Mo containing medium; however, it was observed that there was an increase in total hydrogen production up to 30% accompanied with elevated molybdenum concentrations (1.65 μM and 16.5 μM) [9,76]. Hydrogen evolution rate of *Rb. sphaeroides* RV cells is also slightly influenced by the presence of magnesium and phosphorus [76].

Light

Light intensity strongly affects the hydrogen production in photosynthetic purple non-sulfur bacteria. The rate of hydrogen production increases with increasing light intensity and reaches a value called the "light saturation point" which is the lowest light intensity supporting maximal hydrogen production. Light saturation point ranges from 65 to 200 W/m^2 for many strains of phototrophic bacteria. Very high light intensities (higher than 400 W/m^2) may cause photoinhibition of hydrogen gas production in some bacterial strains [69].

Bacteria absorb light at specific wavelengths based on its bacteriochlorophyll and carotenoid pigments (Fig. **5**). Lack of infrared light (wavelength 750–950 nm) was found to decreases photoproduction of hydrogen by 40% indicating that the infrared region of light spectrum where the bacteriochlorophyll *a* absorption maxima exist is very important for hydrogen production. Light sources which do not emit light in the red-infrared region (750-950 nm) and the bioreactor construction materials that absorb/reflect these wavelengths of light are not suitable for an efficient hydrogen production process [35].

In the photobioreactors that are exposed to light-dark cycles similar to the diurnal cycle, hydrogen production and cell growth stop during the dark periods and restores when illumination starts again. These inactive periods correspondingly results in low average hydrogen production rate for the process compared to the rate obtained in continuously illuminated photobioreactors. The total hydrogen produced also decreases compared to the continuously illuminated photobioreactor, possibly due to the consumption of available substrates during dark periods by the bacteria [34,35,76]. For outdoor operations, artificial illumination during night might be considered based on the gain reported and the cost of such an installment.

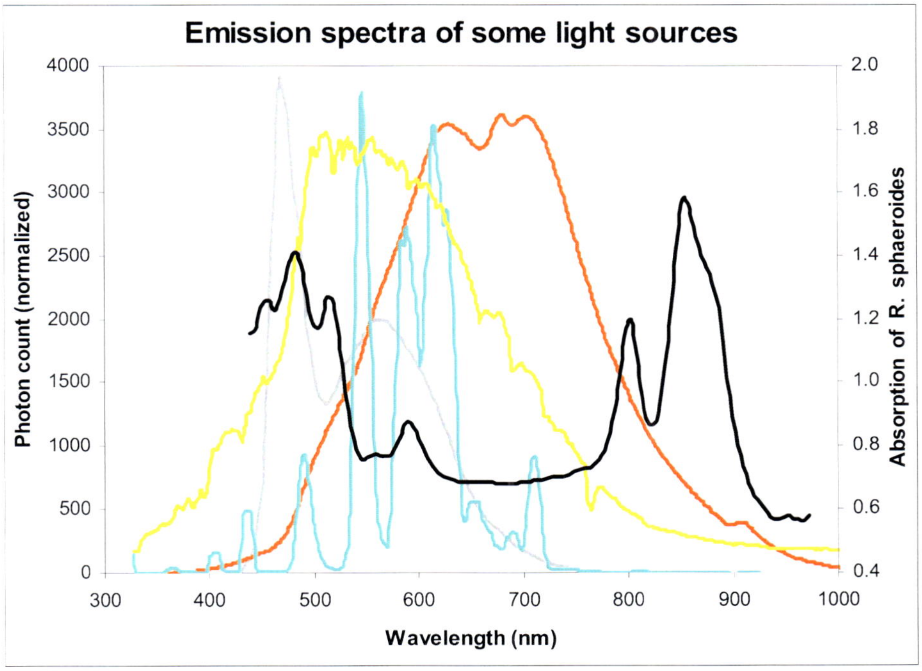

Figure 5: Absorption spectra of *Rb. sphaeroides* O.U. 001 (—) and emission spectra of different light sources; (—): sunlight, (—): white LED, (—): fluorescent, (—): tungsten.

Immobilization of Microorganisms

Using immobilized microorganisms is an alternative to using suspension cell cultures. Immobilizing microorganisms can result in diffusion limitations of substrates and product (H_2), on the other hand, cells can be reused, bioreactor effluent becomes cell-free, and higher cell concentrations in the bioreactor can be achieved which may result in a higher rate of hydrogen production.

Indeed, immobilized cells usage is reported to improve hydrogen production. In a photobioreactor containing a porous glass sheet on which *Rb. sphaeroides* RV was immobilized (11.2 mg cdw/mL porous glass), the maximum rate of hydrogen evolution was 260 mL H_2/L.h [77]. This is 8-10 times higher compared to the rates obtained by free cultures of various strains of *Rhodobacter* cells [4]. Photoproduction of hydrogen, were studied in free and alginate immobilized whole cells of a purple non-sulfur photosynthetic bacterium *Rb. sphaeroides* O.U.001. Four-fold increase in hydrogen production were observed in immobilized cells compared to free cells [78].

In addition, the immobilization within specific gels may improve hydrogen production by protecting the bacteria from the inhibitory effect of ammonium ion. This is particularly useful for wastewater containing significant amounts of ammonium. Hydrogen production from the wastewater of tofu factory was examined by using anoxygenic phototrophic bacterium *Rb. sphaeroides* immobilized in agar gels. The immobilization improved the capability of the bacterium to resist against the inhibitory effect of ammonium ion [79]. Moreover, hydrogen production in the presence of NH_4^+ by *Rb. sphaeroides* RV cells entrapped in the complex gel containing chitosan and agar improved considerably compared to that in the control containing only agar. The addition of chitosan increased the resistance to the diffusion of positively charged NH_4^+, but had no effect on negatively charged lactate thereby enhancing hydrogen production in the presence of ammonia. A buffer effect in the complex gels was also revealed [80].

Strain Improvement

Genetic improvement of purple non-sulfur bacteria is concentrated in the following major areas:

1. Elimination of uptake hydrogenase,
2. Eliminating CO_2 fixation,
3. The construction of mutants that have altered pigmentation,
4. Ammonium insensitive mutants,
5. The blocking of alternative pathways such as poly hydroxybutyrate synthesis.

Nitrogenase is activated by the RegB/RegA global regulatory system during photosynthetic growth. The terminal cytochrome *cbb*$_3$ oxidase serves as a redox signaling to the RegB/RegA regulatory systems in *Rhodobacter*. In the absence of the *cbb*$_3$ oxidase, electron flow is blocked, which renders RegB in the kinase-active mode. This leads to phosphorylation of RegA, resulting in the induction of nitrogenase and repression of uptake hydrogenase. Indeed, *Rb. capsulatus* mutant lacking the cytochrome *cbb*$_3$ oxidase had a hydrogen production rate increased by about 2-fold compared to the wild type strain [20].

The uptake hydrogenase is encoded by the *hupSL* genes and catalyzes the conversion of molecular hydrogen to electrons and protons and thus decreasing the hydrogen production efficiency of the nitrogenase as explained in "Enzymes" section [6]. *hupSL* deleted mutant strains lacking uptake hydrogenase (*hup*$^-$) enzyme has been constructed and their hydrogen productivity was compared to that of wild type cells. *Rb. sphaeroides* KD131 and its mutant strain lacking uptake hydrogenase (*hup*$^-$) have been studied in media containing 30 mM of malate as the carbon source. The rate of hydrogen production by *hup*$^-$ mutant strain was 83% higher compared to the rate obtained by wild-type strain [81]. In a similar study, the hydrogen production potential of the *Rb. sphaeroides* O.U.001 and its *hupSL* deleted mutant strain was evaluated in 15 mM malate containing media. The *hupSL*$^-$ *Rb. sphaeroides* produced 2.42 L H_2/L as compared to the wild type cells which evolved 1.97 L H_2/L [82]. *Rb. capsulatus* ST410, a mutant of the wild strain B100 lacking hydrogenase activity, converted 60 mM malate to hydrogen at a yield of 68%, compared to only 25% with the wild strain under the same conditions [83].

Pigment reduction is also effective for the enhancement of hydrogen production in photosynthetic bacteria. A mutant strain of *Rb. sphaeroides* with reduced photosynthetic pigment content (constructed by removing the light harvesting complex II genes) lessens the photoinhibition under high intensity light, improve light utilization efficiency and helps the volumetric productivity of the photosynthetic microorganism [84].

In a similar study, a mutant strain of *Rb. sphaeroides* RV was created by UV radiation. The contents of bacteriochlorophylls and carotenoids were reduced to 41% and 49% of those from the wild type strain RV, respectively. As a result, the amount of light absorbed by the mutant strain was lower than that by the wild-type strain enabling the illumination of cells in the deeper parts of a reactor. Although mutant had less pigments, its growth rate was equivalent to that of wild type over a wide range of light intensities and mutant produced hydrogen with a stable manner. Using a plate-type reactor, it produced 50% more hydrogen than wild type [85].

It may be possible to increase the fraction of substrate utilized for hydrogen production if PHB synthesis is disabled. *Rb. sphaeroides* KD131 and its mutant strain lacking PHB synthase (phb^-) have been studied on hydrogen production and cell growth. The rate of hydrogen production by phb^- mutant strain was 34% higher compared to the hydrogen production rate of wild-type strain [81].

It is also possible to combine genetic modifications for a cumulative increase in hydrogen production: *Rb. sphaeroides* KD131 and its mutant strain lacking both uptake hydrogenase (hup^-) and PHB synthase (phb^-) have been studied on hydrogen production. The rate of hydrogen production was 153% higher in hup^-/phb^- mutant strain, compared to the rate obtained by wild-type strain [81].

POTENTIAL APPLICATIONS

Techno-Economics of Photofermentative Hydrogen Production

Complicated techno-economic analyses are necessary to compare biologically produced hydrogen and the various other conventional fossil fuels due to the intervening large number of other parameters. The socially relevant costs of bringing any fuel to market must also include such factors as pollution and other short-term and long-term environmental costs, as well as direct and indirect health costs. When these factors are taken into consideration, hydrogen is surely the most logical choice as a worldwide energy medium [86].

Biologically produced hydrogen is currently more expensive than other fuel options. This high cost is one of the most critical issues for commercial application of biohydrogen. In 1998, it was suggested that the cost of large scale photofermentative hydrogen production would be US$20/kg. This is much higher than the costs of conventional hydrogen production based on fossil fuels *via* reforming processes, but less than the costs of a combined photovoltaic/electrolysis process [41]. A newer report in 2004 suggested that the cost of hydrogen produced by a process in which photofermentation coupled with dark fermentation was US$6/kg. On the other hand, 2020 estimates are in the range of US$2/kg for a similar coupled process [87].

The 6th EU Framework Programme Integrated Project "Non-Thermal Production of Pure Hydrogen from Biomass" (HYVOLUTION) started on 1 January 2006 and will end on 31 December 2010. Its aim, "Development of a blueprint for an industrial bioprocess for decentral hydrogen production from locally produced biomass," adds to the number and diversity of hydrogen production routes giving greater security of supply at the local and regional level. Moreover, this "Integrated Project" provides a complementary strategy to fulfill the increased demand for renewable hydrogen expected in the transition to an "Hydrogen Economy". The novel approach adopted in the project is based on a combined bioprocess employing thermophilic and phototrophic bacteria, to provide the highest hydrogen production efficiency from multiple biomass feedstocks in small-scale, cost-effective industries [88].

A first cost evaluation has addressed different scenarios concerning substrate concentration, yield, and productivities based. The cost calculations showed that the cost of the feedstock is of major importance. The objective in HYVOLUTION is about €6/GJ, contributed by biomass production and pretreatment, for the total cost for hydrogen from an industrial HYVOLUTION plant. The current estimations range from €4 to €12.7/GJ (corresponding to US$0.8 to US$2.5/kg H_2) for wheat bran and sugar beet in various regions, respectively. Preliminary cost evaluation of the overall process also showed the decisive effect of the cost of enzymes for pretreatment, yeast extract, and the material for the photobioreactor [88].

For comparison, as of 2010, commercial price of hydrogen is approximately US$2.1/kg when produced by steam reforming of natural gas and US$6.8/kg when produced by water electrolysis [89]. The price of natural gas, another gas fuel, in international market is approximately US$0.24/kg (equivalent to

US$0.64/kg hydrogen) [90]. Those prices show that the cost of the hydrogen gas produced by photofermentation is ten times higher than natural gas and three times higher than hydrogen gas produced by other technologies. If technology improvements succeed in bringing down costs, photofermentative hydrogen production is likely to play a major role in the economy.

Another major bottleneck for the commercialization of the photofermentative processes is low rate of hydrogen production. Considering the actual hydrogen production rates, photofermentative hydrogen production plants would require extremely large areas for production of large amounts of hydrogen. Pilot plant trials of the photo-fermentation processes also require more attention.

Use of cheaper raw materials and development of genetically modified microorganisms will surely make photofermentative hydrogen production processes more competitive with the conventional hydrogen generation processes in near future.

Integration with other Technologies

Another potential application of photofermentative hydrogen production is the integration with other process technologies.It is already demonstrated that photofermentation can be combined with dark fermentation of carbohydrates for enhanced hydrogen production. In such a combined process, carbohydrates are converted to hydrogen, carbon dioxide and intermediates during dark fermentation which is followed by a consecutive photofermentation. During photofermentation, all the intermediates are converted to hydrogen and carbon dioxide, as a result of which higher yields are obtained compared to single processes.

Several studies demonstrated the feasibility of using the two-stage process combining dark and photofermentation for simultaneous hydrogen production. In such a combined dark and photofermentation study, cassava starch was fed as the substrate to dark fermentation and the metabolite byproducts of the dark fermentation, which were mainly acetate and butyrate, were reused as the substrates in the photofermentation inoculated with the *Rp. palustris* bacteria. The maximum hydrogen yield dramatically increased from 240 mL H_2/g starch only in the dark fermentation to 402 mL H_2/g starch in the combined dark and photofermentation, while the energy conversion efficiency increased from 17.5–18.6% to 26.4–27.1% [91].

In a similar study, the soluble metabolites resulting from dark fermentation by *Clostridium pasteurianum* CH4, consisting of butyric and acetic acid, were further used for hydrogen production in the subsequent photofermentation by *Rp. palustris* WP3-5, elevating the total sucrose to hydrogen conversion efficiency from 16% (dark fermentation) to 42% (dark/photofermentation) [60].

In another study, an extreme thermophile *Caldicellulosiruptor saccharolyticus* was used for the dark fermentation, and several photosynthetic bacteria (*Rb. capsulatus* wild type, *Rb. capsulatus hup*$^-$, *Rp. palustris*) were used for the photofermentation, in batch operated bioreactors. The overall sucrose to hydrogen conversion efficiency increased from 18% in dark fermentation to 57% by sequential dark and photofermentation [92,93].

Photofermentative hydrogen production can also be coupled to biodiesel production. Biodiesel manufacturing processes inevitably produce a crude glycerol side fraction. The photosynthetic bacterium *Rp. palustris* is capable of the photofermentative conversion of glycerol, both pure and a crude glycerol fraction, to hydrogen with relatively high efficiencies, up to 75%. Even the crude glycerol fraction was readily converted to hydrogen with no apparent evidence of inhibition or toxicity [94].

Photofermentative hydrogen production can also be combined with a fuel cell for light energy conversion to electricity. An 11 L volume photobioreactor containing *Rb. sphaeroides* RV culture was directly connected to a polymer electrolyte fuel cell and it was shown that evolved gas did not contain any inhibitory gas for the fuel cell. The output power of the fuel cell was 1 W and the conversion efficiency of light to electricity reached to 1.9% [64]. In a comparable study, the biogas produced by different wild type and mutant strains of *Rb. capsulatus* in a 3 L volume photobioreactor was directly introduced in a small proton exchange membrane fuel cell, without prior purification. The maximal current value closed to 80mA for an electrode area of approximately 16 cm^2. At this working point, the cell potential was about 0.8V. The maximal current was maintained during a period of 11-32 h during the batch [95].

Different physicochemical methods including chemical oxidation with ozone and Fenton's reagent, photodegradation by UV radiation and adsorption with clay or zeolite were used for treatment of olive mill wastewater for the utilization of their liquid effluents for biological hydrogen production. Among these different 2-stage processes, the clay treatment method was selected as the optimum process that allows fast and low cost treatment as well as its effluent found to have the highest hydrogen production potential [96].

Biohydrogen production by dark and photo-fermentation of ground wheat starch has been concentrated on sequential and combined dark and light fermentations. Effects of dark/light bacteria ratio, substrate and cell concentrations on hydrogen yield and formation rate were investigated [97]. Sequential fermentation was found to be more advantageous as compared to combined fermentation. Combined fermentation operated in fed-batch mode was also investigated as an alternative to batch operation [98].

Biohydrogen production from cheese whey was conducted in two stages; I) dark fermentation and ii) photofermentation. Thermophilic conditions were determined to be better and more stable than mesophilic and the co-cultures of two pure species of *Clostridium* were found to more beneficial [99].

A pilot-scale tubular photobioreactor (80 L) was developed for photobiological hydrogen production and it was operated in outdoor conditions in fed-batch mode for 30 days using *Rhodobacter capsulatus* wild type on synthetic and real dark fermenter effluents. The average hydrogen productivity was 6.9 ml H_2/L.h, or 2.5 L H_2/day.m^2 ground area. The overall substrate conversion efficiency was 15% [65].

REFERENCES

[1] Reith JH, Wijffels RH, Barten H, Eds. Bio-methane & Bio-hydrogen: status and perspectives of biological methane and hydrogen production. The Netherlands: Dutch Biological Hydrogen Foundation 2003.

[2] Asada Y, Miyake J. Photobiological hydrogen production. J Biosci Bioeng 1999; 88(1): 1-6.

[3] Kars G, Gündüz U. Towards a super H_2 producer: improvements in photofermentative biohydrogen production by genetic manipulations. Int J Hydrogen Energy 2010; 35: 6646-56.

[4] Koku H, Eroglu I, Gündüz U, Yücel M, Türker L. Aspects of the metabolism of hydrogen production by *Rhodobacter sphaeroides*. Int J Hydrogen Energy 2002; 27: 1315-29.

[5] Das D, Veziroglu TN. Hydrogen production by biological processes: a survey of literature. Int J Hydrogen Energy 2001; 26: 13-28.

[6] Vignais PM, Colbeau A, Willison JC, Jouanneau Y. Hydrogenase, nitrogenase, and hydrogen metabolism in the photosynthetic bacteria. Adv Microb Physiol 1985; 26: 155-234.

[7] Akköse S, Gündüz U, Yücel M, Eroglu I. Effects of ammonium ion, acetate and aerobic conditions on hydrogen production and expression levels of nitrogenase genes in *Rhodobacter sphaeroides* O.U.001. Int J Hydrogen Energy 2009; 34(21): 8818-27.

[8] Vignais PM, Magnin JP, Willison JC. Increasing biohydrogen production by metabolic engineering. Int J Hydrogen Energy 2006; 31(11): 1478-83.

[9] Kars G, Gündüz U, Yücel M, Türker L, Eroglu I. Hydrogen production and transcriptional analysis of *nifD*, *nifK* and *hupS* genes in *Rhodobacter sphaeroides* O.U.001 grown in media with different concentrations of molybdenum and iron. Int J Hydrogen Energy 2006; 31(11): 1536-44.

[10] Willison JC, Jouanneau Y, Colbeau A, Vignais PM. H_2 metabolism in photosynthetic bacteria and relationship to N_2 fixation. Ann Microbiol (Paris) 1983; 134(1): 115-35.

[11] Goldberg I, Nadler V, Hochman A. Mechanism of nitrogenase switch-off by oxygen. J Bacteriol 1987; 169: 874-9.

[12] Hochman A, Burris RH. Effect of oxygen on acetylene reduction by photosynthetic bacteria. J Bacteriol 1981; 147(2): 492-9.

[13] Schneider K, Müller A, Schramm U, Klipp W. Demonstration of a molybdenum- and vanadium-independent nitrogenase in a *nifHDK*-deletion mutant of *Rhodobacter capsulatus*. Eur J Biochem 1991; 195(3): 653-61.

[14] Sasikala K, Ramana CV, Raghuveer RP, Subrahmanyam M. Effect of gas phase on the photoproduction of hydrogen and substrate conversion efficiency in the photosynthetic bacterium *Rhodobacter sphaeroides* O.U. 001. Int J Hydrogen Energy 1990; 15(11): 795-7.

[15] Jouanneau Y, Wong B, Vignais PM. Stimulation by light of nitrogenase synthesis in cells of *Rhodopseudomonas capsulata* growing in N-limited continuous cultures. BBA-Protein Struct M 1985; 808(1): 149-55.

[16] Meyer J, Kelley BC, Vignais PM. Effect of light nitrogenase function and synthesis in *Rhodopseudomonas capsulata*. J Bacteriol 1978; 136(1): 201-8.

[17] Vignais PM, Billoud, B. Occurrence, classification, and biological function of hydrogenases: an overview. Chem Rev 2007; 107: 4206-72.

[18] Zorin NA. Redox properties and active center of phototrophic bacteria hydrogenases. Biochimie 1986; 68(1): 97-101.

[19] Kars G, Gündüz U, Rakhely G, Yücel M, Eroglu I, Kovacs KL. Improved hydrogen production by uptake hydrogenase deficient mutant strain of *Rhodobacter sphaeroides* O.U.001. Int J Hydrogen Energy 2008; 33(12): 3056-60.

[20] Oztürk Y, Yücel M, Daldal F, *et al.* Hydrogen production by using *Rhodobacter capsulatus* mutants with genetically modified electron transfer chains. Int J Hydrogen Energy 2006; 31: 1545-52.

[21] Hillmer P, Gest H. H_2 metabolism in the photosynthetic bacterium *Rhodopseudomonas capsulata*: H_2 production by growing cultures. J Bacteriol 1977; 129(2): 724-31.

[22] Koku H. Hydrogen metabolism and factors affecting hydrogen production in *Rhodobacter sphaeroides*. MSc Thesis. Ankara: Middle East Technical University 2001.

[23] Tabita FR. The biochemistry and metabolic regulation of carbon metabolism and CO_2 fixation in purple bacteria. In: Blankenship RE, Madigan MT, Bauer CE, Eds. Anoxygenic photosynthetic bacteria. The Netherlands: Kluwer Academic Publishers 1995; pp. 885-914.

[24] Wiessner W. Photometabolism of organic substrates. In: Halldal P, Ed. Photobiology of microorganisms. Great Britain: John Wiley and Sons Ltd 1970; pp. 95-133.

[25] Meister M, Saum S, Alber BE, Fuchs G. L-Malyl-coenzyme A/β-methylmalyl-coenzyme A lyase is involved in acetate assimilation of the isocitrate lyase-negative bacterium *Rhodobacter capsulatus*. J Bacteriol 2005: 1415-25.

[26] Alber BE, Spanheimer R, Ebenau-Jehle C, Fuchs G. Study of an alternate glyoxylate cycle for acetate assimilation by *Rhodobacter sphaeroides*. Mol Microbiol 2006; 61(2): 297-309.

[27] Uyar B. Hydrogen production by microorganisms in solar bioreactor. PhD Thesis. Ankara: Middle East Technical University 2008.

[28] Gurgun V, Kirschner G, Pfennig N. Fermentation of pyruvate by 7 species of phototrophic purple bacteria. Z Allg Mikrobiol 1976; 16(8): 573-86.

[29] Uyar B, Eroglu I, Yücel M, Gündüz U. Photofermentative hydrogen production from volatile fatty acids present in dark fermentation effluents. Int J Hydrogen Energy 2009; 34: 4517-23.

[30] Pellerin NB, Gest H. Diagnostic features of the photosynthetic bacterium *Rhodopseudomonas sphaeroides*. Curr Microbiol 1983; 9: 339-44.

[31] Conrad R, Schlegel HG. Influence of aerobic and phototrophic growth conditions on the distribution of glucose and fructose carbon into the Entner- Doudoroff and Embden-Meyerhoff pathways in *Rhodopseudomonas sphaeroides*. J Gen Microbiol 1977; 101: 277-90.

[32] Conrad R, Schlegel HG. An alternative pathway for the degradation of endogeneous fructose during the catabolism of sucrose in *Rhodopseudomonas capsulata*. J Gen Microbiol 1978; 105: 305-13.

[33] Akkerman I, Janssen M, Rocha J, Wijffels RH. Photobiological hydrogen production: photochemical efficiency and bioreactor design. Int J Hydrogen Energy 2002; 27: 1195-208.

[34] Miyake J, Wakayama T, Schnackenberg J, Arai T, Asada Y. Simulation of the daily sunlight illumination pattern for bacterial photo-hydrogen production. J Biosci Bioeng 1999; 88: 659-63.

[35] Uyar B, Eroglu I, Yücel M, Gündüz U, Türker L. Effect of light intensity, wavelength and illumination protocol on hydrogen production in photobioreactors. Int J Hydrogen Energy 2007; 32(18): 4670-7.

[36] Nakada E, Nishikata S, Asada Y, Miyake J. Light penetration and wavelength effect on photosynthetic bacteria culture for hydrogen production. In: Zaborsky OR, Ed. Biohydrogen. New York: Plenum Press 1998; pp. 345-352.

[37] Yoon JH, Shin JH, Kim MS, Sim SJ, Park TH. Evaluation of conversion efficiency of light to hydrogen energy by *Anabaena variabilis*. Int J Hydrogen Energy 2006; 31(6): 721-7.
[38] Wakayama T, Miyake J. Light shade bands for the improvement of solar hydrogen production efficiency by *Rhodobacter sphaeroides* RV. Int J Hydrogen Energy 2002; 27: 1495-500.
[39] Zabut B, El-Kahlout K, Yücel M, Gündüz U, Türker L, Eroglu I. Hydrogen gas production by combined systems of *Rhodobacter sphaeroides* O.U.001 and *Halobacterium salinarum* in a photobioreactor. Int J Hydrogen Energy 2006; 31(11): 1553-62.
[40] Tredici MR. Mass production of microalgae: photobioreactors. In: Richmond A, Ed. Handbook of microalgal culture, biotechnology and applied phycology. London: Blackwell Scientific 2004; pp. 178-214.
[41] Modigell M, Holle N. Reactor development for a biosolar hydrogen production process. Renew Energ 1998; 14: 421-6.
[42] Hoekema S, Bijmans M, Janssen M, Tramper J, Wijffels RH. A pneumatically agitated flat-panel photobioreactor with gas re-circulation: anaerobic photoheterotrophic cultivation of a purple non-sulfur bacterium. Int J Hydrogen Energy 2002; 27: 1331-8.
[43] Sasikala CH, Ramana CHV, Rao PR. Regulation of simultaneous hydrogen photoproduction during growth by pH and glutamate in *Rhodobacter sphaeroides* O.U.001. Int J Hydrogen Energy 1995; 20: 123-6.
[44] Eroglu E, Gündüz U, Yücel M, Türker L, Eroglu I. Photobiological hydrogen production by using olive mill wastewater as a sole substrate source. Int J Hydrogen Energy 2004; 29: 163-71.
[45] Younesi H, Najafpour G, Syahidah K, Ismail K, Mohamed A, Kamaruddin AH. Biohydrogen production in a continuous stirred tank bioreactor from synthesis gas by anaerobic photosynthetic bacterium: *Rhodospirillum rubrum*. Bioresource Technol 2008; 99: 2612-9.
[46] Sierra E, Acien FG, Fernandez JM, Garcia JL, Gonzalez C, Molina E. Characterization of a flat plate photobioreactor for the production of microalgae. Chem Eng J 2008; 138(1-3): 136-47.
[47] Eroglu I, Tabanoglu A, Gündüz U, Eroglu E, Yücel M. Hydrogen production by *Rhodobacter sphaeroides* O.U.001 in a flat plate solar bioreactor. Int J Hydrogen Energy 2008; 33(2): 531-41.
[48] Basak N, Das D. Photofermentative hydrogen production using purple non-sulfur bacteria *Rhodobacter sphaeroides* O.U.001 in an annular photobioreactor: a case study. Biomass Bioenerg 2009; 33(6-7): 911-9.
[49] Carlozzi P, Lambardi M. Fed-batch operation for bio-H_2 production by *Rhodopseudomonas palustris* (strain 42OL). Renew Energ 2009; 34(12): 2577-84.
[50] Ogbonna JC, Soejima T, Tanaka H. An integrated solar and artificial light system for internal illumination of photobioreactors. J Biotechnol 1999; 70: 289-97.
[51] Richmond A, Cheng-Wu Z. Optimization of a flat plate glass reactor for mass production of *Nannochloropsis* sp. outdoors. J Biotechnol 2001; 85: 259-69.
[52] Hu Q, Guterman H, Richmond A. A flat inclined modular photobioreactor for outdoor mass cultivation of photoautotrophs. Biotechnol Bioeng 1996; 51: 51-60.
[53] Zhang K, Kurano N, Miyachi S. Outdoor culture of a cyanobacterium with a vertical flat-plate photobioreactor: effects on productivity of the reactor orientation, distance setting between the plates and culture temperature. Appl Microbiol Biotechnol 1999; 52: 781-6.
[54] Ozgür E, Uyar B, Oztürk Y, Yücel M, Gündüz U, Eroglu I. Biohydrogen production by *Rhodobacter capsulatus* on acetate at fluctuating temperatures. Resour Conserv Recy 2010; 54(5): 310-14.
[55] Miyake J, Asada Y. Biological production of hydrogen by environmentally acceptable technologies. In: Ohta T, Homma T, Eds. New energy systems and conversions. USA: Universal Academy Press Inc 1993; pp. 219-23.
[56] Kitajima Y, El-Shistawy RMA, Ueno Y, Otsuki S, Miyake J, Morimoto M. Analysis of compensation point of light using plane-type photosynthetic bioreactor. In: Zaborsky OR, Ed. Biohydrogen. New York: Plenum Press 1998; pp. 359-68.
[57] Rechenberg I. Artificial bacterial algal symbiosis (project ArBAS): sahara experiments. In: Zaborsky OR, Ed. Biohydrogen. New York: Plenum Press 1998; pp. 281-94.
[58] Trambouze P, Euzen JP, Eds. Chemical reactors: from design to operation. Paris: Technip 2004.
[59] Shi XY, Yu HQ. Continuous production of hydrogen from mixed volatile fatty acids with *Rhodopseudomonas capsulata*. Int J Hydrogen Energy 2006; 31(12): 1641-7.
[60] Chen CY, Yang MH, Yeh KL, Liu CH, Chang JS. Biohydrogen production using sequential two-stage dark and photo fermentation processes. Int J Hydrogen Energy 2008; 33(18): 4755-62.
[61] Avcioglu G, Androga DD, Uyar B, *et al.* Continuous hydrogen production by *R. capsulatus* on acetate in panel photobioreactors. Proceedings of the 8th Hydrogen Power Theoretical and Engineering Solutions International Symposium (HYPOTHESIS 2009); 2009 April 1-3; Lisbon, Portugal.

[62] Otsuki T, Uchiyama S, Fujiki K, Fukunaga S. Hydrogen production by a floating type bioreactor. In: Zaborsky OR, Editor. Biohydrogen. New York: Plenum Press 1998; pp. 369-74.

[63] Melnicki MR, Bianchi L, Philippis RD, Melis A. Hydrogen production during stationary phase in purple photosynthetic bacteria. Int J Hydrogen Energy 2008; 33: 6525-34.

[64] Nakada E, Nishikata S, Asada Y, Miyake J. Photosynthetic bacterial hydrogen production combined with a fuel cell. Int J Hydrogen Energy 1999; 24(11): 1053-7.

[65] Boran E, Ozgur E, Gebicki J, *et al.* Investigation of influencing factors for biological hydrogen production by *R. capsulatus* in tubular photo-bioreactors. Chem Eng Trans 2009; 18: 357-62.

[66] Gebicki J, Modigell M, Schumacher M, van der Burg J, Roebroeck E. Development of photobioreactors for anoxygenic production of hydrogen by purple bacteria. Chem Eng Trans 2009; 18: 363-6.

[67] Mathews J, Wang G. Metabolic pathway engineering for enhanced biohydrogen production. Int J Hydrogen Energy 2009; 34(17): 7404-16.

[68] Nath K, Das D. Effect of light intensity and initial pH during hydrogen production by an integrated dark and photofermentation process. Int J Hydrogen Energy 2009; 34(17): 7497-501.

[69] Sasikala K, Ramana CV, Raghuveer RP, Kovacs KL. Anoxygenic phototrophic bacteria: physiology and advances in hydrogen production technology. Adv Appl Microbiol 1993; 38: 211-95.

[70] Favinger J, Stadtwald R, Gest H. *Rhodospirillum centenum*, sp. nov., a thermotolerant cyst-forming anoxygenic photosynthetic bacterium. Antonie Van Leeuwenhoek 1989; 55: 291-6.

[71] Mahakhan P, Chobvijuk C, Ngmjarearnwong M, *et al.* Molecular hydrogen production by a thermotolerant *Rubrivivax gelatinosus* using raw cassava starch as an electron donor. ScienceAsia 2005; 31: 415-24.

[72] Eroglu I, Aslan K, Gündüz U, Yücel M, Türker L. Substrate consumption rates for hydrogen production by *Rhodobacter sphaeroides* in a column photobioreactor. J Biotechnol 1999; 70: 103-13.

[73] Asada Y, Ohsawa M, Nagai Y, *et al.* Re-evaluation of hydrogen productivity from acetate by some photosynthetic bacteria. Int J Hydrogen Energy 2008; 33(19): 5147-50.

[74] Lee JZ, Klaus DM, Maness PC, Spear JR. The effect of butyrate concentration on hydrogen production *via* photofermentation for use in a Martian habitat resource recovery process. Int J Hydrogen Energy 2007; 32: 3301-7.

[75] Zhu H, Fang HHP, Zhang T, Beaudette LA. Effect of ferrous ion on photo heterotrophic hydrogen production by *Rhodobacter sphaeroides*. Int J Hydrogen Energy 2007; 32(17): 4112-8.

[76] Fascetti E, D'Addario E, Todini O, Robertiello A. Photosynthetic hydrogen evolution with volatile organic acids derived from the fermentation of source selected municipal solid wastes. Int J Hydrogen Energy 1998; 23(9): 753-60.

[77] Tsygankov AA, Hirata Y, Miyake M, Asada Y, Miyake J. Photobioreactor with photosynthetic bacteria immobilized on porous glass for hydrogen photoproduction. J Ferment Bioeng 1994; 77(5): 575-8.

[78] Sasikala K, Ramana CV, Rao PR, Subrahmanyam M. Photoproduction of hydrogen, nitrogenase and hydrogenase activities of free and immobilized whole cells of *Rhodobacter sphaeroides* O.U.001. FEMS Microbiol Lett 1990; 72(1-2): 23-8.

[79] Zhu H, Suzuki T, Tsygankov AA, Asada Y, Miyake J. Hydrogen production from tofu wastewater by *Rhodobacter sphaeroides* immobilized in agar gels. Int J Hydrogen Energy 1999; 24(4): 305-10.

[80] Zhu H, Wakayama T, Suzuki T, Asada Y, Miyake J. Entrapment of *Rhodobacter sphaeroides* RV in cationic polymer/agar gels for hydrogen production in the presence of NH_4^+. J Biosci Bioeng 1999; 88(5): 507-12.

[81] Kim MS, Baek JS, Lee JK. Comparison of H_2 accumulation by *Rhodobacter sphaeroides* KD131 and its uptake hydrogenase and PHB synthase deficient mutant. Int J Hydrogen Energy 2006; 31(1): 121-7.

[82] Kars G, Gündüz U, Yücel M, Rakhely G, Kovacs KL, Eroglu I. Evaluation of hydrogen production by *Rhodobacter sphaeroides* O.U.001 and its *hupSL* deficient mutant using acetate and malate as carbon sources. Int J Hydrogen Energy 2009; 34(5): 2184-90.

[83] Ooshima H, Takakuwa S, Katsuda T, *et al.* Production of hydrogen by a hydrogenase-deficient mutant of *Rhodobacter capsulatus*. J Ferment Bioeng 1998; 85(5): 470-5.

[84] Kim NJ, Jeong KL, Lee CG. Pigment reduction to improve photosynthetic productivity of *Rhodobacter sphaeroides*. J Microbiol Biotechnol 2004; 14(3): 442-9.

[85] Kondo T, Arakawa M, Hirai T, Wakayama T, Hara M, Miyake J. Enhancement of hydrogen production by a photosynthetic bacterium mutant with reduced pigment. J Biosci Bioeng 2002; 93(2): 145-50.

[86] Nath K, Das D. Improvement of fermentative hydrogen production: various approaches. Appl Microbiol Biotechnol 2004; 65: 520-9.

[87] Claassen PAM, de Vrije T. Non-thermal production of pure hydrogen from biomass: HYVOLUTION. Int J Hydrogen Energy 2006; 31: 1416-23.

[88] Claassen PAM, de Vrije T, Koukios EG, *et al.* Non-thermal production of pure hydrogen from biomass: HYVOLUTION. In: Stolten D, Ed. Hydrogen energy. Weinheim: Wiley-VCH Verlag GmbH 2010.

[89] Committee on Alternatives and Strategies for Future NRC, National Academy of Engineering. The hydrogen economy: opportunities, costs, barriers, and R&D needs. Hydrogen production and use. Washington, DC: National Academies Press 2004; pp. 52.

[90] New York Mercantile Exchange (NYMEX) Natural Gas Futures Contracts [homepage on the Internet, cited March 1st, 2010]. Available from: http://www.oilnergy.com/1gnymex.htm

[91] Su H, Cheng J, Zhou J, Song W, Cen K. Improving hydrogen production from cassava starch by combination of dark and photo fermentation. Int J Hydrogen Energy 2009; 34(4): 1780-6.

[92] Ozgür E, Afsar N, de Vrije T, *et al.* Potential use of thermophilic dark fermentation effluents in photofermentative hydrogen production by *Rhodobacter capsulatus*. J Clean Prod 2010; 18: S23-8.

[93] Ozgür E, Mars AE, Peksel B, *et al.* Biohydrogen production from beet molasses by sequential dark and photofermentation. Int J Hydrogen Energy 2010; 35(2): 511-7.

[94] Sabourin-Provost G, Hallenbeck PC. High yield conversion of a crude glycerol fraction from biodiesel production to hydrogen by photofermentation. Bioresource Technol 2009; 100(14): 3513-7.

[95] He D, Bultel Y, Magnin JP, Roux C, Willison JC. Hydrogen photosynthesis by *Rhodobacter capsulatus* and its coupling to a PEM fuel cell. J Power Sources 1999; 141(1): 19-23.

[96] Eroglu E, Eroglu I, Gündüz U, Yücel M. Treatment of olive mill wastewater by different physicochemical methods and utilization of their liquid effluents for biological hydrogen production. Biomass Bioenerg 2009; 33: 701-5.

[97] Argun H, Kargi F, Kapdan IK. Effects of the substrate and cell concentrations on biohydrogen production from ground wheat by combined dark and light fermentations. Int J Hydrogen Energy 2009; 34: 6181-8.

[98] Ozmihci S, Kargi F. Effects of starch loading rate on performance of combined fed-batch fermentation of ground wheat for bio-hydrogen production. Int J Hydrogen Energy 2010; 35: 1106-11.

[99] Azbar N, Dokgoz FT, Keskin T. Comparative evaluation of biohydrogen production from cheese whey wastewater under thermophilic and mesophilic anaerobic conditions. Int J Green Energy 2009; 6(2): 192-200.

CHAPTER 5

Integration of Biological H_2 Producing Processes

Anatoly A. Tsygankov* **and Daria N. Tekucheva**

Institute of Basic Biological Problems RAS, Institutskaya, 2, Pushchino, Moscow, Russia

Abstract: Biological methods of H_2 generation are preferable to physico-chemical methods for several reasons: i) biological systems can use renewable sources of energy (sun, organic wastes); ii) biological processes are carrying under ambient pressures and temperatures; that is why they are safer; and iii) biological systems are self-supporting, self-repairing, and self-reproducible in principle. Different biological systems have own advantages and peculiarities. Combining them, the individual strength of each may be exploited and their weaknesses can be overcome. Different strategies of their integration are discussed in this chapter based on literature data. Some methods of integration are promising but still they have not been experimentally supported. The integration of dark fermentative H_2 production using organic wastes in the first stage and H_2 photoproduction by photosynthetic anoxygenic bacteria using an effluent from the fermentation as the second stage attracted much attention last years. This review evaluates published data with attempts to reveal the most important factors affecting the productivity and efficiency of these dual systems.

Keywords: Biohydrogen, biological hydrogen production, integration of biohydrogen processes, integration of dark fermentation and microbial fuel cell, integration of dark fermentation and photofermentation, organic wastes decomposition, alternative energy from organic wastes.

INTRODUCTION

Biological hydrogen production attracts researchers not only as a metabolic peculiarity. Molecular hydrogen (H_2) is an ideal fuel. During H_2 combustion virtually no environmentally dangerous products are released. That is why ecologically clean H_2 production can supply human society with energy, without a negative impact on the environment. Hydrogen can be produced and consumed by microorganisms. All metabolic pathways in microorganisms with H_2 production can be classified on the basis of initial energy source and key participating enzymes (Table **1**). Dark processes (numbers 1-3 in Table **1**) produce H_2 using energy of organic compounds. Among them dark fermentative H_2 production, with hydrogenase as key enzyme, is the most studied process. During fermentation of simple sugars, such as glucose, by granulated biomass, a volumetric H_2 production rate as high as 15 L H_2 per L of reactor per hour has been shown [1]. This is the highest rate reported in the literature, and is close to the rate required for practical applications [2, 3].

All heterotrophic diazotrophic bacteria produce H_2 during nitrogen fixation (number 2 in Table **1**). It an intrinsic property of the key nitrogen fixation enzyme – nitrogenase. However, only few of diazotrophic bacteria release H_2 outside cells, due to the activity of uptake hydrogenase enzymes. This process is not well studied for practical applications. Nevertheless, it is possible that under nitrogen limitation, some *Clostridia* species should be able to produce hydrogen *via* nitrogenase with high rates. Dark anaerobic CO decomposition to H_2 and CO_2 (number 3 in Table **1**) by bacteria from different groups (including some purple bacteria) is potentially valuable process [4, 5], which might be applicable for syngas purification with simultaneous production of H_2 and other useful products.

Light dependent H_2 evolution by anoxygenic photosynthetic bacteria under nitrogen limitation (number 4 in Table **1**) is under active investigations by many research teams for several decades. The key enzyme for this process is nitrogenase. Researchers have showed high rates of H_2 production by purple nonsulfur bacteria. The main drawback of this process is that bacteria need reduced electron donors like organic acids. Other groups of microorganisms (purple sulfur and green sulfur bacteria, heliobacteria) still deserve a thorough study.

***Address correspondence to Anatoly A. Tsygankov:** Institute of Basic Biological Problems RAS, Institutskaya, 2, Pushchino, Moscow, Russia; E-mail: ttt@issp.serpukhov.su

Nuri Azbar and David B. Levin (Eds)

Table 1: Microbial H_2 Producing Processes with Indications of Advantages and Drawbacks for Possible Practical Application.

No	The process	Organisms	Key enzyme	Advantages	Disadvantages
1	Dark anaerobic fermentation	Wide range of anaerobic bacteria	Hydrogenase	Substrates: wide range of organics including wastes; highest rates of the process.	Low yield of the process (not more than 4 moles H_2 per 1 mole of glucose); organic acids, alcohols as by-products; high H_2 concentration inhibits the process
2	Dark aerobic or anaerobic	All chemotrophic diazotrophic bacteria during nitrogen fixation	Nitrogenase	All organic substrates available for aerobic or anaerobic decomposition by diazotrophic bacteria under nitrogen fixing conditions	Low efficiency of the bioconversion, especially under aerobic conditions.
3	Dark anaerobic CO decomposition (water-shift reaction)	Some purple bacteria	Hydrogenase in conjunction with CO-dehydrogenase	The substrate: toxic gas which represents big part of syngas.	High sensitivity to inactivation by oxygen; organic substrates and H_2 can inhibit the process
4	Anaerobic photosynthetic under N deficiency	Nitrogen fixing anoxygenic photosynthetic bacteria	Nitrogenase	High rates, especially by purple nonsulfur bacteria, might be activated by near-infrared light; low sensitivity to high H_2 pressure.	Sensitivity to N sources (due to repression and inactivation of nitrogenase); needs in simple organics as electron donor for photosynthesis; narrow range of organics; inactivation by oxygen; low efficiency (needs in electrons and ATP)
5	Anaerobic photosynthetic under over reduced conditions	Anoxygenic photosynthetic bacteria and cyanobacteria under anaerobic conditions with reversible hydrogenase (*hox YH*)	Hydrogenase	Relatively high efficiency (does not need ATP)	At present knowledge appears only during short-term experiments under over-reduced conditions; high H_2 concentration inhibits the process; under mesophilic conditions a possibility for pathogenic bacteria development
6	Aerobic photosynthetic	Nitrogen fixing cyanobacteria	Nitrogenase	Converts light energy into H_2 fuel in stoichiometry near to $2H_2O \leftrightarrow 2H_2 + O_2$; low sensitivity to high H_2 concentrations; temporal or spatial separation of O_2 and H_2 production	Low rates of the process; low efficiency (needs in electrons and ATP)
7	Temporal separation of oxygenic photosynthesis and light-dependent H_2 production under sulfur deprivation	Microalgae with Fe-Fe hydrogenase; proved for several species including *Chlamydomonas* and *Chlorella*	Hydrogenase	Converts light energy into H_2 fuel; long term process; possibility to repeat the cycle "oxygenic photosynthesis-light-dependent H_2 production".	Low rates; high sensitivity to oxygen inactivation; still need experimental studies to determine efficiency of process
8	Aerobic photosynthetic	Microalgae with Fe-Fe hydrogenase	Hydrogenase	Converts light energy into H_2 fuel in stoichiometry near to $2H_2O \leftrightarrow 2H_2 + O_2$; high efficiency; high rates of the process	Appears only during short-term experiments at the start of anaerobic cultures illumination; under S-deprivation shows decreased rates but for longer period. By-products under S-deprivation are organic acids and ethanol.
9	Temporal separation of oxygenic photosynthesis with accumulation of polysaccharides and dark fermentative H_2 production	Proved for some cyanobacteria with reversible hydrogenase (*hoxYH*) and marine microalgae	Hydrogenase	Converts light energy into H_2 fuel; long term process; possibility to repeat the cycle "oxygenic photosynthesis-dark fermentative H_2 production."	Low rates; high sensitivity to oxygen inactivation; organic acids (especially acetate) as by-products.
10	Microbial electrolytic cell (MEC)	Proved for microbial consortia living in MFC	-	Converts energy of organics with external electricity	Need in catalyst for H_2 production

Several years ago, it was found that under over-reduced conditions, a cyanobacterium *Synechocystis* sp. [6] and the purple sulfur bacterium *Thiocapsa roseopersicina* BBS [7, 8] produce H_2 in the presence of light (number 5 in Table **1**) *via* reversible hydrogenase. Potentially, it is a way for H_2 production using hydrogenase instead of nitrogenase. However, rates and other parameters of this process are still under investigations.

Aerobic photosynthetic H_2 production by heterocystous cyanobacteria *via* nitrogenase (number 6 in Table **1**) is well-known process, which is under active research for decades. Certain cyanobacteria produce H_2 even under elevated concentrations of oxygen in the reaction with nearly stoichiometric to water decomposition to oxygen and hydrogen. Furthermore, the process is already spatially separated since O_2 is produced in vegetative cells, and H_2 is produced in heterocysts.

Some unicellular nitrogen-fixing cyanobacteria have a unique capability to combine the photosynthesis with O_2 production process of nitrogen fixation (with H_2 production; number 7, in Table **1**), which is strictly anaerobic, in one space. This is possible when photosynthesis occurs in the presence of light and nitrogen fixation occurs during the night (in the dark) under anaerobic conditions. In some cases, synchronous cultures of *Synechococcus* produce H_2 with high rates [9, 10]. The practical importance of this physiological peculiarity is still unclear.

Short-term hydrogen evolution by dark adapted microalgae without O_2 in response to illumination (number 8 in Table **1**) was shown more than 65 years ago [11]. This process has a high efficiency of light energy bioconversion [12]. Much later, Melis and coauthors [13] extended this process for several days by depriving the cultures of sulfur. The efficiency of sulfur-deprived microalgae still needs to be determined. However, it is lower than short-term H_2 production due to decreased PSII activity.

During the night time, cyanobacteria and microalgae under anaerobic conditions produce ATP *via* substrate phosphorylation, by degrading accumulated polysaccharides like starch and glycogen. In some cases, H_2 is the product of the fermentation together with acetate and small quantities of other products [14-16]. Thus, the single cell integrates photosynthesis with light energy accumulation, followed by fermentation of accumulated products with H_2 production (number 9 in Table **1**).

Recently, the possibility of H_2 production using organics and external electricity was shown [17]. This process (number 10 in Table **1**) is very attractive, but has not yet been studied in detail. The short summary of different microbial H_2 producing processes presented above shows that every process has its advantages and disadvantages. However, by combining different processes, the individual advantages of each may be explored and their weaknesses may be overcome. This review aims to evaluate published data of integrated systems of H_2 production for possible industrial application.

POTENTIAL INTEGRATION SCHEMES

In general, a combination of different microbial H_2 producing processes aims to use by-products of one stage as the substrate for another stage, to combine energy sources (organic molecules and light), or to use light energy more efficiently.

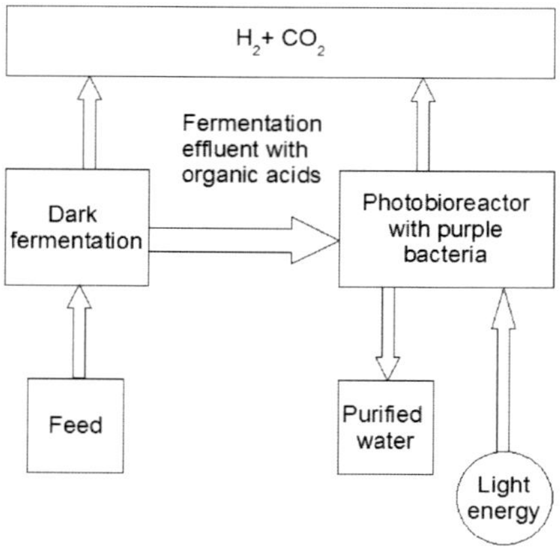

Figure 1: Integration scheme of dark fermentative H_2 production and anaerobic photosynthetic H_2 production.

The most studied scheme of integration is the combination of dark fermentative H_2 production, with organic acids as by-products (number 1 in Table **1**), followed by anaerobic photosynthetic H_2 production under nitrogen deficiency (number 4 in Table **1**) using the organic acids from first stage (Fig. **1**) as substrate. In this case, two drawbacks are eliminated because the by-products of the first stage (first drawback, Table **1**) become substrates for the second stage (second drawback). Potentially, it is possible to get as much as 12 molecules H_2 per mole of glucose. Furthermore, this scheme can be adopted for simultaneous H_2 production and waste purification (Fig. **1**). An additional advantage of this system is that two processes that have the highest rates of H_2 production are combined in one system. This integrated system is under active investigation for last several years and has been reviewed [18, 19].

Another scheme (Fig. **2**) is the integration of the process of temporal separation of oxygenic photosynthesis and dark fermentative H_2 production by single photosynthetic organism (number 9 in Table **1**), and the process of anaerobic photosynthetic H_2 production under nitrogen deficiency (number 4 in Table **1**). Similar to previous scheme, this integration allows the use by-products of the first stage as electron donors for the second stage. The advantage of this system is that it is closed concerning potential carbon emissions. If no biomass accumulates in either stage of the process, no carbon is released as CO_2. The single source of energy for this integrated system is sunlight. After small modifications, this scheme (Fig. **2**) can be combined with microalgal H_2 production under sulfur deprivation (number 7 in Table **1**) in the first stage and anaerobic photosynthetic H_2 production (number 4 in Table **1**) in the second stage.

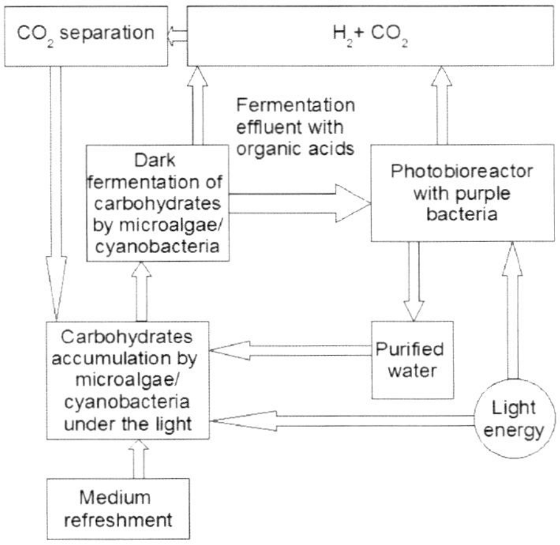

Figure 2: Integration of microalgal/cyanobacterial H_2 production in cyclic mode and anaerobic photosynthetic H_2 Production.

In principle, there is a possibility of integrating two photobioreactors in one unit. In the outer photobioreactor, any process with oxygenic photosynthesis (numbers 5-9 in Table **1**) can be used, whereas the inner photobioreactor would contain anaerobic photosynthetic H_2 production under nitrogen deficiency (number 4 in Table **1**). Due to different absorption spectra of microorganisms, light can be used more efficiently. This scheme of integration can be applied for the system represented in Fig. **2**.

An alternative integrated process combines photoautotrophic biomass accumulation with successive biomass decomposition. In this system, photoautotrophic biomass, which does not produce H_2 by itself, accumulates in the first stage. This is followed by anaerobic dark fermentative H_2 production. This process adds two new properties to the system illustrated in Fig. **1**: the sole source of energy for the whole system is sunlight, and the integrated system is neutral for CO_2 release.

Recently microbial electrolysis cells (MECs) have been investigated as a new way for H_2 production using organic acids (like acetate) and electricity [17]. Since acetate is by-product of dark fermentation, it is interesting to integrate dark fermentation process and microbial electrohydrogenesis.

The integrated systems described above all have one important peculiarity: all stages accumulate biomass during operation, but the biomass excess can inhibit the overall process. In the case of dark fermentation processes, biomass excess results in the limitation of substrates and product diffusion. In the case of photobioreactors, biomass excess causes self-shadowing of microorganisms, which results decreases in the specific rates of substrate utilization and product formation. These facts suggest that excessive biomass should be eliminated from reactors. Excessive biomass decomposition needs more complicated integrated schemes. The most evident additional element to the system depicted in Fig. **1** is the biomass processing (for example by disruption) and feeding dark reactors with processed biomass. Different integrated systems have been studied to different extents. The following sections of this review will give more detailed descriptions of them.

Integration of Dark Fermentative and Phototrophic Anoxygenic Hydrogen Production

The integration of these best studied processes, which have the highest established H_2 production rates in under laboratory conditions, looks very simple. It is necessary to choose organic wastes, select microorganisms for both stages, adjust volumes of dark fermentative reactor and photobioreactor for complete conversion of incoming substrate, efficient light utilization, and connect them together. The choice of organic wastes has several prerequisites. Since the second, anoxygenic photosynthetic stage of this system, is based on the nitrogenase as the key enzyme, the feed of this stage should not have inhibitory or repressive concentrations of nitrogen. Organic wastes must be metabolized during dark fermentative stage with production of by-products which are easily consumed by the second stage. Finally, these wastes must have mineral components in concentrations that are sufficient, but not inhibitory, for both stages. Applying different strategies for selection of microorganisms for the 1st and 2nd stages, we have to take into account that in theory, there are at least 4 different fermentation pathways [18], which give 12 moles of H_2 per mole of glucose in an integrated system:

A:

1 glucose $\rightarrow$ 2 lactate + 0 H_2; lactic fermentation

2 lactate $\rightarrow$ 12 H_2; purple bacteria

B:

1 glucose $\rightarrow$ 4 H_2 + 2 acetate; acetogenic fermentation

2 acetate $\rightarrow$ 8 H_2; purple bacteria

C:

1 glucose $\rightarrow$ 2 H_2 + 1 ethanol + 1 acetate; mixed fermentation

1 ethanol + 1 acetate $\rightarrow$ 10 H_2; purple bacteria

D:

1 glucose $\rightarrow$ 2-4 H_2 + 0-1 butyrate + 0-2 acetate; aceto-butyric fermentation

1 butyrate $\rightarrow$ 8-10 H_2; purple bacteria

Utilizing different metabolic pathways in the first stage produces various quantities of H_2 and different by-products. As a result, the second stage should be adapted to these particular products. Furthermore, the relative volumes of dark reactor and photobioreactor for complete organics utilization depend on the fermentation pathway in the first stage. Development of a technique to combine two stages is also the topic of investigations.

Substrates for the First Stage

Researchers which are studying model compounds as wastes for integrated systems use glucose, sucrose, dextrose, starch, cellulose, and cellulose hydrolysate (Table **2**). Almost all model substrates have been utilized in integrated systems with simultaneous H_2 production on both stages. However, if the synthetic media used contained excessive quantities of nitrogen, no H_2 was produced by the second stage [20]. When the 1st stage is loaded with high concentrations of organics, high concentrations of organic acids are produced in dark reactor. As a result, H_2 production is inhibited in the 2nd stage. Researchers avoided this by diluting the fermentation effluent with water [21]. After that the second stage operated efficiently. However, in some cases the total efficiency was not very high. It was found that after dilution some mineral components like sulfates and Fe were in limiting concentrations (see below). Moreover, to stabilize pH in the second stage, phosphates were added, and these were also stimulatory for H_2 production in the 2nd stage [22]. Addition of mineral components increased rates and efficiency of H_2 production [22].

Among real organic wastes molasses, cow and poultry manure, palm oil mill effluent, fruit and vegetable wastes, olive mill wastewater, potato homogenate, sawdust hydrolysate, tofu wastewater, rice-vine wastewater, corn steep liquor, or potato steam peel hydrolysate have been studied as substrates for integrated systems (Table **3**). All these wastes proved to be suitable for H_2 production by these integrated systems. The efficiencies of substrate conversion into H_2 by integrated systems varied in different studies from 3.1 to 7.1 moles H_2 $mole^{-1}$ glucose (Tables **2**, **3**). Based on presented data, it is difficult to clarify which substrate is more suitable for integrated system. This is because other parameters of integration also play important role in H_2 production.

Table 2: Integrated Systems for H_2 Production with Pure Cultures of Dark Fermentative Bacteria in the First Stage.

Substrate	**Bacterium in the 1st stage**	**Productivity of 1st stage (moles per 1 $mole^{-1}$ hexose)**	**Major products of the 1st stage**	**Bacterium in the 2nd stage**	**Total H_2 productivity (moles $mole^{-1}$ hexose)**	**Ref.**
Cellulose	*Cellulomonas* sp.	-	-	*Rb. capsulatus*	4.3	[23]
Glucose	*C. butyricum*	-	-	*Rb. sphaeroides*	7.0	[24]
Starch	*C. butyricum*	1.9	Acetate, lactate, butyrate	*Rhodobacter* sp.	6.6	[25]
Sweet potato residue (with addition of peptone)	*C. butyricum* and *Enterobacter aerogenes*	2.4	Acetate, lactate, butyrate	*Rhodobacter* sp.	7.0	[26]
Glucose	*Lactobacillus debrueckii*	-	-	*Rb. sphaeroides*	7.1	[27]
Glucose	*Escherichia coli*	0.4	Acetate, lactate, ethanol	*Rb. sphaeroides*	0.4	[20]
Sucrose	*C. pasteurianum*	1.9	Acetate and butyrate	*Rps. palustris*	7.1	[28]
Glucose	*Ethanoligenens harbinense*	1.83	Acetate, ethanol	*Rhodopseudomonas faecalis*	6.32	[29]
Miscanthus hydrolysate	*Thermotoga neapolitana*	-	Acetate, lactate	*Rb. capsulatus*	-	[21]
Glucose	*Clostridium butyricum*	1.72	Acetate and butyrate	*Rps. palustris*	5.48	[30]
Glucose	*C. butyricum*	-	-	*Rps. faecalis*	5.4	[31]
Beet molasses	*Caldicellulosiruptor saccharolyticus*	2.1	Lactate and acetate	*Rhodobacter capsulatus* hup⁻ mutant	6.85	[32]
Glucose	*Eth. harbinens*	-	-	*Rps. faecalis*	3.1	[33]

Table 3: Integrated Systems for H_2 Production with Microbial Consortium in the First Stage.

Substrate	Source of microbial consortium	Productivity of 1st stage (moles mole^{-1} hexose)	Major products of the 1st stage	Bacterium in the 2nd stage	Total H_2 productivity (moles mole^{-1} hexose)	Refs.
Sucrose	Heat treated cattle dung	1.83	Acetate, butyrate	*Rb. sphaeroides*	3.32	[34]
Wheat powder	Heat treated anaerobic sludge	-	-	*Rb. sphaeroides*	156.8 ml per 1 g of starch (1.26)	[35]
Potato homogenate	Heat treated silo pit liquid	1.6 moles H_2 per 1 mole of hexose	Acetate, butyrate	*Rb. capsulatus*	5.6 moles H_2 per 1 mole of hexose	[22, 36]
Wheat powder	Heat treated anaerobic sludge	-	-	*Rb. sphaeroides*	139 ml per 1 g of starch (1.12)	[37]

Different Types of Microorganisms in the Dark Stage

Generally, integrated systems described in literature explore pure cultures or microbial consortia. Among pure heterotrophic cultures for the first stage, different *Clostridium* or *Enterobacter* species are most widely studied (Table **2**). Others are different lactic bacteria, *Cellulomonas*, *Klebsiella pneumoniae*, and thermophilic archae like *Caldicellulosiruptor saccharolyticus* and *Thermotoga neapolitana*. When single bacterial species is used in the first stage of the integrated system and synthetic medium is used, it is possible to calculate the efficiency of glucose conversion to H_2. Table **2** indicates that 1 mole glucose (or 0.5 mole of sucrose) can be converted into the first stage into 1.1-3.3 moles of H_2. Only a single case reported the efficiency of substrate conversion to hydrogen gas as high as 83% of theoretical yield [38]. Other results indicate that the efficiency was not higher than 50% or even lower than theoretical. Total yield of H_2 in the integrated system in many cases reach 5-8 moles H_2 per mole of hexose which is higher than theoretical yield in dark fermentation process (Table **2**). Main portion of H_2 was produced by the second stage during anoxygenic photosynthesis.

Application of pure cultures for the first stage appears to be very useful approach for experimental estimation of H_2 production rates, yields as well as in determination of produced organic acids. Experiments proved that the integrated system is much more efficient than single dark fermentation stage. However for practice, pure culture in the first stage is not applicable due to high cost of waste sterilization before feeding into reactors. Besides, pure cultures are unable to use wide spectrum of organics in wastes for H_2 production.

Application of microbial consortia for the first stage of integrated system is the next step for laboratory research. Using this technology, laboratory systems can be adapted from synthetic wastes with simple organics like glucose, as well as complex mixture of organics in real wastes. In spite of the fact that dark fermentative hydrogen production by microbial consortia is widely studied [39], only few authors use this technology for integrated dark-light H_2 producing system (Table **3**). The total productivity of the integrated system with microbial consortium is much higher than single dark fermentation process and it is close to systems with single bacteria at the first stage. Thus, microbial consortia can be used in the first stage of integrated system. Based on this fact it is easy to predict that in near future papers describing operation of integration systems using microbial consortia and real complex wastes will appear.

Microorganisms for the Second, Anoxygenic Photosynthetic Stage

Potentially, several groups of nitrogen fixing anoxygenic photosynthetic bacteria can be applied for the second stage of integrated system: purple sulfur and non-sulfur bacteria, green sulfur bacteria, and heliobacteria. However, purple and green sulfur bacteria grow slower and show low nitrogenase activity compared to purple non-sulfur bacteria. In addition, oxygen is very toxic to them. The last group, heliobacteria, grows fast under nitrogen fixing conditions, uses organic acids and other complex organics, and looks like promising candidate for the second stage of a H_2 producing integrated system. However, heliobacteria are obligate anaerobes and are difficult to cultivate in the laboratory. In contrast, purple non-sulfur bacteria use wide spectrum of organic acids and other complex organics, can grow aerobically, and have very high nitrogenase activity under anaerobic conditions in the presence of light. That is why all researches use purple non-sulfur bacteria for second stage of the integrated system (Tables **2**, **3**). The most studied species are: *Rb. sphaeroides, Rb. capsulatus, Rps. palustris, Rps. rubrum, R. marinum,* and some other non-identified strains. In all cases, predefined strains were used for the second stage. However, it is not clear still how to adopt this stage for practical operation under non-sterile conditions.

The first, dark stage of integrate system in practice is loaded with wastes at very variable concentrations. Thus, the concentrations of the organic acids may vary widely. That is why it is important to know the range of organic acids concentrations which allow purple bacteria in the second stage to produce H_2 without limitations and inhibition. For *Rb. capsulatus* it was defined that acetate saturated the rate of H_2 production at 1 mM [40]. Same authors showed that in short-term experiments 80 mM of acetate, lactate, butyrate, and isobutyrate inhibited H_2 production by 20-30%. Propionate added to the culture which was not adapted to this organic acid inhibited H_2 production by 70%. In contrast, if culture was pre-grown using propionate the inhibition of H_2 production was only 30%. Additional studies are necessary to clarify the range of organic acid concentrations for stable H_2 production in the second stage.

Estimation of Relative Volumes of Dark Reactor and Photobioreactor

Adaptation of integrated systems for simultaneous H_2 production and waste treatment requires appropriate reactor volumes. Relative volume of the reactors may be determined by the specific rate of H_2 production in each stage and the stoichiometry of the process. Unfortunately, specific rates of H_2 production by different stages presented in the literature are very different. It makes impossible to create general conclusion. That is why in our estimation of reactor volumes, we shall use the stoichiometry of the process and possible connection of two single-stage processes described in the literature with highest rates of H_2 production.

The application of various dark fermentation strategies (see beginning of Section 3, A-D) has direct impact to relative proportions of reactor volume due to different stoichiometries of processes. From theoretical considerations, one could conclude that in the case of lactic acid fermentation (Case A, above), H_2 is produced exclusively in the photobioreactor.

In all other cases, the stoichiometry of complete substrate utilization by the integrated system and the need of steady-state operation of both reactors dictate that the proportion of H_2 production is equal to rates of H_2 production by whole reactors. In the case of acetogenic fermentation (Case B) the rate of H_2 production by a photobioreactor should be 2 times higher than the dark fermentation reactor. In the Case C, this parameter equal to 5, and in Case D, H_2 production by a photobioreactor should be 2-5 times higher than the dark fermentation reactor. Thus, the volume of the photobioreactor is determined by the relative volumetric rates of H_2 production by the dark fermentation reactor and the photobioreactor, by multiplying by a factor of 2 - 5 (excluding Case A).

Alternative information for the integration of the two stages comes from published experiments on each single stage. One could reproduce experiments with the highest rates of H_2 production in the dark and photosynthetic stages, respectively, and connect them together directing the effluent from dark stage to photosynthetic stage and adjusting reactors volume. The adjustment of reactor volumes can be made using two different strategies. The first strategy is the simultaneous H_2 production and wastes treatment in the integrated system without waste generation. In this case, relative volumes of reactors should be calculated on the basis of complete utilization of organic acids derived from complete degradation of wastes. The second strategy is based on experimentally determined maximum rates of H_2 production by stages of integrated system. In both cases, the suitability of fermentation effluent for photosynthetic stage remains unclear.

As an example, one could consider the integration of sucrose dark fermentation with highest H_2 production rates [1] and photobioreactor with immobilized purple bacterium *Rhodobacter sphaeroides* GL1 showing highest rate of H_2 production among purple bacteria [41]. The integration of these processes is simplified because both use continuous flow processes. The rate of H_2 production in the dark stage was 15 L H_2 L^{-1} h^{-1} of reactor, at a hydraulic retention time (HRT) equal to 0.5 [1]. Under these conditions, 80 g of sucrose were consumed L^{-1} h^{-1} and 262 mmoles of butyrate and 42 mmoles of acetate, with minor quantities of propionate (9%) and ethanol (6%) were produced L^{-1} h^{-1}. Efficient operation of the integrated system requires that the photobioreactor in the second stage should consume all produced organic acids. The volume of the photobioreactor, as well as the feed rate, can be calculated using available data [41]. The photobioreactor with immobilized purple bacterium *Rhodobacter sphaeroides* GL1 produced 3.8 L of H_2 with 1 L of matrix, or 0.63 L of H_2 per L^{-1} h^{-1} at a HRT=1 h [41]. One (1) L of the photobioreactor volume with immobilized *Rb. sphaeroides* consumed 10 mmoles of lactate. No data is available about butyrate or acetate consumption rates for this system. One could suggest that the rate of organic acid consumption by purple bacteria can be compared on the basis of the same rate of organic carbon assimilation independently of the nature of organic acid. Thus, butyrate (C_4 compound) consumption rate equals to 0.75 of the lactate (C_3 compound) consumption rate by this bacterium. Similarly, the rate of acetate (C_2 compound) consumption equals to 1.5 of the lactate consumption rate. Strictly speaking, it is not correct. Furthermore, some data indicate that purple bacteria consume organic acids from their mixture selectively [21, 40, 42]. In general, organic acids with higher content of carbon are consumed later. However, in the light of the following experimental optimization results, the preliminary estimations of dark reactor/photobioreactor relative reactor volumes using mentioned relative rates of organic acids, is quite appropriate. Taking into consideration this proportion of lactate/acetate and lactate/butyrate consumption rates, one could suggest that 1 L of the photobioreactor [41] could consume 7.5 mmoles of butyrate or 15 mmoles of acetate per hour. So, for assimilation of 262 mmoles of butyrate and 42 mmoles of acetate (from the dark fermentation rector described in [1]) every hour, the volume of the photobioreactor should be equal to 262/7.5 + 42/15 = 37.7 L. Taking into account minor products, the volume of the photobioreactor should be larger by approximately 15%, whih would be approximately 43.4 L. So, for simultaneous H_2 production and complete utilization of organics every 1 L of dark fermentation reactor effluent must be supplied to a photobioreactor of at least 43.4 L.

If relative volumes of reactors are calculated on the basis of their volumetric H_2 production rates, the volume of the photobioreactor will be calculated as 15/0.63 which is multiplied by the factor 2 - 5. This type of calculation gives the volume of the photobioreactor equal to 47.6 - 119 L. Taking into account that dark fermentation reactor [43] is operated in the mixed mode of fermentation, but predominantly as butyrate fermentation, the volume of the photobioreactor is closer to higher value (approximately 110 L). Thus, different calculation strategies determine significantly different relative volumes of the dark fermentation reactor and photobioreactor.

Other estimations for relative volumes of reactors for dark and light stages exist in literature. Redwood and coauthors [18], based on literature data, assumed that for an 80 L dark fermentation reactor, a photobioreactor of 7, 684 L was necessary (ratio 1:96). Using potato homogenate as an artificial waste, the ratio 1:10 for dark-light stages reactors was calculated [22]. The dramatic differences in these results indicates that this question should be further studied.

Connection of Integration System Stages

The present state of knowledge does not allow feeding of the second stage directly by the effluent from the first stage. Several reasons for this exist:

1. During H_2 production by the first stage, the fermentation effluent has decreased pH. In studies on dark fermentative H_2 production, with the pH control or with indication of fermentation effluent pH, it was shown that optimal H_2 production occurs in the region pH 4.5 - 6.0 [44]. However, purple bacteria grow and produce H_2 at pH 6.5 – 9.0 [45]. Some purple bacteria like *Rps. acidophila* [46] are able to grow and produce H_2 at pH lower than 6.

However, rates of the process are not clear. Thus, pH is a very important parameter of feeding media and the low pH of the dark fermentation reactor effluent does not allow direct feeding of the second stage;

2. In many cases, the efficient first stage produces H_2 with maximum rates when the load by substrate is rather high. This results in elevated concentrations of organic acids in fermentation effluent (more than 100 mM). Although the inhibition of purple bacteria by organic acids is not studied well, in all studied cases organic acid concentrations more than 50 mM inhibited H_2 production in the second stage. This results in the need to dilute the dark fermentation effluent (see Table **4**);

3. Purple bacteria are mesophilic (optimal growth between 30 °C and 40 °C). If the first stage is conducted under thermophilic conditions (optimal growth between > 55 °C), the fermentation effluent need to be cooled before the second stage feeding.

4. Media used for the first and the second stages are quite different. That is why preliminary media adjustment is necessary;

5. The fermentation effluent from the first stage contains bacteria. Furthermore, it is easy to predict that in practice it will contain insoluble (and non-transparent) particles. Excessive bacteria and particles inside the photobioreactor for the second stage will decrease light availability;

6. Production of compounds in the first stage that may be toxic for purple bacteria in the second stage has not been studied well.

These factors together push researchers to use other methods of second stage feeding by fermentation effluent (Table **4**).

One Reactor for Two Stages

Very promising approach is the combination of the first and the second stages in one reactor. Different clostridia strains, *Cellulomonas*, *Lactobacillus*, or even microbial consortium have been mixed with purple non-sulfur bacteria (Table **4**). These co-cultures accumulated biomass and produced H_2 using glucose, starch, cellulose, and ground wheat. The efficiency of the substrate utilization varied from 1.13 to 7.1 moles H_2 per mole of glucose. However, in general, co-immobilized cultures had higher efficiency (Table **4**). This tendency might be explained by the fact that dark fermentative bacteria have pH optimum lower than 6 for H_2 production [44], whereas purple bacteria produce H_2 at neutral or alkaline pH [45]. So, in the co-culture, bacteria are cultivated under non-optimal pH. This fact can affect rate and efficiency of H_2 production. In contrast, co-immobilized cultures have a heterogeneous environment. Different parts of immobilized matrix can have optimal pH for one or another bacterium. Co-cultivation experiments have demonstrated, at least in principal, the possibility for integration of dark fermentation with photofermentation in the same reactor, and allowed some estimation of the overall efficiency. Unfortunately, this system cannot reach the maximal rates of H_2 production due to large differences in optimal conditions for each of the bacterial components.

Table 4: Techniques of Two Stages Connection.

Substrate	Microorganism(s)	Phototrophic bacterium	Total H_2 yield (moles mole^{-1} hexose)	Co-culture technology	Ref.
		Co-cultures			
Cellulose	*Cellulomonas* sp.	*Rb. capsulatus* B100	4.3	Suspension co-culture	[23]
Glucose	*C. butyricum*	*Rb. sphaeroides* RV	7.0	Co-immobilized	[24]
Starch (with yeast extract and glutamate)	*C. butyricum*	*Rhodobacter* sp. M-19	6.4	Suspension co-culture	[25]

Table 4: cont….

Glucose	*C. butyricum*	*Rb. sphaeroides*	-	Co-immobilized	[47]
Glucose	*Lactobacillus debrueckii*	*Rb. sphaeroides RV*	7.1	Co-immobilized	[27]
Ground wheat	Microbial consortium	*Rhodobacter* sp.	1.13	Suspension co-culture	[35]
Sequential cultivation					
Starch (with yeast extract and glutamate)	*C. butyricum*	*Rhodobacter* sp. M-19	3.6	Suspension Dark stage: Polypeptone, yeast extract, cystein addition Light stage: centrifugation, sterilization of fermentation effluent	[25]
Sweet potato	*C. butyricum* and *Eneterobacter aerogenes*	*Rhodobacter* sp.	7.0	Suspension. Dark stage: Polypeptone addition Light stage: Addition of Mo and EDTA. pH control at both stages (5.45 and 7.5)	[26]
Glucose	*E. coli*	*Rb. sphaeroides*	2.4	Dark stage: addition of nutrient broth (Oxoid). Light stage: no H_2 due to high N content	[20]
Sucrose	Microbial consortium	*Rb. sphaeroides*	3.32	Light stage: centrifugation, dilution, pH adjustment of fermentation effluent from dark stage	[34]
Sucrose	*C. pasteurianum*	*Rps. palustris*	7.1	Light stage: centrifugation, dilution, pH adjustment, of fermentation effluent from dark stage	[28]
Glucose	*E. cloaceae*	*Rb. sphaeroides*	4.62	Dark stage: addition of yeast extract and malt extract. Light stage: centrifugation and dilution of the fermentation effluent	[38]
Glucose	*Eth. harbinense*	Immobilized *Rps. faecalis* RLD-53	6.32	Dark stage: Phosphates and vitamins mixture addition. Light stage: Soluble part of fermentation effluent centrifugation, sterilization, pH adjustment (7), dilution	[29]
Miscantus hydrolysate	*T. neapolitana*	*Rb. capsulatus*	-	Light stage: centrifugation, dilution, pH adjustment, buffer addition, sterilization of fermentation effluent from dark stage	[21]
Sugar beet molasses	*Caldicellulosiruptor saccharolyticus*	*Rb. capsulatus* hup$^-$ mutant	6.85	Fermentation effluent was centrifuged, sterilized, diluted with addition of buffer, Fe, and Mo before the 2nd stage	[32]

Sequential Batch Cultures with Conditioning of Fermentation Effluent

Another approach for connecting the components of integrated system is the use of independent stages. In this case, fermentation effluent from the first stage is collected, processed, and then directed to the second stage. The most common procedures for fermentation effluent processing include: sterilization, pH adjustment, and dilution (Table **4**). In some cases, bacteria were separated from the fermentation effluent by centrifugation before sterilization. Depending on the concentration of organic acids in fermentation effluent, the applied dilution was different. When the dilution was significant, an addition of mineral components and buffers was required to restore the appropriate concentrations. This method is very useful when different stages are used in separate laboratories. In this case, H_2 production efficiency was 2.4 - 7.1 moles H_2 per mole of glucose (Table **4**). Unfortunately rates of the process are not available in many publications, prohibiting comparative analysis of different systems as well as techniques of stages' connection.

Integration of Photoautotrophic Biomass Production and its Degradation in Two Stage Dark-Light Integrated System

A three-stage dark-light integrated system (described above), in which the first stage contains the photoautotrophic production of microalgal biomass (followed by dark fermentartion in the second stage and photofermentation in the third stage), may be controlled to capture all energy for the process from sunlight. The possibility of H_2 production using algal biomass was studied by several research teams [48-50]. The efficiency of H_2 production was calculated on the basis of starch from microalgal cells and varied from 7.3 moles H2 per mole of glucose [50] to 8.3 moles H_2 per mole of glucose [48]. Taking into account that microalgal biomass contains other organics, the H_2 yields are rather over-estimated. Nevertheless, this three stage integrated system requires only sunlight and is neutral in terms of CO_2 production. It can use seawater for the operation decreasing the cost of the system.

The problem for the integration of photobioreactors in the system is the technique for the transfer of microalgal biomass to dark fermentation reactor. In the described experiments, the biomass of *Dunaliella* and *Chlamydomonas* was freezed-thawed [50]. This procedure disrupts main part of cells, simplifying biomass processing in dark fermentation reactor. In other cases, the processing of microalgal biomass was not described, but it was concentrated or reconstructed from powder [48]. In contrast, biomass directly from bioreactor is alive and additional research is necessary to understand how this biomass will be processed by dark fermentation reactor.

Integration of Temporal Separation of Photosynthesis with Light Energy Accumulation and Dark Fermentation by One Organism as the First Stage and Anoxygenic Photosynthetic H_2 Production by Purple Bacteria as the Second Stage

That microalgal and cyanobacterial cells to produce H_2 during periods of dark is a well-known phenomenon. During day time, microalgae use sunlight energy for growth and accumulation of storage products like starch or glycogen. During night time (in the dark), under aerobic conditions, they grow due to respiration using these storage materials. Under dark anaerobic conditions, they use energy derived from fermentation. Hydrogen is one of fermentation products. This property of microalgae and general property of purple bacteria to produce H_2 in the presence of light, using organic acids, suggests a scheme of integration depicted on Fig. **2**.

Usually microalgae produce negligible quantities of H_2 during fermentation. Particular strain of *Chlamydomonas* with high ability for dark fermentative H_2 production was isolated [16]. Acetate and ethanol were additional products of fermentation for this strain. The efficiency of starch to H_2 conversion by *Chlamydomonas* was estimated as 2 moles H_2 per mole starch, which was recalculated to hexose (glucose) equivalents. To improve the H_2 yield, the medium after dark fermentation by *Chlamydomonas* was directed to nitrogen-deficient purple bacterium *Rhodovulum sulfidofilum* W1S [51]. Under continuous illumination the efficiency of the integrated system was as high as 10.5 moles H_2 per mole of hexose equivalent. However, since purple bacterium produced H_2 even after the consumption of acetate and ethanol, other non-identified products of microalgal metabolism were present in the medium.

These strains are very attractive in practical systems, since they grow and produce H_2 in sea water. Pilot scale reactors were constructed using these particular strains [52]. The efficiency of this system was 5.8 moles H_2 per mole of hexose equivalent, which was much lower than in the laboratory scale experiments [53]. Nevertheless, pilot-scale experiments show sustainable hydrogen production.

This integrated system can be applied for cyanobacterial/purple bacteria dual systems using established processes of H_2 production by cyanobacteria in light-dark cycles, such as *Mycricystis aerogenosa* [54] or *Gloeocapsa alpicola* [55], and a variety of purple bacteria [45]. Unfortunately, such data are not available in the peer-reviewed literature.

Integrated System for Efficient Light Energy Capture

Integration of H_2 production by microalgae and purple bacteria in one system can expand available sunlight energy due to different absorption spectra. Microalgae consume light energy in the range 400-750 nm, whereas purple bacteria can consume light energy in the range 780-1100 nm. Co-cultivation of *Chlamydomonas reinhartii* and *Rs. rubrum* in one medium demonstrated this possibility [56]. Depending on the light intensity, microalgae or purple bacteria were predominant in the culture. However, simultaneous cultivation of these bacteria for H_2 production seems doubtful, since efficiently growing microalgae produce oxygen, which is a strong repressor of Fe-Fe hydrogenases and nitrogenases. Possibly, two-chamber photobioreactor with microalgae in one chamber and purple bacteria in another chamber can be a model for efficient light utilization. However, no available data on such integrated system of H_2 production exist.

Integration of Microbial Electrolysis Cell (MEC) with Dark Fermentation

To convert dark fermentation by-products to H_2, some additional energy is required. Light energy is used in the case of the photosynthetic second stage. A very interesting alternative was proposed by Liu and colleagues [17]. They applied a usual microbial fuel cell (MFC) in unusual way. MFCs have microbial catalyst (electrogenic bacteria) at the anode and an inorganic (metal) catalyst at the cathode. Electrogenic bacteria oxidize organics at anode surface and generate electricity. The inorganic catalyst at the cathode catalyses oxygen reduction. The presence of oxygen in the cathode chamber creates a potential problem in that diffusion of oxygen to anode chamber will inhibit the bacteria at the anode. If the cathode chamber keep under anaerobic conditions, and a small electrical current (>200 mV) is applied, H_2 is produced at the cathode (instead of electricity). Thus MFC is converted into a microbial electrolysis cell (MEC). In the first experiments, acetate was used as substrate for bacteria [17]. The recovery of electrons as H_2 was over 90%. The Coulombic efficiency, defined as the recovery of total amount of electrons in acetate as current, depended on the applied voltage and ranged from 60 to 78%. It means that at optimum applied voltage the overall H_2 yield was 2.9 moles H_2 per mole of acetate.

By improving experimental methods, H_2 was produced at yields of up to 99%, at an applied voltage 0.8 V, using acetate [57]. H_2 production by MECs was further demonstrated by using glucose, butyric, lactic, propionic and valeric acids, as well as cellulose. The total energy efficiency of the process was as high as 92%.

Further improvement of the system includes the development of membrane-less single chamber MECs [58]. It allows placing electrodes closely to each other decreasing ohmical resistance of the system. This MEC showed high H_2 production rate (up to 3.12 L H_2 per L of reactor in 1 day). The overall energy efficiency relative to both electricity and energy of acetate in average was equal to 78% with a maximum value – 86%. This MEC system with improved electrodes was used for H_2 production with swine wastewater treatment [59]. The efficiency of H_2 production was similar to previously described but the removal of organic carbon was not high (28%). Operation of MEC was successful using fermentation effluent from ethanol-H_2 coproducing reactor [60], and in a two-stage process combining fermentation and electrohydrogenesis [61]. Thus, H_2 production by MECs looks very promising, and the integration of MEC and dark fermentation in one system allows converting virtually any organic substrates into H_2 with high efficiency.

CONCLUSIONS

Data in the published literature suggest that all integrated systems eliminate drawbacks of single stages, and combinations of different single stages do not decrease the advantages of single stages. Furthermore, in some cases, the integration of single stages in one process improves overall efficiency (as in the case of MECs and dark fermentation integration). However, new technologies and investigations open door for new questions. For example, during integration of purple bacteria with dark fermentation, or with microalgal/cyanobacterial cyclic

H_2 production as whole processes, the main question arises is "how can the different stages be connected to each other effectively?" Described technologies use energy-cost approaches. Further investigations are necessary on how to transfer fermentation effluent from first to second stage taking into account cost and energy efficiency. Another unclear question is, "how does the first stage of the process influence the stability and efficient operation of the second stage?" Purple bacteria have optimal range of substrate concentrations when they produce H_2 efficiently, but this range is not studied well.

The very promising integration of dark fermentation and MECs also has many unsolved problems. For example, this process is not studied in terms of available ranges of environmental factors for stable operation. The catalyst for H_2 production by MECs is also not clear. Platinum (Pt) and other metal catalysts are not applicable for microbial community since microbial consortium produces catalytic poisons like H_2S, CO, sulfur containing peptides *etc.* Other approaches should be studied. Finally, in spite of investigations of single stages separately, attention should be paid to particular problems of integrated systems, which may be the only way to maximize H_2 yields from biological H_2 production systems.

REFERENCES

[1] Wu S-Y, Hung C-H, Lin C-N, *et al.* Fermentative hydrogen production and bacterial community structure in high-rate anaerobic bioreactors containing silicone-immobilized and self-flocculated sludge. Biotechnol Bioeng 2006; 93(5): 934-46.

[2] Levin DB, Pitt L, Love M. Biohydrogen production: prospects and limitations to practical application. Int J Hydrogen Energy 2004; 29(2): 173-85.

[3] Levin DB. Re: Biohydrogen production: prospects and limitations to practical application - erratum. Int J Hydrogen Energy 2004; 29(13): 1425-6.

[4] Choi D, Chipman DC, Bents SC, Brown RC. A techno-economic analysis of polyhydroxyalkanoate and hydrogen production from syngas fermentation of gasified biomass. Appl Biochem Biotechnol 2010; 1032-46.

[5] Sipma J, Henstra AM, Parshina SN, *et al.* Microbial CO conversions with applications in synthesis gas purification and bio-desulfurization. Crit Rev Biotechnol 2006; 26(1): 41-65.

[6] Cournac L, Guedeney G, Peltier G, *et al.* Sustained photoevolution of molecular hydrogen in a mutant of *Synechocystis* sp. strain PCC6803 deficient in the type I NADH-dehydrogenase complex. J Bacteriol 2004; 186(6): 1737-46.

[7] Rakhely G, Kovacs AT, Maroti G, *et al.* Cyanobacterial-type, heteropentameric, NAD(+)-reducing NiFe hydrogenase in the purple sulfur photosynthetic bacterium *Thiocapsa roseopersicina*. Appl Environ Microbiol 2004; 70(2): 722-8.

[8] Rakhely G, Laurinavichene TV, Tsygankov AA, *et al.* The role of Hox hydrogenase in the H_2 metabolism of *Thiocapsa roseopersicina*. BBA-Bioenergetics 2007; 1767(6): 671-6.

[9] Suda S, Kumazawa S, Mitsui A. Change in the H_2 photoproduction capability in a synchronously grown aerobic nitrogen-fixing cyanobacterium, *Synechococcus* sp miami bg 043511. Arch Microbiol 1992; 158(1): 1-4.

[10] Borodin VB, Rao KK, Hall DO. Manifestation of behavioural and physiological functions of *Synechococcus* sp. Miami BG 043511 in a photobioreactor. Mar Biol 2002; 140: 455-63.

[11] Gaffron H, Rubin J. Fermentative and photochemical production of hydrogen in algae. J Gen Physiol 1942; 26: 219-40.

[12] Boichenko VA, Satina LY, Litvin FF. Efficiency of hydrogen photoproduction in algae and cyanobacteria. Fiziol Rast (USSR) 1989; 36: 239-47.

[13] Melis A, Zhang LP, Forestier M, *et al.* Sustained photobiological hydrogen gas production upon reversible inactivation of oxygen evolution in the green alga *Chlamydomonas reinhardtii*. Plant Physiol 2000; 122(1): 127-35.

[14] Asada Y, Miyake M, Tomizuka N. Hydrogenase mediated hydrogen metabolism in some cyanobacteria. Photosynth Res 1992; 34(1): 128.

[15] Troshina O, Serebryakova L, Sheremetieva M, *et al.* Production of H_2 by the unicellular cyanobacterium *Gloeocapsa alpicola* CALU 743 during fermentation. Int J Hydrogen Energy 2002; 27(11-12): 1283-9.

[16] Miura Y, Ohta S, Mano M, *et al.* Isolation and characterization of a unicellular green alga exhibiting high activity in dark hydrogen production. Agric Biol Chem 1986; 50(11): 2837-44.

[17] Liu H, Grot S, Logan BE. Electrochemically assisted microbial production of hydrogen from acetate. Environ Sci Technol 2005; 30: 785-93.

[18] Redwood MD, Paterson-Beedle M, Macaskie LE. Integrating dark and light bio-hydrogen production strategies: towards the hydrogen economy. Rev Environ Sci Biotechnol 2009; 8(2): 140-85.

[19] Ren NQ, Wang AJ, Cao GL, *et al.* Bioconversion of lignocellulosic biomass to hydrogen: Potential and challenges. Biotechnol Adv 2009; 27(6): 1051-60.

[20] Redwood MD, Macaskie LE. A two-stage, two-organism process for biohydrogen from glucose. Int J Hydrogen Energy 2006; 31(11): 1514-21.

[21] Uyar B, Schumacher M, Gebicki J, *et al.* Photoproduction of hydrogen by *Rhodobacter capsulatus* from thermophilic fermentation effluent. Bioprocess Biosyst Eng 2009; 32(5): 603-6.

[22] Laurinavichene T, Belokopytov BF, Laurinavichius KS, *et al.* Towards the integration of dark- and photo-fermentative waste treatment. 3. Potato as substrate for sequential dark fermentation and light-driven H_2 production. Int J Hydrogen Energy 2010: 35(16); 8536-43.

[23] Odom JM, Wall J. Photoproduction of H_2 from cellulose by an anaerobic bacterial culture. Appl Environ Microbiol 1983; 45(4): 1300-5.

[24] Miyake J, Mao X-Y, Kawamura S. Hydrogen photoproduction from glucose by a co-culture of a photosynthetic bacteria and *Clostridium butyricum*. J Ferment Technol 1984; 62(6): 531-535.

[25] Yokoi H, Mori S, Hirose J, *et al.* H_2 production from starch by mixed culture of *Clostridium butyricum* and *Rhodobacter* sp M-19. Biotechnol Lett 1998; 20(9): 895-9.

[26] Yokoi H, Saitsu A, Uchida H, *et al.* Microbial hydrogen production from sweet potato starch residue. J Biosci Bioeng 2001; 91(1): 58-63.

[27] Asada Y, Tokumoto M, Aihara Y, *et al.* Hydrogen production by co-cultures of *Lactobacillus* and a photosynthetic bacterium, *Rhodobacter sphaeroides* RV. Int J Hydrogen Energy 2006; 31(11): 1509-13.

[28] Chen CY, Yang MH, Yeh KL, *et al.* Biohydrogen production using sequental two-stage dark and photo fermentation processes. Int J Hydrogen Energy 2008; 33(21): 4755-62.

[29] Liu BF, Ren NQ, Xing DF, *et al.* Hydrogen production by immobilized *R. faecalis* RLD-53 using soluble metabolites from ethanol fermentation bacteria *E. harbinense* B49. Bioresource Technol 2009; 100(10): 2719-23.

[30] Su HB, Cheng J, Zhou JH, *et al.* Combination of dark- and photo-fermentation to enhance hydrogen production and energy conversion efficiency. Int J Hydrogen Energy 2009; 34(21): 8846-53.

[31] Liu BF, Ren NQ, Xie GJ, *et al.* Enhanced bio-hydrogen production by the combination of dark- and photo-fermentation in batch culture. Bioresource Technol 2010; 101(14): 5325-9.

[32] Ozgur E, Mars AE, Peksel B, *et al.* Biohydrogen production from beet molasses by sequential dark and photofermentation. Int J Hydrogen Energy 2010; 35(2): 511-7.

[33] Xie GJ, Feng LB, Ren NQ, *et al.* Control strategies for hydrogen production through co-culture of *Ethanoligenens harbinense* B49 and immobilized *Rhodopseudomonas faecalis* RLD-5. Int J Hydrogen Energy 2010; 35(5): 1929-35.

[34] Tao Y, Chen Y, Wu Y, *et al.* High hydrogen yield from a two-step process of dark- and photo-fermentation of sucrose. Int J Hydrogen Energy 2007; 32(2): 200-6.

[35] Argun H, Kargi F, Kaplan IK. Effects of the substrate and cell concentration on bio-hydrogen production from ground wheat by combined dark and photo-fermentation. Int J Hydrogen Energy 2009; 34(15): 6181-9.

[36] Belokopytov BF, Laurinavichius KS, Laurinavichene TV, *et al.* Towards the integration of dark- and photo-fermentative waste treatment. 2. Optimization of starch-dependent fermentative hydrogen production. Int J Hydrogen Energy 2009; 34(8): 3324-32.

[37] Argun H, Kargi F. Effects of light source, intensity and lighting regime on bio-hydrogen production from ground wheat starch by combined dark and photo-fermentations. Int J Hydrogen Energy 2010; 35(4): 1604-12.

[38] Nath K, Muthukumar M, Kumar A, *et al.* Kinetics of two-stage fermentation process for the production of hydrogen. Int J Hydrogen Energy 2008; 33(4): 1195-203.

[39] Li CL, Fang HHP. Fermentative hydrogen production from wastewater and solid wastes by mixed cultures. Crit Rev Env Sci Tec 2007; 37(1): 1-39.

[40] Laurinavichene TV, Tekucheva DN, Laurinavichius KS, *et al.* Towards the integration of dark and photo fermentative waste treatment. 1. Hydrogen photoproduction by purple bacterium *Rhodobacter capsulatus* using potential products of starch fermentation. Int J Hydrogen Energy 2008; 33(23): 7020-6.

[41] Tsygankov AA, Fedorov AS, Talipova IV, *et al.* Use of immobilized phototrophic microorganisms for wastewater treatment and simultaneous production of hydrogen. Appl Biochem Microbiol 1998; 34(4): 362-6.

[42] Uyar B, Eroglu I, Yucel M, *et al.* Photofermentative hydrogen production from volatile fatty acids present in dark fermentation effluent. Int J Hydrogen Energy 2009; 34: 4517-23.

[43] Wu SY, Hung CH, Lin CN, *et al.* Fermentative hydrogen production and bacterial community structure in high-rate anaerobic bioreactors containing silicone-immobilized and self-flocculated sludge. Biotechnol Bioeng 2006; 93(5): 934-46.

[44] Wang J, Wan W. Factors influencing fermentative hydrogen production: a review. Int J Hydrogen Energy 2009; 34(2): 799-811.

[45] Rocha J, Barbosa HR, Wijffels RH. Hydrogen production by photosynthetic bacteria: culture media, yields and efficiencies. In: Miyake J, Matsunaga T, San Pietro A, editors. Biohydrogen II. An approach to environmentally acceptable technology. Amsterdam: Pergamon 2001; pp. 3-32.

[46] Holt JG, Krieg NR, Sneath PHA, *et al.* Bergey's manual of determinative bacteriology, 9th ed. Baltimore: Williams and Wilkins 1994.

[47] Zhu H, Wakayama T, Asada Y, *et al.* Hydrogen production by four cultures with participation by anoxygenic photosynthetic bacterium and anaerobic bacterium in the presence of NH_4. Int J Hydrogen Energy 2001; 26(11): 1149-54.

[48] Kim MS, Baek JS, Yun YS, *et al.* Hydrogen production from *Chlamydomonas reinhardtii* biomass using a two-step conversion process: Anaerobic conversion and photosynthetic fermentation. Int J Hydrogen Energy 2006; 31(6): 812-6.

[49] Ike A, Kawaguchi H, Hirata K, *et al.* Hydrogen photoproduction from starch in algal biomass. In: Miyake J, Matsunaga T, San Pietro A, Eds. Biohydrogen II: an approach to environmentally acceptable technology. Amsterdam: Pergamon 2001; pp. 53-61.

[50] Kawaguchi H, Hashimoto K, Hirata K, *et al.* H_2 production from algal biomass by mixed culture of Rhodobium marinum A-501 and Lactobacillus amylovorus. J Biosci Bioeng 2001; 91: 277-82.

[51] Miura Y, Saitoh C, Matsuoka S, *et al.* Stably sustained hydrogen production with high molar yield through a combination of a marine green alga and a photosynthetic bacterium. Biosci Biotechol Biochem 1992; 56(5): 751-4.

[52] Akano T, Miura H, Fukatsu K, *et al.* Hydrogen production by photosynthetic microorganisms. Appl Biochem Biotechnol 1996; 57-58(6): 677-88.

[53] Miura Y, Akano T, Fukatsu K, *et al.* Stably sustained hydrogen production by biophotolysis in natural day/night cycle. Energ Convers Manage 1997; 38(SS): S533-7.

[54] Asada Y, Kawamura S, Ho KK. Hydrogenase from the unicellular cyanobacterium, *Microcystis aeruginosa.* Phytochemistry 1987; 26(3): 637-40.

[55] Serebryakova LT, Tsygankov AA. Two-stage system for hydrogen production by immobilized cyanobacterium *Gloeocapsa alpicola* CALU 743. Biotechnol Prog 2007; 23(5): 1106-10.

[56] Melis A, Melnicki R. Integrated biological hydrogen production. Int J Hydrogen Energy 2006; 31(11-12): 1563-73.

[57] Cheng SA, Logan BE. Sustainable and efficient biohydrogen production *via* electrohydrogenesis. Proc Natl Acad Sci USA 2007; 104: 18871–3.

[58] Call DF, Logan BE. Hydrogen production in a single chamber microbial electrolysis cell lacking a membrane. Environ Sci Technol 2008; 42: 3401-6.

[59] Wagner RC, Regan JM, Oh SE, *et al.* Hydrogen and methane production from swine wastewater using microbial electrolysis cells. Water Res 2009; 43(5): 1480-8.

[60] Lu L, Ren NQ, Xing D, *et al.* Hydrogen production with effluent from an ethanol–H_2-coproducing fermentation reactor using a single-chamber microbial electrolysis cell. Biosens Bioelectron 2009; 24(10): 3055-60.

[61] Lalaurette E, Thammannagowda S, Mohagheghi A, *et al.* Hydrogen production from cellulose in a two-stage process combining fermentation and electrohydrogenesis. Int J Hydrogen Energy 2009; 34(15): 6201-10.

CHAPTER 6

Fundamentals of Dark Hydrogen Fermentations: Multiple Pathways and Enzymes

Patrick C. Hallenbeck*

Département de Microbiologie et Immunologie, Université de Montréal, Canada

Abstract: Biohydrogen production through dark fermentation could be a promising route to the generation of a renewable fuel source. Dark fermentation is attractive since its deployment would probably be based on known reactor technology, and, in a first application, could use various carbohydrate rich waste streams, followed by lignocellulosics, a vast largely untapped resource. A thorough understanding of the enzymes and pathways is involved would help develop strategies, especially metabolic engineering, for improving rates and yields of hydrogen production. Here, the various hydrogenases and their active sites and biochemistry are discussed followed by an examination of known hydrogen producing pathways.

Keywords: Hydrogen fermentation, [FeFe] hydrogenase, [NiFe] hydrogenase, *Clostridium, Escherchia coli,* mixed acid fermentation, syntrophic metabolism, carbohydrate degradation, ethanol-acetate cofermentation, redox balance.

INTRODUCTION

All biological energy generating metabolisms are based on coupling oxidation reduction reactions to mechanisms that provide chemical energy in some form. In the absence of exogenous electron acceptors which might permit some variety of respiration, microbes must carry out fermentation to generate the energy required for cell growth and survival. Thus, in fermentations, the required electron acceptors are produced through metabolic conversions of the substrate or intermediates creating various reduced products. The wide range of fermentation products found in nature demonstrate the wonderful diversity possible in microbial metabolism, and, from a more practical perspective, provide a large number of useful products for humans. In nature, fermentative hydrogen production is of potential benefit to a microbe since it provides an avenue to avoid the liquid phase accumulation of a possibly toxic product and the hydrogen which is produced can be a major driving force in many anaerobic and aerobic microbial ecosystems. The majority of biologically produced hydrogen is consumed by these processes, although the magnitude is difficult to estimate. A small fraction escapes into the atmosphere, making hydrogen a relatively abundant trace gas (~550 ppb). Hydrogen producing pathways are found in diverse organisms which can possess one or several different hydrogenase enzymes. Here we will examine the hydrogenase enzymes involved, the range and variety of substrates that lead to hydrogen production, and the various known pathways leading to hydrogen production.

HYDROGENASE ENZYMES

Of course, the key player in biological hydrogen production is hydrogenase and a great deal of effort has gone into understanding its synthesis, biochemistry and physiology. Hydrogenases catalyze one of the simplest chemical reactions, $2H^+ + 2e^- \leftrightarrow H_2$, yet their structure is very complex [1-8]. Until 2004 hydrogenases were classified according to the metals thought to be at their active sites; three classes were recognized: iron-only [FeFe], nickel-iron [Ni-Fe], and "metal-free". In 2004 it was shown that the metal-free hydrogenases in fact contain iron [9], and they are now called [Fe]-hydrogenases, since they contain only a mononuclear Fe active site and no iron-sulfur clusters, in contrast to the [FeFe]-enzymes. Hydrogen production by organisms carrying out a dark fermentation is by either a [Ni-Fe] hydrogenase or a [FeFe] hydrogenase [10].

***Address correspondence to Patrick C. Hallenbeck:** Département de Microbiologie et Immunologie, Université de Montréal, Canada; E-mail: patrick.hallenbeck@umontreal.ca

Nuri Azbar and David B. Levin (Eds)

Different Hydrogenases, Different Active Sites

Thus, there are three types of enzymes uniquely devoted to hydrogen metabolism based on their active centers (Fig. **1**): [NiFe], [FeFe], and Hmd, which has been shown to contain iron, but not, unlike the other two, inorganic sulfur and can now be described as a [Fe] hydrogenase. A fourth enzyme, nitrogenase, can reduce protons to hydrogen in the absence of other reducible substrates, but, as yet, the participation of this enzyme in dark fermentative hydrogen production has not been demonstrated. All of these enzymes contain complex metal centers that are very sensitive to oxygen inactivation. [FeFe] hydrogenases are quickly and irreversibly destroyed by exposure to oxygen, which is thought to be reduced at the di-iron domain ($2Fe_H$) of the H cluster, generating reactive oxygen species which attack and destroy the 4Fe-4S domain [11]. On the other hand, [NiFe] hydrogenases react with oxygen to give intermediates which can be reactivated by reduction.

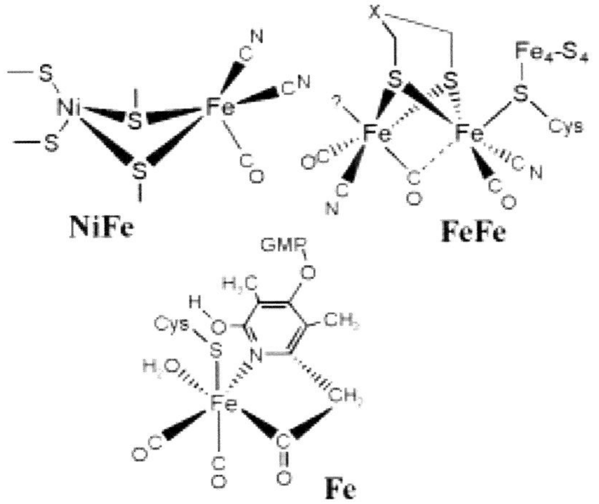

Figure 1: Active site structures of the three hydrogenases. The active sites of the [FeFe] hydrogenase, the [NiFe] hydrogenase, and the [Fe] hydrogenase (Hmd) are shown. Note that the [Fe] hydrogenase does not contain inorganic sulfur or CN, prominent features of the other two active sites. This Figure is reproduced by permission from Heinekey *et al.* [12].

Much recent work has focused on the synthesis and structure of hydrogenase active sites, both to gain insight into the catalytic mechanism as well as for developing model synthetic compounds that might serve as active catalysts in proton reduction or hydrogen activation [12]. A great deal of work over the last two decades has provided many molecular details of the assembly of the [NiFe] center and consequently of active hydrogenases which contain this active site [13,14]. Studies have revealed a large number of genes involved in the expression, processing and assembly of functional [NiFe] hydrogenase [6,13-19]. Some organisms possess the genetic capacity to express multiple [NiFe] hydrogenases, for example, *E. coli*, whose genome encodes four different [NiFe] hydrogenases. However, commonly there are multiple operons encoding the structural genes and some accessory genes, with a common set of genes, *hyp*, being used for core generic functions required for [NiFe] hydrogenase maturation.

Until recently, relatively little was known about the synthesis of the active site of [FeFe] hydrogenases, the H-cluster. Of some interest is how the toxic ligands found in the active sites; CO for all three, CN for NiFe and FeFe, are generated by cellular metabolism and inserted into the complex metal centers. Cyanide is generated by two different routes depending upon the active site to be made. The CN component of the NiFe center comes from enzyme-bound thiocyanate generated from carbamoyl phosphate [20] and thus free cyanide as an intermediate is avoided. The CN found in the FeFe active site, the so-called H cluster, has recently been shown to be derived from a different route, the S-adenosylmethionine–dependent cleavage of tyrosine by HydG to give p-cresol and cyanide [21,22]. It is not known if free cyanide is liberated, or if it is immediately passed to HydF, thought to function as a scaffold for the assembly of at least part of the H-cluster [23]. Evidence that HydF contains a CO and CN- ligated di-iron cluster has recently been presented [24] as well as evidence for the synthesis of the 2Fe subcluster of the H cluster on HydF from a [2Fe-2S]cluster framework in a process requiring HydE, HydG, and GTP [25]. The recent determination of the structure of $HydA^{\Delta EFG}$ (HydA expressed in a host lacking HydE, F, and G) revealed that, in line with previous biochemical studies [26]; the [4Fe-4S] subcluster can be formed by generalized host-cell machinery [27].

Taken together these recent results strongly suggests that the role of the specialized maturation proteins, HydE, HydF, and HydG, is specifically to synthesize the 2Fe subcluster with its CN and CO ligands. How the CO that forms an integral part of the H-cluster is formed is completely unknown. In fact, the origin of the critical CO ligands is unknown for all three hydrogenase active sites, although at least for the NiFe center it is thought to be derived from a different pathway than that used for synthesis of the CN ligand [28]. Recent evidence for dithiomethylamine as the previously unknown ligand has been presented [29] and it is likely that HydG is involved in its synthesis [30].

While all hydrogenases, as other enzymes, are theoretically reversible, in practice a given hydrogenase usually fulfills a specific metabolic role and is thus dedicated to functioning in one direction or the other; either proton reduction or hydrogen activation. In fact, hydrogenases are usually poised to function primarily in one direction, either internally, by the redox potential of auxiliary co-factors, or externally, through the redox potential of their natural electron acceptor or donor. In general [FeFe] hydrogenases usually function to evolve hydrogen and [NiFe] hydrogenases to oxidize hydrogen, however, exceptions are known; *i.e.* the periplasmic [FeFe] hydrogenases of *Desulfovibrio* [31] and probably *Shewanella*, which carry out the oxidation of hydrogen to drive the energy yielding reduction of external electron acceptors, and some of the [NiFe] hydrogenases that fall into the Ech class, the best known example being the hydrogen evolving hydrogenase 3 of *Escherchia coli* (Fig. **2**). Another group of [NiFe] hydrogenases (group3, [4]) are soluble cytoplasmic enzymes that are normally thought oxidize hydrogen and reduce NAD for cellular metabolism. The best known example by far is the SH hydrogenase of *Ralstonia eutropha* [32]. As far as I am aware, there is as yet no evidence for hydrogen evolution by this class of hydrogenase in non-photosynthetic organisms belonging to the *Bacteria* during fermentation whereas they are widespread and appear fully bidirectional in various *Archeae*. In some [NiFe]-hydrogenases one of the Ni-bound cysteine residues is replaced by selenocysteine [33].

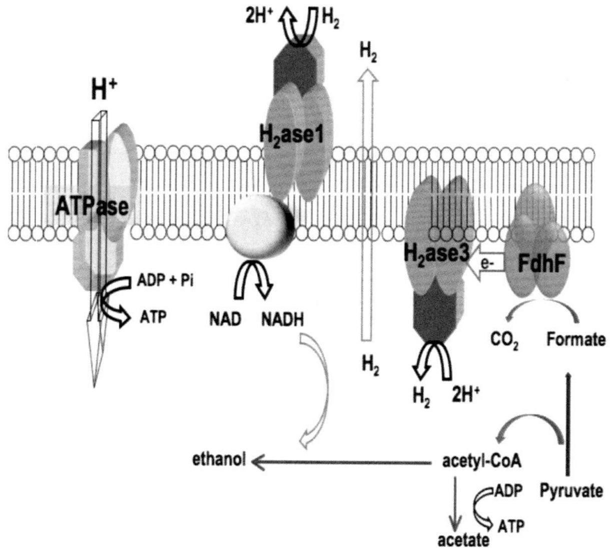

Figure 2: The Ech hydrogenase of *Escherchia coli* and putative hydrogen cycling. The differential localization of different hydrogenases in the same organism creates a potential for hydrogen recycling and the development of a proton gradient across the cytoplasmic membrane. In *E. coli*, Hyd3, the hydrogenase responsible for hydrogen production is oriented towards the cytoplasm, where it reduces protons with electrons derived from formate to hydrogen. Hyd3 and FdhF, the membrane associated formate dehydrogenase that acts under fermentative conditions, along with two membrane bound electron carriers, constitute the FHL complex. The hydrogen that is produced freely diffuses across the cytoplasmic membrane to the periplasm where it encounters Hyd1. Hyd1 oxidizes the hydrogen, leaving protons in the periplasm and the extracted electrons are used to reduce NAD to NADH on the cytoplasmic side. Note that the overall reaction represents the transfer of protons from the cytoplasm to the periplasm. The resultant proton gradient can then be used to drive ATP synthesis or other energized membrane dependent processes, for example chemotaxis. NADH is used to reduce acetyl-CoA to ethanol to maintain redox balance.

Other classes of [NiFe] hydrogenases are wide-spread and function in hydrogen oxidation. Elimination of this activity may be desirable, at least in some cases, as hydrogen uptake by these systems could potentially decrease the overall hydrogen yields of a practical hydrogen fermentation process (Fig. **2**). These systems are fairly well understood and documented and thus, in what follows, emphasis will be placed on hydrogen producing systems. Likewise, Hmd, or the Fe hydrogenase of methanogens almost certainly participates solely in hydrogen oxidation during methanogenesis and will not be considered further here.

Finally, given the complex machinery required for synthesizing the [NiFe] or [FeFe] active site discussed above, it is apparent that not only must an organism possess the structural genes for a particular hydrogenase, there is also an absolute requirement for additional genes for the assembly and proper insertion of these complex metal containing cofactors (reviewed in [4, 6, 8, 13, 34]). Therefore, attempts to increase fermentative hydrogen production through metabolic engineering by the heterologous expression of a hydrogenase must account for these additional factors if it is to be successful. Before these additional requirements were known, many attempts at heterologous expression of hydrogenases were unsuccessful. With this in mind, reports of hydrogenase activity when only the hydrogenase structural genes were expressed in organisms that do not possess the requisite maturation machinery must be viewed with some skepticism.

Until very recently, attempts to express functional [Ni-Fe] hydrogenases in organisms belonging to different genera other than the organism providing the hydrogenase-encoding gene(s) have so far largely failed even when they contain homologs of the required maturation genes, indicating an inability of one or more host maturation genes to correctly process and assemble functional heterologous [NiFe] hydrogenases [34] (PCH unpublished results). However, very recently, the successful expression of a [NiFe] hydrogenase from *Pyrococcus furiousus*, belonging to the Archeae, in *E. coli* was reported [35]. Remarkably, the only *P. furiousus* gene in addition to the hydrogenase structural genes required for high level production of functional protein was a specific protease required for the C-terminal processing step. This indicates that all the other maturation machinery can be supplied by *E. coli* and potentially opens the door for heterologous expression of [NiFe] hydrogenase from diverse organisms. Likewise, the heterologous expression of [FeFe] hydrogenases in *E. coli* was unsuccessful after many (unpublished) attempts until it was shown that the co-expression of the three maturation genes, *hydE*, *hydF*, and *hydG* was both necessary and sufficient [36,37].

DIVERSITY OF HYDROGENASE AND ORGANISMS

In general, hydrogenases, either [FeFe] or [NiFe], or both, are found in a wide variety of organisms. As well, there is also diversity at the enzyme level, with a variety of types of FeFe and NiFe hydrogenases evident from gene sequences and genomic context see below). [NiFe] hydrogenases occur only in Bacteria and Archeae whereas [FeFe] hydrogenases occur only in Bacteria and Eucarya. [NiFe] hydrogenases are more widely distributed than [FeFe] hydrogenases being approximately three times more abundant on the organism as well as gene level [38].

However, the [NiFe] hydrogenases of interest in hydrogen production are only a small subset of the total [NiFe] hydrogenase operons known and rather narrowly distributed. The soluble cytoplasmic [NiFe] hydrogenases (SH hydrogenases, group 3) while relatively abundant in the Archaea have been described for only a few bacteria outside of the cyanobacteria. However, genomic analysis suggests that substantial numbers of as yet uncharacterized hydrogenases of this type might be found in bacteria [7]. The Ech hydrogenases, found in a variety of organisms, are interesting since they appear capable of coupling hydrogen evolution with the generation of a membrane proton gradient [8,39]. In fact, depending upon the organism and growth conditions, Ech hydrogenases are thought to be capable of coupling hydrogen oxidation to ferredoxin reduction using the membrane gradient to drive the necessary reverse electron transport or to energize the membrane during proton reduction to hydrogen. These hydrogenases have also been found in hyperthermophilic archaeon, *Pyrococcus furiosus* [40,41] and the gram-positive *Thermoanaerobacter tengcongensis* [42] and may be relatively widespread as a recent analysis found in 56 different bacterial and 28 different archeael genomes [8]. However, on a cautionary note, not all Ech hydrogenases are thought to function physiologically in proton reduction. For example, Ech in methanogens plays a key role in generating reduced ferredoxin through hydrogen oxidation to drive methanogenesis and carbon fixation [43]. Therefore, caution should be exercised in assigning a role to genomic sequences without further biochemical or physiological evidence. The presence of genes potentially encoding Ech components in some *Clostridial* genomes has recently been noted [44]. It would be interesting to know how these function in the overall metabolism. Since under normal fermentative conditions the ferredoxin pool is probably maintained largely in the reduced state, it is likely that Ech would function in hydrogen evolution, as suggested for *Thermoanaerobacter tengcongensis* [42]. However, in the closely related *Thermoanaerobacter saccharolyticum*, mutagenic analysis shows that the Ech only plays a minor role at best in hydrogen production [45].

[FeFe] hydrogenases seem to be much more narrowly distributed. On the other hand, they show a highly diverse modular structure [38,44] (Fig. **3**). If one only counts what are likely to be functional [FeFe] hydrogenases, that is genomes that contain both a *hydA* homologue as well as *hydE*, *hyd*F, and *hydG*, [FeFe] hydrogenases are widely found amongst the Gram positive phylum Firmicutes, especially the order Clostridia, as well as in the Gram negative γ proteobacteria (*Desulfovibrio*), with one example known in the δ proteobacteria (*Shewanella*) [38], and are absent from the α and β proteobacteria. A surprisingly large number of "orphan" *hydA*s can be found, and what their possible function might be is an interesting question but completely unknown at present.

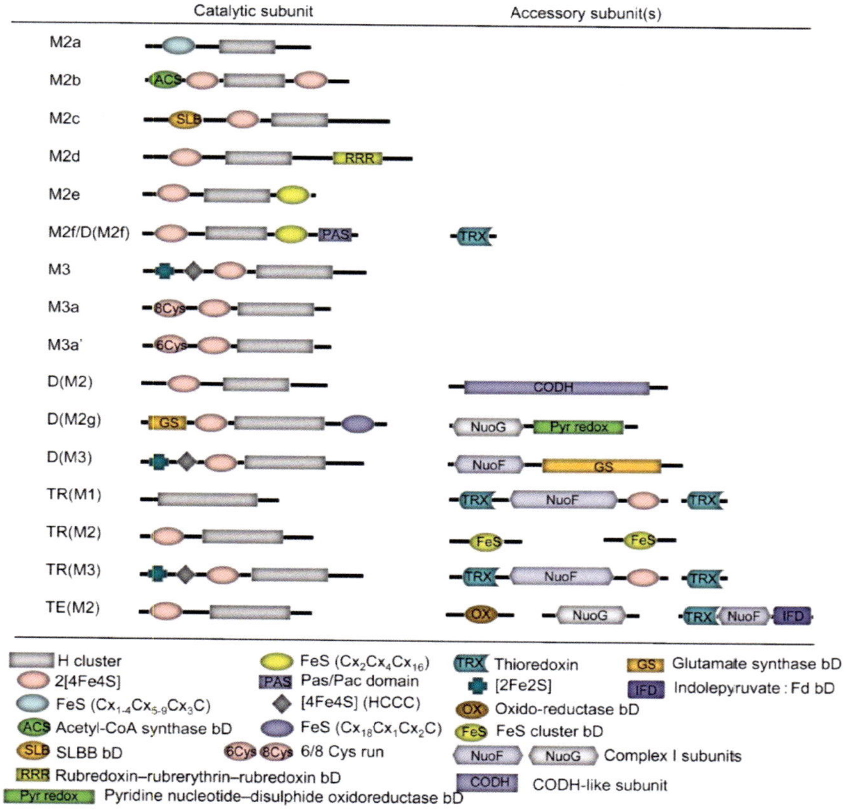

Figure 3: Modular structure and diversity of [FeFe] hydrogenases**.** Shown are different hydrogenases found in genomes of various Clostridia (Calusinska *et al.* [44]), but is essentially the same (with the same nomenclature) as previously found in a bacteria-wide genome search (Meyer 2007). Many different monomeric forms can be seen, some with additional domains that may either be involved in specialized redox reactions or in sensing and signaling. In addition, there appear to be several different trimeric forms as well as a dimeric and tetrameric forms. Some *Clostridial* species can contain up to five different hydrogenase forms. Obviously much molecular and biochemical work remains to be done to elucidate the function of these various hydrogenases. This Figure is reproduced by permission from Calusinska *et al.* [44].

Of some interest is the wide modular structure that is found in general [38], and, perhaps not surprising given the propensity of the Clostridia for fermentation and hydrogen production, amongst these organisms

[44] (Fig. **3**). In fact, many Clostridial genomes encode multiple [FeFe] hydrogenases of different classes, many of which are relatively uncharacterized. The basic underlying module consists of an H domain, necessary for the catalytic site, and an F domain, consisting of the sequence necessary to ligand two [4Fe4S] clusters which serve as a molecular wire to bring electrons from the redox partner to the active site (H cluster). This arrangement is seen in all hydrogenases except that from group M2a which migh coordinate a single FeS cluster (Fig. **3**).

Different classes of hydrogenases can be distinguished based on their having additional domains, or even additional subunits [38,44]. Members of a few different classes are of interest for their potential participation in hydrogen production. Many different [FeFe] hydrogenases are monomeric and include the best characterized hydrogenases, that of *Clostridium pasteurianum* (M3) and *Desulfovibrio desulfuricans* (also from the M2 group), both of which have been extensively studied on the biochemical and structural levels [33,46]. The type of hydrogenase exemplified by that of *C. pasteurianum* are widespread amongst the *Clostridium*, and are likely to be the primary players in fermentative hydrogen evolution, reacting with reduced ferredoxin. These, and perhaps one other, M2e, differ in their content of auxiallry FeS centers, suggesting that they interact with different partners. The other monomeric hydrogenases contain additional domains; either domains suggesting additional redox functions, such as previously unseen runs of Cys, suggesting coordination of an unknown metal center (M3a), and a complex rubredoxin domain (M2d), or domains suggestive of regulatory functions (PAS, SLBB). Nothing is known about these hydrogenases, except their presence in the genome, and additional characterization on the molecular level will be required to determine their function.

Genome analysis indicates the presence in many *Clostridial* species as well as other bacteria of putative multimeric hydrogenases. There appear to be several different classes of dimeric hydrogenases. While some members can readily be identified as the well characterized hydrogenase component of a carbon monoxide dehydrogenase system (DM2), others appear to be novel, and since they contain a glutamate synthase interaction domain and may therefore be directly involved in supplying reductant from H_2 for nitrogen assimilation. These remain to be characterized.

Of considerable interest in understanding hydrogen production by these, and many other organisms (*e.g. Thermotoga* [47], *Thermoanaerobacterium* [42], *Caldicellulosiruptor* [48], are the trimeric and tetrameric hydrogenases (Fig. **3**). In terms of their modular structure there appears to be two major types. One group (TR (M2)) has a basic [FeFe] hydrogenase subunit and two subunits carrying [FeS] clusters and, might be involved in hydrogen production through formate oxidation, as previously suggested for *Eubacterium acidaminophilum* [49]. While in general formate metabolism has been relatively little studied in the Clostridia, some details have been elucidated in *C. thermocellum* [50-52]. Interestingly, this bacterium accumulates considerable amounts of formate during fermentative growth [52], probably due to the fact that it appears to lack both the TR(M2) enzyme system [44] as well as formate dehydrogenase.

Finally, a recent study a hydrogenase from a different class of trimeric enzymes (TR(M3)) demonstrated that it has unique properties that require a re-examination of the bioenergetics of the hydrogen producing fermentations of many bacteria [47]. This enzyme, encoded in the genomes of at least one third of known hydrogen producing bacteria, appears (at least in *Thermotoga*) to use the simultaneous oxidation of NADH and reduced ferredoxin to reduce protons to hydrogen. In this way, the additional free energy available in reduced ferredoxin can drive the thermodynamically unfavorable flow of electrons from NADH to hydrogen. This type of bifurcating enzyme is proposed to be an example of a newly found form of energy conservation in bacteria, one that has already been found in the bifurcating system of *Clostridium kluyveri* that couples the energy gained during butyrate formation to drive ferredoxin reduction by NADH, leading to hydrogen evolution [53]. Although the similar trimeric hydrogenase from *Thermoanaerobacter tengcongensis* was previously proposed to be a NADH-dependent hydrogenase [42], it is in fact quite possible that it too is a bifurcating enzyme under conditions of low P_{H2} [53,54]. In the related *Thermoanaerobacterium saccharolyticum*, hydrogen production is profoundly affected in a strain carrying a knock-out of the trimeric hydrogenase, but not one with a knock-out of the more conventional hnd type[FeFe] hydrogenase, suggesting that this trimeric enzyme is in fact the primary one involved in hydrogen fermentations [45]. Thus, hydrogen production in many Clostridial-type fermentations may be more favorable than previously thought, where it was considered that two hydrogens per hexose would be derived from reduced ferredoxin and an additional maximum of two by the oxidation of NADH which could only occur at very low hydrogen partial pressures. This would help explain at least in part some reports of what seemed like unusually high molar hydrogen yields of some fermentations [10,55,56].

Based on an initial metagenomic survey, the natural occurrence and diversity of [Fe-Fe] hydrogenases in the environment has been considered to be rather low [57], and a more recent blast search of over one billion base pairs of non-redundant sequence found only 10 hits to a [FeFe] hydrogenase bait sequence [38]. This is further borne out by several studies of specific environments. However, some studies did not examine the *hydA* sequences obtained to verify that they are bone fide sequences containing the critical residues for binding of the H-cluster. As well, none are capable of determining in the organism in question also possesses the requisite *hydE*, *hydF*, and *hydG*, or if the observed *hydA* sequence is an "orphan" sequence of unknown function. A study of the upper (photic) region of a saline microbial mat showed that a large fraction of the putative HydA sequences obtained were most closely affiliated with the *Firmicutes* (66.2% of the clones) and the *Verrucomicrobia* (24.6%), with a lower proportion related to the *Bacteroidetes* (9.2%) [58]. Interestingly, many of the putative HydA sequences recovered were found to have substitutions in the L1 motif, which is involved in the coordination of the oxygen-labile [4Fe-4S] subcluster of the H cluster, suggesting that perhaps these substitutions could confer a degree of oxygen stability. In another study [59], genome sequences extracted from an acidic ethanol-hydrogen coproducing system gave 11 phylotypes closely related to *Ethanoligenens* and *Clostridium thermocellum*, in the Clostridiaceae family, well known hydrogen producers as detailed above. This investigation also found sequences associated with lactate-, propionate-, and butyrate-oxidizing bacteria, but these are unlikely to participate in overall hydrogen production as these are mostly syntrophic-type bacteria (see below, Metabolic Pathways). Similarly, hydrogenase gene fragments obtained from fen soil were related to well known hydrogen producers and carboxydotrophs [60].

SUBSTRATE UTILIZATION

Types of Substrates

Dark fermentation to produce hydrogen could potentially use a wide variety of feedstocks, either various waste streams, sugar rich crops (corn, sugar beet, *etc.*) or dedicated energy crops. Use of complex materials requires a wide range of hydrolytic and catabolic activities, and in this regard mixed microbial consortia are particularly useful [61,62]. Of the variety of possible substrates available in these materials, particularly waste streams, practical hydrogen fermentations are restricted to carbohydrate rich materials since little or no hydrogen can be fruitfully extracted from either the protein or lipid fractions. The amino acids obtained from hydrolysis of the protein fraction are principally fermented in pairs by so-called Strickland reactions where the oxidation of one amino acid leads to a reduction of a second amino acid to a keto acid. These reactions thus yield no hydrogen. Some amino acids can be fermented alone, but only a few of these (glutamate, cysteine, and serine) yield hydrogen as well as a variety of other products [63]. Thus hydrogen yields from a complex mixture of proteins are poor at best. Lipids are hydrolyzed to glycerol and long-chain fatty acids (LCFAs). In natural systems, LCFAs are degraded by syntrophic bacteria to acetate and hydrogen (see below, Metabolic Pathways) [64], but this reaction is only possible at the extremely low pH_2s maintained by the associated methanogenic or sulphate-reducing bacteria [65,66]. Much more energy could be obtained from this source by conversion to methane [67]. Thus, some wastes, even though present in relatively large amounts, such as effluents from the secondary treatment of sewage, are poor substrates for net hydrogen production due to their relative lack of fermentable carbohydrates. They are more suitable for anaerobic digestion, where transitory hydrogen production by synthrophs is rendered possible by being directly coupled to hydrogen consumption by methanogens (see below).

Use of Carbohydrates

Based on the considerations detailed above, hydrogen yielding fermentations for biofuel production must therefore be based on carbohydrate rich substrates. As a first step, waste streams could, and should, be targeted. Various food processing wastes fit the bill. This subject is discussed at length elsewhere in this volume. However, if any biofuel is to be produced at the scale required for it to substitute for a significant, or even moderate fraction of the present fossil fuel use, these resources would not be enough.

Efforts in the past few decades have led to the development of biofuel production on a significant level. However, in reality this has mostly been bioethanol production from corn starch, now viewed as creating an unacceptable competition with food production. Producing significant amounts of biofuels without seriously compromising the world food supply requires the ability to utilize biomass resources which contain substantial quantities of energy stored in lignocellulosic compounds [68-71]. Therefore, lignocellulosic biomass, an abundant renewable biological resource that is continually replenished by solar energy driven photosynthetic reduction of carbon dioxide is now being targeted as a replacement for food crops as a biofuel substrate [72].

Although lignocellulose could possibly be a very energy rich substrate for biofuels production in general, actually achieving this potential is difficult due to the almost crystalline state of the cellulose component and the intractability of the lignin. Different strategies have been developed in response to this problem (Table 1) [73]. Current use of this substrate depends upon physical/chemical pretreatment followed by the addition of what are at present cellulose degrading enzymes whose production is costly. This is viewed as the limiting factor in the economics of lignocellulosic ethanol production [74,75] and this of course applies to the production of any biofuel, including hydrogen, from this substrate.

However, being able to use the available chemical energy of the compounds generated by hydrolysis of complex lignocellulosic substrates requires the fermentative conversion of a variety of hexose and pentose sugars into a biofuel. The ability to both use a wide variety of substrates, and to carry out a rapid and high yielding fermentation, do not necessarily occur in the same organism, suggesting useful areas in which to apply metabolic engineering [76]. Two approaches can be suggested. One would be to engineer a capable fermenter to be able to consume a wider range of substrates through the introduction of pathways for the catabolism of a variety of pentoses and hexoses. Another would be to increase the fermentative powers of an organism with an omnivorous appetite, such as *Escherchia coli* [77].

Table 1: Comparison of Strategies for Lignocellulosic Degradation.

***Separate* Hydrolysis and Fermentation**	-Reactor size and operating conditions easily optimized -Enzyme specificity and efficiency can be adjusted to substrate -Newly discovered or engineered enzymes easily incorporated	-Enzyme production costly -Enzymes may need to be cloned from different sources -Two stage system required
***Native* Consolidated Bioprocessing**	-Direct conversion of cellulose to biofuel possible -Single stage process; simpler facility, ease of operation -Avoidance of glucose inhibition of cellulose degradation -Uses existing metabolic machinery	-Optimal temperatures for cellulose degradation and fermentation may be different -Low rates and yields of useful products by native organism -Low titers of active enzymes due to inefficient anaerobic growth
***Engineered* Consolidated Bioprocessing**	-Optimal cellulose degradation capacity in efficient fermenter -Single stage process -Cost-effective production of cellulases -Designer cellulosomes can be constructed	-Need for complex metabolic engineering, expression of multiple components -May lack synergy factors found in native organism

Recently, a new method for biofuels production from lignocellulosics has been proposed, consolidated bioprocessing (CBP), that combines enzyme production, cellulose saccharification, and fermentation into a single process which could potentially dramatically reduce the cost of biofuels production [73,78]. As originally proposed, this process would be carried out by anaerobic bacteria (Clostridia) which are capable of elaborating a complex molecular assembly at their cell surface, the cellulosome containing dozens of enzymes with different specificities. However, the successful use of this approach requires that a number of technical barriers be overcome [50,79] including augmenting volumetric production rates and overcoming inhibition by non-cellulosic components. Another possibility is to incorporate artificial cellulosomes into fermenting organisms already endowed with the capacity to achieve the desired rates and yields of biofuel production, but again there are significant technical barriers to this approach including establishing efficient expression and assembly of the introduced macromolecular machinery. This is further addressed in another chapter.

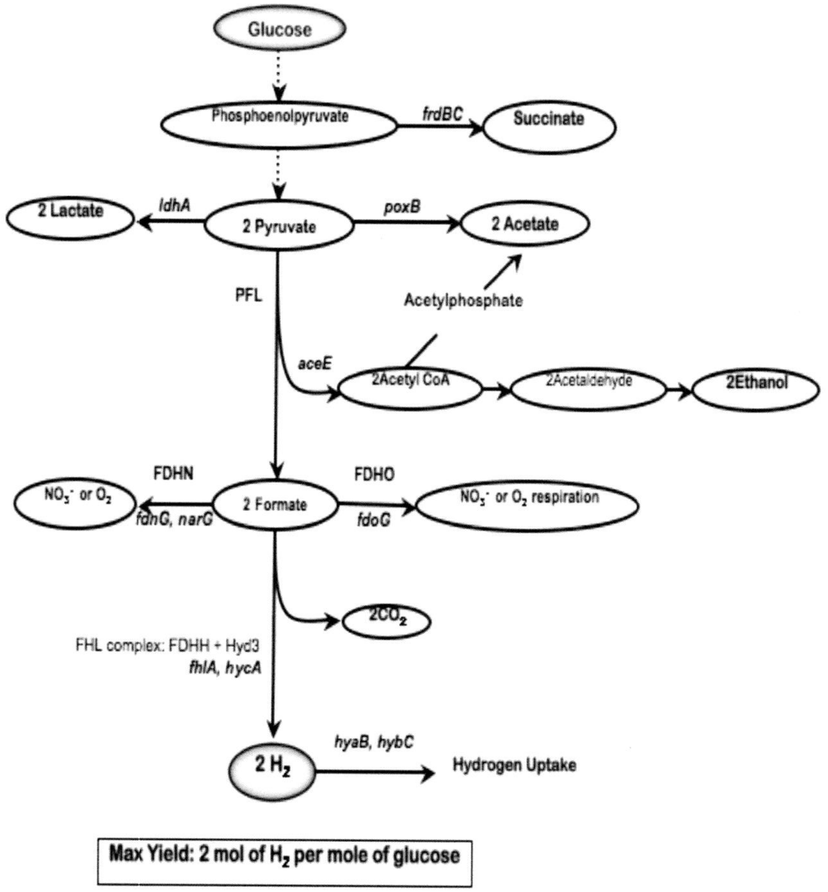

Figure 4: The main fermentative pathways of organisms like *Escherchia coli* carrying out mixed acid fermentations. Fermentative hydrogen production from glucose by *E. coli*, a well-studied facultative anaerobic bacterium, is shown. Hydrogen is produced through the action of the FHL complex, including the Ech [NiFe] hydrogenase (hyrogenase 3). The maximum theoretical hydrogen yield is 2 mol of H_2 per mole of glucose. The metabolic break down of glucose yields succinate, lactate, acetate, ethanol and formate as fermentation end-products. Succinate production is a rather minor pathway. Production of acetate generates ATP. Lactate production and formate dissimilation to CO_2 and H_2 occur primarily upon acidification of the medium. The proteins shown in bold with an asterisk have been studied through metabolic engineering in order to enhance the biohydrogen production. PFL, pyruvate formate lyase; FDH, formate dehydrogenase; FHL, formate hydrogen lyase; Hyd, hydrogenase; CoA, coenzyme A.

HYDROGEN PRODUCING PATHWAYS

Main Metabolic Pathways Leading to Fermentative Hydrogen Production

As discussed in detail previously [5,10], pyruvate generated during the glycolytic breakdown of carbohydrate-derived sugars is the key intermediate in the principal fermentation pathways leading to hydrogen. Pyruvate can be oxidized into intermediates involved in hydrogen production by either of two different enzymes; pyruvate:formate lyase in enteric-type mixed acid fermentation, and pyruvate:ferredoxin oxidoreductase (PFOR) in clostridial-type fermentations. In any case, acetyl-CoA is produced, from which other fermentation products can be derived. The redox neutral conversion to acetate allows ATP synthesis from the acetyl-phosphate intermediate and supports cell growth.

Facultative anaerobes such as *E. coli* carry out a mixed acid fermentation (Fig. **4**), where pyruvate is cleaved to acetyl-CoA and formate without the immediate production of reducing equivalents. Cultures at neutral pH can accumulate large amounts of formate, as there is actually no advantage to the bacterium to breakdown formate to hydrogen (and carbon dioxide) unless active metabolism causes the pH to drop to inhibitory levels. Under these conditions, the *fhl* system (formate:hydrogen lyase) is induced by formate acting on the activator FhlA, and the subsequent conversion of formic acid to hydrogen alleviates the acidification. Hydrogen is formed by an Ech ([NiFe] hydrogenase (hydrogenase 3) driven by formate dehydrogenase (FdhF) (Fig. **2**). This is the reaction that needs to be favoured to maximize hydrogen yields. Hydrogen recycling through hydrogenase 1 can decrease overall hydrogen yields and thus is a target for metabolic engineering. The advantage to the cell provided by hydrogen recycling is that a proton gradient is created by the outward diffusion of hydrogen and its subsequent oxidation in the cytoplasm.

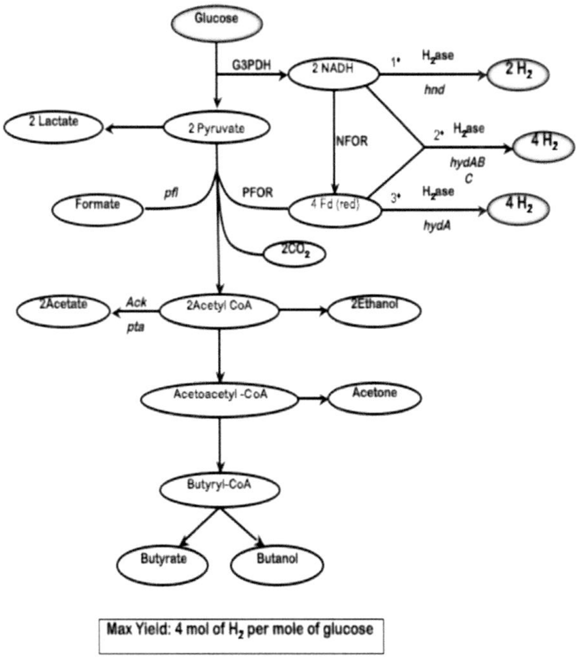

Figure 5: The main fermentative pathways of organisms like *Clostridium*. Fermentative hydrogen production from glucose by *C. acetobutylicum*, a strict anaerobic bacterium, is shown. Hydrogen can be produced through the action of PFOR and NFOR. The maximum theoretical hydrogen yield is 4 mol of H_2 per mole of glucose, with acetate or acetone as the fermentation end-product. The metabolic breakdown of glucose potentially yields lactate, acetate, ethanol, acetone, butanol and butyrate as fermentation end-products. Production of acetate generates ATP. Formation of lactate, ethanol, butanol, or butyrate consumes NADH, reducing the overall hydrogen yield. The proteins shown in bold with an asterisk have been studied in *Clostridium* species through metabolic engineering in order to enhance biohydrogen production. G3PDH, glyceraldehyde-3-phosphate dehydrogenase; PFOR, pyruvate ferredoxin oxidoreductase; NFOR, NADH:ferredoxin oxidoreductase; NADH, nicotinamide-adenine dinucleotide; red, reduced. Three different types of [FeFe] hydrogenases have been described in Clostridia. HydA, the classical clostridial hydrogenase that reacts with reduced ferredoxin, Hnd, a NAD(P)H reactive hydrogenase, and HydABC a novel bifurcating hydrogenase that reacts simultaneously with NADH and reduced ferredoxin.

Note that acidification also favors increased conversion of pyruvate to lactic acid, since this leads to the formation of one molecule of acid (lactic) instead of potentially two (acetic and formic). Obviously, this reaction diverts reductant from the hydrogen production pathway and presents another target for metabolic engineering. Since hydrogen is derived from formate, and a maximum of two formates are formed per glucose, maximum hydrogen yields can be predicted to be $2H_2$/glucose. As in all fermentations, the NADH produced during the glycolytic production of pyruvate must be re-oxidized to regenerate the NAD necessary for continued glycolysis. Here NADH is regenerated by the production of a reduced organic compound, principally ethanol. *E. coli* has the capacity to metabolize and produce hydrogen from a variety of sugars and sugar derivatives at different redox states [77], due to its wide range of catabolic systems and the flexibility accorded by mixed acid fermentation. Different redox states of substrates are reflected in the ratio of ethanol/acetate, which is one for glucose (actually a little higher due to hydrogen recycling, Fig. **2**).

On the other hand, Clostridia (strict anaerobes) are thought to principally use the PFOR (pyruvate:ferredoxin oxidoreductase) system to convert pyruvate to acetyl-CoA, and CO_2, producing reduced ferredoxin (Fig. **5**). The reduced ferredoxin transfers electrons to a [FeFe] hydrogenase, driving hydrogen evolution. This assures the production of two moles of hydrogen per mole of glucose consumed. However, *C. thermocellum* has been shown to produce formate, suggesting that it has an active Pfl system as well [52]. This drain on the pyruvate pool very likely reduces hydrogen production, although this has not been formally demonstrated.

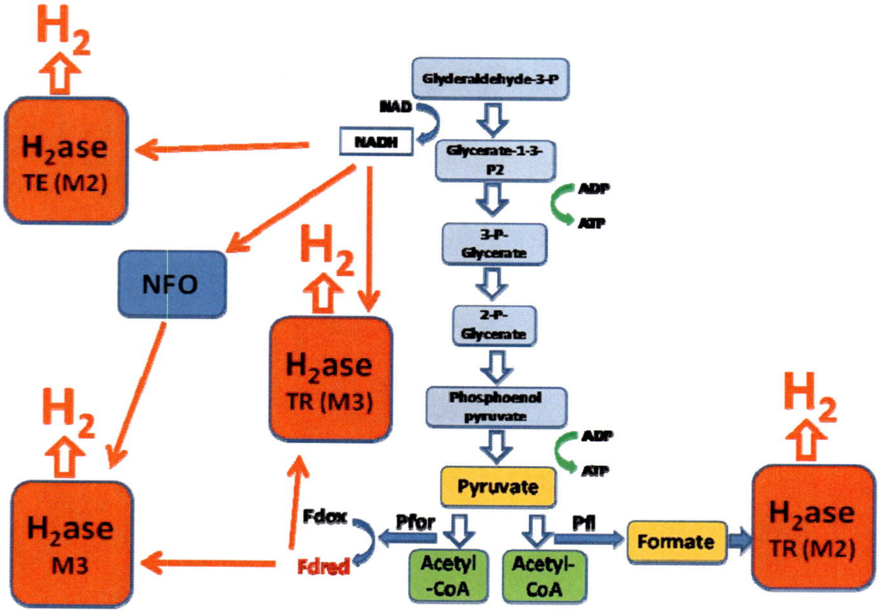

Figure 6: Various hydrogenases and the pyruvate node of metabolism. Shown are the various possibilities for connection of different hydrogenases in the Clostridia to glycolysis and pyruvate degradation suggested by the present evidence. Some aspects will require further verification. Hydrogenases are as noted in Fig. **3** and discussed in the text. Some organisms can degrade at least part of the pyruvate to formate. If present, a trimeric hydrogenase may, perhaps anchored to the cytoplasmic membrane by an associated membrane protein, couple formate oxidation to proton reduction. The well known monomeric *C. pasteurianum* type hydrogenase (Type M3) catalyzes hydrogen production from reduced ferredoxin, either produced directly from pyruate, or produced by NADH oxidation by Nfo, NADH:ferredoxin oxidoreductase. Nfo could be the soluble activity described several decades ago, or the more recently described membrane comlex of the Rnf family. Some multimeric hydrogenases (TE (M2)), similar to Hnd from the sulfate reducers, could possibly evolve hydrogen directly from NAD(P)H at very low P_{H2}s. Finally, the newly described timeric bifurcating hydrogenase (TR (M3)) could simultaneously use NADH and reduced ferredoxin to evolve hydrogen with the excess energy available from ferredoxin oxidation driving the unfavorable oxidation of NADH.

It is interesting to note that many organisms actually contain genes encoding both pyruvate:formate lyase (Pfl) and pyruvate:ferredoxin (flavodoxin) oxidoreductase (PFOR). Usually one of these plays a major role in fermentation (*e.g.* Pfl, in gram negative organism such as *E. coli*) whereas the second one is involved in minor pathways (*e.g.* Pfo, *E. coli,* which is involved in reduction of the flavodoxin necessary for formation of the glycyl radical active site of Pfl) [80]. In fact, organisms may contain multiple homologues of these genes. For example, *E. coli* has five *pfl* homologues, one of which is necessary for anaerobic threonine degradation. Obviously, it is therefore impossible to predict the fermentation pathways of a particular organism from its complement of Pfl and/or PFOR encoding genes.

In addition, unlike the organisms carrying out mixed-acid fermentation, hydrogen fermenting anaerobes like Clostridia possess pathways that allow them, under the proper conditions (low P_{H2}), to couple NADH oxidation to proton reduction to hydrogen. The pathways(s) by which this occurs is not totally clear, some possible routes suggested by published reports are shown in Fig. **6**. Soluble NADH:ferredoxin oxidoreductase activity in crude extracts was reported early on [81]. However, the corresponding enzyme has yet to be purified or described in molecular detail, nor has a gene encoding this activity been identified. The reduction of ferredoxin from NADH could actually be carried out by a membrane-bound NADH:ferredoxin oxidoreductase, the so-called Rnf complex, known to be present in some *Clostridial* species [82]. The Rnf complex, encoded by seven genes where it appears to be essential for photoheterotrophic nitrogen fixation, and probably drives ferredoxin reduction from NADH using an ion gradient [83]. Genomic analysis shows that Rnf complexes are widespread in nature and are found in a variety of Gram-negative and Gram-positive organisms. In most cases it is thought that they function primarily to create ion gradients using the energy liberated by reduced ferredoxin reduction of NAD. However, it is unknown to what extent these are reversible, and it remains an intriguing possibility that in some fermentative cases ferredoxin reduction by NADH could be driven by an Rnf complex using an existing ion gradient.

Recent work suggests a number of other ways in which NADH can be used to drive hydrogen evolution. Some organisms have been shown to contain heteromeric [FeFe] hydrogenases, Hnd hydrogenases, which can couple directly with NAD(P)(H) since they contain subunit homologues of some Complex I proteins. However, they may serve different functions depending upon the organism. In *Desulfovibrio fructosovorans* the Hnd [FeFe] hydrogenase appears to function physiologically in reduction of NADP with hydrogen [84]. However, in the Firmicutes, this enzyme may catalyze H_2 evolution from NADH, as has been show in *Thermoanaerobacter tengcongensis* [42]. The report that the trimeric [FeFe] hydrogenase from *Thermotoga maritima* is a bifurcating enzyme that uses NADH and reduced ferredoxin synergistically and that homologues of this enzyme are widespread [46], requires that the energetics of hydrogen production from NADH be re-examined. This would greatly increase the capacity to produce hydrogen at a given P_{H2} and therefore may be the major route for this reaction in the (many) Clostridia that encode this enzyme.

Thus, unlike organisms such as *E. coli* which rely on mixed acid fermentation, organisms using the PFOR fermentation pathway are potentially capable of producing additional hydrogen by reoxidizing the NADH generated during glycolysis. Given that two moles of NADH are produced during the glycolysis of glucose, up to a maximum of two additional molecules of H_2 can be generated by this pathway. Thus, these organisms appear to be capable in theory of producing up to $4H_2$/glucose, and indeed yields approaching this value have been obtained in some studies, albeit at low hydrogen partial pressures. How close an organism can get to this maximum depends upon the specific mechanism used for ferredoxin reduction (see above for Rnf and bifurcating hydrogenases) since ferredoxin reduction with NADH is unfavourable given that the midpoint potential of ferredoxin (~450 mV) is appreciably lower than that of the NADH/NAD couple (~320mV). Thus, production of H_2 with electrons derived from NADH is only possible at greatly reduced P_{H2} or with the input of energy in some form. Failure to reoxidize NADH by hydrogen production leads to its oxidation by pathways that produce reduced organic compounds; commonly, butyric acid and ethanol, but also butanol and acetone in solventogenic Clostridia. Some studies with these organisms have also noted that as the medium acidifies, butyric acid production is increased at the expense of hydrogen although the molecular mechanism is unclear.

It would appear that much fruitful work remains to be done to understand the molecular details of hydrogen production in the *Clostridia*, some of which might be useful in designing more efficient and more effective strains for biohydrogen production. Never the less, these organisms with their naturally endowed abilities cannot be expected to surpass 4 mol H_2/mol glucose.

Other Pathways that Yield Hydrogen

Although the main pathways to practical hydrogen are those mentioned above, it is worth noting that there exist in nature several mechanisms for the conversion of fermentation products to hydrogen, not all of which could be useful in the production of hydrogen as a biofuel. The interesting thing is that in these cases, hydrogen is produced from the end products of fermentation by other bacteria.

Syntrophic Hydrogen Production

Some organisms can carry out "impossible" fermentations of some fermentation end products to hydrogen. These might be regarded as "impossible" since the free energy change is positive under standard conditions (Table **2**), and only become possible at very low hydrogen partial pressures (Tables **2** and **3**). Even under these conditions these organisms are living close to the thermodynamic limits [85]. Syntrophic bacteria have been fairly well studied in terms of systematics and physiology [86], in spite of the difficulty of obtaining some of them in pure culture. In fact, these bacteria are intimately involved in natural anaerobic degradative processes as well as anaerobic digestion. In these systems hydrogen partial pressures are kept at the very low concentrations required for syntrophic growth (Table **3**) by methanogenic or sulfate reducing bacteria. Such a tight relationship can be observed in defined co-cultures of a syntrophic bacterium and a methanogen, and can even be used to isolate new species of syntrophic bacteria [65,87]. The maintenance of low hydrogen partial pressures by hydrogen consuming bacteria or archeae permits anaerobic growth on aromatic compounds such as benzoate and LCFAs (long chain fatty acids) as well as butyrate and propionate. However, their metabolic pathways generate primarily NADH and not the reduced ferredoxin for hydrogenase reduction. Even at 1 kPa P_{H2} ferredoxin reduction from NADH is unfavorable. Various membrane complexes are postulated to create ion gradients, which can then be used to drive ferredoxin reduction from NADH, by a Rnf complex [88].

Table 2. Coupled Metabolic Reactions Permitting Syntrophic Growth.

Reactions	$\Delta G^{0'}$	$\Delta G'^{*}$
Hydrogen producing bacteria		
$Propionate^- + 3H_2O \rightarrow Acetate^- + HCO_3^- + H^+ + 3H_2$	+72 kJ	-21kJ
$Butyrate^- + 2H_2O \rightarrow 2\ Acetate^- + H^+ + 2H_2$	+48 kJ	-22kJ
Methanogenic archeae		
$4\ H_2 + CO_2 \rightarrow CH_4 + 2\ H_2O$	-131kJ	-15kJ
$Acetate^- + H^+ \rightarrow CO_2 + CH_4$	-36kJ	-36kJ

* $\Delta G'$(standard Gibbs free energy change) is expressed in kJ mol^{-1} and was calculated for H_2 in the gaseous state at 1 Pa, and CH_4 and CO_2 in the gaseous state at 10^4 Pa. All other compounds were at 10mM. Values taken from Field *et al.* [69].

Table 3. The Thermodynamics of some Reactions Involved in Syntrophic Metabolism.

Reactions	$\Delta G^{0'}$	pH_2 for $-\Delta G'^{*}$
$Ethanol + H_2O \rightarrow Acetate- + H^+ + 2H_2$	+9.6	$<10^{-1}$
$Propionate^- + 3H_2O \rightarrow Acetate^- + HCO_3^- + H^+ + 3H_2$	+76.1	$<10^{-4}$
$Butyrate^- + 2H_2O \rightarrow 2\ Acetate^- + H^+ + 2H_2$	+48.3	$<10^{-4}$

*The partial pressure of hydrogen needed for the reaction to be thermodynamically favorable ($-\Delta G'$), which was calculated with concentrations of substrate and acetate of 0.1mM and a bicarbonate concentration of 100mM. Values taken from McInerney *et al.* [86].

Fermentation of Acetate and Ethanol or Lactate to Hydrogen by Some Clostridia

C. kluyveri can grow on ethanol and acetate producing butyrate, 1 caproate, and hydrogen, with the formation of 1 mol of ATP by substrate-level phosphorylation per 2 H_2 formed [89].

$$6\ \text{ethanol} + 3\ \text{acetate}^- \rightarrow 3\ \text{butyrate}^- + 1\ \text{caproate}^- + 1\ H^+ + 4\ H_2O + 2\ H_2$$

While this might appear to provide a pathway to increase overall hydrogen yields of a mixed culture fermenting glucose to hydrogen plus ethanol and acetate, this would only give at most 0.5 mol of additional hydrogen per glucose, this would only really be advantageous if the glucose fermenter was an organism such as *E. coli* since direct fermentation of glucose to butyrate by an organism that can carry this out (*i.e.* a *Clostridium*) would be more effective (2 mol H_2 per glucose). Likewise, the fermentation of lactate and acetate to butyrate and hydrogen by some *Clostridial* strains has long been recognized [90,91]. In particular this metabolism is carried out by *C. tyrobutyricum*, a potential source of problems in the dairy industry. These pathways are shown schematically in Fig. **7**. They both depend ultimately on the ability of the butyryl-CoA dehydrogenase/Etf complex to function as a bifurcating enzyme and use the energy gained in the reduction of crotonyl-CoA to drive ferredoxin reduction from NADH. Over 10% of the soluble protein in *C. kluyveri* is estimated to consist of this complex [54].

$$2\ \text{NADH} + 1\ \text{crotonyl-CoA} + 1\text{Fd}_{ox} \rightarrow 2\ \text{NAD} + 1\ \text{butyryl-CoA} + 1\text{Fd}_{red}$$

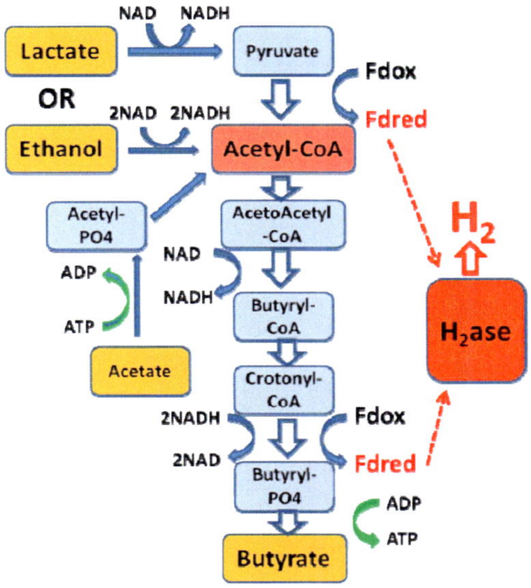

Figure 7: Fermentation of acetate and ethanol or lactate. Schematic diagram of the metabolic steps involved in the formation of hydrogen during co-fermentation of acetate and either ethanol or lactate. Stoichiometires are not indicated, normally ethanol or lactate are in excess in these fermentations.

CONCLUSIONS

Dark hydrogen fermentation, a naturally occurring process for a variety of microbes, could potentially form the basis for production of hydrogen as a biofuel. Underlying this possibility is a great deal of fundamental knowledge that has been gained over the past few decades. Much is now known about the molecular details of hydrogenases, their active sites, biochemistry, and in some cases three dimensional structures. The relatively recent discovery of [FeFe] hydrogenase diversity, as well as abundant content of different types in the same organism, demonstrates that further research is necessary in order to understand the mechanism and control of hydrogen production on a molecular level. Further information might easily be useful in designing more productive biohydrogen strains. As well, the main metabolic pathways involved are well understood in general, giving defined targets for metabolic engineering to increase rates and yields from suitable substrates, such as lignocellulosics.

ACKNOWLEDGEMENTS

Biohydrogen research in the laboratory of PCH is supported by FQRNT and NSERC (Natural Sciences and Engineering Research Council of Canada). Dipankar Ghosh and Mona Abo-Hashesh are thanked for helpful discussions. M. A.-H. is thanked for help with Figure presentation.

REFERENCES

[1] Thauer RK. Biochemistry of methanogenesis: a tribute to Marjory Stephenson. Microbiology 1998; 144: 2377-406.

[2] Adams MWW, Stiefel EI. Biological hydrogen production: Not so elementary. Science 1998; 282: 1842–3.

[3] Nicolet Y, Lemon BJ, Fontecilla-Camps JC, Peters JW. A novel FeS cluster in Fe-only hydrogenases. Trends Biochem Sci 2000; 25: 138–43.

[4] Vignais PM, Billoud B, Meyer J. Classification and phylogeny of hydrogenases. FEMS Microbiol Rev 2001; 25: 455–501.

[5] Frey M. Hydrogenases: hydrogen-activating enzymes. Chembiochem 2002; 3:153–60.

[6] Vignais PM, Colbeau A. Molecular Biology of Microbial Hydrogenases. Curr Issues Mol Biol 2004; 6: 159-88.

[7] Vignais PM, Billoud B. Occurrence, classification, and biological function of hydrogenases: an overview. Chem Rev 2007; 107: 4206-72.

[8] Vignais PM. Hydrogenases and h[+]-reduction in primary energy conservation. Results Probl Cell Differ 2008; 45: 223-52.

[9] Lyon EJ, Shima S, Buurman G *et al.* UV-A/blue-light inactivation of the 'metal-free' hydrogenase (Hmd) from methanogenic archaea - The enzyme contains functional iron after all. Eur J Biochem 2004; 271: 195-204.

[10] Hallenbeck PC. Fermentative hydrogen production: Principles, progress, and prognosis. Int J Hydrogen Energy 2009; 34(17): 7379-89.

[11] Stripp ST, Goldet G, Brandmayr C, *et al.* How oxygen attacks [FeFe] hydrogenases from photosynthetic organisms. Proc Natl Acad Sci USA 2009; 106: 17331–6.

[12] Heinekey DM. Hydrogenase enzymes: Recent structural studies and active site models. J Organomet Chem 2009; 694: 2671-80.

[13] Böck A, King PW, Blokesch M, Posewitz MC. Maturation of hydrogenases. Adv Microb Physiol 2006; 51: 1–71.

[14] Forzi, L Sawers, G. Maturation of [NiFe]-hydrogenases in *Escherichia coli.* Biometals 2007; 20: 565–78.

[15] Cauvin B, Colbeau A, Vignais PM. The hydrogenase structural operon in *Rhodobacter capsulatus* contains a third gene, hupM, necessary for the formation of a physiologically competent hydrogenase. Mol Microbiol 1991; 5: 2519–27.

[16] Vignais PM, Toussaint B. Molecular biology of membrane-bound H_2 uptake hydrogenases. Arch Microbiol 1994; 161: 1–10.

[17] Olson JW, Fu C, Maier RJ. The HypB protein from *Bradyrhizobium japonicum* can store nickel and is required for the nickel-dependent transcriptional regulation of hydrogenase. Mol Microbiol 1997; 24: 119–28.

[18] Durmowicz MC, Maier RJ. Roles of HoxX and HoxA in biosynthesis of hydrogenase in *Bradyrhizobium japonicum.* J Bacteriol 1997; 179: 3676–82.

[19] Schwartz E, Gerischer U, Friedrich,B. Transcriptonal regulation of *Alcaligenes eutrophus* hydrogenase genes. J Bacteriol 1998; 180: 3197–204.

[20] Reissmann S, Hochleitner E, Wang H, *et al.* Taming of a poison: biosynthesis of the [NiFe]-hydrogenase cyanide ligands. Science 2003; 299: 1067–70.

[21] Driesener RC, Challand MR, McGlynn SE, *et al.* [FeFe]-Hydrogenase Cyanide Ligands Derived From S-Adenosylmethionine-Dependent Cleavage of Tyrosine. Angew Chem Int Ed 2010; 49(9): 1687-90.

[22] Kuchenreuther JM, Stapleton JA, Swartz JR. Tyrosine, Cysteine, and S-Adenosyl Methionine Stimulate *In Vitro* [FeFe] Hydrogenase Activation. PLoS ONE 2009; 4(10): e7565.

[23] McGlynn SE, Shepard EM, Winslow MA, *et al.* HydF as a scaffold protein in [FeFe] hydrogenase H-cluster biosynthesis. FEBS Lett 2008; 582(15): 2183-7.

[24] Czech I, Silakov A, Lubitz W, Happe T. The [FeFe]-hydrogenase maturase HydF from *Clostridium acetobutylicum* contains a CO and CN-ligated iron cofactor. FEBS Letters. Feb; 584(3): 638-642

[25] Shepard EM, McGlynn SE, Bueling AL, *et al.* Synthesis of the 2Fe subcluster of the [FeFe]-hydrogenase H cluster on the HydF scaffold. Proc Natl Acad Sci USA 2010; 107 : 10448-53.

[26] Mulder DW, Ortillo DO, Gardenghi DJ, *et al.* Activation of HydA(Delta EFG) Requires a Preformed [4Fe-4S] Cluster. Biochemistry 2009; 48: 6240-8.

[27] Mulder DW, Boyd ES, Sarma R, *et al.* Stepwise [FeFe]-hydrogenase H-cluster assembly revealed in the structure of HydA(Delta EFG). Nature 2010; 465: 248-52.

[28] Forzi L, Hellwig P, Thauer RK, Sawers RG. The CO and CN- ligands to the active site Fe in NiFe-hydrogenase of *Escherichia coli* have different metabolic origins. FEBS Lett 2007; 581(17): 3317-21.

[29] Ryde U, Greco C, De Gioia L. Quantum Refinement of [FeFe] Hydrogenase Indicates a Dithiomethylamine Ligand. J Am Chem Soc 2010; 132(13): 4512.

[30] Pilet E, Nicolet Y, Mathevon C, Douki T, Fontecilla-Camps JC, Fontecave M. The role of the maturase HydG in [FeFe]-hydrogenase active site synthesis and assembly. FEBS Letters 2009; 583 : 506-11.

[31] Pohorelic BKJ, Voordouw JK, Lojou E, Dolla A, Harder J, Voordouw G. Effects of Deletion of Genes Encoding Fe-Only Hydrogenase of *Desulfovibrio vulgaris* Hildenborough on Hydrogen and Lactate Metabolism. J Bacteriol 2002; 184: 679-86.

[32] Burgdorf T, Lenz O, Buhrke T, *et al.* NiFe -hydrogenases of Ralstonia eutropha H16: Modular enzymes for oxygen-tolerant biological hydrogen oxidation. J Mol Microbiol Biotechnol 2005; 10(2-4): 181-96.

[33] Fontecilla-Camps JC, Volbeda A, Cavazza C, Nicolet, Y. Structure/function relationships of [NiFe]- and [FeFe]-hydrogenases. Chem Rev 2007; 107: 4273–303.

[34] English CM, Eckert C, Brown K, Seibert M, King PW. Recombinant and *in vitro* expression systems for hydrogenases: new frontiers in basic and applied studies for biological and synthetic H2 production. Dalton Trans 2009; 45: 9970-8.

[35] Sun JS, Hopkins RC, Jenney FE, McTernan PM, Adams MWW. Heterologous expression and maturation of an NADP-dependent NiFe -hydrogenase: a key enzyme in biofuel production. PLoS ONE 2010; 5(5): e10526.

[36] Posewitz MC, King PW, Smolinski SL, Zhang LP, Seibert M, Ghirardi ML. Discovery of two novel radical S-adenosylmethionine proteins required for the assembly of an active Fe hydrogenase. J Biol Chem 2004; 279(24): 25711-20.

[37] King PW, Posewitz MC, Ghirardi ML, Seibert M. Functional studies of FeFe hydrogenase maturation in an *Escherichia coli* biosynthetic system. J Bacteriol 2006; 188(6): 2163-72.

[38] Meyer J. [FeFe] hydrogenases and their evolution: a genomic perspective. Cell Mol Life Sci 2007; 64: 1063-84.

[39] Hedderich R, Forzi L. Energy-converting [NiFe] hydrogenases:more than just H2 activation. J Mol Microbiol Biotechnol 2005; 10: 92–104.

[40] Sapra R, Verhagen M, Adams MWW. Purification and characterization of a membrane-bound hydrogenase from the hyperthermophilic archaeon *Pyrococcus furiosus*. J Bacteriol 2000; 182: 3423–8.

[41] Silva PJ, van den Ban EC, Wassink H, *et al.* Enzymes of hydrogen metabolism in *Pyrococcus furiosus*. Eur J Biochem 2000; 267: 6541–51.

[42] Soboh B, Linder D,Hedderich R. A multisubunit membrane-bound [NiFe] hydrogenase and an NADH-dependent Fe-only hydrogenase in the fermenting bacterium *Thermoanaerobacter tengcongensis*. Microbiology 2004; 150**:** 2451-63.

[43] Meuer J, Kuettner HC, Zhang JK, Hedderich R, Metcalf WW. Genetic analysis of the archaeon *Methanosarcina barkeri* Fusaro reveals a central role for Ech hydrogenase and ferredoxin in methanogenesis and carbon fixation. Proc Natl Acad Sci USA 2002; 99(8): 5632-7.

[44] Calusinska M, Happe T, Joris B, Wilmotte A. The surprising diversity of clostridial hydrogenases:a comparative genomic perspective. Microbiology 2010; 156: 1575–88.

[45] Shaw AJ, Hogsett DA, LyndLR. Identification of the [FeFe]-Hydrogenase Responsible for Hydrogen Generation in *Thermoanaerobacterium saccharolyticum* and Demonstration of Increased Ethanol Yield *via* Hydrogenase Knockout. J Bacteriol 2009; 191: 6457–64.

[46] Fontecilla-Camps JC, Amara P, Cavazza C, Nicolet Y, Volbeda A. Structure–function relationships of anaerobic gas-processing metalloenzymes. Nature 2009; 460: 814-22.

[47] Schut GJ, Adams MWW. The Iron-Hydrogenase of *Thermotoga maritima* Utilizes Ferredoxin and NADH Synergistically: a New Perspective on Anaerobic Hydrogen Production. J Bacteriol 2009; 191(13): 4451-7.

[48] van de Werken HJG, Verhaart MRA, VanFossen AL, *et al.* Hydrogenomics of the Extremely Thermophilic Bacterium *Caldicellulosiruptor saccharolyticus*. Appl Environ Microbiol 2008; 74: 6720–9.

[49] Graentzdoerffer A, Rauh D, Pich A, Andreesen JR. Molecular and biochemical characterization of two tungsten and selenium-containing formate dehydrogenases from *Eubacterium acidaminophilum* that are associated with components of an iron-only hydrogenase. Arch Microbiol 2003; 179: 116–30.

[50] Carere CR, Sparling R, Cicek N, Levin DB. Third generation biofuels via direct cellulose fermentation. Int J Mol Sci 2008; 9(7): 1342-60.

[51] Rydzak T, Levin DB, Cicek N, Sparling R. Growth phase-dependant enzyme profile of pyruvate catabolism and end-product formation in *Clostridium thermocellum* ATCC 27405. J Biotechnol 2009; 140(3-4): 169-75.

[52] Sparling R, Islam R, Cicek N, Carere C, Chow H, Levin DB. Formate synthesis by *Clostridium thermocellum* during anaerobic fermentation. Can J Microbiol 2006; 52(7): 681-8.

[53] Herrmann G, Jayamani E, Mai G, Buckel W. Energy conservation *via* electron-transferring flavoprotein in anaerobic bacteria. J Bacteriol 2008; 190: 784-91.

[54] Li F, Hinderberger J, Seedorf H, Zhang J, Bucke W, Thauer RK. Coupled Ferredoxin and Crotonyl Coenzyme A (CoA) Reduction with NADH Catalyzed by the Butyryl-CoA Dehydrogenase/Etf Complex from *Clostridium kluyveri*. J Bacteriol 2008; 190: 843–50.

[55] Hallenbeck PC. Fundamentals of the fermentative production of hydrogen. Water Sci Technol 2005; 52(1-2): 21-9.

[56] Zeidan AA, van Niel EWJ. A quantitative analysis of hydrogen production efficiency of the extreme thermophile *Caldicellulosiruptor owensensis* OLT. Int J Hydrogen Energy 2010; 35: 1128-37.

[57] Meyer J. Miraculous catch of iron–sulfur protein sequences in the Sargasso Sea. FEBS Lett 2004; 570: 1–6.

[58] Boyd ES, Spear JR, Peters JW. [Fe-Fe]-hydrogenase genetic diversity provides insight into molecular adaption in a saline microbial mat community. Appl Environ Microbiol 2009; 75: 4620-3.

[59] Xing D, Ren N, Rittmann BE. Genetic diversity of hydrogen-producing bacteria in an acidophilic ethanol-H2-coproducing system, analyzed using the [Fe]-hydrogenase gene. Appl Environ Microbiol 2008; 74: 1232-9.

[60] Schmidt O, Drake HL, Horn MA. Hitherto Unknown [Fe-Fe]-Hydrogenase Gene Diversity in Anaerobes and Anoxic Enrichments from a Moderately Acidic Fen. Appl Environ Microbiol 2010; 76: 2027–31.

[61] Hallenbeck PC, Ghosh D. Advances in fermentative biohydrogen production: the way forward? Trends Biotechnol 2009; 27(5): 287-97.

[62] Valdez-Vazquez I, Poggi-Varaldo HM. Hydrogen production by fermentative consortia. Renew Sust Energ Rev 2009; 13: 1000–13.

[63] Mechichi T, Fardeau M-L, Labat M, Garcia J-L, Verhé F, Patel BKC. *Clostridium peptidivorans* sp. nov., a peptide-fermenting bacterium from an olive mill wastewater treatment digester. Int J Syst Evol Microbiol 2000; 50: 1259–64.

[64] Sousa DZ, Pereira M A, Stams AJM, Alves MM, Smidt H. Microbial Communities Involved in Anaerobic Degradation of Unsaturated or Saturated Long-Chain Fatty Acids. Appl Environ Microbiol 2007; 73: 1054–64.

[65] Sousa DZ, Smidt H, Alves MM, Stams AJM. Syntrophomonas zehnderi sp. nov., an anaerobe that degrades long-chain fatty acids in co-culture with Methanobacterium formicicum. Int J Syst Evol Microbiol 2007; 57: 609–15.

[66] Stams AJM, Plugge CM. Electron transfer in syntrophic communities of anaerobic bacteria and archaea. Nat Rev Microbiol 2009; 7: 568-77.

[67] Alves MM, Alcina Pereira MA, Sousa DZ, *et al.* Waste lipids to energy: how to optimize methane production from long-chain fatty acids (LCFA). Microb Biotechnol 2009; 2(5): 538–50.

[68] Sagar AD, Kartha S. Bioenergy and sustainable development? Ann Rev Env Resour 2007; 32: 131–67.

[69] Field CB, Campbell JE, Lobell DB. Biomass energy: the scale of the potential resource. Trends Ecol Evol 2008; 23: 65-72.

[70] Martindale W, Trewavas A. Fueling the 9 billion. Nature Biotechnol 2008; 26:1068-9.

[71] Tilman D, Socolow R, Foley JA, *et al.* Beneficial Biofuels: The Food, Energy, and Environment Trilemma. Science 2009; 325: 270-1.

[72] Sims, REH, Mabee W, Saddler JN, Taylor M. An overview of second generation biofuel technologies. Bioresource Technol 2010; 101: 1570–80.

[73] Lynd LR, Weimer PJ, van Zyl WH, Pretorius IS. Microbial Cellulose Utilization: Fundamentals and Biotechnology. Microbiol Mol Biol Rev 2002; 66: 506–77.

[74] Margeot A, Hahn-Hagerdal B, Edlund M, Slade R, Monot F. New improvements for lignocellulosic ethanol. Curr Opin Biotechnol 2009; 20: 372–80.

[75] Wilson DB. Cellulases and biofuels. Curr Opin Biotechnol 2009; 20: 295–9.

[76] Hallenbeck PC, Ghosh D, Abo-Hashesh M, Wang R. Metabolic engineering for enhanced biofuels production with emphasis on the biological production of hydrogen. In: Taylor JC, Ed. Advances in chemistry research volume 6. Nova Science Publishers 2011; pp.125-54.

[77] Ghosh D, Hallenbeck PC. Fermentative hydrogen yields from different sugars by batch cultures of metabolically engineered *Escherichia coli* DJT135. Int J Hydrogen Energy 2009; 34(19): 7979-82.

[78] Lynd LR, van Zyl WH, McBride JE, Laser M. Consolidated bioprocessing of cellulosic biomass: an update. Curr Opin Biotechnol 2005; 16(5): 577-83.

[79] Levin DB, Carere CR, Cicek N, Sparling R. Challenges for biohydrogen production *via* direct lignocellulose fermentation. Int J Hydrogen Energy 2009; 34(17): 7390-403.

[80] Sawers G, Watson G. A glycyl radical solution: oxygen-dependent interconversion of pyruvate formate-lyase. Mol Microbiol 1998; 29: 945-54.

[81] Jungermann K, Thauer RK, Leimenstoll G, Decker K. Function of reduced pyridine nucleotide-ferredoxin oxidoreductases in saccharolytic *Clostridia*. BBA-Bioenergetics 1973; 305: 268–80.

[82] Boiangiu CD, Jayamania E, Brügelet D, *et al.* Sodium ion pumps and hydrogen production in glutamate fermenting anaerobic bacteria. J Mol Microbiol Biotechnol 2005; 10: 105–19.

[83] Schmehl M, Jahn A, Meyer zu Vilsendorf A, *et al.* Identification of a new class of nitrogen fixation genes in Rhodobacter capsulatus: a putative membrane complex involved in electron transport to nitrogenase. Mol Gen Genet 1993; 241: 602–15.

[84] Malki S, Saimmaime I, De Luca G, Rousset M, Dermoun Z, Belaich JP. Characterization of an operon encoding an NADP-reducing hydrogenase in *Desulfovibrio fructosovorans*. J Bacteriol 1995; 177: 2628–36.

[85] Jackson BE, McInerney MJ. Anaerobic microbial metabolism can proceed close to thermodynamic limits. Nature 2002; 415: 454-6.

[86] McInerney MJ, Struchtemeyer CG, Sieber J, *et al.* Physiology, Ecology, Phylogeny, and Genomics of Microorganisms Capable of Syntrophic Metabolism. Ann NY Acad Sci 2008; 1125: 58–72.

[87] de Bok FAM, Harmsen HJM, Plugge CM, *et al.* The first true obligately syntrophic propionate-oxidizing bacterium, *Pelotomaculum schinkii* sp nov., co-cultured with *Methanospirillum hungatei*, and emended description of the genus *Pelotomaculum*. Int J Syst Evol Microbiol 2005; 55: 1697-703.

[88] McInerney MJ, Rohlin L, Mouttaki H, *et al.* Life at the thermodynamic limit of microbial growth: The genome of *Syntrophus aciditrophicus*. Proc Natl Acad Sci USA 2007; 104: 7600–05.

[89] Thauer RK, Junermann K, Henninger H, Wenning J, Decker K. The Energy Metabolism of *Clostridium kluyveri*. Eur J Biochem 1968; 4: 173-80.

[90] Bhat JV, Barker HA. *Clostridium lacto-acetophilum* nov. spec. and the role of acetic acid in the butyric acid fermentation of lactate. J Bacteriol 1947; 54: 381-91.

[91] Diez-Gonzalez F, Russell JB, Hunter JB. The role of an NAD-independent lactate dehydrogenase and acetate in the utilization of lactate by *Clostridium acetobutylicum* strain P262. Arch Microbiol 1995; 164: 36-42.

CHAPTER 7

Biohydrogen Production *via* Fermentation of Biowastes by Microorganisms

Vikineswary Sabaratnam[1]* and Mohammad Ali Hassan[2]

[1]*Institute of Biological Science, Faculty of Science, University of Malaya, Kuala Lumpur, Malaysia and* [2]*Universiti Putra Malaysia, Universiti Putra Malaysia, Serdang, Malaysia*

Abstract: The production of hydrogen which is predicted to be the fuel of the future has a number hurdles to overcome. One of the constraints is cost of production. However, cost of production can be significantly reduced considering biowastes as a potential and cheap renewable substrate for hydrogen production. The wastes are a result of the many activities of humans leading to wastes generation and includes household, municipal, agricultural and industrial wastes. Further, the microbial diversity available enables us to select the right microorganisms to ferment a particular biowaste in obligate anaerobic or facultative anaerobic single stage or hybrid system to covert the carbonaceous components to biohydrogen. The systems can be operated with an axenic culture or a consortia of microbial cultures. Factors that affect microbial growth need to be optimized and the systems can be operated in a continuous mode. Further, this approach leads to dual benefits, that is, a simultaneous biohydrogen – a green fuel production and pollution load reduction for the industry. Cost saving technologies are always attractive to industry where cost of waste management is high. The aim of this review is to summarise the possible technologies that combine utilisation of biowastes with production of biohydrogen. This will be attractive to industry as the cost of management of the biowastes can be off set by the alternative green fuel produced during the process. When biowastes are utilised as energy substrates for biohydrogen producing microbes.

Keywords: Biohydrogen, photofermentation, heterotrophic bacteria, biowastes, hybrid systems, phototrophic bacteria, biomass, agroindustrial residues, *Rhodopseudomonas palustris,* palm oil mill effluent, biofuel.

INTRODUCTION

The worldwide energy demand has been increasing exponentially. Today, about 80% of the world energy demand is being fulfilled by fossil fuels [1]. This will eventually lead to the depletion of limited fossil energy resources. In addition, the overwhelming scientific evidence shows that the combustion of fossil fuels has serious negative effects on the environment that causing the world's climate to change, with potential disastrous effect due to the emission of pollutants such as CO_x, NO_x, SO_x, C_xH_x, soot, ash, droplets of tars and other organic compounds, release into the atmosphere as a result of their combustion. For these reasons, new sustainable energy sources that could substitute fossil fuel has extensively been explored by many researchers. Hence, the security of energy supply has raised the demand towards the establishment of hydrogen economy. The transition of global energy utilization is illustrated in Fig. **1**. Hydrogen is considered as a promising and ideal source of energy in the future.

It is regarded as a clean non-polluting fuel because its combustion only releases water vapour and a little of nitrogen oxide [2]. However, the major problem in utilization of hydrogen as a fuel is its unavailability in nature, as it is available in combination with other molecules. Hydrogen is considered as a promising and ideal source of energy for the future. This makes it the ninth most abundant element with about 0.9% by weight of earth's surface is hydrogen [3]. Therefore, development of cost effective and efficient hydrogen production technologies has gained significant attention in recent years.

***Address correspondence to Vikineswary Sabaratnam:** Institute of Biological Science, Faculty of Science, University of Malaya, Kuala Lumpur, Malaysia; E-mail: viki@um.edu.my

Nuri Azbar and David Levin (Eds)

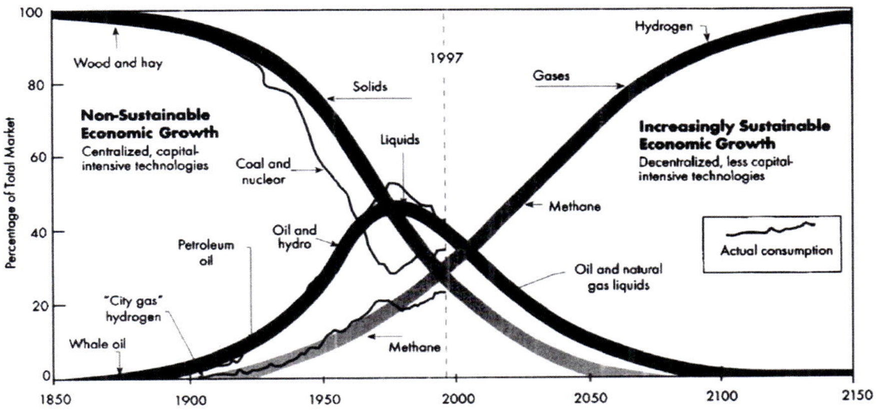

Figure 1: An illustration of the global energy system transition from petroleum oil to hydrogen, 1850-2150 [4].

At present, hydrogen is mainly produced from fossil fuels, biomass and water. Fig. **2** shows the principal hydrogen production routes and 96% is produced from fossil fuels. Currently, the annual production of hydrogen by steam reforming of fossil fuels is about 0.1 Gton [5]. However, these processes are energy intensive as they are operated at high temperatures (> 850°C). Further, it may eventually lead to the foreseeable depletion of limited fossil energy resources.

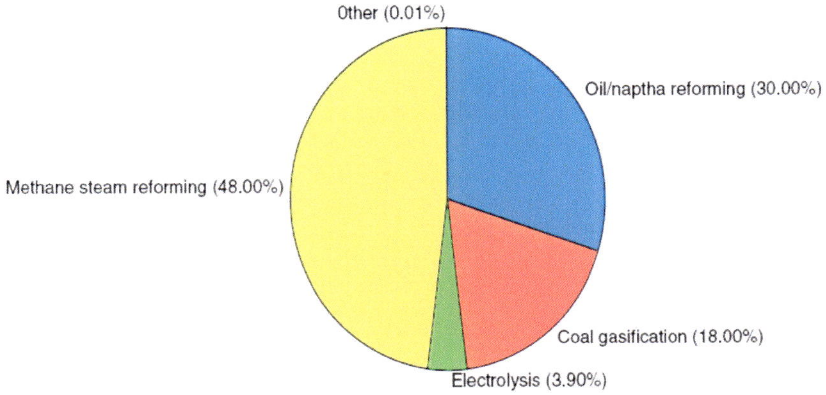

Figure 2: The distribution of primary energy sources and hydrogen production methods [6].

The aim of this chapter is to provide a concise and comprehensive account of the production of hydrogen by selected microorganisms using renewable substrates such as agro-industrial residues. The chapter highlights that the simultaneous treatment/utilisation of the residues to reduce organic load and production of biohydrogen may be an attractive solution to not only manage the wastes but also harness clean and efficient fuel –hydrogen. However, the constraints are that the selected biowaste should be cheap, available all year round and in a form that requires minimum pre- treatments prior to use in the fermentation. In this chapter, palm oil mill effluent (POME) will be presented as a type study for the potential use of a sustainable substrate from the palm oil industry as a renewable and organic resource for biohydrogen production.

TECHNOLOGIES FOR HYDROGEN PRODUCTION

In the preceding chapters of this book several technologies for hydrogen production are discussed. In brief, hydrogen can be obtained from water by electrolysis, photolysis, thermochemical processes and thermolysis (direct thermal decomposition). Electrolysis of water may be the cleanest technology available currently for hydrogen production. However, the electricity consumed in this process is far more valuable than the hydrogen produced as the electricity costs account for 80% of the operating cost of hydrogen production. Further, the feed water used in this process has to be demineralised to avoid deposits on the electrode and corrosion [7]. The thermochemical and electrochemical hydrogen generation processes are considered impractical, too. Not only are these processes energy intensive, the cost of production is high and can lead to environmental pollution. Recently, biomass is being considered favourably as a substrate for biological hydrogen (biohydrogen) production which can be less energy and capital intensive as ainablewell as less polluting to the environment. Various aspects including the choice of the substrates, the microogranisms and culture systems have to be considered to ensure successful, sustainable and cost attractive production of biological hydrogen.

Biowastes as Substrates for Biohydrogen Production

Simple sugar such as glucose, sucrose and lactose are readily biodegradable and are the preferred substrates for hydrogen production. However, pure carbohydrate sources are expensive raw materials to be considered for hydrogen production. Biowastes that can be considered are by-products of agricultural activities (cellulosic biomass) and/or processing of agricultural produce such as cassava for starch (starchy-cellulosic residues). Biowastes can also be from food industries, carbohydrate - rich industrial wastewaters and waste sludges from organic wastewater treatment plants. The crucial criteria for the selection of organic wastes including biomass as substrates for biohydrogen production are its availability, cost, carbohydrate content and biodegradability. Renewable resources including agroindustrial wastewaters such as POME qualify as they meet all prerequisites for choice as substrates for fermentation by microbes.

Biomass

Biomass is a versatile, clean and renewable energy source and contributes to net CO2 emission. Further, biomass is organic matter formed by photosynthetic capture of solar energy and fixation of carbon dioxide which is then stored by plants. The chemical components which are available in biomass are cellulose and hemicellulose in the cell walls and starch as storage product. Further, in an agricultural activity to produce food, large amounts (40-80%) end up as lignocellulosics either in the field or as by-products of processing. Therefore, crops, such as agricultural and forestry products and processing residues, livestock effluents and the organic fraction of domestic and industrial wastes are potential energy – rich biomass for alternative fuel [8]

On the other hand, wet biomass substrates (organic wastes or biomass and water) are favourable for biological or photobiological hydrogen production [9]. Biological processes are carried out largely at ambient temperatures and pressures. The operations are less energy intensive than chemical or electrochemical methods [10]. In addition, they do not contribute to pollution, so that it can be considered as a viable alternative for hydrogen production.

Selection of Biowastes

There have been several studies, reports and reviews to evaluate renewable resources as suitable substrates to produce biohydrogen *via* dark fermentation. Selected examples of biowastes that have been considered for biohydrogen production are chemical wastewater [11], food processing and domestic wastewater [12, 13], sugar refinery wastewater [14], tofu wastewater [15] and palm oil mill effluent [16]. In Malaysia, the potential utilisation of palm oil mill effluent (POME) for hydrogen production substrate has been studied. Currently, from the approximately 300 mills operating throughout Malaysia generating the wastewater (POME) which is highly polluting and posing problems in environmental friendly management. Current treatment strategies include anaerobic digestion. For example, a large scale methane production facility has been implemented at a palm oil mill, Felda Palm Industries Sdn. Bhd. which is located at Serting Hilir, Negeri Sembilan, Malaysia with reactor capacity of 500 m^3 [17]. However, POME which is high in organic carbon content and is available throughout the year is a potential substrate for hydrogen production [18].

METHODS FOR BIOHYDROGEN PRODUCTION

The phenomenon of biohydrogen production was observed one century ago. The technology started to receive attention when the oil crisis broke out in 1970s. Today, there are several methods for hydrogen production from biomass and it can be divided into two general categories: thermochemical and biological methods. Combustion, pyrolysis, liquefaction and gasification are the four thermochemical methods. The biological hydrogen production methods are biophotolysis of water by algae and cyanobacteria (direct and indirect biophotolysis), biological water-gas shift reaction, photo-fermentation and dark-fermentation. The advantages and disadvantages of these processes as well as the microbes involved in biological hydrogen production are given in Table **1**.

Table 1: Biohydrogen Production Process and their Relative Advantages and Disadvantages.

Process	General reactions	Type of microorganisms	Advantages	Disadvantages
Photo fermentation	$CH_3COOH + 2H_2O +$ light $\rightarrow 4H_2 + 2CO_2$	Photosynthetic bacteria *Rhodobacter sphaeroides, Rhodobacter capsulatus, Rhodobacter sulidophilus, Rhodopseudomonas sphaeroides, Rhodopseudomonas palustris, Rhodopseudomonas capsulate, Rhodospirillum rubrum, Chromatium* sp. *Miami PSB 1071*	A wide spectral light energy and different waste materials: *e.g.* distillery effluents, waste *etc.* can be used by these bacteria	Light conversion efficiency is very low, only 1-5% O_2 is a strong inhibitor of hydrogenase
Dark fermentation	$C_6H_{12}O_6 + 2\ H_2O \rightarrow 4\ H_2 + 2\ CH_3COOH + 2\ CO_2$ (acetate) $C_6H_{12}O_6 \rightarrow 2\ H_2 + 2\ CH_3CH_2CH_2COOH + 2\ CO_2$ (butyrate)	Fermentative bacteria *Enterobacter aerogenes, Enterobacter cloacae, Clostridium butyricum, Clostridium pasterianum, Desulfovibrio vulgaris, Magashaera elsdenii, Citrobacter intermedius* and *Escherichia coli*	It can produce H_2 all day long without light from varieties of carbon sources It produces valuable metabolites such as butyric, lactic and acetic acids as by products It is anaerobic process, so there is no O_2 limitation problem	Relatively lower achievable yields of H_2 As yields increase H_2 fermentation becomes thermodynamically unfavourable Product gas mixture contains CO_2 which has to be separated
Biological water-shift reaction	$CO + H_2O \rightarrow CO_2 + H_2$, $\Delta G^o = -20$ kJ mol^{-1}	Photoheterotrophic bacteria (Superfamily of Rhodospirilillaceae, *Rhodopseudomonas gelatinosa, Rhodopspirillum rubrum)*	Minimum energy is required and low process cost compared to chemical catalysts	The used of CO/synthesis gas for hydrogen production is a drawback compared to other biological processes
Hybrid system	a. $C_6H_{12}O_6 + 2H_2O \rightarrow 4H_2 + 2CH_3COOH + 2\ CO_2$ b. $2CH_3COOH + 4H_2O \rightarrow 8H_2 + 4CO_2$	Fermentative bacteria Photosynthetic bacteria	Can achieve high overall yield of H_2	Suitable microbial strain, process modification, efficient bioreactor are needed

On-going research have shown that biohydrogen can be produced from variety of renewable resources. However, many of these studies have been demonstrated in laboratory scale and the practical, large scale application needs to be demonstrated. The concept of biohydrogen production and its utilization is given in Fig. **3**.

Microorganisms for Biohydrogen Production

A large number of different taxonomic and physiological different microbial species can produce hydrogen (Table **1**). The biological species involved in hydrogen can be divided into two categories: photosynthetic and dark fermentators, namely, green algae, cyanobacteria heterocystous, cyanobacteria non-heterocystous, photosynthetic bacteria and fermentative bacteria (refer to related chapters in this eBook)

Biological systems provide a wide range of approaches to generate hydrogen that include biophotolysis (direct and indirect biophotolysis) photo-fermentation, dark-fermentation and two stage dark/photofermentation (hybrid system). These approaches also involve many taxonomically diverse species, a variety of enzymes and several metabolic pathways and processes (Tables **1** and **2**) [1]. The favoured microorganisms are facultative or obligate anaerobic and phototrophic bacteria depending on the fermentation type selected. The microorganisma are either utilised as axenic or monocultures or in consortia. Co-culture and sequential culture systems are also possible. For example, biohydrogen production can be realized by anaerobic and photosynthetic microorganisms using carbohydrate and non-toxic raw materials. All biohydrogen production processes are controlled by hydrogen-producing enzymes, such as hydrogenase and nitrogenase.

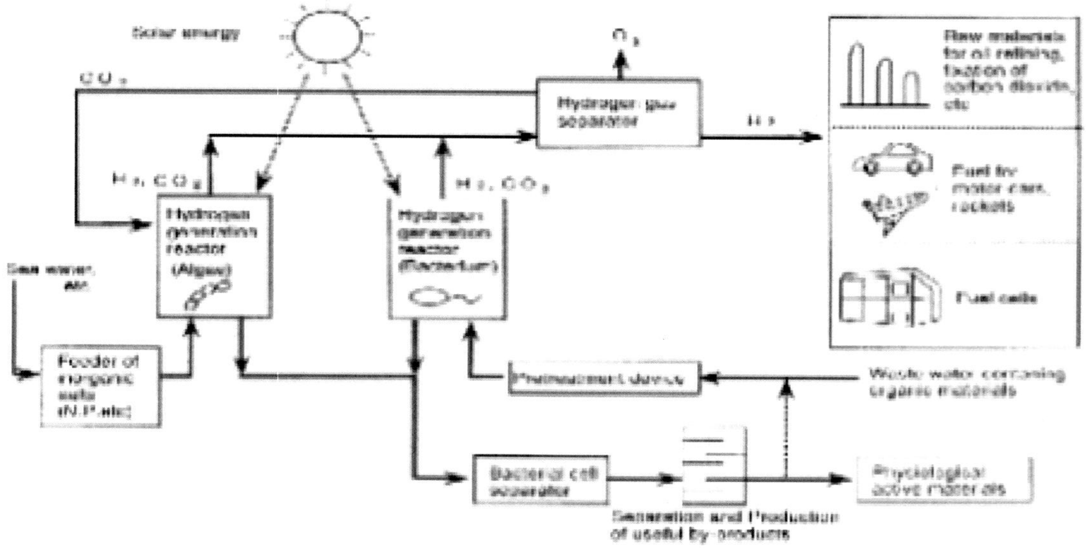

Figure 3: Schematic concept of biological hydrogen production and utilization [19].

Table 2: Classification and Examples of Microorganisms that have been Studied for Production of Hydrogen [1].

Classification	**Microorganisms**
Green algae	*Scenedesmus obliquus, Chlamydomonas reinhardii* and *Chalmydomonas moewusii*
Cyanobacteria (heterocystous)	*Anabaena azollae, Anabaena CA, Anabaena cylindrical, Nostoc muscorum, Nostoc spongiaeforme* and *Westiellopsis prolifica*
Cyanobacteria (non-heterocystous)	*Plectonema boryanum, Oscillotoria limnetica, Synechococcus* sp., *Aphanothece halophytico, Mastidocladus laminosus,* and *Phormidium valderianum*
Photosynthetic bacteria	*Rhodobacter sphaeroides, Rhodobacter capsulatus, Rhodobacter sulfidophilus, Rhodopseudomonas sphaeroides, Rhodopseudomonas palustris, Rhodopseudomonas capsulata, Rhodospirillum rubrum, Chromatium* sp., *Miami PSB 1071, Chlorobium licola, Thiocapsa roseopersicina* and *Halobacterium halobium*
Fermentative bacteria	*Enterobacter aerogenes, Enterobacter cloacae, Clostridium butyricum, Clostridium pasterianum, Desulfovibrio vulgaris, Magashaera elsdenii, Citrobacter intermedius* and *Escherichia coli*

ENZYMES INVOLVED IN BIOHYDROGEN PRODUCTION

Fundamentally, the processes for biohydrogen depends on the presence of hydrogen-producing enzymes. The quantity or inherent activities of these enzymes could limit the overall process of hydrogen production. There are two types of enzymes known as hydrogenase and nitrogenase that mediate biohydrogen production. Both nitrogenase and hydrogenase are sensitive to oxygen, hence, anaerobic condition is favourable for biohydrogen production.

Nitrogenase: Nitrogenase produces hydrogen as by-product of nitrogen fixation by reducing molecular nitrogen into ammonia. The reaction is also associated with the consumption of 16 mol of Adenosine triphosphate (ATP) as shown in Eq. (1) [20]. With the intention of maintaining the nitrogen-fixing reaction in the desired direction, nitrogenase produces hydrogen in excess, which then diffuses out of the cells,

$$N_2 + 8H^+ + 8e- + 16\ ATP \longrightarrow 2NH_3 + H_2 + 16\ ADP + 16\ P_i \quad \mathbf{(1)}$$

where, ADP and P_i refer to adenosine diphosphate and inorganic phosphate, respectively.

In the absence of nitrogen, nitrogenase can produce molecular hydrogen from protons (hydrogen ions, H^+) under anaerobic conditions with associates of ATP consumption (Eq.2). The chemical reaction that yields hydrogen by nitrogenase-based system [21] is

$$2e^- + 2H^+ + 4\ ATP \longrightarrow H_2 + 4\ ADP + 4\ P_i \quad \mathbf{(2)}$$

where, ADP and P_i refer to adenosine diphosphate and inorganic phosphate, respectively.

Hydrogenase: Another, key enzyme in biohydrogen production is hydrogenase. This enzyme catalyses the formation and decomposition of hydrogen [22]. Hydrogenases exist in dark fermentative and also a number of photosynthetic microorganisms. Hydrogenases can be classified into two categories, uptake hydrogenases and reversible hydrogenases. The enzymes are responsible for reversibly catalyzing the reaction (Eq.3) as follows:-

$$\text{Hydrogenase } H_2 \leftrightarrow 2e^- + 2H^+ \quad \mathbf{(3)}$$

Hydrogenase is generally considered to be responsible for dark hydrogen production alone. However, it also may contribute to photobiological hydrogen production. The properties of nitrogenase and hydrogenase are shown in Table **3**.

Table 3: Comparison between the Properties of Nitrogenase and Hydrogenase Enzymes of Hydrogen Production [23]

Property	**Nitrogenase**	**Hydrogenase**
Substrates	ATP, H^+ or N_2, electrons	H^+, H_2
Products	H_2 or NH_4^+, CO_2	ATP, H^+, H_2, electrons, CO_2
Number of proteins	Two (Mo-Fe and Fe)	One
Metal components	Mo, Fe	Ni, Fe, S
Optimum temperature	30°C	55°C(*Rs. rubrum*) 70°C (*R. capsulatus*),
Optimum pH	7.1 – 7.3	6.5- 7.5 (*R. sulfidophilus*)
Inhbitors	N_2 (of H_2 production) NH_4^+, O_2 high N to C ratio	CO,EDTA (ethylenediaminetetraacetic acid), O_2, presence of organic compound
Stimulators	Light	H_2 (*R. sphaeroides*), Absence of organic compounds (*Rs. rubrum, R. capsulatus*)

BIOHYDROGEN PRODUCTION *VIA* FERMENTATION

Biohydrogen production by dark (facultative and obligate anaerobic bacteria) or light (phototrophic bacteria) fermentation and a hybrid of the two using suitable biomass as substrate are possible. These systems have their strengths and weakness (Table **1**).

Dark Fermentation

Biohydrogen production from biomass and industrial wastes by dark fermentation was reviewed by Chong *et al.* [24]. Hydrogen can be produced by anaerobic fermentative bacteria (strict anaerobes and facultative anaerobes) as well as some microalgae (green algae), grown in the dark on carbohydrate-rich substrate [10]. They have dual functions of simultaneously generating hydrogen and reducing the pollution load of the substrate used in the production. In this process, carbon dioxide and a small amount of methane, carbon monoxide and hydrogen sulphide are produced [25]. The dark fermentation processes for hydrogen production share many common features with methanogenic anaerobic digestion, which is widely used in wastewater treatment. The most common feature is the separation of the gaseous products from the treated water [26].

Dark fermentation processes produce a mixed biogas containing primarily hydrogen and carbon dioxide, CO2; and may contain lesser amounts of methane, CH_4; carbon monoxide, CO and hydrogen sulphide, H_2S. Dark microbial hydrogen production is driven by the anaerobic metabolism of the key intermediate, pyruvate. Carbohydrates are the preferred substrates for hydrogen producing fermentation. Glucose yields different quantities of hydrogen depending on the fermentation pathway and end product(s). In dark fermentation using strict anaerobic bacteria, a theoretical maximum of 4 moles H_2 per mole of glucose is obtained:

$$C_6H_{12}O_6 + 2H_2O \longrightarrow 2CH_3COOH + 4H_2 + 2CO_2$$

However, if the main producers are facultative anaerobes such as *E. coli* a maximum of 2 moles of H_2 per mole of glucose is obtained:

$$C_6H_{12}O_6 + 2H_2O \longrightarrow 2CH_2CH_2CH_2COOH + 2H_2 + 2CO_2$$

In a typical anaerobic digestion process, methane (Fig. **4**) and several solvents (*e.g.* acetone, butanol, ethanol) are produced as the end products. The process generally involves three groups of microorganisms which coexist and work as a consortia. They are acetogens which degrade various carbon sources to produce volatile fatty acids (VFA) and hydrogen; acetogens that degrade some VFA (*e.g.* propionate, butyrate) to produce acetate and hydrogen and finally, acetate and hydrogen are used to produce methane by methanogens [27]. Hydrogen is produced as an intermediate product of methane production. Thus, to produce hydrogen as the main product, methanogenesis has to be blocked. Pretreatment such as heat treatment, chemical treatment and pH treatment will block or repress methanogenic activity during biohydrogen production [27-30].

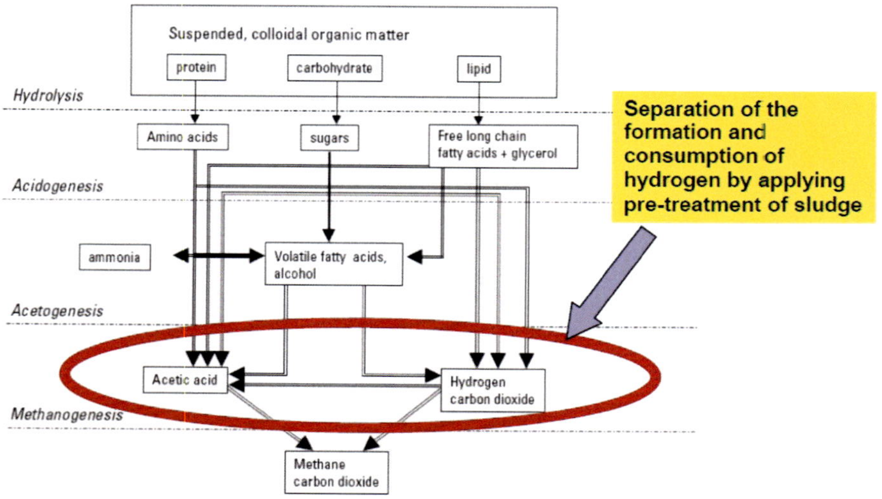

Figure 4: Simplified schematic representation of the anaerobic degradation by facultative and obligate anaerobic bacteria [31].

Anaerobic fermentation can be operated at mesophilic (25-40 °C), thermophilic (40-65°C), extreme thermophilic (65-80°C) or hyperthermophilic (> 80°C) temperatures. Extreme thermophilic anaerobic bacteria have been selected because of their high yield with respect to hydrogen production. The yield is about 83-100% of the theoretical maximum value of 4 moles H_2/mole glucose [32]. Further, biohydrogen production at high temperatures using mixed thermophilic bacteria has been identified as a potential process favorable to reaction kinetics while avoiding contamination by hydrogen - consuming bacteria. *Thermoanaerobacterium* sp has been identified as an effective hydrogen producing bacteria [33]. O-Thong *et al.* [34] isolated *Th. thermosaccharolyticum* PSU-2 from anaerobically digested palm oil mill sludge. This bacterium showed a high yield of hydrogen at optimum pH 6.25 and at 60°C. Maximum hydrogen production with a yield of 2.53 mol H_2/ mol hexose was obtained during late exponential phase at 8th to 16th hour and decreased when fermentation time was prolonged. This could be due to accumulation of organic acid.

In contrast, the maximum of 2 moles of H_2/mole of glucose is produced by the strict anaerobic bacteria such as *Clostridia* and the facultative anaerobes (e.g *E. coli*) gives a hydrogen yield of less than two (2). Among the fermentative microorganisms, *Clostridium butyricum* has been studied most extensively because it is a fast growing bacteria. But it is extremely sensitive to oxygen (strict anaerobe). A study carried out by Lin *et al.* [35] on biohydrogen production by an obligate anaerobic *Clostridium beijerinckii* L9 gave the highest hydrogen production yield of 2.81 mmol H_2/ mmol glucose. However, Chen *et al.* [36] reported about 209 mL H_2/L/h was produced by another strain of obligate anaerobic *Clostridium* - *Clostridium butyricum* CGS5 grown in a synthetic media (DMI). Facultative anaerobes are less sensitive to oxygen, and sometimes recover the hydrogen production activity from accidental oxygen damage for themselves by rapidly depleting the oxygen in the broth [37]. A good example of facultative bacteria is *Enterobacter cloacae* IIT-BT 08, which has been reported to have a high specific H_2 production rate of 29.6 mmol H_2/g cell/h and a high conversion yield of 2.2 mmol H_2/mmol glucose [38].

Photofermentation

Purple non sulfur (PNS) bacteria constitute a non-taxonomic group of versatile microorganisms which can grow as photoheterotrophs, photoautotrophs or chemoheterotrophs-switching from one mode to another depending on available conditions such as degree of anaerobiosis, availability of carbon source (CO_2 for autotrophic growth, organic compounds for heterotrophic growth) and availability of a light source (needed for phototrophic growth) [39]. Photosynthetic bacteria offer the advantage combining organic waste disposal problem with the production of clean fuel hydrogen. Many studies are focusing on purple non sulphur species such as *Rhodobacter sphaeroides* [22,40], *Rhodobacter capsulatus* [41] and *Rhodopseudomonas palustris* [16,36].

Photosynthetic bacteria have long been studied for their capacity to produce significant amounts of hydrogen through the action of their nitrogenase system [42] using solar energy and organic acids or biomass. Under anaerobic condition, these bacteria specifically photoherotrophic bacteria (purple non sulphur bacteria) and cyanobacteria, are able to use simple organic acids like acetic acid as electron donor. These electrons are transported to the nitrogenase by ferredoxin (Fd) using energy in the form of ATP. On the other hand, these bacteria are also able to fix nitrogen. The overall reaction of hydrogen production can be given as follow:

$$C_6H_{12}O_6 + 12H_2O \xrightarrow{\text{light energy}} 12H_2 + 6CO_2$$

In the presence of nitrogen, nitrogenase can also reduce proton (hydrogen ion, H^+) into hydrogen gas again using extra energy in the form of ATP. The overall energy consuming reaction:

$$N_2 + 8H^+ + 8e- + 16\ ATP \longrightarrow 2NH_3 + H_2 + 16\ ADP + 16\ P_i$$

Nitrogenase enzyme catalyses nitrogen fixation by photoheterotrophic bacteria (purple non sulphur bacteria) and cyanobacteria as shown in Eq. (1). Nitrogenase also catalyses the evolution of hydrogen, particularly in the absence of molecular nitrogen and oxygen [39,43]. This explains why photobioreactors are usually operated under anaerobic conditions free of nitrogen, with illumination and limiting concentrations of nitrogen sources [20,44]. The overall biochemical pathways for the photofermentation process in the absence of molecular nitrogen can be expressed as shown in Fig. **5**; where ferredoxin (Fd) acts as electron carrier in the presence of nitrogenase in the cell membrane [1,45]. Further, it has been reported that four mutant strains of the photosynthetic bacterium *Rhodopseudomonas palustris* produced hydrogen constitutively, even in the presence of ammonium, a condition in which wild-type cells do not accumulate detectable amounts of hydrogen [46].

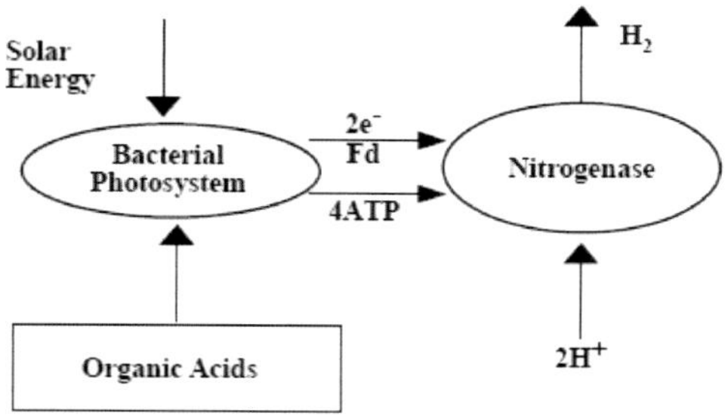

Figure 5: Schematic overall biochemical pathways for photofermentation process [47].

There are several studies on photofermentation biohydrogen production by purple non sulfur bacteria (PNB)namely *Rhodobacter sphaeroides*, *Rhodobacter capsulatus, Rhodopseudomonas palustris etc.* These studies were carried out by growing selected PNB in synthetic medium or wastewater as substrate. According to Li and Fang [48], *Rubrivivax gelatinosus* L31could produce hydrogen from carbohydrates, including glucose, sucrose and starch as well as from lactate and malate. Fang *et al.* [49] investigated photobiological hydrogen production from individual substrate of acetate and butyrate by *Rhodobacter capsulatus*. The maximum hydrogen yield produced by *Rhodobacter capsulatus* was 2.5 mol-H_2/mol-acetate and 3.7 mol-H_2/mol-butyrate. Sunita and Mitra [50] studied the production of hydrogen from sewage by *Rhodopseudomonas* sp. and the maximum production rate was 0.003 L H_2/L/h. The production of photofermentation hydrogen by using *Rhodobacter sphaeroides* O.U.001 grown in olive mill wastewater [51], sugar refinery wastewater [52] and distillery wastewater [53] has been reported. The maximum hydrogen production rate was 0.009 L H_2/L/h, 0.005 L H_2/L/h and 0.022 L H_2/L/h respectively. The production rate of 0.016-0.024 L H_2/L/h, 0.012-0.02 L H_2/L/h and 0.008-0.002 L H_2/L/h were obtained from lactic acid production wastewater, yogurt waste and whey wastewater respectively by using *Rhodosprillum rubrum* S1 [54]. However, on comparing the various studies on photofermentation hydrogen production, the highest production rate recorded at 0.059 L H_2/L/h was by *Rhodobacter sphaeroides* RV grown in tofu wastewater as the substrate [55].

Biological Water-Gas Shift Reaction

Certain photoheterotrophic bacteria within superfamily Rhodospirillaceae, such as *Rhodospirillum rubrum* can grow in the dark using carbon monoxide, CO as the sole carbon source to generate ATP (adenosine triphosphate). This reaction is coupled to the oxidation of CO and the reduction of H^+ to H_2. Thermodynamics of the reactions are very favourable to CO-oxidation and hydrogen synthesis. The reaction, however, is mediated by proteins coordinated in an enzymatic pathway.

Biohydrogen production *via* biological water-gas shift reaction is still under laboratory scale and only a few studies have been reported. The common objectives of these studies were to identify the suitable microorganisms that had high CO uptake and to estimate the hydrogen production rate [47]. Younesi *et al.* [56] had reported that bioconversion of synthesis gas (syngas, 55% CO, 20% H_2, 15% Ar, 10% CO_2) to hydrogen *via* water-gas shift reaction by *Rhodopirillum rubrum* produced 16 mmol H_2/g cell/h and the hydrogen yield was 87%. In a similar study, the same species when grown in a medium with an acetate concentration of 1.5 g/L produced 2.35 mmoles H_2 and the yield of hydrogen was 98% [57].

Hybrid System

Methods for producing hydrogen by anaerobic dark fermentation or photofermentation of wastewater are known for not only producing clean energy but for simultaneous reduction in the pollution load of the biowastes. The dark fermentation is, however, an incomplete oxidation as the organic matter is not completely oxidized to CO_2 but to intermediate compounds like acetic acid or lactate. Further, many anaerobic bacteria cannot utilize these organic compounds as carbon sources. The complete oxidation of one mole of glucose in fermentation process yields 12 moles of hydrogen.

$$C_6H_{12}O_6 + 6H_2O \longrightarrow 2H_2 + 6CO_2 \qquad \Delta G^o = +3.2\ kJ$$

In the absence of external energy (dark fermentation) the glucose will oxidize into other by-products namely acetate (organic acid) together with 4 moles of hydrogen and carbon dioxide by fermentative bacteria (Table **1**). Thereafter, the accumulated organic acids can be further treated phototrophically using photosynthetic bacteria to produce more hydrogen. Thus, an approach of coupling the dark fermentation and photofermentation processess in two subsequent stages are recommended to enhance the hydrogen production and become an efficient way for biological hydrogen production [32]. The schematic diagram of these two stages are shown in Fig. **6**. Several studies were carried out by applying hybrid systems in order to produce hydrogen from different types of substrates; olive mill effluent [58], municipal solid wastes [59], sucrose [60], starch-manufacturing wastes [61] and sugar [62].

Palm Oil Mill Effluent - A Potential Substrate for Biohydrogen Production: A Case Study

The oil palm tree *Elaeis guinessis* which originated from West Africa is currently a commercial oil yielding plantation crop that is grown in large hectares across the tropical belt of three continents – Asia, Africa and South America. It was first introduced to Malaysia in early 1870's and since then has been domesticated as an agricultural crop. The phenomenal growth of palm oil industry since 1960's has turned Malaysia as one of the largest producers and exporters of palm oil in the world. In 2006, about 15.9 million tonnes of palm oil was produced from its 4.17 million hectares of oil palm [63]. In many countries, the processing for the production of palm oil is carried out throughout the year. The biomass residues and wastewaters arising from this industry are renewable, cheap and available in abundance and has been extensively researched for various applications including as a substrate biohydrogen production. In this chapter, the potential of palm oil mill effluent (POME) as a suitable substrate will be highlighted.

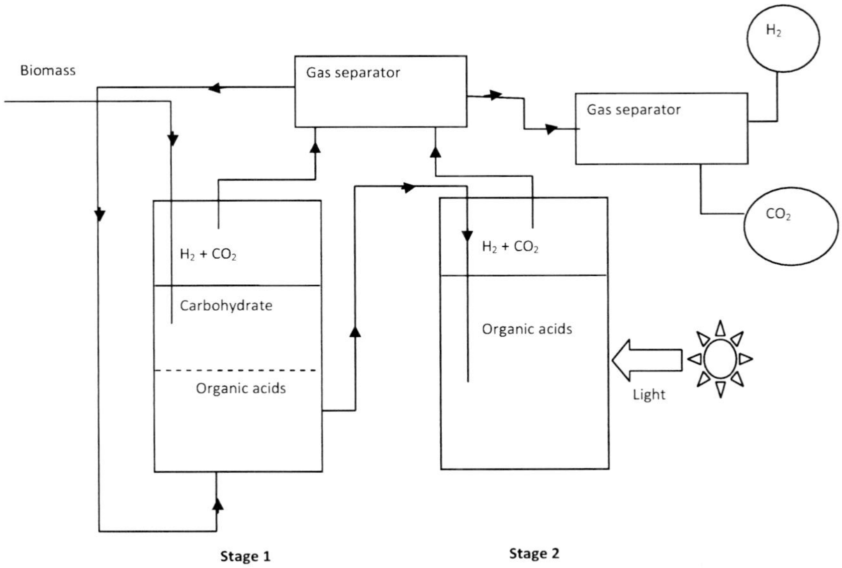

Figure 6: Hybrid system concept for biohydrogen production [32].

Palm Oil Mill Effluent (POME) for Hydrogen Production

In 2007, there were 434 mills operating with the production capacity of 93.6 million tonnes fresh fruit bunches (FFB) per year [64]. Palm oil mill effluent (POME) is generally produced during a series of palm oil milling processess to yield crude palm oil (CPO): steam sterilization of fresh fruit bunches (FFB), fruit-stalk separation by stripping, digestion of stripped fruits, oil expression by means of screw press and clarification of palm oil. It is estimated that for very tonne of CPO produced, 5 -7.5 tonnes of water are required, and more than 50% of the water will end up as POME [65]. POME contains mainly cellulose and lignocellulosic material with mixture of carbohydrates and oil. The biological oxygen demand (BOD) of POME is approximately 20 g/L and the chemical oxygen demand (COD) is at the range of 60 to 100 g/L. The total carbohydrate content is from 12 to 30 g/L [66]. Therefore, the processing of 20 tonnes of FFB, releases 100 tonnes of POME that yield about 400 m^3 of biogas [67].

The POME is treated by anaerobic digestion to reduce the pollution load and meet the Department of Environment's Standards for discharge into waterways. Biogas/methane is produced. Further, oil palm biomass is converted into various types of value-added products *via* several conversion technologies that are readily available. Oil palm biomass can be transformed into three types of biomass energies; *i.e.* bio-products (materials and adsorbent), bio-fuels (bioethanol, biodiesel and biomethanol) and bio-power (direct-fired of biomass, hydrogen, methane *etc.*) [67]. Empty Fruit Bunches (EFB) and fruit mesocarp fibres are mostly sent back to the plantation areas for mulching to release nutrients for the palm as well as for soil conservation purposes. Some of the fruit fibres and the kernel shell are burnt in boilers to generate steam and electricity for the mills. This system is called cogeneration system which uses biomass residue as fuel in the boiler. The lignocellulosic biomass, *i.e.* the EFB, can be converted to hydrogen gas through gasification process [67].

POME can be an ideal substrate for hydrogen production. In the Malaysian scenario, research and development on producing hydrogen from renewable sources began in early 2000. The research in this area continues until today by taking advantage of abundant waste, particularly POME [18,31,68-70]. In these studies, the dark fermentation method was used with microflora and axenic cultures. Biohydrogen production through photofermentation with photosynthetic bacteria is also possible because of their capacity to produce significant amounts of hydrogen. Photosynthetic bacteria are found to be as promising as other microbial system due to their high substrate to product conversion yield, lack of oxygen-evolving activity (which is desirable for biohydrogen production), ability to use a wide wavelength of light, and capability to consume organic substrates (derived from wastes) for hydrogen production that also helps in bioremediation processes (wastewater treatment) [60]. Photosynthetic bacteria carry out anoxygenic photosynthesis using organic compounds and reduced sulphur compounds electron donors, which are categorized as non-sulfur and sulfur photosynthetic bacteria respectively [71].

Biohydrogen production using POME as substrate and microflora from a variety of sources has been reported by several researchers: natural anaerobic microflora from cow dung [31], microflora from the POME sludge, sludge compost, CREST compost [18,72] and *Thermoanaerobacterium*-rich sludge [17,34]. Further, high COD is reduced during the production of hydrogen. Vijaraghavan and Ahmad [31] in their studies reported that POME treatment was coupled with biohydrogen during anaerobic degradation process by a consortia of microflora from cow dung in an anaerobic contact filter reactor. The microflora was initially pretreated by pH adjustment coupled with heat treatment to kill or suppress the methanogenic and non-sporulating bacteria. The biogas generated was free from methane. At pH 5, maximum COD removal, biogas generation and hydrogen content were observed. The average biogas generation was found to be 0.42 L/g COD removal, with hydrogen content of 57 % at 7 days hydraulic retention time (HRT).

Morimoto *et al.* [72] studied the capability of three different microflora isolated from POME sludge, sludge compost containing a mixture of POME sludge and empty fruit bunch (EFB) and CREST compost containing a mixture of product of municipal wastes (kitchen and urban wastes) and cow manure. Hydrogen production was tested in an artificial wastewater containing 1% glucose, 0.2% yeast extract and 0.018 % magnesium chloride hexahydrate under anaerobic fermentation in a batch culture. All three microflora produced significant amounts of hydrogen. The maximum production yield of hydrogen was 2.1 mol/ mol glucose at a conversion rate of 0.137 L/ L medium/ h at 50°C obtained in sludge compost. O-Thong *et al.* [17,34] studied hydrogen production from POME using *Thermoanaerobacteriuma*-rich sludge in anaerobic sequencing batch reactor. The maximum hydrogen production was achieved at 6.33 L H_2/ L POME along with 55% COD reduction at 257 mg Fe^{2+}/L, a C/N ratio of 74 and C/P ratio of 559.

Biohydrogen was successfully produced by *Clostridium butryicum* E6, a strict anaerobic bacteria growing in POME [34]. The hydrogen gas percentage in the biogas produced ranged from 60 to 75 with an estimated 298 ml of hydrogen produced per gram of carbohydrate consumed by the bacteria under optimised conditions of growth. It was observed that environmental factors such as pH, temperature and COD loads significantly influenced biohydrogen production.

In another study, Jamil *et al.* [16] studied hydrogen production from raw POME *via* photofermentation process by *Rhodopseudomonas palustris* strain PBUM 001. The production of hydrogen was low because photosynthetic bacteria specifically purple non sulphur bacteria require organic acids to produce hydrogen. The raw POME contained low levels of organic acids and pre-treatment *via* dark fermentation can increase the organic acids required by these bacteria.

CONCLUSIONS

The production of biohydrogen in biowastes *via* fermentation using suitable microbes have been shown to be feasible in small scale studies. However, there is a need for more scale-up and optimisation studies to take these potential technologies from bench to commercial scale. The wastes utilised need to be identified based on locality – in other words, scout for major organic polluters in your neighbourhood. For example, POME has vast potential for tropical developing countries producing palm oil as a economic earner. The hydrogen production by dark fermentation shows a promising application comparing to the photo fermentation. However, the low hydrogen production can be solved with serial fermentation process from dark fermentation then followed by photofermentation process in a hybrid system. This will eventually increase the efficiency of hydrogen production from the organic wastes and thus pave the way for biohydrogen generated from biowastes to be a major contributor of renewable energy.

ACKNOWLEDGEMENTS

The authors thank their respective Universities and the Ministry of Science, Technology and Innovation, Malaysia and University of Malaya and University Putra Malaysia for the grants and scholarships for staff and postgraduate students that made this study possible.

REFERENCES

[1] Das D, Veziroglu TN. Hydrogen production by biological process: a survey of literature. Int J Hydrogen Energy 2001; 26: 13-28.
[2] Contreras A, Yigit S, Ozay K, Veziroglu TN. Hydrogen as aviation fuel: A comparison with hydrocarbon fuels. Int J Hydrogen Energy 1997; 10: 1053-60.
[3] Penner SS. Steps toward the hydrogen economy. Energy 2006; 31: 33-43.
[4] Momirlan M, Veziroglu TN. The properties of hydrogen as fuel tomorrow in sustainable energy system for a cleaner planet. Int J Hydrogen Energy 2005; 30: 795-802.
[5] Marban G, Valdes-Solis T. Towards the hydrogen economy? Int J Hydrogen Energy 2007; 32:1625-37.
[6] Ewan BCR, Allen RWK. A figure of merit assessment of the routes to hydrogen. Int J Hydrogen Energy 2005; 30: 809-19.
[7] Kapdan IL, Kargi F. Bio-hydrogen production from waste materials. Enzyme Microb Tech 2006; 38: 569-82.
[8] Florin N, Harris A. Hydrogen production from biomass. Environmentalist 2007; 27: 207-15.
[9] Vijayaraghavan K, Soom AM, D. Trends in biohydrogen generation – a review. J Integr Environ Sci 2006; 3: 255-71.
[10] Nath K, Das D. Improvement of fermentative hydrogen production: various approaches. Appl Microbiol Biotechnol 2004; 65: 520-9.

[11] Mohan SV, Bhaskar YJ, Krishna PM, Rao NC, Babu VL, Sarma PN. Biohydrogen production from chemical wastewater as substrate by selectively enriched anaerobic mixed consortia: Influence of fermentation pH and substrate composition. Int J Hydrogen Energy 2007; 32: 2286-95.

[12] Fang HHP, Li CL, Zhang T. Acidophilic biohydrogen production from rice slurry. Int J Hydrogen Energy 2006; 31: 683-92.

[13] Ginkel SV, Oh SE, Logan BE. Biohydrogen gas production from food processing and domestic wastewaters. Int J Hydrogen Energy 2005; 30: 1535-42.

[14] Boligger R, Zurrer H, Bachofen R. Photoproduction of molecular hydrogen from waste water of a sugar refinery by photosynthetic bacteria. Appl Microbiol Biotechnol 1985; 23:147-51.

[15] Zhu H, Ueda S, Asada Y, Miyake J. Hydrogen production as a novel process of wastewater treatment – studies on tofu wastewater with entrapped *Rhodobacter sphaeroides* and mutagenesis. Int J Hydrogen Energy 2002; 27: 1349-57.

[16] Jamil Z, Annuar MSM, Ibrahim S, Vikineswary S. Optimization of phototrophic hydrogen production by *Rhodopseudomonas palustris* PBUM 001 *via* statistical experimental design. Int J Hydrogen Energy 2009; 34: 7502-12.

[17] Yacob S, Shirai Y, Hassan MA, Wakisaka M, Subash S. Start-up operation of semi-commercial closed anaerobic digester for palm oil mill effluent treatment. Process Biochem 2006; 41: 962-4.

[18] Atif AAY, Fakru'l-Razi A, Ngan MA, Morimoto M, Iyuke SE, Veziroglu TN. Fed batch production of hydrogen from palm oil mill effluent using anaerobic microflora. Int J Hydrogen Energy 2005; 30: 1393-7.

[19] Miyake J, Miyake M, Asada. Biotechnological hydrogen production: research for efficient light energy conversion. J Biotechnol 1999; 70: 89-101.

[20] Akkerman I, Janssen M, Rocha J, Wijffels RH. Photobiological hydrogen production: photochemical efficiency and bioreactor design. Int J Hydrogen Energy 2002; 27: 1195-208.

[21] Hallenback PC, Benemann JR. Biological hydrogen production: fundamentals and limiting processes. Int J Hydrogen Energy 2002; 27: 1185-93.

[22] Kovacs KL, Maroti G, Rakhely G. A novel approach for biohydrogen production. Int J Hydrogen Energy 2006; 31: 1460-8.

[23] Koku H, Eroğlu İ, Gündüz U, Yücel M, Türker L. Aspects of the metabolism of hydrogen production by *Rhodobacter sphaeroides.* Int J Hydrogen Energy (2002); 27: 1315-29.

[24] Chong ML, Vikineswary S, Shirai S, Hassan MA. Biohydrogen production from biomass and industrial wastes by dark fermentation. Int J Hydrogen Energy 2009; 34. 3277-87.

[25] de Vrije T, Claassen PAM. Dark hydrogen fermentations. In: ReithJH, Wijffels RH Barten H, Eds. Bio-methane and Bio-hydrogen: Status and Perspectives of Biological Methane and Hydrogen Production. The Netherlands: Dutch Biological Hydrogen Foundation, 2003; pp. 103-23.

[26] Angenent LT, Karim K, Al-Dahhan MH, Wrenn BA, Espinosa RD. Production of bioenergy and biochemicals from industrial and agricultural wastewater. Trends Biotechnol 2004; 22: 477-85.

[27] Hu B, Chen S. Pretreatment of methanogenic granules for immobilized hydrogen fermentation. Int J Hydrogen Energy 2007; 32: 3266-73.

[28] Zhu H, Beland M. Evaluation of alternative methods of preparing hydrogen production seeds from digested wastewater sludge. Int J Hydrogen Energy 2006; 31: 1980-8.

[29] Mohan SV, Bhaskar YV, Krishna PM, Rao NC, Babu VL, Sarma PN. Biohydrogen production from chemical wastewater as substrate by selectively enriched anaerobic mixed consortia: Influence of fermentation pH and substrate composition. Int. J Hydrogen Energy 2007; 32: 2286-95.

[30] Mohan SV, Babu VL, Sarma PN. Effect of various pretreatment methods on anaerobic mixed microflora to enhance biohydrogen production utilizing dairy wastewater as substrate. Bioresource Technol 2008; 99: 59-67.

[31] de Mes TZD, Stams AJM, Reith JH, Zeeman G. Methane production by anaerobic digestion of wastewater and solid wastes. In: Reith JH, Wijffels RH, Barten H, Eds. Bio-methane & Bio-hydrogen: Status and Perspectives of Biological Methane and Hydrogen Production. The Netherlands: Dutch Biological Hydrogen Foundation, 2003; pp. 58-102.

[32] Vijayaraghavan, K., and Ahmad, D. (2006) Biohydrogen generation from palm oil mill effluent using anaerobic contact filter. Int J Hydrogen Energy 2006; 31: 1284-91.

[33] O-Thong S, Prasertsan P, Intrasungkha N, Dhamwichukorn S, Birkeland NK. Improvement of biohydrogen production and treatment efficiency on palm oil mill effluent with nutrient supplementation at thermophilic condition using an anaerobic sequencing batch reactor. Enzyme Microb Tech 2007; 41: 583-90.

[34] O-Thong S, Prasertsan P, Karakashev D, Angelidaki I. Thermophilic fermentative hydrogen production by the newly isolated *Thermoanaerobacterium thermosaccharolyticum* PSU-2. Int J Hydrogen Energy 2008; 33: 1204-14.

[35] Lin PY, Whang LM, Wu YR, Ren WJ, Hsiao CJ, Li SL, Chang JS. Biological hydrogen production of the genus *Clostridium*: Metabolic study and mathematical model stimulation. Int J Hydrogen Energy 2007; 32: 1728-35.

[36] Chen CH, Lu WB, Wu JF, Chang JS. Enhancing phototrophic hydrogen production of *Rhodopseudomonas palustris via* statistical experimental design. Int J Hydrogen Energy, 2007; 32: 940-9.

[37] Oh YK, Seol EH, Lee EY, Park S. Fermentative hydrogen production by a new chemoheterotrophic bacterium *Rhodopseudomonas palustris* P4. Int J Hydrogen Energy 2002; 27: 1373-9.

[38] Kumar N, Das D. Enhancementof hydrogen production by *Enterobacter cloacae* IIT-BT 08. Process Biochem 2000; 35: 589-93.

[39] Basak N, Das D. The prospect of purple non sulfur (PNS) photosynthetic bacteria for hydrogen production: the present state of the art. World J Microbiol Biotechnol 2007; 23: 31-42.

[40] Koku H, Eroğlu İ, Gündüz U, Yücel M, Türker L. Kinetics of biological hydrogen production by photosynthetic bacterium *Rhodobacter sphaeroides* O.U. 001. Int J Hydrogen Energy 2003; 28: 381-8.

[41] He D, Bultel Y, Magnin JP, Roux C, Willison JC. Hydrogen photosynthesis by *Rhodobacter capsulatus* and its coupling to a PEM fuel cell. Int. J Hydrogen Energy 2005; 30: 19-23.

[42] Tan JW, Cheah WL, Lai YW, Chua KH, Thong KL, Nirmala Devi A, Raha Abdul Rahim and Vikineswary S. Development of a PCR assay for the detection of *nifH* and *nifD* genes in indigenous photosynthetic bacteria. Int J Hydrogen Energy 2009; 34: 7538-41.

[43] Boligger R, Zurrer H, Bachofen R. Photoproduction of molecular hydrogen from waste water of a sugar refinery by photosynthetic bacteria. Appl Microbiol Biotechnol 1985; 23:147-51.

[44] Chen CY, Lee CM, Chang JS. Hydrogen production by indigenous photosynthetic bacterium *Rhodopseudomonas palustris* WP3-5 using optical fiber-illuminating photobioreactors. Biochem Eng J 2006; 32: 33-42.

[45] Hallenback PC, Benemann JR. Biological hydrogen production: fundamentals and limiting processes. Int J Hydrogen Energy 2002; 27: 1185-93.

[46] Rey FE, Heiniger EK, Harwood CS. Redirection of metabolism for biological hydrogen production. Appl Environ Microbiol 2007; 73: 1665-71.

[47] Ni M, Leung DYC, Leung MKH, Sumathy K. An overview of hydrogen production from biomass. Fuel Process Technol 2006; 87: 461-72.

[48] Li RY, Fang HHP. Hydrogen production characteristics of photoheterotrophic *Rubrivivax gelatinosus* L31. Int J Hydrogen Energy 2008; 33: 974-80.

[49] Fang HHP, Liu H, Zhang T. Phototrophic hydrogen production from acetate and butyrate in wastewater. Int J Hydrogen Energy 2005; 30: 785-93.

[50] Sunita M, Mitra CK. Photoproduction of hydrogen by photosynthetic bacteria from sewage and wastewater. J Biosci 1993; 18: 155-60.

[51] Eroglu E, Gunduz U, Yucel M, Turker M, Eroglu I. Photobiological hydrogen production by using olive mill wastewater as a sole substrate source. Int J Hydrogen Energy 2004; 29: 163-71.

[52] Yetis M, Gunduz U, Eroglu I, Yucel M, Turker L. Photoproduction of hydrogen from sugar refinery wastewater by *Rhodobacter sphaeroides* O.U. 001. Int J Hydrogen Energy 2000; 25: 1035-41.

[53] Sasikala K, Ramana CV, Rao PR. Photoproduction of hydrogen from the wastewater of a distillery by *Rhodobacter sphaeroides* O.U.001. Int J Hydrogen Energy 1999; 17: 23-7.

[54] Zurrer H, Bachofen R. Hydrogen production by the photosynthetic bacterium *Rhodospirillum rubrum*. Appl Environ Microbiol 1979; 37: 789-93.

[55] Zhu H, Suzuki T, Tsygankov AA, Asada Y, Miyake J. Hydrogen production from tofu wastewater by *Rhodobacter sphaeroides* immobilized in agar gels. Int J Hydrogen Energy 1999; 24: 305-10.

[56] Younesi H, Najafpour G, Ismail KSK, Mohamed AR, Kamaruddin AZ. Biohydrogen production in a continuous stirred tank bioreactor from synthesis gas by anaerobic photosynthesis bacterium: *Rhodopirillum rubrum*. Bioresource Technol 2008; 99: 2612-9.

[57] Najafpour G, Younesi H, Mohamed AR. Effect of organic substrate on hydrogen production from synthesis gas using *Rhododspirillum rubrum* in batch culture. Int J Hydrogen Energy 2004; 21: 123-30.

[58] Eroglu E, Eroglu I, Gunduz U, Turker L, Yucel M. Biological hydrogen production from olive oil wastewater with two-stage processes. Int J Hydrogen Energy 2006; 31: 1527-35.

[59] Fascetti E, D'Addario E, Todini O, Robertiello A. Photosynthetic hydrogen evolution with volatile organic acids derived from the fermentation of source selected municipal solid wastes. Int J Hydrogen Energy 1998; 23: 753-60.

[60] Tao Y, Chen Y, Wu Y, He Y, Zhou Z. High hydrogen yield from a two-step process of dark- and photo-fermentation of sucrose. Int J Hydrogen Energy 2007; 32: 200-6.

[61] Yokoi H, Maki R, Hirose J, Hayashi, S. Microbial production of hydrogen from starch-manufacturing wastes. Biomass Bioenerg 2002; 22: 389-95.

[62] Lee CM, Chen PC, Wang CC, Tung YC. Photohydrogen production using purple non sulfur bacteria with hydrogen fermentation reactor effluent. Int J Hydrogen Energy 2002; 27; 1309-13.

[63] MPOC (Malaysian Palm Oil Council). Oil palm-tree of life. Available online at <http://www.mpoc.org.my/) 2006) (accessed on May 2009).

[64] MPOB (Malaysia Palm Oil Board). Malaysian Oil Palm Statistics 2007. 27th Edn. MPOB. 2008.

[65] Ahmad AL, Ismail S, Bhatia S. Water recycling from palm oil mill effluent (POME) using membrane technology. Desalination 2003; 157: 87-95.

[66] Basiron Y, Chan KW. The oil palm and its sustainability. J Oil Palm Res 2004; 16: 1-10.

[67] Sumanthi S, Chai SP, Mohamed AR. Utilization of oil palm as a source of renewable energy in Malaysia. Renew Sust Energ Rev 2008; 12: 2404-21.

[68] Chung WC. Biohydrogen production from palm oil mill effluent as part of zero emission system in the palm oil industry. PhD Dissertation, 2005, University Putra Malaysia.

[69] Chong ML, Rahim AR, Shirai Y, Hassan MA. Biohydrogen production by *Clostridium butyricum* EB6 from palm oil mill effluent. Int J Hydrogen Energy 2009, 34: 764-71.

[70] Yusoff ZM, Hassan MA, Abd-Aziz S, Rahman AA. Start-up of biohydrogen production from palm oil mill effluent under non-sterile condition in 50 L continuous stirred tank reactor. Int J Agric Res 2009; 4: 163-8.

[71] Asada Y, Miyake J. (1999). Photobiological hydrogen production. J Biosci Bioeng.1999; 8: 1-6.

[72] Morimoto M, Atsuko M, Atif AAY, *et al.* Biological production of hydrogen from glucose by natural anaerobic microflora. Int J Hydrogen Energy 2004; 29: 709-13.

CHAPTER 8

Kinetics of Biohydrogen Production by Dark Fermentation Processes

Kaustubha Mohanty[1] and Debabrata Das[2*]

[1]*Indian Institute of Technology Guwahati, India and* [2]*Indian Institute of Technology Kharagpur, India*

Abstract: Various kinetic models were developed to describe the fermentative biohydrogen production in a batch system. These models took into account the effects of substrate degradation (and inhibition), temperature and pH on biomass growth as well as on the rate of biohydrogen production. This chapter summarizes various such models describing the effect of different parameters on the biological hydrogen production process. It was found that the modified Gompertz model was most suitable to describe the progress of biohydrogen formation process. Substrate inhibition was well described by the Andrew model, which is the modified form of the classical Monod model as well as by the Han-Levenspiel model. The modified Logistic model was used to describe the growth of the biomass as well as the rate of biohydrogen production. The effect of temperature was well described by the Arrhenius model, but up to the critical temperature only. The Ratkowsky model was able to describe the effect of both temperature as well as pH on fermentative biohydrogen production.

Keywords: Biohydrogen, dark fermentation, Gompertz model, Andrew model, Han-Levenspiel model, Ratkowsky model, Monod model, inhibition, Luedeking-Piret model, activation energy, *Enterobacter cloacae*.

NOMENCLATURE

X	biomass concentration (g l^{-1})
X_0	Initial cell mass concentration (g l^{-1})
S	substrate concentration (g l^{-1})
S_{crit}	critical substrate concentration (g l^{-1})
N	sampling number
K_S	saturation constant (g l^{-1})
D	dilution rate
T	time of fermentation (h)
$Y_{X/S}$	yield coefficient for cell growth [g cell mass (g substrate degraded)$^{-1}$]
K_i	Inhibition constant (g l^{-1})
$H(t)$	cumulative volume of hydrogen production (ml)
P	gas production potential (ml)
r_{fp}	rate of hydrogen production (g l^{-1} h^{-1})
r_{fx}	rate of cell mass production (g l^{-1} h^{-1})
A	Arrhenius constant
R	rate of hydrogen production (ml h^{-1})
R_m	the maximum production rate (ml h^{-1})
R_g	ideal gas constant
E_a	activation energy (kcal/mol)
T	absolute temperature (K)

***Address correspondence to Debabrata Das:** Indian Institute of Technology Kharagpur, India; E-mail: ddas.iitkgp@gmail.com

Nuri Azbar and David B. Levin (Eds)

Greek Symbols	
α	growth associate coefficient
β	non-growth associate coefficient (h^{-1})
ν	specific product formation rate (h^{-1})
μ	specific growth rate (h^{-1})
μ_{max}	maximum specific growth rate (h^{-1})
λ	the lag time (h)

INTRODUCTION

Developing countries need advanced production and processing technology to secure their energy needs. The environmental pollution generated due to the use of fossil fuels and their gradual shortfall has focused the attention of researchers worldwide on alternative-renewable energy sources. Hydrogen is a promising fuel and of great potential to become alternative to fossil fuels. Moreover, it is a clean fuel as it produces only water when combusted or converted to electricity. The liabilities like emissions, spills, CO_2, global warming and acid rain *etc.* which are common to fossil fuels are not associated with hydrogen [1]. Hydrogen can be produced in a number of ways, such as steam reforming of natural gas, thermal cracking, electrolysis of water and *via* biological methods (using microorganisms). The production of hydrogen by biological methods is less energy intensive and therefore gaining a lot of attention these days as a sustainable alternative to the conventional methods. Biohydrogen can be produced by two ways, first, in presence of light by photosynthetic organisms (such as photosynthetic bacteria, cyanobacteria, green algae and photofermentative bacteria), and second, in absence of light by dark fermentative organisms (strict anaerobes and facultative anaerobes). It is well known that the rate of hydrogen production by dark fermentative organisms is always faster than the photosynthetic route. The efficiency of photosynthetic hydrogen production is low as it cannot be operated in the absence of light, whereas fermentative hydrogen production can produce hydrogen all day long without light [1,2]. Moreover, the fermentative hydrogen production utilizes various industrial and agricultural wastes thus facilitating waste recycling. In addition, this route has other advantages such as higher hydrogen production stability, simpler control requirements, lower operating costs and higher feasibility for industrialization [3]. Anaerobic fermentation, commonly termed as dark fermentation is more feasible and widely used for biohydrogen production. However, the major drawback of this process is the lower achievable yield of hydrogen per mol of substrate utilized which is restricted to 4 mol of hydrogen per mole of glucose if all the substrates would be converted *via* acetic acid pathway. More in-depth research is required to increase the theoretical maximum yield of hydrogen which will make the process economically viable.

Study of kinetics of any fermentative process is essential for the analysis, design and operation of the process [4,5]. In fermentative biohydrogen process, the growth of the hydrogen-producing bacteria and the production of hydrogen take place simultaneously while the substrate is continuously degraded producing some soluble metabolites. The substrate degradation rate and the bacterial growth rate were related with hydrogen production rate by many earlier researchers. Several kinetic models are proposed to predict the substrate degradation, bacterial growth, hydrogen production and some soluble metabolite formation in a batch fermentative hydrogen production [6,7]. Apart from these, various other parameters like substrate concentration, temperature, pH, dilution rate and inhibitors also can influence the biohydrogen production [8]. There are some kinetics models reported in literature which considered the effect of these parameters on biohydrogen production. The present paper attempts to summarize the kinetics models describing the effect of various parameters on the biological hydrogen production process.

MATHEMATICAL MODELING OF BIOLOGICAL HYDROGEN PRODUCTION PROCESSES

The primary objective of a kinetic model developed for fermentation is the prediction of kinetic behavior of hydrogen fermentation performance. An appropriate model of fermentation, with the technical, economic and physiological implications would be a powerful means to predict and control problem of fermentations, and be helpful to understand the fermentation process [9]. Among the numerous models developed, the majority of the models are biochemically knowledge-based models, which consist of a set of mathematical equations describing the phenomena occurring during fermentation. The main advantage of this type of model is that they account for biological phenomena. The model parameters with some biological significance can be obtained, but their structures may be strongly non-linear, complex and difficult to verify and validate, and can pose problems in terms of parameters identification. In general, the fermentation kinetic model can be subdivided into a growth model, a substrate model and a product model. This chapter attempts to summarize the various kinetic models, which have been proposed to describe the biohydrogen production in batch system. Apart from these the kinetic models describing the effect of pH, temperature and dilution rates are also discussed.

Kinetic Modeling based on Growth of Cells

The biomass concentration in the batch hydrogen production experiments depends on the concentration in limiting substrate. The classical Monod model and it's various modified forms were used to describe the cell growth kinetics satisfactorily.

Determination of cell growth kinetic parameters: The rate of biomass production and substrate degradation can be determined by the differential technique as follows:

$$\frac{dx}{dt} = \frac{X_{n+1} - X_{n-1}}{t_{n+1} - t_{n-1}}, \quad \frac{ds}{dt} = \frac{S_{n+1} - S_{n-1}}{t_{n+1} - t_{n-1}} \tag{1}$$

where X = biomass concentration (gl^{-1}); S = substrate concentration (gl^{-1}); n = sampling number. The specific growth rate (μ) of biomass can be written as:

$$\mu = \frac{1}{x}\frac{dx}{dt} \tag{2}$$

where dx/dt are calculated for different experimental data points.

Assuming that cell growth follows Monod equation [10] specific cell growth rate can be approximated as:

$$\mu = \frac{\mu_{max} S}{K_S + S} \tag{3}$$

Equation (3) may be linearized to find out the kinetic parameters with the help of Lineweaver-Burk plot:

$$\frac{1}{\mu} = \frac{Ks}{\mu_{max} S} + \frac{1}{\mu_{max}} \tag{4}$$

Regression analysis was used to find the best fit for a straight line on a plot $1/\mu$ vs. $1/S$ to determine the values of maximum specific growth rate (μ_{max}) and saturation constant (K_S). The values of μ_{max} and K_S were 0.4 h^{-1} and 5.5 g/l respectively using glucose as substrate using *Enterobacter cloacae* IIT-BT [11]. Both the values of μ_{max} and K_S were found lower than those reported (1.12 h^{-1} and 8.9 g/l respectively).

In case of chemostat (CSTR), under steady state condition and sterile feed, D = μ. Then the Equation (4) may be written:

$$\frac{1}{D} = \frac{Ks}{\mu_{max} S} + \frac{1}{\mu_{max}} \tag{5}$$

Kinetic models based on material balances for continuous hydrogen production process were constructed to describe the steady state behavior of the concentrations of substrate (sucrose), biomass and products (H_2 and VFA) in the anaerobic culture adapted to the enhanced production of hydrogen [12]. The proposed model was able to interpret the trends of the experimental data. The maximum specific growth rate (μ_{max}), saturation constant (K_S), and yield coefficient for cell growth ($Y_{X/S}$) were reported as 0.17 h^{-1}, 68 mg COD/l, and 0.1 g/g respectively. Table **1** summarizes the kinetic parameters of biomass growth in batch systems for different seed and substrate.

Table 1: Cell Growth Kinetic Parameters of Different Hydrogen Producing Microorganisms in a Batch Process.

Seed	**Substrate**	**Maximum specific growth rate, μ_{max} (h^{-1})**	**Saturation constant, K_S ($g\ l^{-1}$)**	**Biomass yield coefficient, $Y_{X/S}$ [g cell mass (g substrate degraded)$^{-1}$]**	**References**
Enterobacter cloacae IIT-BT 08	Glucose	0.568	3.658	0.837	[1]
Enterobacter cloacae DM11	Glucose	0.4	5.51	ND	[11]
Enterobacter cloacae IIT-BT 08	Glucose	1.12	8.9	0.097	[13]
Ruminococcus albus	Glucose	0.765	0.654	0.139	[14]
Caldicellulosiruptor saccharolyticus	Sucrose	0.13	0.81	NA	[4]
Anaerobic bacteria	Glucose	0.03	16.1	ND	[15]
Cl. butyricum CGS5	Xylose	0.15	0.67	ND	[16]
Cl. pasteurianum CH4	Sucrose	0.31	4.39	ND	[16]
Escherichia coli BL-21	Glucose	0.4	11.11	ND	[17]
Th. thermosaccharolyticum PSU-2	Sucrose	0.31	1.47	0.3	[18]

ND: Not determined; NA: Not available

Development of mathematical models to correlate substrate and biomass concentration with time: The effect of time and substrate concentration on the biomass growth can be calculated using the following equations, written in the form:

$$q = ap + b + \frac{c}{p} \tag{6}$$

Now, when a data set of p and q is available, the coefficients a, b, and c can be calculated by solving the following three equations simultaneously:

$$an + b\,\Sigma\frac{1}{p_i} + c\,\Sigma\frac{1}{p_i^{\,2}} = \Sigma\frac{q_i}{p_i} \tag{7}$$

$$a\,\Sigma p_i + bn + c\,\Sigma\frac{1}{p_i} = \Sigma q_i \tag{8}$$

$$a\,\Sigma p_i^2 + b\,\Sigma p_i + cn = \Sigma q_i p_i \tag{9}$$

Here, n = number of data of p and q, i refers to the ith data of p or q. Once values for a, b, and c are obtained, values of kinetic parameters K_S, μ_{max}, and K_i can be calculated by using equations (16), (17), and (18).

Simulation of biomass concentration and substrate concentration profiles as functions of time: Now, the modified Andrew equation for the substrate inhibition as mentioned in eqn. (13) may be written as

$$\frac{dx}{dT} = \mu x = \frac{\mu_{max} S}{K_S + S - K_i S^2} x \quad \textbf{(10)}$$

Thus,

$$\frac{dS}{dt} = \frac{-1}{Y_{x/S}} \frac{\mu_{max} S}{K_S + S - K_i S^2} x \quad \textbf{(11)}$$

x in equation (26) is replaced by an expression involving S, S_0, x_0, and $Y_{x/S}$ (from eqn. 9). It can then be integrated to give an expression for simulated values of S as a function of t:

$$\mu_{max}(x_0+Y_{x/S} S_0)t = [x_o+Y_{x/S}(S_0+K_S)]\ln(\frac{x_0 + Y_{x/S}(S_0 - S)}{x_0})$$

$$K_i(x_0+Y_{x/S} S_0)[S-S_0+\frac{(x_0 + Y_{x/S}S_0)}{Y_{x/S}}\ln(\frac{x_0 + Y_{x/S}(S_0 - S)}{x_0})]$$

$$K_S Y_{x/S} \ln\frac{S}{S_0} \quad \textbf{(12)}$$

Hence, from the above expression, simulated values for substrate concentration at different times are calculated.

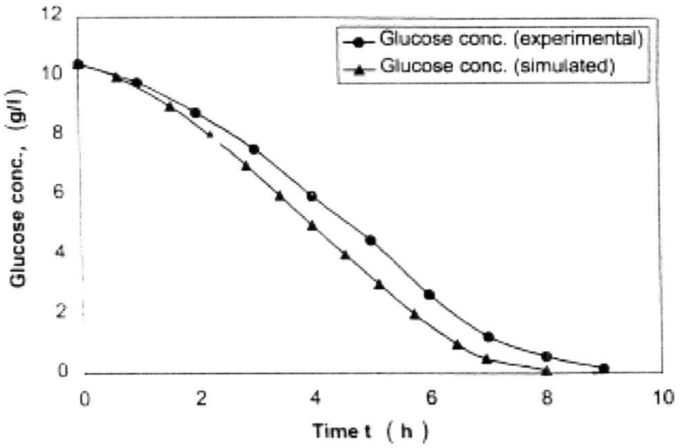

Figure 1: Comparison between experimental and simulated substrate concentration profiles using substrate inhibition model.

Simulated values for biomass concentration (x) are calculated by substituting the values of the substrate concentration in equation (11). Substrate, biomass concentration and hydrogen production profiles (Figs. **1** to **4**) have good resemblance as compared to the experimental results [1]. The little deviation between the experimental and theoretical curves might be due to the assumption that $Y_{x/S}$ remained constant over the system irrespective of time of fermentation. This was assumed by ignoring the effect of maintenance energy on cell growth and metabolism.

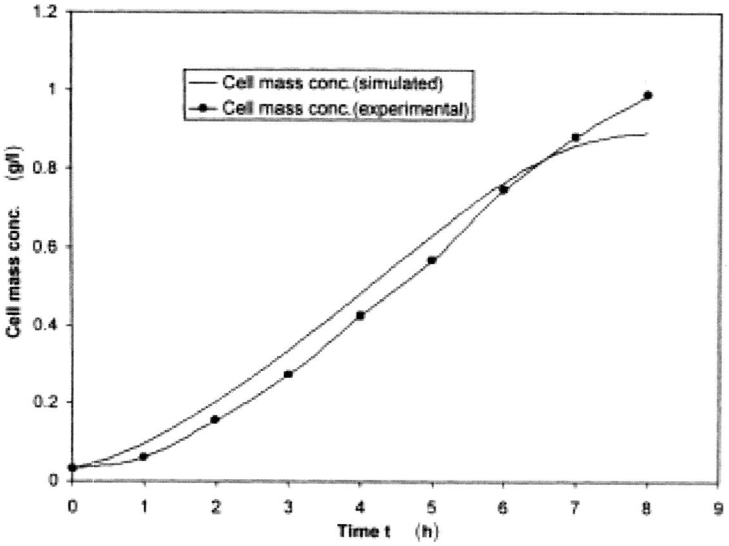

Figure 2: Comparison between experimental and simulated cell mass concentration profiles using substrate inhibition model.

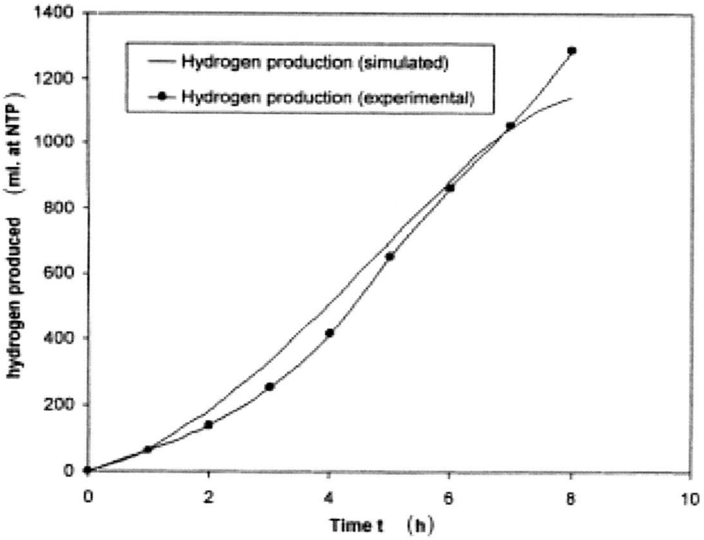

Figure 3: Comparison between experimental and simulated product generation profiles.

KINETIC MODELING BASED ON PRODUCT FORMATION

Various kinetic models were used to describe the biohydrogen formation mechanism during fermentation. The rate of hydrogen production was affected by the substrate concentration, biomass growth and formation of soluble metabolites [19]. Some kinetic models considered these parameters. The modified Gompertz model, the Logistic model and the Luedeking-Piret model are the most widely used models for product formation kinetics.

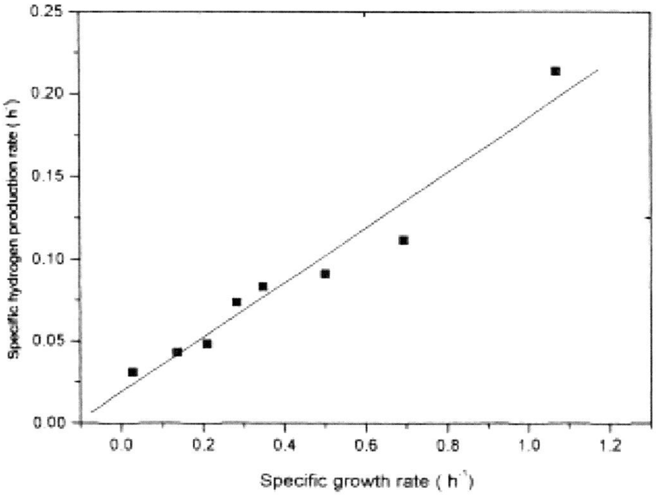

Figure 4: Plot of specific hydrogen production rate versus specific growth rate using Leudeking-Piret model.

Modified Gompertz Equation

The Cumulative hydrogen production in the anaerobic reactor can be modeled by modified Gompertz equation [20], which can be written as:

$$H(t) = P.\exp\{-\exp[\frac{R_m.e}{P}(\lambda - t) + 1]\} \quad (13)$$

This was found to be a suitable equation for describing the progress of cumulative gas production obtained from the experiment, where H (t) represents cumulative volume of hydrogen production (ml), P the gas production potential (ml), R_m the maximum production rate ($ml.h^{-1}$), λ (h) the lag time, t incubation time (h), e.the exp(1) = 2.718.

Gompertz equation can be simplified as:

$$Y = a\exp\{-\exp(b - cx)\} \quad (14)$$

Where a = P; b = $\frac{R_m.e}{P}(\lambda + 1)$; and c = $\frac{R_m e}{P}$

The typical cumulative hydrogen production curve was nonlinearly modeled by the above equation. The values of P, R_m and (for each set of experiment) were determined by best fitting the cumulative hydrogen production data for the above equation using a software program, Curve expert 1.3. The spent medium was obtained from the batch of dark fermentation by *E. cloacae* using MYG medium originally containing 1% (w/v) of glucose. The values of P, R_m and λ were 615.65 ml., 16.15 ml h^{-1} and 14.32 h respectively determined by best fitting the cumulative hydrogen production data for the above equation [11].

Luedeking-Piret Model

Product formation kinetics can be modeled on the basis of two-parameter kinetic expression, termed as Luedeking-Piret equation of the following form [10]:

$$r_{fp} = \alpha r_{fx} + \beta x \quad (15)$$

Dividing both sides by x, we get,

$$\frac{1}{x}r_{fp} = \alpha\frac{1}{x}r_{fx} + \beta$$

or

$$\nu = \alpha\mu + \beta \tag{16}$$

ν is the specific productivity = (1/x) (dP/dt), where P = product formed after time interval *t*, α and β are constants. The volume of hydrogen produced was noted from time to time and the values of γ calculated. The values of ν and μ obtained at various time intervals were plotted on a graph. The nature of the best-fit curve tells us whether the product formation is growth-associated or not. In case of growth-associated product formation, the graph passes through the origin. Fig. **4** indicates that the hydrogen is a purely growth associated product [1]. The value of growth associate coefficient (α) was 0.166. From Eqn. (35), values of dP/dt at different points of time can be found out from dx/dt. From dP/dt profile, a simulated P profile can be obtained.

Modified Logistic Model

The modified logistic model is of the following form:

$$H = \frac{P}{1+\exp\left[\frac{4R_m(\lambda-t)}{P}+2\right]} \tag{17}$$

where H represents cumulative volume of hydrogen production (ml), P the gas production potential (ml), R_m ($ml.h^{-1}$) the maximum production rate, λ (h) the lag time and t incubation time (h).

Although this model was not that widely used as the modified Gompertz, it has similar properties like the modified Gompertz model. Recently, Wang and Wan [21] have used the modified logistic model to describe the formation of hydrogen using glucose as the substrate in batch tests satisfactorily.

Kinetic Modeling Describing the Effect of pH and Temperature

Temperature and pH are the most important factors which influences the fermentative biohydrogen production by affecting the growth of the biomass. A drastic change in temperature affects the activity of some of the essential enzymes thereby hampering the growth of the biomass. Similarly, extreme change in pH affects the ionization states of the active components of the biomass as well as substrates [19]. Till date the Arrhenius model has been used to describe the effect of temperature of the rate of biohydrogen production.

$$R = A\exp\left(\frac{-E_a}{R_g T}\right) \tag{18}$$

Where R is the rate of hydrogen production, A is Arrhenius constant, E_a is activation energy, R_g is the ideal gas constant and T is the absolute temperature.

Kumar and Das, 2000 [13] studied the rate of hydrogen production by *E. cloacae* IIT-BT 08 by varying temperature. They reported that the hydrogen yields increased from 15 to 36 °C while afterwards it decreased. The maximum hydrogen yield was obtained at 36 °C (2.25 mol H_2/mol glucose). However, 45 °C was found as critical temperature for hydrogen production because at this temperature although there was cell growth but no hydrogen production. This was found slight different from the reported work on *E. aerogens* [22] where hydrogen production increases proportionately up to 40 °C. The activation energy (E_a) for hydrogen production was calculated as 12.84 kcal/mol considering glucose as the substrate. Though from these reported works, *R* can be calculated with increase in temperature, however, the Arrhenius model cannot account for the decrease in *R* with increasing temperature beyond the critical temperature thus making it unsuitable for a wide operating temperatures. The Ratkowsky model, however, takes into account both the increasing and the decreasing trend of *R* along with increasing temperature:

$$R = [A(T - T_{\min})]^2 \{1 - \exp[B(T - T_{\max})]\}^2 \quad \textbf{(19)}$$

where, A, B are constants, *T* is the absolute temperature, T_{min} and T_{max} are the minimum and maximum temperatures. According to Wang and Wan, 2008 [23] the Ratkowsky model could describe the effect of temperature on the hydrogen production potential, maximum hydrogen production rate and the lag time during fermentative hydrogen production of very well.

Like temperature, pH is also a very sensitive parameter of biohydrogen production. Tanisho *et al.*, 1987 [24] reported that the hydrogen producing ability of *E. aerogenes* strain E82005 was strongly influenced by pH, with a hydrogen production rate at pH 7.0 which was 50% lower than the one at the optimum pH of 6.0–6.5. The Andrew model was widely used to describe the effects of H^+ concentration on the rate of biohydrogen production:

$$\mathrm{R} = \frac{R_{\max}[\mathrm{H}^+]}{K_a + [\mathrm{H}^+] + [\mathrm{H}^+]^2 / K_b} \quad \textbf{(20)}$$

where *R* is the rate of hydrogen production, R_{max} is the maximum rate of hydrogen production, K_a and K_b are constants. However, it is convenient to use the value of pH directly in the equation rather than H^+ concentration. The Ratkowsky model (Eq. 41) is a better choice is this aspect.

$$R = [A(pH - pH_{\min})]^2 \{1 - \exp[B(pH - pH_{\max})]\}^2 \quad \textbf{(21)}$$

CONCLUSIONS

The present paper reviewed the various kinetics models that were proposed to describe the progress of the batch fermentative biohydrogen production processes. The effects of substrate concentration, pH and temperature on the growth of the biomass as well as rate of biohydrogen production were discussed. It was found that the widely used modified Gompertz model was the best to describe the progress of a batch fermentative hydrogen production process, while the Andrew model was widely used to describe the effects of substrate inhibition on the biomass growth and hydrogen production rate. To describe the effect of temperature the Arrhenius model was widely used. The effect of pH was suitably described by the Ratkowsky model. In addition, further in-depth study is required which will incorporate the effect of hydrogen partial pressure and regulatory mechanisms like $NADH/NAD^+$.

REFERENCES

[1] Kumar N, Monga PS, Biswas AK, Das D. Modeling and simulation of clean fuel production by *Enterobacter cloacae* IIT-BT 08. Int J Hydrogen Energy 2000; 25(10): 945–52.

[2] Benemann J. Hydrogen biotechnology: progress and prospects. Nat Biotechnol 1996; 14: 1101-3.

[3] Wang J, Wan W. Kinetic models for fermentative hydrogen production: A review. Int J Hydrogen Energy 2009; 34: 331323.

[4] van Niel EWJ, Claassen PAM, Stams AJM. Substrate and product inhibition of hydrogen production by the extreme thermophile, *Caldicellulosiruptor saccharolyticus*. Biotechnol Bioeng 2003; 81(3): 255–62.

[5] Mu Y, Wang G, Yu HQ. Kinetic modeling of batch hydrogen production process by mixed anaerobic cultures. Bioresource Technol 2006; 97(11): 1302–7.

[6] Wang B, Wan W, Wang JL. Inhibitory effect of ethanol, acetic acid, propionic acid and butyric acid on fermentative hydrogen production. Int J Hydrogen Energy 2008; 33(23): 7013–9.

[7] Wang JL, Wan W. Experimental design methods for fermentative hydrogen production: a review. Int J Hydrogen Energy 2009; 34(1): 235–44.

[8] Wang JL, Wan W. Factors influencing fermentative hydrogen production: a review. Int J Hydrogen Energy 2009; 34(2): 799–811.

[9] Marin MR. Alcoholic fermentation modelling: current state and perspectives. Am J Enol Vitic 1999; 50:166-78.

[10] Bailey JE, Ollis DF. In: Biochemical engineering fundamentals. 2nd ed. New York: McGraw-Hill, 1986.

[11] Nath K, Muthukumar M, Kumar A, Das D. Kinetics of two-stage fermentation process for the production of hydrogen. Int J Hydrogen Energy 2008; 33(4): 1195–203.

[12] Chen CC, Lin C-Y, Chang JS. Kinetics of hydrogen production with continuous anaerobic cultures utilizing sucrose as the limiting substrate. Appl Microbiol Biotechnol 2001; 57: 56–64.

[13] Kumar N, Das D. Enhancement of hydrogen production by *Enterobacter cloacae* IIT-BT 08. Process Biochem. 2000; 35(6): 589–93.

[14] Ntaikou I, Gavala HN, Kornaros M, Lyberatos G. Modeling of fermentative hydrogen production from the bacterium *Ruminococcus albus*: Definition of metabolism and kinetics during growth on glucose, Int J Hydrogen Energy 2009; 34: 3697-709.

[15] Sharma Y, Li B. Optimizing hydrogen production from organic wastewater treatment in batch reactors through experimental and kinetic analysis. Int J Hydrogen Energy 2009; 34: 6 171-80.

[16] Lo YC, Chen WM, Hung CH, Chen SD, Chang JS. Dark H_2 fermentation from sucrose and xylose using H_2-producing indigenous bacteria: feasibility and kinetic studies. Water Res 2008; 42(4–5): 827–42.

[17] Chittibabu G, Nath K, Das D. Feasibility studies on the fermentative hydrogen production by recombinant *Escherichia coli* BL-21. Process Biochem 2006; 41(3): 682–8.

[18] O-Thong S, Prasertsan P, Karakashev D, Angelidaki I. Thermophilic fermentative hydrogen production by the newly isolated *Thermoanaerobacterium thermosaccharolyticum* PSU-2. Int J Hydrogen Energy 2008; 33(4): 1204–14.

[19] Mu Y, Yu HQ, Wang G. A kinetic approach to anaerobic hydrogen-producing process. Water Res 2007; 41(5): 1152–60.

[20] Lay JJ, Li Y-Y, Noike T. Influences of pH and moisture content on the methane production in high solids sludge digestion. Water Res 1997; 3: 1518–24.

[21] Wang JL, Wan W. The effect of substrate concentration on biohydrogen production by using kinetic models. Sci China Ser B 2008; 51(11): 1110–7.

[22] Tanisho S. Feasibility study of biological hydrogen production from sugar cane by fermentation. Proceedings of the 11th World Hydrogen Energy Conference Stuttgart, Germany, Hydrogen Energy Progress XI, 1996: 2601–6.

[23] Wang JL, Wan W. Effect of temperature on fermentative hydrogen production by mixed cultures. Int J Hydrogen Energy 2008; 33(20): 5392–7.

[24] Tanisho S, Suzuki Y, Wakao N. Fermentative hydrogen evolution by *Enterobacter aerogenes* strain E82005. Int J Hydrogen Energy 1987; 12(9): 623–7.

CHAPTER 9

Hydrogen Production by Thermophilic Fermentation

Ed W. J. van Niel[1*], Karin Willquist[1], Ahmad A. Zeidan[1], Truus de Vrije[2], Astrid E. Mars[2] and Pieternel A. M. Claassen[2]

[1]Lund University, Lund, Sweden and [2]Food and Biobased Research, Wageningen, Netherlands

Abstract: Of the many ways hydrogen can be produced, this chapter focuses on biological hydrogen production by thermophilic bacteria and archaea in dark fermentations. The thermophiles are held as promising candidates for a cost-effective fermentation process, because of their relatively high yields and broad substrate palette. Yet many challenges remain to be faced, including improving productivity, tolerance to high osmolality and growth inhibitors, and reactor configuration. This review consolidates current insights in the quest for high yields and productivities within thermophilic hydrogen production. Important is to understand how environmental parameters affect the redox- and energy metabolism of the microorganism(s) involved. This knowledge is required for designing an optimal bioreactor configuration and operation.

Keywords: Central carbon metabolism, hydrogenase, hydrogen productivity, hydrogen yield, hydrogen partial pressure, hydrolysate, medium composition, osmolality, physiology, reactor design, thermodynamics, thermophile.

INTRODUCTION

Interest in sustainable hydrogen production has been regained in the last couple of decades for both its application as an energy carrier and as a chemical. Hydrogen can be produced in many ways, but in this chapter we focus on biological hydrogen production in dark fermentations by thermophilic bacteria and archaea. The following advantages for using thermophilic hydrogen producers have been recognized: 1) they usually digest a wide range of sugars, including hexoses, pentoses, oligosaccharides and polysaccharides such as cellulose; 2) they produce hydrogen yields close to the theoretical maximum; 3) at elevated temperature there is a significant reduction in contamination problems; 4) the high temperature eliminates the need to cool pretreated feeds to a significant extent.

A critical aspect for a cost-efficient biohydrogen production is to obtain a high substrate-conversion efficiency due to the central cost of the raw material and the requirement to reduce the chemical oxygen demand (COD) in the waste stream [1]. The stoichiometrically maximal yield of 12 moles of hydrogen per mole of glucose is not thermodynamically feasible since the reaction is endergonic:

$$C_6H_{12}O_6 + 12\ H_2O \rightarrow 12\ H_2 + 6\ HCO_3^- + 6\ H^+ \quad (1)$$

($\Delta G^{0'}$: +3.2 kJ mol^{-1} at 25°C, but ΔG estimated to be -90 kJ mol^{-1} at 70°C)

Therefore, input of energy is required to obtain this high yield under standard conditions, but the reaction might become more favorable at elevated temperatures. In practice, one strategy to improve the overall process yield is to apply a hybrid two-stage system where sugars are converted in two steps. The first step consists of a dark fermentation yielding maximally ⅓ of the hydrogen *via* the reaction:

$$C_6H_{12}O_6 + 2\ H_2O \rightarrow 4\ H_2 + 2\ HCO_3^- + 2\ CH_3COO^- + 4\ H^+ \quad (2)$$

($\Delta G^{0'}$: -216.1 kJ mol^{-1} at 25°C, but ΔG estimated to be -260 kJ mol^{-1} at 70°C)

***Address correspondence to Ed W. J. van Niel:** Lund University, Lund, Sweden; E-mail: Ed.van_Niel@tmb.lth.se

Nuri Azbar and David B. Levin (Eds)

Thus the empirical maximum yield is 4 mol H_2/mol hexose, which is also called the Thauer limit [2]. Obviously, acetate is the preferred by-product of dark fermentation, which can be converted to H_2 and CO_2 by either photofermentation or microbial electrohydrogenesis [1,3], using input of energy in the form of sunlight or electricity, respectively. These topics are treated elsewhere in this eBook (chapter 4 and 12).

Research on thermophilic hydrogen production is relatively recent, and has mainly focused on applying undefined consortia to fermentations of waste material, or isolation and characterization of new species. So far several genomes have been sequenced of thermophilic hydrogen producers (including *Pyrococcus furiosus* [4], *Clostridium thermocellum* [5], *Thermotoga maritima* [6], (*Thermoanaerobacter tengcongensis* [7], and *Caldicellulosiruptor saccharolyticus* [8], but more are on the way), allowing more in depth studies of their physiology. The latter three are among the best fermentative H_2-producers in terms of yield. However, maximum yields have been mostly gained in artificial media, whereas yields on real substrates, such as hydrolysates, are quite variable, depending on their composition, and presence of metabolic inhibitors.

The interest in biohydrogen research is rapidly increasing, as exemplified through the explosion of reviews published in the last few years. Whereas most of these reviews focused on the kind of microorganisms involved or their redox-related physiology [9-11], this chapter will put special emphasis on other aspects such as performance on biomass substrates, energy-related metabolism and designed co-cultures as opposed to undefined consortia. From the work on metabolic networks it is known that the redox- and energy metabolism are highly interlinked, and their optimal regulation may benefit high hydrogen yields and productivities. Optimal metabolic regulation is in turn dependent on environmental conditions, hence it is important to understand the physiology of the thermophiles in relation to environmental parameters to improve the overall fermentation process such that it becomes cost-effective. In the post-genomic era this field has been coined 'enviromics' [12], and will be used as an underlying theme that runs through this chapter.

ORGANISMS INVOLVED

Several promising extreme thermophilic candidates will be summarized here as opposed to thermophilic undefined consortia. Attention will be given also to 'designed co-cultures', a relatively new concept for hydrogen production.

Pure Cultures

To find an efficient microorganism is vital for successful commercialization of biohydrogen production. In addition, an industrially successful hydrogen producer should have the following characteristics: 1) high hydrogen yields and productivities; 2) ability to tolerate high concentrations of sugars and fermentation products; 3) ability to convert both hexose and pentose sugars; 4) resistance to growth inhibitors in biomass hydrolysates; and 5) good growth in inexpensive medium formulations. From all pure culture studies it is clear that the make-up of the genome (chapter 10), and thus the metabolic network, plus the thermodynamic barriers have an influence on the maximum H_2 yield. However, there is no microbe or combination of microbes studied that possess all these traits, yet some promising candidates are at hand. Moreover, there is still a lack of complete information about their behavior in various environmental conditions, especially including different feedstocks.

Pure cultures are mainly used for studying the effect of environmental parameters on fermentation profiles and carbon metabolism, using lab-scale reactors. So far, only *Caldicellulosiruptor saccharolyticus* has been successfully tested as a pure culture in a pilot-scale bioreactor using non-sterilized feedstock [13]. It has been confirmed that most of the hydrogen-producing thermophiles can achieve yields >80% of the Thauer limit (Table **1**). The characteristics of the various genera were recently reviewed [9,11], and will not be treated here.

Table 1: A selection of Thermophilic and Extreme Thermophilic Hydrogen Producing Species, Showing their Optimal Growth Temperature (Opt T; °C) and Maximum Hydrogen Yields (Max Yield; mol H_2 mol Hexose^{-1}).

Organism	Opt T (°C)	Cultivation	Substrate	Max Yield	Reference
Thermophiles					
Clostridium thermocellum	60	Batch		2.4	[14]
Extreme & Hyper Thermophiles					
Thermotoga maritima	80	Batch	Glucose	4.0	[15]
Thermotoga neopolitana	77	Batch	Glucose	3.9	[16]
Thermotoga elfii	65	Batch	Glucose	3.3	[17]
Caldicellulosiruptor saccharolyticus	70	Continuous	Glucose	3.6	[18]
Caldicellulosiruptor owensensis	70	Batch	Glucose	4.0	[19]
Thermoanaerobacter tengcongensis	75	Batch	Glucose	4.0	[20]
Thermoanaerobacter pseudoethanolicus	70	Continuous	Xylose	1.2	[21]
Pyrococcus furiosus	100	Batch	Glucose	3.5	[22]
Thermococcales kodakaraensis	88	Continuous	Starch	3.3	[23]

Undefined Consortia

Mixed cultures can sometimes add a new dimension to many biotechnological processes that could, otherwise, be carried out using a single organism. Elimination of sterilization, enhancing biofilm formation as well as broadening the substrate spectrum is among the main motives for utilizing mixed culture in biotechnology. Applying continuous enrichment techniques for isolating H_2-producing organisms can provide a stable microbial consortium adapted to continuous culture conditions, which is especially valuable in a continuous-flow commercial process [24]. This approach also can result in the enrichment of H_2-producing organisms not readily recovered by batch enrichment or purification on solid media. Within this context, microflora from various sources has been used as inocula for H_2 production in different types of reactors. The process is always coupled to biowaste treatment and the starting culture is usually derived from samples of compost or anaerobic treatment sludge [25]. Several strategies have been proven efficient for eliminating H_2-consuming organisms from the reactors, and are described elsewhere [26,27]. However, as most studies using mixed cultures have so far been carried out under mesophilic conditions, H_2 yields hardly exceeded two moles per mol glucose [28,29], but high volumetric rates of hydrogen production could be achieved [1]. The former is mainly due to the variety of reduced end products formed during mixed-culture fermentations, which results in draining some of the reducing power of pyruvate. Fewer studies have described thermophilic mixed consortia, of which the highest H_2 yield reported with continuous operation under extreme thermophilic conditions (70°C) was 2.5 mol mol glucose^{-1} [30], meaning that there is still room for improvement. Similar results were obtained (2.6 ± 0.1 mol mol hexose^{-1}) with an undefined, thermophilic compost microflora under batch operation conditions at 70°C [31]. PCR-based analytical techniques targeting the 16S rRNA gene, such as denaturing gradient gel electrophoresis (DGGE) [32-34] and cloning-sequencing [35], have been employed for analyzing the microbial communities used for H_2 production. *Thermoanaerobacterium* spp. as well as thermophilic *Clostridium* spp. are usually the predominant organisms in thermophilic H_2-producing consortia [36]. Changes in process parameters, such as pH, hydraulic retention time and organic loading rate, were found to significantly influence the composition of the microbial community in a mixed population [36], which results in undesirable variations in H_2-production efficiency. It should be clear that in addition to the low H_2 yields associated with mixed-culture fermentations, optimizing the operational parameters is conceptually challenging due to the complex nature of microbial interactions. In conclusion, applying undefined consortia for H_2 production appears to be beneficial primarily when it is coupling biowaste treatment, where high H_2 yields are not crucially essential. However, if H_2 is to be produced from dedicated energy crops, maximum yields must be aimed for, which might not be practically attainable with a complex microbial community.

Designed Co-Cultures

In recent years, there has been a renewed interest in using designed co-cultures, consisting of two or more microorganisms intentionally combined together, for improving yields and productivities of various products, such as bioethanol [37-39] and organic acids [40-41], as well as extending the range of substrate utilization. This is in addition to the long history of mixing different bacterial species in food fermentations [42,43]. Designed co-cultures may offer a better performance than mixed-population enrichments and in addition can overcome some of the shortcomings of pure cultures, based on synergies among the microorganisms involved. Only a few studies have reported on the use of designed thermophilic co-cultures for H_2 production. Liu *et al.* [44] combined two thermophilic organisms, *i.e. Clostridium thermocellum* and *Thermoanaerobacterium thermosaccharolyticum*, in a co-culture for improving H_2 yields and substrate conversion efficiency. The bacteria were co-cultured in batch mode in uncontrolled conditions at 60ºC, and achieved a maximum hydrogen yield of only 1.8 mol mol glucose^{-1}. On the other hand, superior hydrogen yields (3.8 mol mol glucose^{-1}) were obtained in a co-culture of *Caldicellulosiruptor saccharolyticus* and *C. kristjanssonii* under batch-operation conditions on glucose and xylose [31]. Interestingly, these yields were significantly higher than that obtained by either organism alone under the same conditions, which suggested the existence of synergy on H_2 production between the two organisms. A variety of interesting interactions can be expected by combining microorganisms from different habitats, some of which might be potentially useful for H_2 production, such as tolerance to higher hydrogen partial pressure (P_{H2}) and higher substrate concentration and improved chance of biofilm formation. Although optimizing a co-culture of two or more organisms might be challenging and requires a thorough knowledge on the physiology of and interaction between the involved organisms, it still can be more readily achievable than with undefined consortia. A bioreactor system that allows for real-time quantitative assessment of individual cell activities in a mixed culture system has been previously developed [45]. In addition, with the continued increase in the number of sequenced microbial genomes, unraveling different mechanisms of interspecies interaction can be facilitated by the use of microarray-based functional genomics approaches [46] and metabolic modeling approaches [47]. These approaches can offer essential tools for constructing, optimizing and stabilizing tailor-made co-cultures for H_2 production.

High Yields *Versus* High Productivities

As pointed out by others, the low cell density of the thermophilic cultures leads to a low volumetric productivity [10]. Overall, there is a trade-off between yield and productivity; at higher dilution rates the productivity increases with the expense of a lower yield [18,31]. The same is true when comparing different microorganisms, *i.e.*, *E. coli* produces much lower yields than *C. saccharolyticus,* but the volumetric productivity is about 100-fold higher in *E. coli* [48] compared to *C. saccharolyticus* [18]. The specific productivities of these organisms are similar, but the reason for a higher volumetric productivity is the 200-fold higher cell mass in *E. coli* cultures than in *C. saccharolyticus* cultures. A similar trade-off scenario is present when comparing performances of different bioreactor types [49]. Interestingly, it has been observed that when the volumetric productivity increases, the specific productivity decreases [48]. This phenomenon could be due to that an increased cell mass decreases the mass transfer flux of substrate and products, hence leading to lower conversion efficiencies in each cell [48].

SUBSTRATE SOURCES: HYDROLYSATES AND WASTE

Hydrogen Production from Hydrolysates

Many thermophilic microorganisms can produce hydrogen during growth on a variety of carbohydrates, from simple sugars to complex polysaccharides like starch, hemicellulose, cellulose and pectin (listed by [50]). In these organisms, various hydrolytic enzymes involved in polysaccharide degradation have been identified [9,51]. Subsequently, the substrates seem to be co-utilized because for a number of these thermophiles simultaneous consumption of sugars in mixtures of monosaccharides or mixtures of mono- and di- or oligosaccharides has been observed and catabolite repression by glucose appeared to be absent. Because of these features, thermophiles are potentially excellent candidates for hydrogen production from a variety of biomass feedstocks with high biomass conversion efficiencies. Table **2** shows an overview of present data on thermophilic fermentations with substrates from biomass. These real substrates were produced from sugar- and starch-based feedstock, feedstock with mixed carbohydrate composition and lignocellulosic biomass rich in cellulose and hemicellulose. The reported hydrogen yields of thermophilic fermentations with these kinds of substrates varied greatly from 0.5 – 3.8 mol mol hexose^{-1}.

Table 2: Hydrogen Yield (Y_{H2}) and Productivity (Q_{H2}) by Thermophilic Bacteria on Real Substrates in Batch Fermentations or in Continuous Cultures (CC).

Feedstock	Culture	Y_{H2}	Q_{H2}	Substrate***		Conditions			References
		mol mol $C6^{-1}$	mmol L^{-1} h^{-1}	[S] g L^{-1}	Used (%)	Temp (°C)	Time (h)	pH	
Sugar-based feedstocks									
Waste water sugar factory	Mixed culture	2.5*	8.3	9.9	70	60	CC	6.8	[52]
Juice of:									
Sugar beet	*C. saccharolyticus*	3.0	7.8	10.6	99	72	26	7.0	[53]
Sugar beet	*C. saccharolyticus*	2.5	5.1	10.8	99	72	32	7.0	[54]
		1.5	3.9	21.4	99	72	50	7.0	
		1.2	3.2	31.5	93	72	72	7.0	
Sweet sorghum	*C. saccharolyticus*	2.6	n.d.	2.0	n.d.	70	n.d.	n.a.	[55]
Sweet sorghum	*C. saccharolyticus*	2.3	4.7	24.2	93	70	63	7.1	[56]
Starch-based feedstocks									
Hydrolysate of:									
Potato peels	*C. saccharolyticus*	2.9	1.9	11.3	72	70	69	7.0	[57]
Mixed feedstocks									
Household solid waste	Mixed culture	0.6**	0.8*	20	n.d.	70	CC	4.7 ± 0.2	[58]
Household solid waste	Mixed culture	2.0	0.06	0.5	n.d.	70	96	7.0	[59]
Palm oil mill effluent	Mixed culture	2.3*	16.9	20	92	60	CC	5.5	[60]
Palm oil mill effluent	Mixed culture	2.6*	1.1	20	92	60	49	n.a.	[61]
Maize leaves - treated	*C. saccharolyticus*	3.7	n.d.	6	n.d.	70	1056	n.a.	[55]
Maize leaves	*C. saccharolyticus*	1.8	n.d.	9	n.d.	70	984	n.a.	[55]
Sweet sorghum plant	*C. saccharolyticus*	1.8	n.d.	20	n.d.	70	552	n.a.	[55]
Silphium trifoliatum leaves	*C. saccharolyticus*	0.5	n.d.	10	n.d.	70	912	n.a.	[55]
Lignocellulosic biomass									
Filter paper	*Cl. thermocellum*	0.7	n.d.	4.5	n.d.	60	264	n.a.	[14]
Delignified wood fibres	*Cl. thermocellum*	1.0	n.d.	4.5	n.d.	60	264	n.a.	[14]
DDGS	*Cl. thermocellum*	1.3	0.23	5	n.d.	60	72	n.a.	[62]
Barley hulls	*Cl. thermocellum*	1.2	0.24	5	n.d.	60	72	n.a.	[62]
Sugar cane bagasse	*C. saccharolyticus*	2.3	n.d.	10	n.d.	70	744	n.a.	[55]
Wheat straw	*C. saccharolyticus*	3.8	n.d.	10	n.d.	70	744	n.a.	[55]
Hydrolysate of:									
Wheat straw	Mixed culture	2.6*	n.d.	0.8	97	70	80	n.a.	[63]
Corn stover (acid)	*Th. thermosaccharolyticum*	2.2	n.d.	11.4	86	60	50	n.a.	[65]
Corn stover (alkaline)	*Th. thermosaccharolyticum*	2.7	4.3*	10	73	60	25	n.a.	[65]
Miscanthus	*C. saccharolyticus*	3.3	4.6	17.3	99	72	69	6.8	[66]
Miscanthus	*Tm. neapolitana*	3.2	7.3	16.1	97	80	39	6.8	[66]
Paper sludge	*C. saccharolyticus*	3.7	4.1	8.4	76	70	32	7.0	[67]

* Calculated from data; ** Y_{H2} based on added substrate; ***Dry matter for solid substrates, carbohydrates for liquid substrates. n.a. not applicable; n.d. not determined.

Juice collected from sugar-based feedstock contains sucrose as the main sugar. Fast and almost complete consumption of the substrate was observed in batch cultures of *C. saccharolyticus*. Hydrogen yields at low substrate concentrations were 2.5 – 3.0 mol mol hexose^{-1} and dropped substantially to 1.2 – 2.3 at higher substrate loadings, which was accompanied by more lactate production at the cost of acetate and hydrogen production. Hydrogen productivity reached values of circa 8 mmol L^{-1} h^{-1} in batch fermentations as well as in a continuous fermentation [51-55].

Hydrogen production was reported from a real starch-based feedstock, *i.e.* potato peelings [56]. The starch was first fully hydrolyzed to glucose by commercial enzyme preparations prior to fermentation. The hydrogen yield was relatively high with almost 3 mol mol glucose^{-1}. The fermentation was performed under limiting N-source resulting in incomplete substrate consumption and low volumetric hydrogen productivity.

Hydrogen production from lignocellulosic biomass was reported for fermentations with the cellolulytic bacteria *Cl. thermocellum* and *C. saccharolyticus*. *Cl. thermocellum* is known for its very high rate of cellulose degradation. For this, it uses a multi-protein complex, the cellulosomes (reviewed by [68]). *C. saccharolyticus* does not possess a cellulosome but genes encoding cellulases and other enzymes involved in the complete degradation of cellulose have been identified. Additionally, in *C. saccharolyticus* other specific β-glycoside hydrolysases for the degradation of hemicellulose have been identified [8,50].

Hydrogen was produced by *Cl. thermocellum* from filter paper, delignified wood fibers, Distillers Dried Grains with Solubles (DDGS) and barley hulls at low biomass concentrations of circa 5 g L^{-1} [14,62]. The hydrogen yields varied from 0.7 – 1.3 mol mol hexose^{-1}. The low yields were due to the mixed product fermentation by this organism. Other major products besides acetate were the reduced end products ethanol and lactate, as well as formate. Yields on sugar cane bagasse and wheat straw in fermentations with *C. saccharolyticus* were higher, *i.e.* 2.3 and 3.8 mol mol hexose^{-1}, respectively, but in this case the duration of the fermentations was much longer and fermentations lasted for more than 30 days [54].

To improve the hydrogen production rate and also the biomass conversion efficiency lignocellulosic biomass is often first pretreated and/or hydrolyzed. Promising technologies for biomass pretreatment have been described by Mosier *et al.* [69]. As a result of the biomass treatment a soluble fraction, the hydrolysate, is obtained that consists of soluble carbohydrates. The hydrolysates used for thermophilic fermentations were produced from wheat straw [64], corn stover [64,65], Miscanthus [66] and paper sludge [67] by hydrothermal or mild acid pretreatment (mainly hemicellulose hydrolysis), or by enzymatic hydrolysis of untreated or alkaline pretreated biomass (cellulose and hemicellulose hydrolysis). The reported hydrogen yields of the thermophilic fermentations varied from 2.2 to as high as 3.7 mol mol hexose^{-1}. The substrates were consumed for more than 70% within 1 to 3 days. Hydrogen productivities of 4 to 7 mmol L^{-1} h^{-1} were reported.

Hydrogen from Waste

Mixed cultures adapted to the mixed feedstocks, household solid waste (HSW) and palm oil mill effluent (POME) were used for hydrogen production in combination with waste treatment [58-61]. Hydrogen yields and productivities on milled household solid waste were both rather low. Data on substrate consumption were not given, but consumption might be low and rate limiting. Reported hydrogen yields on POME are between 2.3 – 2.6 mol mol hexose^{-1}. For this the authors assumed that the COD reduction can be fully ascribed to carbohydrate utilization. The hydrogen productivity of 16.9 mmol L^{-1} h^{-1} in a sequencing batch reactor with a HRT of 2 days was much higher than for a batch cultivation without pH control. *C. saccharolyticus* was able to produce hydrogen from untreated whole plants or plant leaves [55]. The yield was low, *i.e.* 0.5 – 1.8 mol mol hexose^{-1}, and the fermentations continued for 23 – 41 days on substrate concentrations of 9 – 20 g L^{-1} of biomass. Because of this, the hydrogen production rate is most likely very low. Treatment of maize leaves with an aerobic cellulase-producing bacterium, *Bacillus amyloliquefaciens*, increased the hydrogen yield substantially [55], but not necessarily the production rate since the fermentation lasted still for 44 days.

Overall Conclusions on Substrate Usage

The results show that thermophilic microorganisms are able to produce hydrogen from energy crops, agricultural by-products, by-products from the food industry and from waste. Different aspects determine the efficiency of hydrogen production. Pure cultures of extreme thermophiles, *e.g. C. saccharolyticus* and *Tm. neapolitana*, producing mainly acetate as the soluble end product and varying amounts of lactate have the potential of higher hydrogen yields than undefined consortia or pure cultures of the moderate thermophiles, such as *Cl. thermocellum* and *Th. thermosaccharolyticum* that have all a broader end-product profile. Higher yields were also observed on soluble substrates as compared to solid substrates. Additionally, solubilization improved the hydrogen productivity substantially. Further improvement of the productivity by higher substrate concentrations appeared to be as yet unsuccessful because the hydrogen yield dropped significantly. Other potential effects of higher substrate concentration are the inhibition of fermentations by increasing concentrations of inhibitors. This might be especially the case with hydrolysates that are produced from pretreated lignocellulosic biomass. The sometimes severe conditions (high temperature, low pH) and applied chemicals may introduce fermentation inhibitors. It has been observed that extreme thermophiles are inhibited by the sugar degradation products furfural and hydroxymethylfurfural (HMF) [66]. More research is needed on the nutrient requirements of fermentations on natural substrates. The culture media of most of the fermentations listed in Table **2** were supplemented with nutrients like yeast extract and/or peptone, and other chemicals like salts and trace elements. To reduce the hydrogen production costs the use of these compounds should be limited.

PHYSIOLOGY OF THERMOPHILIC HYDROGEN PRODUCERS

To improve hydrogen yields and productivities it is important to understand the pertaining physiology of the microorganism(s) involved. Microorganisms respond and try to adapt to their immediate environmental conditions. Environmental parameters will influence the redox- and energy metabolism of the cell, *via* which they affect growth, yields and productivities (Fig. **1**). This concept is referred to as 'enviromics' [12]. Here the effect of the most important environmental parameters on thermophilic hydrogen production will be discussed, but before that the central carbon pathways are introduced through which the protons and electrons are generated for hydrogen production.

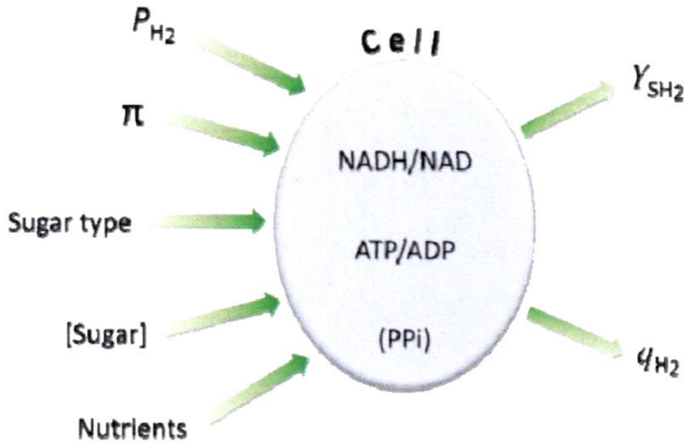

Figure 1: Enviromics of dark fermentation of sugars to hydrogen. Environmental factors (P_{H2} = hydrogen partial pressure, and π = osmolality) influencing the hydrogen yields (Y_{SH2}) and productivities (q_{H2}) *via* the redox- and energy metabolism of the cell (including PPi = pyrophosphate).

CENTRAL CARBON PATHWAYS

Glycolytic Pathways and Pentose Phosphate Pathway

The majority of microorganisms convert hexose sugars *via* the Embden-Meyerhof pathway (EMP) or the Entner-Douderoff pathway (EDP). The EDP is present in Gram-negative species and Archaea, and its widespread occurrence suggested that it may have predated the EMP [70]. However, it is not common in (strict) anaerobes - with the exception of the mesophilic *Zymomonas mobilis* [71], because substrate level phosphorylation yields only 1 ATP glucose^{-1}, half as much as can be gained *via* the EMP [72]. Hence, the EMP is the usual pathway for anaerobic conversion of hexoses, providing sufficient amounts of ATP to enable anaerobic growth. Interestingly, there are hyperthermophilic archaea and bacteria that use two central carbon pathways simultaneously. For instance, the hydrogen producing bacterium *Tm. maritima* ferments glucose to acetate and lactate with a flux distribution of 85% *via* the EMP and 15% *via* the EDP [73].

When focusing on the EMP among thermophilic hydrogen producers, several variations of the 'classical' EMP do exist with regard to hexose-kinases, glyceraldehydes-3-phosphate (GAP)-oxidizing enzymes and kinases converting phospho-enolpyruvate (PEP; Fig. **2**). The archaeon *P. furiosus* [74] and *Thermococcus* species [73] use ADP instead of ATP to donate phosphate-groups to the hexoses. GAP is oxidized in one step to 3-phosphoglycerate in *P. furiosus* and *T. litoralis* by the ferredoxin-dependent GAPOR [75], thereby omitting one ATP gaining step. In total, one net ATP is being produced by substrate-level phosphorylation, *i.e.* two ATP are produced from two PEP but one ATP is invested to convert AMP to ADP in the reaction:

$$ATP + AMP \rightarrow 2\ ADP \qquad \textbf{(3)}$$

Both ADPs can then be invested in the first two glycolytic reactions (Fig. **2**).

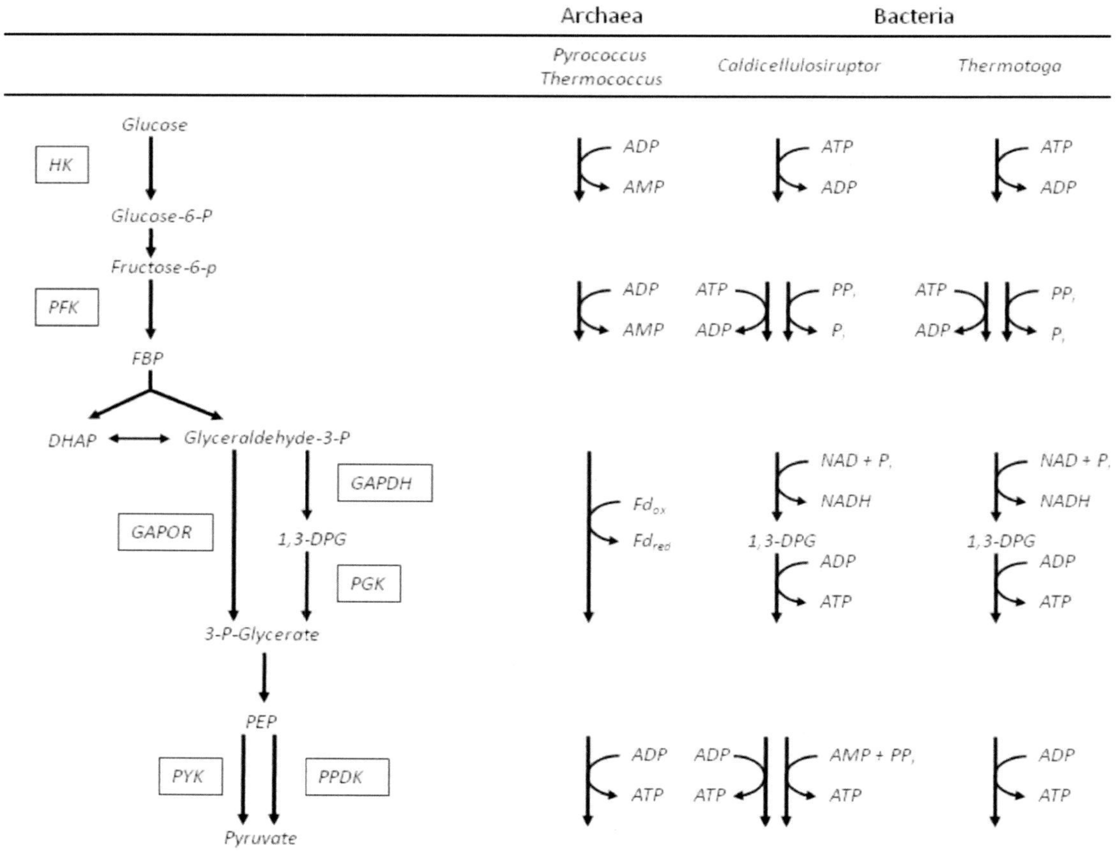

Figure 2: Comparison of EMPs in hyperthermophilic archaea and thermophilic bacteria. The net ATP production for these alternative EMPs are per mol glucose: 1 mol ATP, and 2-3 mol ATP for *Pyrococcus/Thermococcus*, and *Caldicellulosiruptor/Thermotoga*, respectively. The energy carriers (ATP, ADP, and PP_i) of hexokinase (HK) and 6-phosphofructokinase (PFK), and the enzymes for glyceraldehyde-3-phosphate oxidation (GAPOR = glyceraldehyde-3-phosphate:ferredoxin oxidoreductase; GAPDH = glyceraldehydes-3-phosphate dehydrogenase; PGK = phosphoglycerate kinase; PYK = pyruvate kinase; PPDK = pyruvate phosphate dikinase; DHAP = dihydroxyacetone phosphate; 1,3-DPG = 1,3-diphosphoglycerate; PEP = phospho enolpyruvate; Fd_{OX} = oxidized ferredoxin; Fd_{RED} = reduced ferredoxin). For references see in the text.

Tm. maritima possesses an EMP having a slight variation to the classical one, in that it has a pyrophosphate-dependent phosphofructokinase (PP_i-PFK) in addition to an ATP-dependent phosphofructokinase (ATP-PFK) [76]. In the case of *C. saccharolyticus*, it has been recently found that it uses PP_i as a major energy carrier in addition to ATP [77]. The annotated genome and enzyme assays provided information of the presence of a PP_i-PFK and an ATP-PFK. Likewise, there are two enzymes for the conversion of phospho-enolpyruvate (PEP) to pyruvate, *i.e.* an ADP-dependent pyruvate kinase (ADP-PYK) and a pyruvate phosphate dikinase (PPDK) deriving ATP from AMP and PP_i [8]. Both *Tm. maritima* and *C. saccharolyticus* obtain 2-3 mol ATP net production in the glycolysis, depending whether or not PP_i is used in the step catalyzed by PFK.

Pentose phosphate pathway. In general, pentose sugars are converted by microorganisms *via* the pentose phosphate pathway (PPP) or the phosphoketolase pathway (PKP), the latter being mainly used by lactic acid bacteria. In *Tm. maritima* and *C. saccharolyticus* pentoses are converted to fructose-6-P and GAP *via* the non-oxidative PPP [6,8]. Like with archaea [78], they lack the oxidative branch of the PPP, which excludes the obvious production site of NADPH required for biosynthesis. Several alternative enzymes that may produce NADPH have been suggested, but no serious investigations have been performed as yet.

Pathways at the Pyruvate Node

Pyruvate, the end-product of glycolysis, can be catabolically further converted to various fermentation products. The control of the fluxes to each of these products depends on the environmental conditions (see section Enviromics). The strict anaerobic hydrogen producers normally have a selection of pathways to reoxidize NADH, a necessary requirement to cope with variations in these environmental conditions. Examples of the pathways of several hydrogen producers are shown in Fig. **3**. Obviously, the preferred pathway is *via* pyruvate:ferredoxin oxidoreductase (PFOR), with acetate as the oxidized end-product and reduced ferredoxin as the electron donor of a NiFe-hydrogenase, to deliver H_2 as the reduced end-product. The conversion of acetyl-CoA to acetate includes an extra gain of ATP, which is the aim of the cell when it is in the mode of maximizing its energy carrier production.

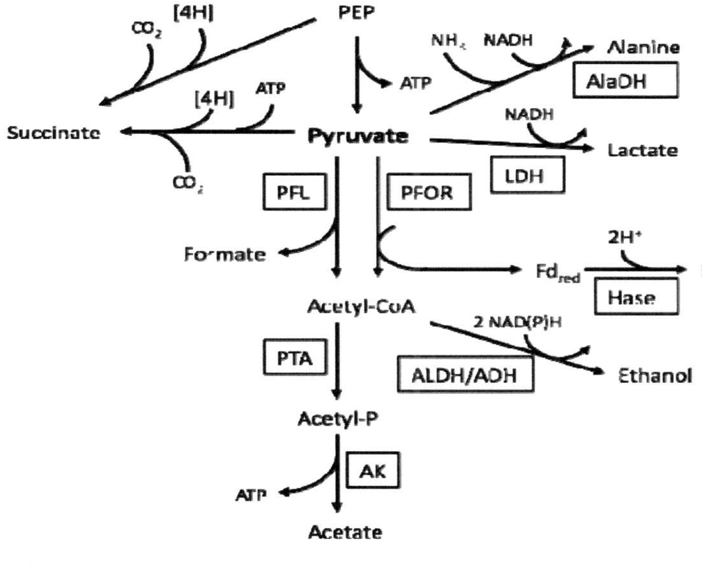

Figure 3: Metabolic pathways at the pyruvate node to: lactate (*Cl. thermocellum, C. saccharolyticus, Tm. maritima*), alanine *via* alanine dehydrogenase (AlaDH) (*P. furiosus, Thermotoga* species), Acetate, ethanol, & formate *via* PFL (*Cl. thermocellum*), or acetate, H_2, and ethanol *via* PFOR (*Cl. thermocellum, C. saccharolyticus, Tm. maritima, P. furiosus*) and succinate (*C. saccharolyticus*) [6,8,74,79-81].

Hydrogenases and Thermodynamics

The hydrogenases of thermophilic hydrogen producers have been well reviewed recently (*e.g.* [10,11]) and will thus not be discussed here in great detail. In general, two main types of hydrogenases have been described based on their metal cofactors, *i.e.* the Fe-only hydrogenases and the NiFe-hydrogenases. Strict anaerobes contain both types, whereas the NiFe-hydrogenases are more widespread among the (facultative) anaerobes and aerobes. Ferredoxin, NADH and NADPH function as their natural electron donors, and are reduced in the central carbon pathways. The two main reduction steps are at the conversion of GAP to 3-P-glycerate and the conversion of pyruvate to acetyl-CoA.

The low H_2 yields observed in mesophilic cultivation have primarily been attributed to the energetically unfavorable H_2-evolving reaction at $P_{H2} > 0.039$ kPa and at ambient temperatures [82,83]. The maximum P_{H2} depends on the difference in the midpoint redox potential of the reduced cofactor and H_2, as was described by Angenent *et al.* [82]:

$$P_{\mathrm{H2,max}} \leq \exp\left\{\frac{2F(E_{H_2}^{0'} - E_{NADH}^{0'})}{RT}\right\} \tag{4}$$

Where F is the Faraday's constant, R is the ideal gas constant, T is the absolute temperature, and E^0 is the midpoint redox potential, which is -320 mV for NADH and approx. -400 mV for Fd_{red}/Fd_{ox} [82]. The actual redox potential of the Fd_{red}/Fd_{ox} couple depends on the organism and temperature, *i.e.* the redox potential of this couple in *P. furiosus* and *Tm. maritima* at their optimal temperature is -600 mV and -453 mV, respectively [84]. It has been shown that even when the reactor is sparged with an inert gas to reduce P_{H2}, the H_2 concentration in the liquid exceeds that value 1000-fold [85], which prevents H_2 generation to be spontaneous. However, the hydrogen production reaction in these organisms will be spontaneous in their natural environment where methanogens and sulphate reducers consume the hydrogen.

In mesophilic clostridia it is therefore the NADH:Fd oxidoreductase (NOR), converting GAP-derived NADH to Fd_{red}, that is inhibited. Due to the lower E^0 of Fd_{red}, the critical P_{H2} for electron transfer from Fd_{red} to H_2 is significantly higher (30 kPa and 25°C; Eq. 4). Therefore, mesophilic anaerobes normally form other reduced compounds such as butanol, butyrate, ethanol, propionate and propanol in addition to H_2 resulting in a maximum yield of 2 mol H_2 mol hexose^{-1}. Several researchers have suggested that from a thermodynamic perspective it is much more favorable for the cell to use a NADPH:H_2 pathway than a NADH:H_2 pathway [83,86], although NADPH turnover is limited by the low flux through the PPP.

Another strategy to overcome this thermodynamic barrier is to increase the cultivation temperature and hence decrease the Gibbs free energy of conversion (ΔG) according to the second law of thermodynamics (provided that ΔS is positive):

$$\Delta G = \Delta H - T\Delta S \tag{5}$$

Where ΔH is the change in enthalpy, T is the absolute temperature and ΔS is the change in entropy. Indeed, extreme and hyperthermophiles reach higher hydrogen yields than mesophiles. However, the optimum growth temperature of *e.g. C. saccharolyticus* is 70°C which would mean that the critical P_{H2} is 0.173 kPa being still well below the actual P_{H2} in the liquid [81]. One possible explanation for this contradiction could be the recently characterized novel type of bifurcating hydrogenase [87] (see below).

Based on its genome sequence, *C. saccharolyticus* possesses two distinct hydrogenases, *i.e.* one NADH-dependent Fe-only hydrogenase and one Fd_{red}-dependent, membrane-associated NiFe hydrogenase [8]. Given that the redox potential of Fd_{red} (E^0 approx. -400 mV) is close to that of H_2 (E^0_{H2}: -414 mV), the Fd-dependent hydrogenase reaction is energetically favorable even at P_{H2} below 39 kPa at 70°C. This calculation is based on the value given by Angenent *et al.* [82]; however, the actual midpoint redox potential of Fd_{red}/Fd_{ox} depends on the strain and the temperature [84]. The actual midpoint redox potential for ferredoxin in *C. saccharolyticus* is not known but should be at the same magnitude as in other related thermophilic bacteria. Nevertheless, it is obvious from thermodynamics that it is primarily the NADH-dependent Fe-only hydrogenase that is inhibited by elevated P_{H2}. Consistently, the NADH-dependent Fe-only hydrogenase in *Th. tengcongensis* was downregulated at high P_{H2}, whereas the Fd-dependent NiFe-hydrogenase is constitutively expressed, independently of the P_{H2} [20].

Recently, Schut and Adams [87] characterized a novel type of Fe-only hydrogenase in *Tm. maritima* that uses NADH and Fd_{red} simultaneously. This novel bifurcating hydrogenase could therefore catalyze the unfavorable oxidation of NADH to H_2 by using the exothermic oxidation of Fd_{red} as a driving force. Interestingly, many of the thermophiles listed in Table **1** and several mesophilic clostridia possess a Fe-only hydrogenase with sequence similarity to the bifurcating hydrogenase in *Tm. maritima*, but it remains to be investigated whether these hydrogenases possess a bifurcating function. This observation could give a plausible explanation to the ability of these thermophiles to produce high H_2 yields in well-sparged bioreactors. However, an intriguing question that arises is why these thermophiles have evolved optimized H_2-production. Knowledge about this subject would aid metabolic engineers to construct high H_2-yielding cell factories. A plausible explanation is discussed below.

Sugar Transport

Bacterial and archaeal ABC transporters consist of five subunits, *i.e.* a sugar binding protein, two transmembrane domains and two nucleotide-binding domains [88]. The substrate-binding protein has been characterized in several thermophilic hydrogen producers, including *P. furiosus* [89], *Tm. maritima* [90,91] and *Thermococcus litoralis* [92]. A common characteristic of these proteins is their ability to capture the sugar outside the cell *via* a cap-lock mechanism, which determines the high affinity for the substrate [93].

Genome annotation has revealed that *Tm. maritima* and *C. saccharolyticus* possess an unusually high number of ABC transporters [6,8], but none of these transporters have been characterized so far. These transporters of *Tm. maritima* seem to be controlled by local regulators responsive to either the transport substrate or a key metabolic degradation product [94]. Based on transcriptional analysis, it has been concluded for *C. saccharolyticus* that the same transporter can be used for translocating different sugars with varying affinities [8,95].

From a cell perspective, one pronounced drawback of ABC transporters is their relatively high energy demand [93]. At least one mole of ATP is required for sugar translocation and a second mole for sugar phosphorylation. This is a relatively high energy investment compared to the more 'economical' PEP-dependent phosphotransferase system (PTS)-based transport that consumes one PEP molecule for both translocation and phosphorylation. Given that 1 PEP is usually equivalent to 1 ATP, makes the PEP-dependent PTS at least twice as economical as the ABC-transport system. Some organisms might be compensating for this high energy cost through a more energy-efficient metabolism. One of such energy-conserving strategies is to oxidize glucose to acetate, resulting in two more ATP per hexose. In this process, H_2 is an electron sink formed in the re-oxidation of the reduced cofactors Fd_{red} and NADH.

This kind of physiology might well be a consequence of the adaptation to the original habitat of organisms such as *C. saccharolyticus*. Being isolated from wood floating in a hot spring in a lake in New Zealand [96], the free sugar concentration will be low and only slowly released through breakdown of (hemi)cellulose. The high number of (hemi)cellulolytic enzymes [8, 97,98], its ability to co-metabolize a large spectrum of sugars [56,66,95] and the lack of glucose repression [8] are further indications of this adaptation.

ENVIROMICS

pH and Temperature

At industrial scale, it will be hardly possible to operate hydrogen fermentations under aseptic conditions. Under those conditions, the pH and temperature are two environmental parameters usually applied as a means to prevent methanogenesis. The effect of these two parameters on hydrogen production is briefly discussed here.

pH. Methane production is limited to a small window between pH 5.5 and 8.2. Indeed, fermenting household waste with mixed cultures at 70ºC, methane production was observed between pH 6 and 7, but was absent at pH 5.5 and lower [59]. As long as methanogens were present, all hydrogen produced was consumed, but hydrogen production could be measured only in low quantities at pH ≤ 5.5. Instead, lactate became a major fermentation product at that pH [59], but whether this was produced by the hydrogen producers remains to be determined. In many studies, it was found that a neutral to slightly acidic pH (6-7) is optimal for hydrogen production [17,58,99,100]. Apparently, at elevated temperatures optimal hydrogen production shares (part of) the same pH window as methanogenesis, thus process operation at low pH is not a practical solution.

Temperature. Few studies have investigated the effect of the temperature on the same inoculum, and drawing the same conclusion: the specific rate of hydrogen production increased with the temperature [49,101]. For instance, the specific hydrogen-production rates were about 4-10 times higher at 55-60ºC than at 35-37ºC. This could be mainly due to the lower biomass production at elevated temperature (≥55ºC) [49]. Yokoyama *et al.* [101] observed two peaks of high hydrogen production from a mixture of manure and urine, *i.e.* one at 60 and one at 75ºC. These two temperature islands corresponded to the optimal growth temperatures of the thermophilic hydrogen producers present in the cow waste slurry used as the inoculum. Indeed, a DGGE analysis revealed that at 60˚C species similar to *Bacteroides xylanolyticus*, *Cl. stercorarium* and *Cl. thermocellum* were present, whereas at 75˚C species related to *Caldanaerobacter subterraneus* were found [102]. Ren *et al.* [100] showed that the hydrogen yield of *Th. thermosaccharolyticum* W16 is correlated to the growth temperature with the highest yield at the optimum growth temperature. It is expected that this will be the case for each hydrogen producer.

Interestingly, a proteomic study of *Th. tengcongensis* cultured at different temperatures revealed that the temperature-sensitive proteins can be broadly divided into two groups: i) specific expression at certain temperatures and (ii) consistent temperature-dependent changes of expression [103]. Especially enzymes involved in redox-regulation capacity are uniquely detected under sub-optimal growth temperatures, but decline strongly at higher temperatures. It is reasoned that this weakens growth at higher temperatures, especially beyond the optimal growth temperature.

Partial Hydrogen Pressure

Thermodynamics have been used to determine the critical P_{H2} for the hydrogenase reaction presenting a negative ΔG [82]. However, the ΔG of the reaction (Eq. 3) is a function of: i) the temperature, ii) the type of cofactor and iii) the ratio of reduced and oxidized cofactors [83]. Different cofactors present varying electric midpoint potentials, *i.e.* the E^0 for Fd_{red}/Fd_{ox} < NADPH/NADP < NADH/NAD [82-84] and unless the cofactor specificity is known, it is not possible to determine the thermodynamics of the system. Moreover, the E^0 of the ferredoxin couple depends on the enzyme and the temperature. The ratio of reduced and oxidized cofactors could also be problematic to determine since it varies with the growth rate and the organism [104,105]. Therefore, this lack of information makes it not possible to use the thermodynamic constraint to estimate the tolerance to hydrogen.

For practical reasons, the P_{H2} is generally used as a measure for hydrogen tolerance. The currently often quoted critical P_{H2}, defined as at which value lactate formation is initiated in *C. saccharolyticus*, is 10-20 kPa [106]. However, further studies have established that the value of the critical P_{H2} varies with environmental conditions. For instance, this quoted critical P_{H2} was determined in batch cultivation on sucrose using no stripping gas in a closed bioreactor. However, in similar experiments, using the same conditions but with xylose as the substrate, the P_{H2} in the headspace reached 60 kPa before the metabolism shifted to lactate [81]. Interestingly, *C. owensensis* is more sensitive to P_{H2}, producing significantly more lactate, when it is growing on xylose rather than on glucose [19]. A possible explanation for these different values for the critical P_{H2} could be that the productivities differed with the two substrates, resulting in different concentrations of dissolved H_2 [81], or it could be because of differences in metabolic stoichiometries between the two substrates.

In addition, during batch growth of *C. saccharolyticus* on glucose (5 g L-1) with continuous N_2 sparging, lactate formation was triggered at the transition to the stationary phase, even though the *P*H2 was 6.3 kPa [81], whereas without stripping exponentially growing cells initiated lactate formation at a *P*H2 of 30 kPa. Moreover, growth and H_2 production were still observed in continuous cultures at a low dilution rate (0.05 h-1) at a P_{H2} of 67 kPa. However, at this high P_{H2}, the metabolism had shifted predominantly (57%) to lactate, thereby decreasing the H_2 productivity and thus the dissolved hydrogen concentration substantially [81].

Due to a lack of a pattern in the critical values for P_{H2} found for each fermentation conditions, it may not be an accurate parameter. Indeed Pauss *et al.* [107] demonstrated in various bioreactor systems that hydrogen is, due to its low solubility, easily supersaturated in the liquid and thus not in equilibrium with the gas phase. A study with *C. saccharolyticus* in a continuous stirred bioreactor confirmed this conclusion that the dissolved H_2 is a more true parameter [81], being a function of H_2 productivity and the mass transfer rate. The cells being in the liquid phase are confronted with the dissolved H_2. The mass transfer rate is, in turn, a function of the gas stripping rate and the stirring rate. In addition, cultivations performed in closed systems without stripping have lower mass transfer rates. Thus, in that case the dissolved H_2 concentration with P_{H2} is even further from equilibrium (as depicted by Henry's law) than in a sparged and open system. This means that the critical P_{H2} determined in closed systems gives a more underestimated value of the tolerance to H_2 than when determined in an open system.

To counterbalance the H_2 hold up in the liquid, the cells will decrease their H_2 productivity to direct away from adverse conditions that may lead to growth inhibition, for instance by shifting the metabolism to lactate to re-oxidize NADH. However, recent studies [81,104,105] revealed that a critical P_{H2} is not always the trigger for lactate formation. Instead, in *C. saccharolyticus* lactate formation is triggered by a combination of the dissolved H_2 concentration and an increased osmolarity through the accumulation of acetate (2 mol acetate per mol glucose). Therefore, lactate can be triggered even though the dissolved hydrogen concentration is low.

In conclusion, the tolerance of *C. saccharolyticus* for H_2 is directly related to the dissolved H_2 concentration, but an increasing osmotic pressure will proportionally lower the H_2 tolerance level. Nonetheless, for practical reasons, since data on dissolved H_2 concentration are scarce in the literature and the P_{H2} is usually the determined parameter, the P_{H2} will be used in the continuation of this chapter.

Osmolality

Hydrogen fermentations have been observed to be affected by higher sugar or by-product concentrations [99,100,106,108], although the effects have not yet been widely studied. Usually, the organic acids formed in the fermentation act as uncouplers thereby creating sub-optimal conditions increasing cellular maintenance requirements and thus inhibit growth. This effect is enhanced at lower pH, due to the dissociation constant of the acids produced. It could be argued that this may be one of the major causes why growth of thermophilic hydrogen producers have a limited pH range around neutral pH.

Alternatively, the osmotic potential can act as a parameter of inhibition since both sugars and organic acids have the same inhibiting effect on growth and hydrogen production [109]. Regarding the osmotic potential, the most advanced study has been carried out with *C. saccharolyticus*, being sensitive to osmotic pressures beyond 0.218 ± 0.005 osm kg H_2O^{-1}, corresponding to the production of about 39 mM acetate and HCO_3^- under N_2-sparged conditions [81,110]. At this critical osmotic potential cell lysis is induced, as evidenced by increases of protein and DNA in the culture supernatant [110]. The critical molarity at which growth is completely inhibited is estimated to be 400-425 mM [106]. To avoid growth inhibition, stripping with CO_2 should be avoided and glucose concentrations of 5 g L^{-1} should be used, giving acetate and bicarbonate concentrations of 55 mM and 70 μM, respectively [110]. The low glucose concentration has obviously a negative effect on hydrogen productivity.

It remains to be established why *C. saccharolyticus* is not able to withstand increased osmolarities, It would be interesting to produce genetic variants that have increased osmotolerance. On the other hand, (moderate) halophilic thermophiles, such as *Tm. neapolitana* or *P. furiosus*, could be an interesting alternative [66]. Indeed, the osmotolerance of *Tm. neapolitana* and *P. furiosus* are much higher than for *C. saccharolyticus*, growing optimally in the presence of 0.46 M and 0.5 M NaCl, respectively [111,112]. The higher osmotolerance of these two thermophiles is due to that they produce compatible solutes such as β-mannosylglycerate, di-myo-inositol-phosphate and glutamate. It remains to be investigated whether *C. saccharolyticus* is able to efficiently synthesis these or other solutes when exposed to osmotic stress. The low salt concentration in its natural habitat [113] may have resulted in a loss of this ability [114]. However, the presence of genes for osmoprotectant biosynthesis such as for proline and glutamate [110] could be an opportunity for directed evolution to render *C. saccharolyticus* more osmotolerant.

Medium Components

Carbon and energy source. The majority of studies used thermophilic hydrogen producers originating from environments that are pervaded with polysaccharides, such as hot springs, waste-water treatment systems, compost heaps and cattle manure. Therefore, a lot of them are capable of not only metabolizing monosaccharides (hexoses and pentoses), but also oligosaccharides and polysaccharides (starch, inulin, cellulose, xylan, and pectin). Recently, there have been several reviews on the hydrolases involved in the breakdown of these polysaccharides [9,115,116] thus it will not be treated here.

In nature, but also in the hydrogen fermentation process running on *e.g.* hydrolysates, the microorganisms will be exposed to (complex) sugar mixtures. Several of the thermophilic hydrogen producers described prioritize glucose over other sugars, and for some of them catabolite repression has been proven [100,117]. *Tm. neapolitana,* although a Gram-negative, possesses a cAMP-independent repression mechanism but may have a more ancient mode of regulation instead [117]. Interestingly, *Caldicellulosiruptor* species lack catabolite repression [8,31,95]; but even though all sugars are taken up simultaneously, they seem to prefer xylose. Lack of catabolite repression is thus a welcome characteristic for fermentation of sugar mixtures.

Growth activation. A very intriguing topic, not profoundly investigated yet, is growth activation. It has been found recently that, in contrast to glucose, growth of *C. saccharolyticus* on xylose and arabinose did not initiate in the absence of sufficient CO_2 [81]. This observation was made in reactor systems sparged with N_2, but was absent when using CO_2 as sparging gas and in closed bottles where self-generated CO_2 is accumulating. Adequate amounts need to be present in the cell to be converted to bicarbonate, which is the substrate for 'CO_2-fixing' pathways. No specific CO_2-fixation pathway was being upregulated with growth on pentoses, as confirmed by a combination of bioinformatic, ^{13}C-NMR and available microarray data analysis. Since addition of acetate had the same positive effect as sparging with CO_2, it was suggested that intracellular acidification of self-generated CO_2 and acetate is the usual trigger of growth. So far, growth activation through intracellular acidification has been studied extensively in *Saccharomyces cerevisiae* only [118], but it remains an elusive process. Alternatively, with pentoses the CO_2/bicarbonate concentrations remain far below the affinity of one or more of the CO_2 fixing enzymes, which include phospho-ribosyl aminoimidazole carboxylase, acetyl-CoA carboxylase and pyruvate carboxylase.

Per molecule of pentose less CO_2 and acetate is produced than per molecule of hexose, apparently keeping their intracellular concentrations below a critical triggering threshold, which is further maintained when the liquid phase is sparged [82]. This phenomenon is most probably widespread; for instance, growth initiation of *Rhodopseudomonas* sp. is also facilitated by accumulated CO_2 through temporarily ceasing gas stripping [119].

Other medium components. O-Thong *et al.* [99] compared inorganic (NH_4Cl) and organic (peptone) nitrogen sources using a similar C:N ratio, showing that on the latter twice more hydrogen could be produced in a culture of *Th. thermosaccharolyticum* PSU-2 with glucose, probably because of a three-fold lower ethanol production. In addition, with peptone two-fold less acetate and 10-fold more butyrate was produced and the biomass increased two-fold. In total more fermentation products were produced, indicating that peptone was partly concurrently metabolized.

So far, no dedicated studies have been performed on other nutrients that are essential to optimize thermophilic hydrogen production. It is possible that certain nutrients are required at higher concentrations than expected. For instance, usually iron concentrations in the order of 25-40 µmol L^{-1} should be not limiting fermentative growth, yet up to 6 mM was required for optimal hydrogen production by *Cl.*

acetobutylicum [120]. Iron is an important nutrient for hydrogenases and ferredoxin, which are essential for optimal hydrogen production. One reason for such a high iron concentration in the culture could be the increasing immobilization of iron through chelation by accumulating fermentation products such as acetate, butyrate and lactate [121]. Obviously, the same can be said for the other trace metal ions in the culture.

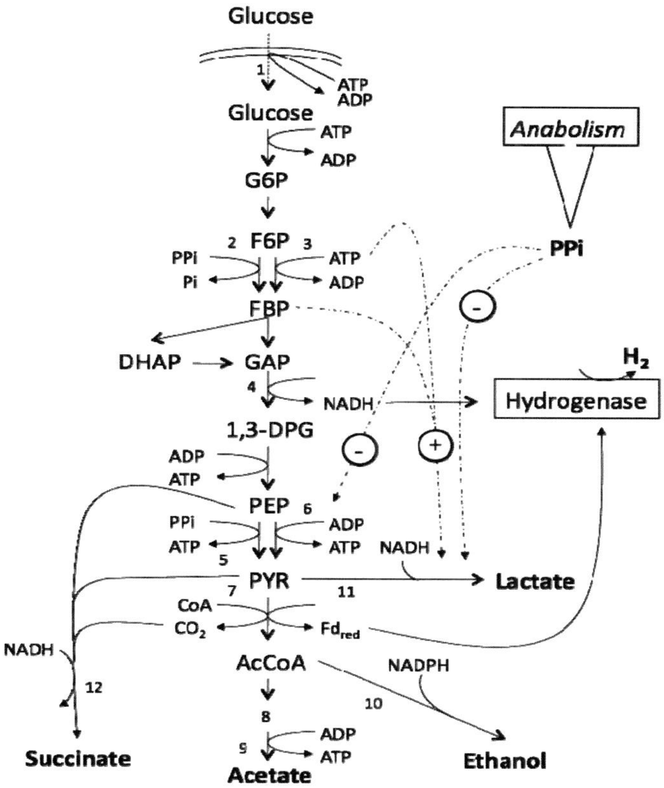

Figure 4: Proposed catabolic network in *C. saccharolyticus*. Glucose is transported over the membrane by an ABC-transport system (1; [8]) and oxidized to pyruvate in the EMP pathway [18] with the exception that PPi can potentially replace ATP (PPi-PFK (2) and ATP-PFK (3)) or ADP (PPDK (5) and PYK (6); [78]). NADH generated from GAPDH (4) and Fd_{red} generated from PFOR (7) can be converted to hydrogen through the catalysis of the cytosolic NADH-dependent Fe-only hydrogenase and membrane-bound Fd-dependent NiFe-hydrogenase, respectively. Acetate is formed from PTA (8) and AK (9), which yields an extra ATP. Alternative routes for NADH reoxidation are formation of ethanol (ADH; (10)), lactate (LDH; (11)), and succinate ((12); [81]). Filled lines show metabolic routes and dashed lines show metabolite activation (+) or inhibition (-) on enzyme level. PPi generated from anabolic reactions is a strong inhibitor of both PYK and LDH activity [77,104].

Regulation of Carbon Flow and Hydrogen Production

Regulation of the classical glycolysis and the flux distribution at the pyruvate node is a very complex and highly tuned process, since it is the principle provider of energy carriers and redox equivalents that need to be precisely balanced under all growth conditions. Regulation takes place at various metabolic levels, but mostly at enzyme and gene expression levels. Enzyme activities normally are regulated *via* direct mass action law (concentrations of substrates or effectors) or *via* allosteric inhibition or activation. Each regulation step has its own relaxation time frame, *i.e.* mass action law < 10^{-4} sec, allosteric control 10^{-5}-100 sec, and gene expression level that affect enzyme concentration >10^{3} sec [122]. Thus, with good allosteric regulation the metabolic system will adapt promptly and smoothly to fluctuations in environmental conditions. Yet, regulation of the glycolytic flux in *P. furiosus* and other archaea has been observed to be mainly at the level of gene expression, such that of GAPOR, and less pronounced *via* allosteric regulation [75]. Apparently, not even the ADP-GLK and ADP-PFK, catalyzing irreversible reactions, are allosterically regulated by any of the known allosteric effectors.

Recently, insights are beginning to be made into the regulation of the flux of glucose to fermentation products in *C. saccharolyticus* and is used as an example here. Under optimal growth conditions, *C. saccharolyticus* produces H_2 yields close to the Thauer limit and it uses the EMP [18] to oxidize glucose to pyruvate, which is further oxidized to acetate with the generation of reduced Fd. Molecular H_2 is then obtained from NADH (generated in the EMP) and Fd_{red} (Fig. **4**). During catabolism, both ATP and PPi are used as energy carriers, the latter being produced entirely in anabolic pathways. PPi is usually a 'waste product', but many thermophiles use it to phosphorylate sugars (via PPi-PFK) and retrieve ATP (via PPDK) (Fig. **2**; [76,77,123]), thereby linking catabolism and anabolism more intensively. This further means that PPi has more influence at higher biosynthetic activities, *i.e.* at higher growth rates. This is supported by two different observations: (i) the PPi concentration is significantly higher in the exponential phase (4 ± 2 mmol L^{-1}) than in the stationary phase (0.6 ± 0.1 mmol L^{-1}) [77], thus having more mass action power; and (ii) PPi is a strong inhibitor of the ATP-dependent PK in *C. saccharolyticus* [77] and most probably also of the ATP-PFK, like in *Tm. maritima* [76].

At high growth rates sugars are converted to acetate obtaining the maximum hydrogen yield. As soon as the exponential growth phase transitions to stationary phase, lactate is formed as well. Lactate formation is accompanied by a rise in the NADH/NAD ratio, decline of the PPi concentration and slight rise in the ATP concentration [77,104]. It was found that lactate dehydrogenase is allosterically activated by ATP and NADH, but was strongly hampered by NAD and PPi through competitive inhibition [104]. Thus it is clear that hydrogen yields are also dependent on energy metabolism in addition to the known redox metabolism. More research is needed how other reduced products such as ethanol and succinate are regulated and whether similar mechanisms are present in other thermophilic hydrogen producers.

REACTOR DESIGN

Reactor Types and Operation

The majority of tests of different types of reactors and modes of operation for hydrogen production made use of mesophilic mixed cultures [1]. The few studies using thermophiles will be discussed and evaluated here with respect to cost-effectiveness.

Usually these studies aim for high volumetric productivities, while putting less emphasis on hydrogen yields, but the opposite is true when working with thermophiles (Table **3**). Indeed, higher yields can be obtained with thermophiles, although optimizations of the process are still needed. In most cases undefined cultures were applied, and so far only one report on a pure culture in a pilot plant-scale reactor has been tested [13]. The hydrogen yield obtained with *C. saccharolyticus* is higher than obtained with mesophilic or undefined thermophilic consortia. Magnusson *et al.* [62] succeeded in producing hydrogen from α-cellulose (4 g L^{-1}) by *Cl. thermocellum* for 125 days (Table **3**). Yet, more comparisons between pure cultures, undefined consortia and designed co-cultures are needed in these reactor systems before the best organisms can be selected [31].

Table 3: Types of Thermo-Bioreactors in Studies on Hydrogen Production, with Productivity (Q_{H2} mmol L Working Volume^{-1} h^{-1}), and Yield (Y_{H2} mol H_2 mol Hexose^{-1}).

Reactor	Inoculum	Reactor size (L)	T (°C)	Q_{H2}	Y_{H2}	Reference
CSTR	*Cl. thermocellum*	5	60	1.0	1.29	[124]
Trickle bed	*C. saccharolyticus*	400	70	22	2.80	[13]
Trickle bed	sludge	2	60	44	1.11	[125]
UASB	Enriched sludge	3	55	5.0	2.14	[126]
UASB	Mixed culture	0.22	70	2.1	2.47	[30]

The proof of principles have been made, revealing that both trickle bed and UASB reactors seem to be promising concepts for obtaining desired yields. Yet many challenges are awaiting to improve productivities while preserving proper yields. For instance, the lower hydrogen yields observed are coupled to reduced by-product formation, such as lactate and ethanol. Metabolic shifts to these undesired products are triggered by higher P_{H2}, growth phase and/or sugar concentrations. These are critical parameters that should be taken into account in the optimization of the reactor configuration and operation mode. One of the critical points is whether systems without sparging gases can be developed (see next section).

Exposing the unsterile feed to 70ºC for *e.g.* 1 hour is a means to prevent development of methanogenesis in the reactor. Thus, it may be expected that operating the fermentation at this elevated temperature is even better, yet thermophilic methanogens eventually manifest when the hydraulic retention time (HRT) is above 1 day [59]. Alternative start-up strategies were also successful, such as bromoethanosulfonate (BrES) and flow recycle [30]. The latter was also convenient to achieve higher hydrogen yields. The interesting outcome of the study by van Groenestijn *et al.* [13] that even though the reactor was run under non-sterile conditions, the originally introduced strain remained dominant during 3 months operation. Therefore, for practical reasons it might be better to use a defined (co-)culture as inoculum, not only to avoid a high probability of methanogenesis, but also reduce the risk of development of satellite populations not contributing to hydrogen production. Preservation of the applied cultures could be even further secured by making the thermophilic strains involved more osmotolerant.

Gas Removal

For a cost-effective industrial-scale operation of a dark fermentation, it is important that the H_2 gas is removed from the liquid phase as efficiently as possible to prevent inhibition of its production. One possibility is to strip the culture with an appropriate gas to increase the mass transfer of H_2 from the liquid to the gas phase, which generally results in higher yields [127]. Although N_2 is commonly used as a stripping gas for lab-scale H_2 production, it is not cost-efficient since it is inert and, therefore, difficult to remove from the effluent gas stream. Instead, it has been suggested that CO_2 can be an appropriate alternative since it can be more easily separated from H_2 and its costs are low because it is a product of the fermentation process [128]. However, it has been shown that sparging with CO_2 can negatively influence H_2 productivity, as reported for *C. saccharolyticus* [110], due to the inevitable increase of osmotic pressure. Higher P_{CO2} increases the concentration of dissolved CO_2, which is hydrolyzed to bicarbonate and protons, the latter which needs to be neutralized with additional caustic agent to maintain the pH. Secondary disadvantages of CO_2 are increase of costs for caustic agent and substantial contribution to environmental burden.

Other strategies, such as applying partial vacuum to remove H_2 and CO_2 from the gas phase, have been shown to efficiently increase H_2 yields at lab scale [129]. However, on an industrial scale under-pressure might not be economically feasible since it is coupled to a high energy cost and may increase the risk of contamination. Instead, researchers are left with at least two alternatives, i) improvements of reactor design, focusing on how to decrease the levels of dissolved H_2 and CO_2, and ii) metabolic engineering and directed evolution of the microorganism to enable high H_2 yields at high dissolved H_2 concentration as far as it is thermodynamically possible.

OUTLOOK

In the past decade thermophilic biohydrogen production has gained increasing interest. Many new candidates have been isolated, characterized and their optimal process conditions have been determined. Further progress in understanding how to improve hydrogen yields and productivities has to come from investigating other physiological and process aspects, of which the most important are treated shortly below:

Biofilm and bioreactors. It has been observed that immobilization on granulated activated carbon, wood shavings or perlite improved hydrogen production by *C. saccharolyticus* [130]. Immobilized culture systems may indeed improve productivity considerably compared to suspended cultures, but often this is compensated by a lower hydrogen yield [49]. Here lies a challenge, how to further improve bioreactor system configurations to gain productivity without jeopardizing the hydrogen yield. The focus has to be on efficient hydrogen removal from the liquid. Also the concept of immobilized designed co-cultures may deserve further scrutiny [31].

Deepen understanding of physiology. Understanding the physiology of hydrogen production is vital to improve yields and productivities. For this, genetic engineering and sophisticated 'omics' tools need to be applied. Omics-based technologies are increasingly applied for thermophilic hydrogen producers [8, 103,131], but so far genetic protocols have been developed only for a few thermophilic hydrogen producers (*e.g.* [132]). Together with genome-wide metabolic models, now being developed for several thermophiles (*e.g.* [133]), the systems biology approach can eventually become manifest for thermophilic hydrogenomic physiology.

Apply thermophilic hydrolase capacity. Many of the thermophilic hydrogen producers are able to breakdown lignocellulosic feedstocks through their possession of cellulases and xylanases. The status and prospects of exploiting these thermophiles for this task in biomass conversion has been recently reviewed [9], but it still needs intensive effort before it can be commercially applied. If successful, it might replace the commercial cellulases and thus positively affect the cost price of hydrogen.

ACKNOWLEDGEMENTS

This work was supported by the EU FP6-SES IP HYVOLUTION (contract no. 019825) and the Swedish Energy Agency (Energimyndigheten) project Biohytan (project no. 31090-1).

REFERENCES

[1] Hallenbeck PC, Ghosh D. Advances in fermentative biohydrogen production: the way forward? Trends Biotechnol 2009; 27(5): 287-97.

[2] Thauer RK, Jungermann K, Decker K. Energy conservation in chemotrophic anaerobic bacteria. Bacteriol Rev 1977; 41(1): 100-80.

[3] Claassen PAM, de Vrije T. Non-thermal production of pure hydrogen from biomass: HYVOLUTION. Int J Hydrogen Energy 2006; 31(11): 1416-23.

[4] Robb FT, Maeder DL, Brown JR *et al.* Genome sequence of hyperthermophile, *Pyrococcus furiosus*: implications for physiology and enzymology. Methods Enzymol 2001; 330: 134-57.

[5] Roberts SB, Gowen CM, Brooks JP, Fong SS. Genome-scale metabolic analysis of *Clostridium thermocellum* for bioethanol production. BMC Syst Biol 2010; 4: 31.

[6] Nelson KE, Clayton RA, Gill SR *et al.* Evidence for lateral gene transfer between Archaea and Bacteria from genome sequence of *Thermotoga maritima*. Nature 1999; 399(6734): 323-9.

[7] Bao Q, Tain Y, Li W *et al.* A complete sequence of the *T. tengcongensis* genome. Genome Res 2002; 12(5): 689-700.

[8] Van de Werken HJG, Verhaart MRA, VanFossen AL *et al.* Hydrogenomics of the extremely thermophilic bacterium *Caldicellulosiruptor saccharolyticus*. Appl Environ Microbiol 2008; 74(21): 6720-9.

[9] Blumer-Schuette SE, Kataeva I, Westpheling J, Adams MWW, Kelly RM. Extremely thermophilic microorganisms for biomass conversion: status and prospects. Curr Opin Biotechnol 2008; 19(3): 210-7.

[10] Chou C-J, Jenney Jr FE, Adams MWW, Kelly RM. Hydrogenesis in hyperthermophilic microorganisms: implications for biofuels. Metab Eng 2008; 10(6): 394-404.

[11] Kengen SWM, Goorissen HP, Verhaart M, van Niel EWJ, Claassen PAM, Stams AJM. Biological hydrogen production by anaerobic microorganisms. In: Soetaert W, Vandamme EJ, Eds. Biofuels. Chichester: John Wiley & Sons 2009; pp. 197-221.

[12] Lidén G. Understanding the bioreactor. Bioproc Biosyst Eng 2002; 24(5): 273-9.

[13] Van Groenestijn JW, Geelhoed JS, Goorissen HP, Meesters KPM, Stams AJM, Claassen PAM. Performance and population analysis of the non-sterile trickle bed reactor inoculated with *Caldicellulosiruptor saccharolyticus*, a thermophilic hydrogen producer. Biotechnol Bioeng 2008; 102(5): 1361-7.

[14] Levin DB, Islam R, Cicek N, Sparling R. Hydrogen production by *Clostridium thermocellum* 27405 from cellulosic biomass substrates. Int J Hydrogen Energy 2006; 31(11): 1496-503.

[15] Schröder C, Selig M, Schönheit P. Glucose fermentation to acetate, CO_2 and H_2 in the anaerobic hyperthermophilic eubacterium *Thermotoga maritima* - Involvement of the Embden-Meyerhof Pathway. Arch Microbiol 1994; 161(6):460-70.

[16] Munro SA, Zinder SH, Walker LP. The fermentation stoichiometry of *Thermotoga neapolitana* and influence of temperature, oxygen, and pH on hydrogen production. Biotechnol Progr 2009; 25(4):1035-42.

[17] Van Niel EWJ, Budde MAW, de Haas GG, van der Wal FJ, Claasen PAM, Stams AJM. Distinctive properties of high hydrogen producing extreme thermophiles, *Caldicellulosiruptor saccharolyticus* and *Thermotoga elfii*. Int J Hydrogen Energy 2002; 27(12):1391-8.

[18] De Vrije T, Mars AE, Budde MAW *et al.* Glycolytic pathway and hydrogen yield studies of the extreme thermophile *Caldicellulosiruptor saccharolyticus*. Appl Microbiol Biotechnol 2007; 74(6):1358-67.

[19] Zeidan AA, van Niel EWJ. A quantitative analysis of hydrogen production efficiency of the extreme thermophile *Caldicellulosiruptor owensensis* OL^T. Int J Hydrogen Energy 2010; 35(3):1128-37.

[20] Soboh B, Linder D, Hedderich R. A multisubunit membrane-bound [NiFe] hydrogenase and an NADH-dependent Fe-only hydrogenase in the fermenting bacterium *Thermoanaerobacter tengcongensis*. Microbiology 2004; 150(Pt 7): 2451-63.

[21] Hild HM, Stuckey DC, Leak DJ. Effect of nutrient limitation on product formation during continuous fermentation of xylose with *Thermoanaerobacter ethanolicus* JW200 Fe(7). Appl Microbiol Biotechnol 2003; 60(6): 679-86.

[22] Kengen SWM, Stams AJM. Growth and energy-conservation in batch cultures of *Pyrococcus furiosus*. FEMS Microbiol Lett 1994; 117(3): 305-9.

[23] Kanai T, Imanaka H, Nakajima A *et al.* Continuous hydrogen production by the hyperthermophilic archaeon, *Thermococcus kodakaraensis* KOD1. J Biotechnol 2005; 116(3): 271-82.

[24] Stanbury PF, Whitaker A, Hall SJ. The isolation, preservation, and improvement of industrially important microorganisms. In: Principles of Fermentation Technology, 2nd ed. Oxford: Butterworth-Heinmann 2000; pp. 35-91.

[25] Hallenbeck PC. Fundamentals of the fermentative production of hydrogen. Water Sci Technol 2005; 52(1-2): 21-9.

[26] O-Thong S, Prasertsan P, Birkeland NK. Evaluation of methods for preparing hydrogen-producing seed inocula under thermophilic condition by process performance and microbial community analysis. Bioresource Technol 2009; 100(2): 909-18.

[27] Zhu HG, Beland M. Evaluation of alternative methods of preparing hydrogen producing seeds from digested wastewater sludge. Int J Hydrogen Energy 2006; 31(14): 1980-8.

[28] Kleerebezem R, van Loosdrecht MCM. Mixed culture biotechnology for bioenergy production. Cur Opin Biotechnol 2007; 18(3): 207-12.

[29] Li CL, Fang HHP. Fermentative hydrogen production from wastewater and solid wastes by mixed cultures. Crit Rev Env Sci Tec 2007; 37(1): 1-39.

[30] Kotsopoulos TA, Zeng RJ, Angelidaki I. Biohydrogen production in granular up-flow anaerobic sludge blanket (UASB) reactors with mixed cultures under (hyper)thermophilic temperature (70 degrees C). Biotechnol Bioeng 2006; 94(2): 296-302.

[31] Zeidan AA, van Niel EWJ. Developing a thermophilic hydrogen-producing co-culture for efficient utilization of mixed sugars. Int J Hydrogen Energy 2009; 34(10): 4524-8.

[32] Ahn Y, Park EJ, Oh YK, Park S, Webster G, Weightman AJ. Biofilm microbial community of a thermophilic trickling biofilter used for continuous biohydrogen production. FEMS Microbiol Lett 2005; 249(1): 31-8.

[33] Liu WT, Chan OC, Fang HHP. Microbial community dynamics during start-up of acidogenic anaerobic reactors. Water Res 2002; 36(13): 3203-10.

[34] Ueno Y, Haruta S, Ishii M, Igarashi Y. Microbial community in anaerobic hydrogen-producing microflora enriched from sludge compost. Appl Microbiol Biotechnol 2001; 57(4): 555-62.

[35] Fang HHP, Zhang T, Liu H. Microbial diversity of a mesophilic hydrogen-producing sludge. Appl Microbiol Biotechnol 2002; 58(1): 112-8.

[36] Valdez-Vazquez I, Poggi-Varaldo HM. Hydrogen production by fermentative consortia. Renew Sust Energ Rev 2009; 13(5): 1000-13.

[37] Fu N, Peiris P, Markham J, Bavor J. A novel co-culture process with *Zymomonas mobilis* and *Pichia stipitis* for efficient ethanol production on glucose/xylose mixtures. Enzyme Microb Tech 2002; 45(3): 210-7.

[38] Mamma D, Koullas D, Fountoukidis G, Kekos D, Macris BJ, Koukios E. Bioethanol from sweet sorghum: Simultaneous saccharification and fermentation of carbohydrates by a mixed microbial culture. Process Biochem 1996; 31(4): 377-81.

[39] Szambelan K, Nowak J, Czarnecki Z. Use of *Zymomonas mobilis* and *Saccharomyces cerevisiae* mixed with *Kluyveromyces fragilis* for improved ethanol production from Jerusalem artichoke tubers. Biotechnol Lett 2004; 26(10): 845-8.

[40] Collet C, Schwitzguebel JP, Peringer P. Improvement of acetate production from lactose by growing *Clostridium thermolacticum* in mixed batch culture. J Appl Microbiol 2003; 95(4): 824-31.

[41] Collet C, Gaudard O, Peringer P, Schwitzguebel JP. Acetate production from lactose by *Clostridium thermolacticum* and hydrogen-scavenging microorganisms in continuous culture - Effect of hydrogen partial pressure. J Biotechnol 2005; 118(3): 328-38.

[42] Pfeiler EA, Klaenhammer TR. The genomics of lactic acid bacteria. Trends Microbiol 2007; 15(12): 546-53.

[43] Sieuwerts S, de Bok FA, Hugenholtz J, van Hylckama Vlieg JE. Unraveling microbial interactions in food fermentations: from classical to genomics approaches. Appl Environ Microbiol 2008; 74(16): 4997-5007.

[44] Liu Y, Yu P, Song X, Qu Y. Hydrogen production from cellulose by co-culture of *Clostridium thermocellum* JN4 and *Thermoanaerobacterium thermosaccharolyticum* GD17. Int J Hydrogen Energy 2008; 33(12): 2927-33.

[45] Tanaka H, Ebata T, Kuwahara I, Matsuo M, Ogbonna JC. Development and application of a system for analysis of mixed cultures of microorganisms. Appl Biochem Biotechnol 1999; 80(1): 51-64.

[46] Maligoy M, Mercade M, Cocaign-Bousquet M, Loubiere P. Transcriptome analysis of *Lactococcus lactis* with *Saccharomyces cerevisiae.* Appl Environ Microbiol 2008; 74(2): 485-94.

[47] Stolyar S, van Dien S, Hillesland KL *et al.* Metabolic modeling of a mutualistic microbial community. Mol Syst Biol 2007; 3: 92.

[48] Yoshida, A, Nishimura T, Kawaguchi H, Inui M, Yukawa H. Efficient induction of formate hydrogen lyase of aerobically grown *Escherichia coli* in a three-step biohydrogen production process. Appl Microbiol Biotechnol 2007; 74(4): 754-60.

[49] Gavala HN, Skiadas IV, Ahring BK. Biological hydrogen production in suspended and attached growth anaerobic reactor systems. Int J Hydrogen Energy 2006; 31(9): 1164-75.

[50] Goorissen HP, Stams AJM. Biological hydrogen production by moderately thermophilic anaerobic bacteria. Proceedings of the 16th World Hydrogen Energy Conference; 2006 Jun 13-16; Lyon, France.

[51] VanFossen AL, Lewis DL, Nichols JD, Kelly RM. Polysaccharide degradation and synthesis by extremely thermophilic anaerobes. Ann NY Acad Sci 2008; 1125(1): 322-37.

[52] Ueno Y, Otsuka S, Morimoto M. Hydrogen production from industrial wastewater by anaerobic microflora in chemostat culture. J Ferment Bioeng 1996; 82(3): 194-7.

[53] Panagiotopoulos IA, Bakker RR, de Vrije T, Urbaniec K, Koukios EG, Claassen PAM. Prospects of utilization of sugar beet carbohydrates for biological hydrogen production in the EU. J Clean Prod 2010; 18:S9-14.

[54] de Vrije T, unpublished results.

[55] Ivanova G, Rákhely G, Kovács KL. Thermophilic biohydrogen production from energy plants by *Caldicellulosiruptor saccharolyticus* and comparison with related studies. Int J Hydrogen Energy 2009; 34(9): 3659-70.

[56] Claassen PAM, de Vrije T, Budde MAW, Koukios EG, Glynos A, Réczey K. Biological hydrogen production from sweet sorghum by thermophilic bacteria. 2nd World Biomass Conference, Biomass for Energy, Industry and Climate Protection; 2004 May 10-14; Rome, Italy. van Swaaij WPM, Fjällström T, Helm P, Grassi A, Eds. ETA-Florence, Florence and WIP-Munich, Munich, 2004; pp. 1522-5.

[57] Claassen PAM, Budde MAW, van Niel EWJ, de Vrije T. Utilization of biomass for hydrogen fermentation. In: Biofuels for Fuel Cells: Renewable energy from biomass fermentation. P. Lens, P. Westerman, M. Haberbauer, A. Moreno, Eds. London: IWA Publishing 2005; pp. 221-30.

[58] Liu D, Min B, Angelidaki I. Biohydrogen production from household solid waste (HSW) at extreme-thermophilic temperature (70ºC) – Influence of pH and acetate concentration. Int J Hydrogen Energy 2008; 33(23): 6985-92.

[59] Liu D, Zeng RJ, Angelidaki I. Effects of pH and hydraulic retention time on hydrogen production versus methanogenesis during anaerobic fermentation of organic household solid waste under extreme-thermophilic temperature (70ºC). Biotechnol Bioeng 2008; 100(6): 1108-14.

[60] Prasertsan P, O-Thong S, Birkeland N-K. Optimization and microbial community analysis for production of biohydrogen from palm oil mill effluent by thermophilic fermentative process. Int J Hydrogen Energy 2009; 34(17): 7448-59.

[61] O-Thong S, Prasetsan P, Intrasungkha N, Dhamwichukorn S, Birkeland N-K. Optimization of simultaneous thermophilic fermentative hydrogen production and COD reduction from palm oil mill effluent by *Thermoanaerobacterium*-rich sludge. Int J Hydrogen Energy 2008; 33(4): 1221-31.

[62] Magnusson L, Islam R, Sparling R, Levin D, Cicek N. Direct hydrogen production from cellulosic waste materials with a single-step dark fermentation process. Int J Hydrogen Energy 2008; 33(20): 5398-403.

[63] Kongjan P, O-Thong S, Kotay M, Min B, Angelidaki I. Biohydrogen production from wheat straw hydrolysate by dark fermentation using extreme thermophilic mixed culture. Biotechnol Bioeng 2010; 105(5): 899-908.

[64] Cao G, Ren N, Wang A, *et al.* Acid hydrolysis of corn stover for biohydrogen production using *Thermoanaerobacterium thermosaccharolyticum* W16. Int J Hydrogen Energy 2009; 34(17): 7182-8.

[65] Ren N, Cao G, Guo W *et al.* Biological hydrogen production from corn stover by moderately thermophile *Thermoanaerobacterium thermosaccharolyticum* W16. Int J Hydrogen Energy 2010; 35(7): 2708-12.

[66] De Vrije T, Bakker RR, Budde MAW, Lai MH, Mars AE, Claassen PAM. Efficient hydrogen production from the lignocellulosic energy crop *Miscanthus* by the extreme thermophilic bacteria *Caldicellulosiruptor saccharolyticus* and *Thermotoga neapolitana*. Biotechnol Biofuels 2009; 2: 12.

[67] Kádár Z, De Vrije T, van Noorden GE *et al.* Yields from glucose, xylose, and paper sludge hydrolysate during hydrogen production by the extreme thermophile *Caldicellulosiruptor saccharolyticus*. Appl Biochem Biotechnol 2004; 113-16: 497-508.

[68] Gilbert HJ. Cellulosomes: microbial nanomachines that display plasticity in quaternary structure. Mol Microbiol 2007; 63(6): 1568-76.

[69] Mosier N, Wyman CE, Dale B *et al.* Features of promising technologies for pretreatment of lignocellulosic biomass. Bioresource Technol 2005; 96(6): 673-86.

[70] Moat AG, Foster JW, Spector MP. Central pathways of carbohydrate metabolism. In: Microbial Physiology 4th ed. New York: Wiley-Liss Inc 2002; pp. 350-67.

[71] Seo J-S, Chong H, Park HS *et al.* The genome sequence of the ethanologenic bacterium *Zymomonas mobilis* ZM4. Nat Biotechnol 2005; 23(1): 63-8.

[72] Gottschalk G. Bacterial metabolism. New York: Springer-Verlag 1979.

[73] Selig M, Xavier KB, Santos H, Schönheit P. Comparative analysis of Embden-Meyerhof and Entner-Doudoroff glycolytic pathways in hyperthermophilic archaea and the bacterium *Thermotoga.* Arch Microbiol 1997; 167(4): 217-32.

[74] Kengen SWM, Stams AJM. Formation of L-alanine as a reduced end product by the hyperthermophilic archaeon *Pyrococcus furiosus*. Arch Microbiol 1994; 161(2): 168-75.

[75] Verhees CH, Kengen SWM, Tuininga JE *et al.* The unique features of glycolytic pathways in Archaea. Biochem J 2003; 375(Pt 2): 231-46.

[76] Ding Y-HR, Ronimus RS, Morgan HW. *Thermotoga maritima* phosphofructokinases: expression and characterization of two unique enzymes. J Bacteriol 2001; 183(2): 791-4.

[77] Bielen AAM, Willquist K, Engman J, van der Oost J, van Niel EWJ, Kengen SWM. Pyrophosphate as a central energy carrier in the hydrogen producing extremely thermophilic *Caldicellulosiruptor saccharolyticus*. FEMS Microbiol Lett 2010; 307(1): 48-54.

[78] Makarova KS, Aravind L, Galperin MY *et al.* Comparative genomics of the archaea (euryarchaeota): evolution of conserved protein families, the stable core, and the variable shell. Genome Res 1999; 9(7): 608-28.

[79] Rydzak T, Levin DB, Cicek N, Sparling R. Growth phase-dependant enzyme profile of pyruvate catabolism and end-product formation in *Clostridium thermocellum* ATCC 27405. J Biotechnol 2009; 140(3-4): 169-75.

[80] Silva PJ, van den Ban ECD, Wassink H *et al.* Enzymes of hydrogen metabolism in *Pyrococcus furiosus*. Eur J Biochem 2000; 267(22): 6541-51.

[81] Willquist K. Physiology of *Caldicellulosiruptor saccharolyticus*: a hydrogen cell factory. PhD thesis, Lund University, Lund, Sweden. 2010.

[82] Angenent LT, Karim K, Al-Dahhan MH, Domiguez-Espinosa R. Production of bioenergy and biochemicals from industrial and agricultural wastewater. Trends Biotechnol 2004; 22(9): 477-85.

[83] Veit A, Akhtar MK, Mizutani T, Jones PR. Constructing and testing the thermodynamic limits of synthetic NAD(P)H:H_2 pathways. Microb Biotechnol 2008; 1(5): 382-94.

[84] Smith ET, Blamey JM, Zhou ZH, Adams MWW. A variable-temperature direct electrochemical study of metalloproteins from hyperthermophilic microorganisms involved in hydrogen production from pyruvate. Biochemistry 1995; 34(21): 7161-9.

[85] Kraemer JT, Bagley DM. Supersaturation of dissolved H_2 and CO_2 during fermentative hydrogen production with N_2 sparging. Biotechnol Lett 2006; 28(18): 1485-91.

[86] Malki S, Saimmaime I, De Luca G, Rousset M, Dermoun Z, Belaich JP. Characterization of an operon encoding an NADP-reducing hydrogenase in *Desulfovibrio fructosovorans*. J Bacteriol 1995; 177(10): 2628-36.

[87] Schut GJ, Adams MWW. The iron-hydrogenase of *Thermotoga maritima* utilizes ferredoxin and NADH synergistically: a new perspective on anaerobic hydrogen production. J Bacteriol 2009; 191(13): 4451-57.

[88] Jones PM, George AM. The ABC transporter structure and mechanism: perspectives on recent research. Cell Mol Life Sci 2004; 61(6): 682-99.

[89] Koning SM, Elferink MG, Konings WN, Driessen AJ. Cellobiose uptake in the hyperthermophilic archaeon *Pyrococcus furiosus* is mediated by an inducible, high-affinity ABC transporter. J Bacteriol 2001; 183(17): 4979-84.

[90] Nanavati D, Noll KM, Romano AH. Periplasmic maltose- and glucose-binding protein activities in cell-free extracts of *Thermotoga maritima*. Microbiology 2002; 148(Pt 11): 3531-7.

[91] Wassenberg D, Liebl W, Jaenicke R. Maltose-binding protein from the hyperthermophilic bacterium *Thermotoga maritima*: stability and binding properties. J Mol Biol 2000; 295(2): 279-88.

[92] Xavier KB, Martins LO, Peist R, Kossmann M, Boos W, Santos H. High-affinity maltose/trehalose transport system in the hyperthermophilic archaeon *Thermococcus litoralis*. J Bacteriol 1996; 178(16): 4773-7.

[93] Linton KJ, Higgins CF. Structure and function of ABC transporters: the ATP switch provides flexible control. Pflug Arch Eur J Phy 2007; 453(5): 555-67.

[94] Conners SB, Montero CI, Comfort DA *et al.* An expression-driven approach to the prediction of carbohydrate transport and utilization regulons in the hyperthermophilic bacterium *Thermotoga maritima*. J Bacteriol 2005; 187(21): 7267-82.

[95] VanFossen AL, Verhaart MRA, Kengen SMW, Kelly RM. Carbohydrate utilization patterns for the extremely thermophilic bacterium *Caldicellulosiruptor saccharolyticus* reveal broad growth substrate preferences. Appl Environ Microbiol 2009; 75(24): 7718-24.

[96] Rainey FA, Donnison AM, Janssen PH *et al.* Description of *Caldicellulosiruptor saccharolyticus* gen-nov., sp. nov. an obligately anaerobic, extremely thermophilic, cellulolytic bacterium. FEMS Microbiol Lett 1994; 120(3): 263-6.

[97] Dwivedi PP, Gibbs MD, Saul DJ, Bergquist PL. Cloning, sequencing and overexpression in *Escherichia coli* of a xylanase gene, xynA from the thermophilic bacterium Rt8B.4 genus *Caldicellulosiruptor*. Appl Microbiol Biotechnol 1996; 45(1-2): 86-93.

[98] Gibbs MD, Reeves RA, Farrington GK, Anderson P, Williams DP, Bergquist PL. Multidomain and multifunctional glycosyl hydrolases from the extreme thermophile *Caldicellulosiruptor* isolate Tok7B.1. Curr Microbiol 2000; 40(5): 333-40.

[99] O-Thong S, Prasertsan P, Karakashev D, Angelidaki I. Thermophilic fermentative hydrogen production by the newly isolated *Thermoanaerobacterium thermosaccharolyticum* PSU-2. Int J Hydrogen Energy 2008; 33(4): 1204-14.

[100] Ren N, Cao G, Wang A, Lee D-J, Guo W, Zhu Y. Dark fermentation of xylose and glucose mix using isolated *Thermoanaerobacterium thermosaccharolyticum* W16. Int J Hydrogen Energy 2008; 33(21): 6124-32.

[101] Yokoyama H, Waki M, Moriya N, Yasuda T, Tanaka Y, Haga K. Effect of temperature on hydrogen production from cow waste slurry by using anaerobic microflora within the slurry. Appl Microbiol Biotechnol 2007; 74(2): 474-83.

[102] Yokoyama H, Moriya N, Ohmori H, Waki M, Ogino A, Tanaka Y. Community analysis of hydrogen-producing extreme thermophilic anaerobic microflora enriched from cow manure with five substrates. Appl Microbiol Biotechnol 2007; 77(1): 213-22.

[103] Wang J, Zhao C, Meng B *et al.* The proteomic alterations of *Thermoanaerobacter tencongensis* cultured at different temperatures. Proteomics 2007; 7(9): 1409-19.

[104] Willquist K, van Niel EWJ. Lactate formation in *Caldicellulosiruptor saccharolyticus* is regulated by the energy carriers pyrophosphate and ATP. Metab Eng 2010; 12(3): 282-90.

[105] Payot S, Guedon E, Cailliez C, Gelhaye E, Petitdemange H. Metabolism of cellobiose by *Clostridium cellulolyticum* growing in continuous culture: evidence for decreased NADH reoxidation as a factor limiting growth. Microbiology 1998; 144(2): 375-84.

[106] Van Niel EWJ, Claassen PAM, Stams AJM. Substrate and product inhibition of hydrogen production by the extreme thermophile, *Caldicellulosiruptor saccharolyticus*. Biotechnol Bioeng 2003; 81(3): 255-62.

[107] Pauss A, Andre G, Perrier M, Guiot SR. Liquid-to-gas mass transfer in anaerobic processes: inevitable transfer limitations of methane and hydrogen in the biomethanation process. Appl Environ Microbiol 1990; 56(6): 1636-44.

[108] Hawkes FR, Hussy I, Kyazze G, Dinsdale R, Hawkes DL. Continuous dark fermentative hydrogen production by mesophilic microflora: principles and progress. Int J Hydrogen Energy 2007; 32(2): 172-84.

[109] Van Niel EWJ, Martens AA, Claassen PAM. Effect of carbon dioxide and temperature on H_2 producing cultures of the extreme thermophile, *Caldicellulosiruptor saccharolyticus*. Proceedings of the 15th World Hydrogen Energy Conference; 2004 Jun 27-30; Yokohama, Japan.

[110] Willquist K, Claassen PAM, van Niel EWJ. Evaluation of the influence of CO_2 on hydrogen production by *Caldicellulosiruptor saccharolyticus*. Int J Hydrogen Energy 2009; 34(11): 4718-26.

[111] Martins LO, Carreto LS, da Costa MS, Santos H. New compatible solutes related to di-myo-inositol-phosphate in members of the order Thermotogales. J Bacteriol 1996; 178(19): 5644-51.

[112] Martins LO, Santos H. Accumulation of mannosylglycerate and di-myo-inositol-phosphate by *Pyrococcus furiosus* in response to salinity and temperature. Appl Environ Microbiol 1995; 61(9): 3299-303.

[113] Sissons CH, Sharrock KR, Daniel RM, Morgan HW. Isolation of cellulolytic anaerobic extreme thermophiles from New Zealand thermal sites. Appl Environ Microbiol 1987; 53(4): 832-8.

[114] Santos H, da Costa MS. Compatible solutes of organisms that live in hot saline environments. Environ Microbiol 2002; 4(9): 501-9.

[115] Conners SB, Mongodin EF, Johnson MR, Montero CI, Nelson KE, Kelly RM. Microbial biochemistry, physiology, and biotechnology of hyperthermophilic *Thermotoga* species. FEMS Microbiol Rev 2006; 30(6): 872-905.

[116] Levin DB, Carere CR, Cicek N, Sparling R. Challenges for biohydrogen production *via* direct lignocellulosic fermentation. Int J Hydrogen Energy 2009; 34(17): 7390-403.

[117] Vargas M, Noll KM. Catabolite repression in the hyperthermophilic bacterium *Thermotoga neapolitana* is independent of cAMP. Microbiology 1996; 142(1): 139-44.

[118] Thevelein JM. Fermentable sugars and intracellular acidification as specific activators of the RAS-adenylate cyclase signalling pathway in yeast: the relationship to nutrient-induced cell cycle control. Mol Microbiol 1991; 5(6): 1301-7.

[119] Hoekema S, Bijmans M, Janssen M, Tramper L, Wijffels RH. A pneumatically agitated flat-panel photobioreactor with gas re-circulation: anaerobic photoheterotrophic cultivation of a purple non-sulfur bacterium. Int J Hydrogen Energy 2002; 27(11-12): 1331-8.

[120] Lee YJ, Miyahara T, Noike T. Effect of iron concentration on hydrogen fermentation. Bioresource Technol 2001; 80(3): 227-31.

[121] Shock EL, Koretsky CM. Metal-organic complexes in geochemical processes: estimation of standard partial molal thermodynamic properties of aqueous complexes between metal cations and monovalent organic acid ligands at high pressures and temperatures. Geochim Cosmochim Ac 1995; 59(8): 1497-532.

[122] Roels JA. Energetics and kinetics in biotechnology. Amsterdam: Elsevier 1983.

[123] Tjaden B, Plagens A, Dörr C, Siebers B, Hensel R. Phosphoenolpyruvate synthetase and pyruvate, phosphate dikinase of *Thermoproteus tenax*: key pieces in the puzzle of archaeal carbohydrate metabolism. Mol Microbiol 2006; 60(2): 287-98.

[124] Magnusson L, Cicek N, Sparling R, Levin D. Continuous hydrogen production during fermentation of α-cellulose by the thermophilic bacterium *Clostridium thermocellum*. Biotechnol Bioeng 2009; 102(20): 759-66.

[125] Oh Y-K, Kim SH, Kim M-S, Park S. Thermophilic biohydrogen production from glucose with trickling biofilter. Biotechnol Bioeng 2004; 88(6): 690-8.

[126] Yu HQ, Zhu ZH, Hu WR, Zhang HS. Hydrogen production from rice winery wastewater in an upflow anaerobic reactor using mixed anaerobic cultures. Int J Hydrogen Energy 2002; 27(11-12): 1359-65.

[127] Kraemer JT, Bagley DM. Improving the yield from fermentative hydrogen production. Biotechnol Lett 2007; 29(5): 685-95.

[128] Van Groenestijn JW, Hazewinkel JHO, Nienoord M, Bussmann PJT. Energy aspects of biological hydrogen production in high rate bioreactors operated in the thermophilic temperature range. Int J Hydrogen Energy 2002; 27(11): 1141-7.

[129] Mandal B, Nath K, Das D. Improvement of biohydrogen production under decreased partial pressure of H_2 by *Enterobacter cloacae*. Biotechnol Lett 2006; 28(11): 831-5.

[130] Ivanova G, Rákhely G, Kovács KL. Hydrogen production from biopolymers by *Caldicellulosiruptor saccharolyticus* and stabilization of the system by immobilization. Int J Hydrogen Energy 2008; 33(23): 6953-61.

[131] Chhabra SR, Shockley KR, Conners SB, Scott KL, Wolfinger RD, Kelly RM. Carbohydrate-induced differential gene expression patterns in the hyperthermophilic bacterium *Thermotoga maritima*. J Biol Chem 2003; 278(9): 7540-52.

[132] Shaw AJ, Podkaminer KK, Desai SG *et al.* Metabolic engineering of a thermophilic bacterium to produce ethanol at high yield. Proc Natl Acad Sci USA 2008; 105(37): 13769-74.

[133] Zhang Y, Thiele I, Weekes D, *et al.* Three-dimensional structural view of the central metabolic network of *Thermotoga maritima*. Science 2009; 325(5947): 1544-9.

CHAPTER 10

Thermodynamic and Biochemical Aspect of Hydrogen Production by Dark Fermentation

Richard Sparling[1*], R. Carlo Carere[2], Thomas Rydzak[1], John Schellenberg[1] and David B. Levin[2]

[1]Department of Microbiology and [2]Department of Biosystems Engineering, University of Manitoba, Winnipeg, Canada

Abstract: The present chapter discusses the opportunities and challenges faced by microorganisms when they produce molecular hydrogen (H_2) as a major fermentative electron sink. We will focus on sugar fermentation to look at the thermodynamic implications of selecting hydrogen as a main electron sink, how competing fermentation pathways compete with hydrogen for electrons and their effect on both flux and yields of hydrogen. The signatures of these pathways can be observed in the genomes of these organisms. We will contrast the putative enzymes and pathways available to different fermentative organisms on the basis of an ever-increasing collection of available genomes. A description of the molecular toolbox available to various phyla and specific organisms will lead to a better understanding of the key reactions involved in electron flow and will lead to rational strategies of molecular engineering to optimize hydrogen concentrations and yields from dark fermentation.

Keywords: Dark fermentation, glycolysis, bio-hydrogen, comparative functional genomics, thermodynamics, Gibb's free energy, electron transferring reactions, hydrogenomics, metabolic pathways, ferredoxin, nicotinamide adenine dinucleotide (NAD), hydrogenase, facultative anaerobes, strict anaerobes.

INTRODUCTION

The benefits of biological hydrogen (H_2) production have been convincingly articulated in other chapters of this book. With respect to dark fermentation, the ultimate technological goal would be to achieve a fermentation that would provide maximal yield of H_2 from the carbon substrate through complete oxidation of the carbon to CO_2 at the fastest rate possible and, for ease of H_2 recuperation and purification, at the highest partial pressure possible. The present chapter will discuss key biochemical factors in dark fermentation that influence specific aspects of H_2 production in microorganisms using fermentative pathways for energy conservation (*i.e.* ATP synthesis and maintenance of a transmembrane ion motive force).

H_2 production itself can be described using 3 different measures, each under the influence of a different set of parameters: 1) concentration of H_2, controlled by the thermodynamics of the chemical reactions involved; 2) rate of production, determined by the catalysts (enzymes) involved in the reaction; 3) yield of H_2, ie. the amount of H_2 generated per molecule of substrate, determined by a combination of thermodynamics and the complement of enzymatic pathways available to a given microorganism.

Human interest in enhancing H_2 production (*i.e.* rate, yield and concentration) may conflict with the specific metabolic "interests" of the microorganisms. In heterotrophic fermentation, the key is to modify organic substrates through metabolic pathways so as to conserve usable energy, primarily through substrate level phosphorylation, as well as to generate biosynthetic intermediates. For this, a microorganism must choose, on the basis of its genetic toolbox, thermodynamically favourable enzymatic reactions and pathways that allow energy conservation under specific growth conditions. While some growth conditions favour H_2 production, others do not, resulting in the release of other reduced end products [1].

***Address correspondence to Richard Sparling:** Department of Microbiology, University of Manitoba, Winnipeg, Canada; E-mail: sparlng@cc.umanitoba.ca

Nuri Azbar and David B. Levin (Eds)

A SURVEY OF ORGANISMS PRODUCING BIOHYDROGEN BY ANAEROBIC FERMENTATION

Measurement of H_2 Evolution by Microbes: Molar Yield *vs.* Production Rate

A clear understanding of how efficiently microbes produce H_2 is complicated by the wide variety of culture methods, substrates and measures employed by researchers. Most studies report batch cultures in tubes or reactor vessels over a period of hours to days, while others use continuous approaches with periodic carbohydrate feeding and pH adjustment over a period of days. Glucose, sucrose and xylose are the most frequently reported carbohydrate substrates, as components of either defined or complex media with varying concentrations of yeast extract or other carbon containing media components. Many studies use undefined waste or cellulosic substrates such as municipal solid waste, industrial effluents, or agricultural or forestry residues as substrates. Tables **1** and **2** summarize a broad cross section of data taken from papers studying H_2 evolution from bacterial and archaeal cultures.

Measures of H_2 evolution under diverse culture conditions are often difficult to compare across studies (Table **2**). While the most frequently reported parameters are *molar yield* and *production rate,* these are not reported systematically. In many studies, yield is defined as the number of moles H_2 produced per mole of substrate, indicating the overall efficiency of the organism, under the experimental conditions tested, to extract the maximum amount of H_2 from the minimum amount of substrate. In others, yields are reported as volume of H_2 per mass substrate, which requires the reader to convert volumes into moles and the mass into moles or, in the case of polymers such as cellulose, into moles of monomer. We have, when the data were available, converted the yields to moles of H_2 per mole of hexose and the rates as moles H_2 per gram dry weight of cells per hour in order to permit direct comparisons between literature reports (Table **2**). Such molar yield calculations are important in order to appreciate how close reported H_2 yields are to the "Thauer limit" [1] of 4 moles H_2 per mole hexose. Most studies report yields ranging from 1 to 3.

In fermentative hydrogen production, H_2 is never the sole product (Table **1**). Oxidized products, for example formate or CO_2 must also be produced. Other more reduced organic compounds such as ethanol or butyrate may also compete against H_2 for electrons. Reporting of carbon and redox balances (for example, [2]) is therefore very useful in order to appreciate the quality of data reported on H_2 production, but not always available in the literature. Excess carbon in the end products can indicate that other carbon compounds, for example added yeast extract, may have contributed significantly to the fermentation. With respect to carbon as CO_2, frequently reported as H_2/CO_2 ratio, gaseous CO_2 is in equilibrium with carbonate and bicarbonate in solution, therefore, dissolved CO_2 represents a significant proportion of the total at pH values near and above pH 7. This can cause a significant under-reporting of the production of CO_2. In *Clostridium thermolacticum,* for example, the H_2/CO_2 ratio in the gas phase increased from 1.26 at pH 6.4 to 3.95 at pH 7.5, however, when also taking into account bicarbonate as well as dissolved CO_2, total H_2/CO_2 remained stable at the different pH values tested [3].

In contrast to yields, production rate indicates how much and how quickly H_2 is liberated under experimental conditions regardless of amount of substrate consumed. Rates are expressed as the amount of H_2 evolved (in volume or molar terms) per unit culture (volume of culture media or cell-based measures such as protein weight, dry cell weight or optical density) per unit time (hours or minutes). Amount of H_2 can be interconverted between volume and molar rates by adjusting to the known volume per mole H_2 at the temperature of assay (for example 24.45 litres at room temperature) [4]. Rates expressed per litre of culture are often described as "volumetric productivity" values, while those expressed per gram dry cell weight (gdcw) are described as "specific productivity" values. Volumetric productivity can be converted to specific productivity by adjusting the rate according to the total dry cell weight per litre, but only if this value is reported. Rate of H_2 evolution may also be expressed as a maximum (*i.e.*, the highest amount of H_2 produced during a time course), or overall (*i.e.* the average amount of H_2 produced over the entire time course).

There are further exceptions. One study of engineered *E. coli* measured H_2 production using a platinum electrode, expressing production rate as volts of oxidation-reduction potential (mV ORP) per minute per gdcw [5]. Another study in engineered *E. coli* reports a production "rate" that does not include a term for unit time, therefore this value is more appropriately reported as a final concentration [6]. Nevertheless, in pure culture studies, the specific rate is the most informative expression, since it is a characteristic of the organism studied under the growth conditions tested.

Yields and rates are complementary measures and should be reported and considered together in order to evaluate experimental conditions designed to maximize H_2 evolution. Conditions that result in an optimal yield may have a low production rate and vice-versa. Therefore, optimization of both parameters is critical to improving H_2 evolution from diverse substrates in experimental and industrial settings.

Diversity of H_2-Producing Microbes

Several studies describe fermentative H_2 production in eukaryotic Protists. While these organisms are of significant ecological importance, for example the cellulolytic *Pseudotrichonympha grassii* in the termite gut [7] or in human parasitic *Trichomonas vaginalis* [8], few if any studies have looked at their potential for H_2 production on a larger scale and will not be described further.

Several prokaryotic groups have been extensively studied for H_2 production under anaerobic fermentative conditions (also known as dark fermentation). These include mesophiles from the Enterobacteriaceae division of the Proteobacteria (*Escherichia, Klebsiella, Citrobacter, Enterobacter*), various species of saccharolytic and cellulolytic, mesophilic to hyperthermophilic Firmicutes (*Clostridium, Ethanoligenens, Thermoanaerobacter, Caldicellulosiruptor*), syntrophs from various taxa (*Syntrophus, Syntrophobacter*), and thermophiles to extreme thermophiles from the bacterial phyla *Coprothermobacter* and Thermotogae (*Thermotoga*) and domain Archaea (*Pyrococcus, Thermococcus*) (Fig. **1**). The following sections highlight the differences in H_2 production within these different groups. The possibility that H_2 producing fermentative heterotrophs of interest may exist in other phyla awaits further bioprospecting.

Proteobacteria: Formate-Dependent H_2 Production

The enzyme system resulting in H_2 production by the "colon-typhoid" group of bacteria (Enterobacteria) was first described in the 1930s [9,10]. Anaerobically, *E. coli* degrades glucose to acetate, ethanol, succinate, lactate and formate *via* pyruvate and acetyl-CoA from glycolysis [11,12]. Large amounts of formate are produced from pyruvate by pyruvate formate-lyase (PFL) [13], which is subsequently metabolized by the formate hydrogenlyase (FHL) complex to H_2 and CO_2 [14]. The FHL reaction is accomplished by two enzymes: formate dehydrogenase (FDH) and hydrogenase (Hyd), the first oxidizing formate and the second reducing protons to produce H_2 gas [13]. End products from pyruvate/acetyl-CoA include acetate, lactate, ethanol, and small amounts of succinate *via* carboxylation of phosphoenolpyruvate (PEP) by PEP carboxylase (PEPC) (Fig. **2**). Many studies examine increased H_2 production in *E. coli* mutants, shutting off enzymes related to the succinate or lactate pathways [11,15], however a wild-type *E. coli* strain has been shown to produce H_2 at a rate of 9.5 mmol/gdcw·h and a yield of 1.08 H_2/glucose (Table **2**).

Other *Enterobacteriaceae* also use the FHL complex to produce H_2 from formate. Several *Klebsiella* sp. produce H_2, CO_2, acetate, lactate and ethanol from carbohydrates, with 2-3, butanediol (a potential jet fuel) produced as the major end product for some strains (Table **1**) [16,17]. At lower glucose concentrations (5g/l), *K. pneumoniae* ECU-15 had the highest yield (2.1 H_2/glucose), while highest production rate was achieved at 30g/l glucose (7.0 mmol/gdcw·h) after 6h incubation (Table **2**) [16]. Similar production rates were observed for another *K. pneumoniae* strain, however yield values increased with increased sucrose concentration, and were highest (0.92 H_2/hexose) at 30g COD/l [17]. This value was similar for the ECU-15 strain at that concentration of glucose, but about half of the value for ECU-15 at 5g/l glucose (Table **2**). A hot-springs isolate, *K. oxytoca*, had similar values for production rate and yield on glucose in batch culture (Table **2**), but a yield of 3.6 H_2/sucrose (1.8 H_2/hexose) was observed in continuous culture [18].

A similar range of end products is observed for *Enterobacter aerogenes* HU-1 (Table **1**), although yield of the wild-type strain was only 0.8 H_2/glucose (Table **2**) [19]. Another *E. aerogenes* strain was found to produce butyrate, and increased H_2 production rate and yields were observed in this strain when formate was added to the media [20]. This increase may have been due to the direct oxidation of the added formate to CO_2 and release of H_2 by FHL. At 10g/l glucose and pH 7, *Enterobacter asburiae* was found to produce H_2 at a rate of 175ml/l·h. Since cell dry weight was determined to be 1.75g/l in this experiment, this rate could be compared to other studies by dividing it by 1.75 to give a rate of 100ml/gdcw·h, and transforming the volume of H_2 to mmol by dividing by 24.45. Therefore, the production rate in this study was 4.0 mmol/gdcw·h, about 1/3 the rate reported for *E. aerogenes* E.82005, while yield was about half (Table **2**). Highest production rates for this strain were observed at 25g/l glucose [21]. The specific genomic and regulatory differences in these closely related strains would also be useful for understanding why these organisms produce distinct yields of H_2 at different rates.

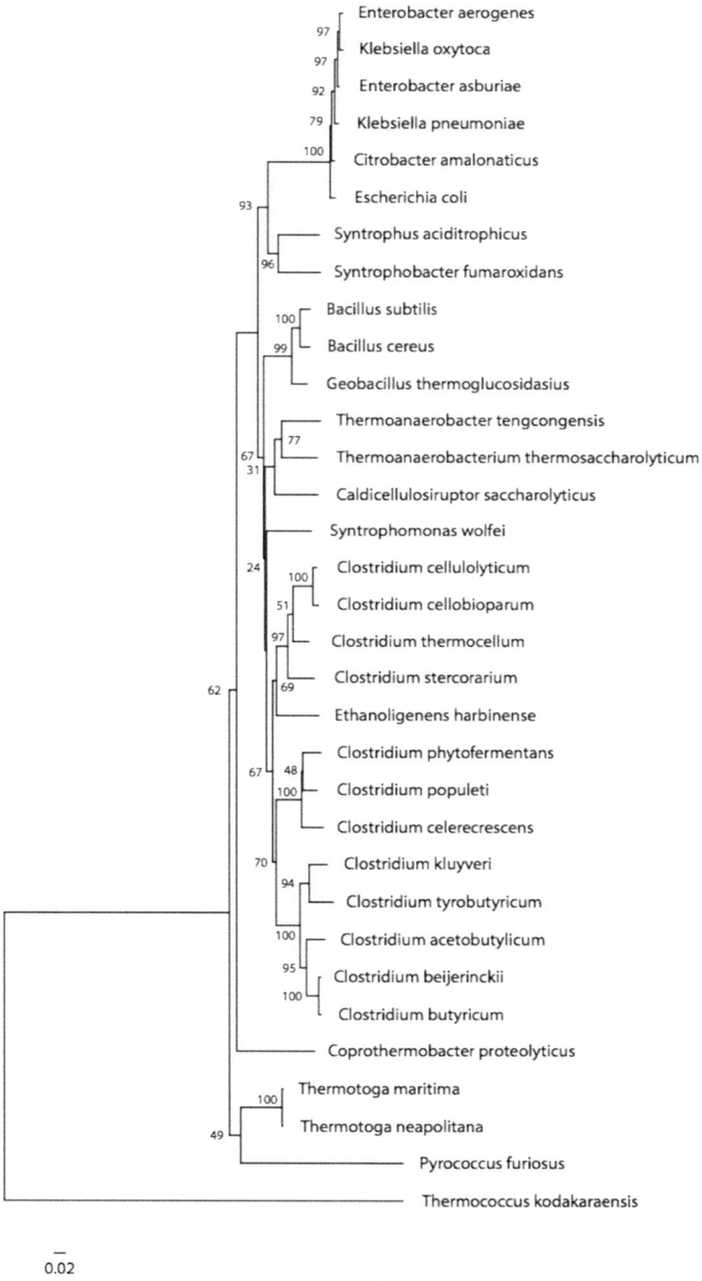

Figure 1: Phylogenetic diversity of H_2-producing organisms.16S rRNA sequences for selected organisms were copied from Genbank and aligned using ClustalW (MEGA 4.0). Bootstrap values were calculated in MEGA based on 500 replicates.

Citrobacter amalonitica Y19 produces a very similar range of end products compared to other enterobacteria (Table **1**) [22]. As in other studies, H_2 yield decreased with increasing glucose concentration, from 1.27 H_2/glucose at 4g/l to 0.7 at 10g/l, concomitant with increased lactate and reduced acetate and ethanol [23]. This strain was also shown to produce H_2 chemoautotrophically using only carbon monoxide and water (known as the water-gas shift) [24]. Facultative heterofermentative anaerobes related to *E. coli* produce a wide range of organic acids and solvents, especially acetate, lactate and ethanol. In all of these organisms, H_2 production is dependent of the presence of formate *via* the PFL system. In general, H_2 yield is increased at lower substrate concentrations, which could also represent lower organic end product concentrations or, in batch cultures, higher pH.

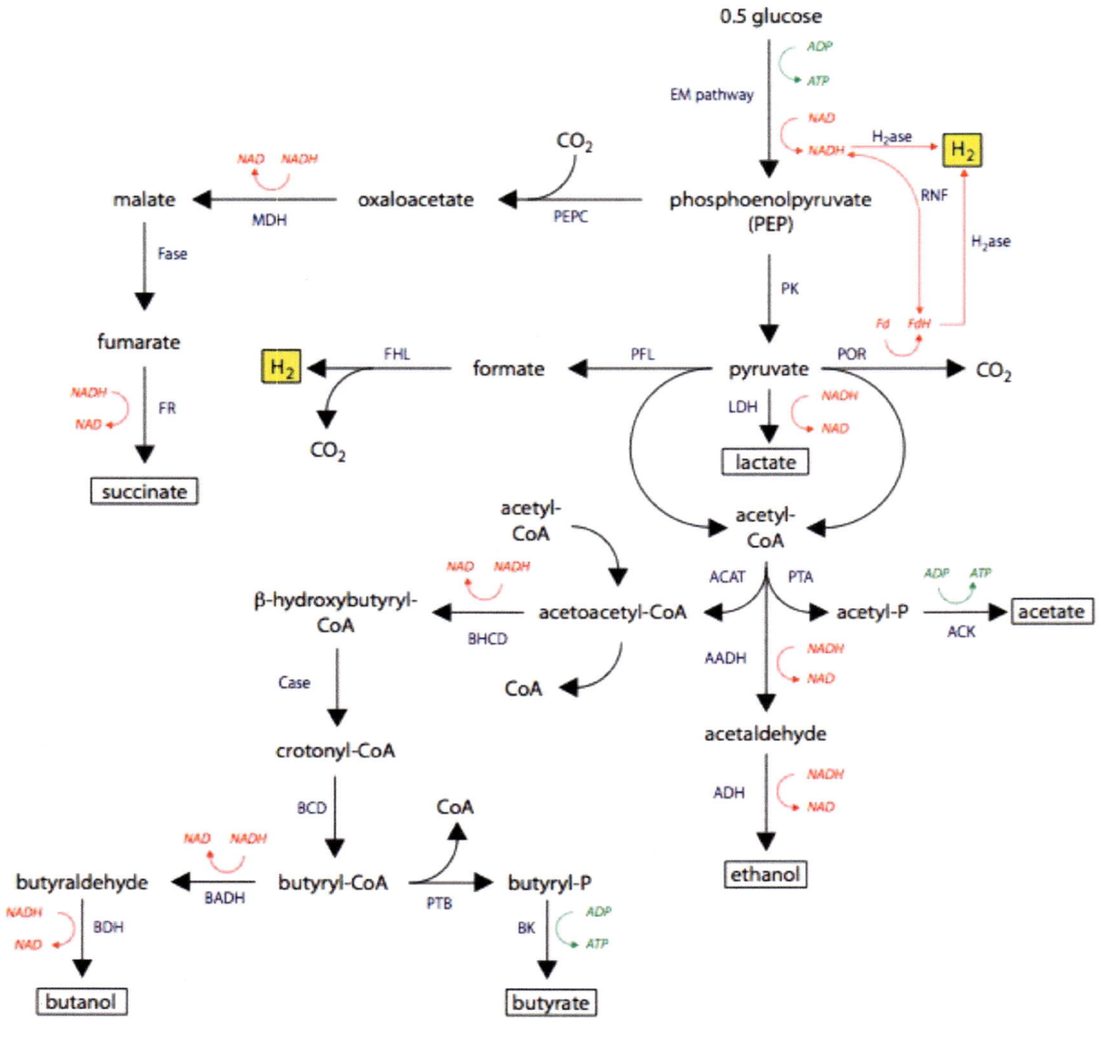

Figure 2: Major pathways of central metabolism involved in H_2 production and competing end products.

Other Proteobacteria are also capable of H_2 evolution during fermentation without going through formate or PFL. For example, *Desulfovibrio vulgaris* donates its electrons to protons forming H_2 during the fermentation of lactate or pyruvate to acetate plus CO_2, in the absence of an alternate terminal electron acceptor [25]. In this case, H_2 production is enhanced through interspecies H_2 transfer, when this organism is placed in co-culture with a methanogen. Indeed, several syntrophic organisms are also members of the Proteobacteria (see below). There are few studies of these organisms as potential H_2 producers because of their poor growth and low production of H_2 in pure culture.

Firmicutes: Facultative Anaerobes

Within the facultative anaerobic Firmicutes, many have been shown not to produce H_2 [26-28]. Rather, electrons associated with the transformation of pyruvate to acetyl-CoA are sequestered in formate through PFL activity, while electrons from the Embden-Meyerhoff pathway are released in ethanol.

Despite this, H_2 production has been reported in *Bacillus coagulans* strain IIT-BT S1, with a yield of 2.3 H_2/glucose [29], and in *Bacillus licheniformis* strain JK1, with a yield of 1.5 H_2/glucose [30]. It should be noted that H_2 production appears to be strain specific, since putative hydrogenase sequences are apparently lacking in the annotation of the sequenced strain *B. lichenoformis* DSM 13 [31], while *B. coagulans* strain 36D1 is primarily known for its high lactate yields [32]. Likewise, in the genus *Geobacillus*, reports have focused on ethanol production [33], with few if any reports of H_2 production. Since some strains are amenable to molecular engineering and are known to produce formate through PFL, acquisition of FHL enzymes through the development of engineered strains may result in significant H_2 production.

H_2 Production by Mesophilic Clostridia

Several mesophiles in order Clostridiales produce H_2, CO_2 and acetate during anaerobic fermentation of various carbohydrate sources *via* pyruvate and acetyl-CoA. However, other more reduced end products are also produced such as butyrate, lactate, ethanol or butanol, depending on the species (Table **1**) [34]. The pathway to butyrate production results in only 2 moles of H_2 per mole of glucose, compared to 4 moles of H_2 per mole glucose when acetate is produced [34].

H_2 yield for the fully sequenced *C. acetobutylicum* ATCC824 strain in pH-controlled batch cultures fed glucose at regular intervals was 2.0 H_2/glucose and final concentration was 650 mmol per litre (Table **2**), with butyrate as the main end product. Production rate was highest (14mmol/OD_{600}·h) at early timepoints when the OD_{600} was lowest and declined to approximately 6 as the OD_{600} increased between 12-14 hours. At low pH, butyrate production shifted to butanol and no H_2 was evolved [34]. Interestingly, CO in the headspace also shifts metabolism away from H_2, acetate and butyrate production towards butanol and ethanol production, likely by inhibition of hydrogenase activity [35].

Another study of *C. acetobutylicum* ATCC824 found a similar yield of H_2 (Table **2**), however no production rate was reported and final concentration was reported as percent headspace, making it difficult to compare to the previous study. Interestingly, a strain that had been cured of a megaplasmid responsible for solvent (butanol and acetone) production had reduced production of biomass and increased H_2 yield relative to the wild-type strain [36].

Table 1: Diversity of End Products Formed by Selected Hydrogen-Producing Organisms.

Group	Phylogeny	Organism	H_2	CO_2	Formate	Acetate	Butyrate	Lactate	EtOH	Other	Ref.
FHL-using	γ-Proteo	*Escherichia coli* W3110	+	+	+	+		+	+	succinate	[12]
		Klebsiella pneumoniae ECU-15	+	+	+	+		+	+	succinate, 2-3,butanediol	[16]
		Enterobacter aerogenes HU-101	+	+	+	+		+	+	succinate, 2,3-butanediol	[19]
		E. aerogenes E82005	+	+	+	+	+	+	+	2,3-butanediol	[20]
		E. asburiae SNU-1	+	+	+	+		+	+	succinate, 2,3-butanediol	[21]
		Citrobacter amalonaticus Y19	+	+	+	+		+	+	succinate	[105]
Syntrophs	δ-Proteo	*Syntrophobacter fumaroxidans* MPOB	+	+		+				succinate	[64]
		Syntrophus aciditrophicus SB	+	+	+	+	+			caproate	[65]
	Clostridiales	*Syntrophomonas wolfei* Gött.	+	+		+				propionate	[68]
Butyrate-producers	Clostridiales	*Clostridium acetobutylicum* ATCC 824	+	+		+	+		+		[41]
		C. acetobutylicum ATCC 824	+	+		+	+		+	butanol	[34]
		C. tyrobutyricum ATCC 25755	+	+		+	+				[38]
		C. butyricum CGS5	+	+		+	+		+	propionate	[42]
		C. beijerinckii FanP3	+	+		+	+		+	butanol	[106]
		C. beijerinckii ATCC 8260	+	+	+	+	+		+	propionate	[44]
		C. kluyveri DSM 555	+	+			+			caproate	[107]

		C. cellobioparum DSM1351	+	+		+	+		+		[41]
		C. populeti DSM5832	+	+		+	+				[41]
		C. celerecrescens DSM5628	+	+		+	+		+	acetone	[41]
		Thermoanaerobacterium thermosaccharolyticum W16	+	+		+	+		+	butanol	[108]
Non-butyrate producers	Clostridiales	*Clostridium cellulolyticum* ATCC 35319	+	+		+			+	n-propanol	[41]
		C. cellulolyticum ATCC 35319	+	+		+		+	+		[109]
		C. phytofermentans ATCC700394	+	+	+	+			+		[41]
		C. thermocellum ATCC27405	+	+	+	+			+		[52]
		C. thermocellum YS	+	+		+			+		[110]
		C. thermolacticum DSM 2910	+	+		+		+	+		[3]
Non-butyrate producers (cont.)	Clostridiales (cont.)	*C. stercorarium* NCIB 11754	+	+		+		+	+	alanine	[111]
		Caldicellulosiruptor saccharolyticus DSM 8903	+	+		+		+			[55]
		Ca. saccharolyticus DSM 8903	+	+		+		+			[58]
		Thermoanaerobacter tengcongensis MB4	+	+		+			+	propionate	[112]
		Ethanoligenens harbinense B49	+	+		+		+	+		[40]
		Coprothermobacter platensis 3R	+	+		+				alanine	[73]
	Thermotoga	*Thermotoga maritima* MSB8	+	+		+		+			[71]
		T. neapolitana DSM 4359	+	+		+		+			[55]
		T. neapolitana DSM 4359	+	+		+		+		alanine	[72]
		T. neapolitana DSM 4359	+	+		+	+				[113]
	Archaea	*Pyrococcus furiosus* DSM3638	+	+		+	+			propionate	[76]
		Thermococcus kodakaraensis KOD1	+	+		+				alanine	[114]
Non-H_2 producers	Firmicutes	*Bacillus cereus* 03BB102		+	+	+		+	+	succinate	[28]
		Bacillus subtilis JH642				+		+	+	acetoin, pyruvate	[26]
		Geobacillus thermoglucosidasius M10EXG			+	+		+	+		[27]

Table 2: Production of Hydrogen Under Various Culture Conditions by Selected Organisms.

Group	Phylogeny	Organism	Method	Media	Substrate	Final conc.	Production rate[a]	Molar yield[b]	Ref.
FHL-using	γ-Proteo.	*Escherichia coli* W3110	Batch, stirred, pH-controlled	defined	glucose	24 mmol in headspace	9.5 mmol/gdcw·h	1.08/glucose	[15]
		Klebsiella pneumoniae ECU-15	Batch, stirred, pH-controlled	complex	glucose		7.0 mmol/gdcw·h	2.1/glucose	[16]
		K. pneumoniae HE-1	Batch, shaken	defined	sucrose	2.7 l	5.2 mmol/gdcw·h[a]	0.48/hexose[b]	[17]
		Klebsiella oxytoca HP1	Batch, shaken	complex	glucose		8.0 mmol/gdcw·h	1.0/glucose	[18]
		Enterobacter aerogenes E82005	Batch, stirred, pH-controlled	complex	glucose	49 mmol/l	12 mmol/gdcw·h (max)	0.89/glucose	[20]
		E. aerogenes HU-101	Batch, stirred	complex	glucose	40 mmol		0.8/glucose	[19]
		E. asburiae SNU-1	Batch, stirred	complex	glucose	29.7 mmol	175 ml l h (max)	0.54/glucose	[21]
		Citrobacter amalonaticus Y19	Batch, stirred	complex	glucose			1.27/glucose	[23]
Syntrophs	δ-Proteo.	*Syntrophus aciditrophicus* SB	Batch	complex	crotonate			1.38/crotonate	[65]
Butyrate-producers	Clostridiales	*Clostridium acetobutylicum* ATCC 824	Batch, pH-controlled, stirred, substrate fed	complex	glucose	650 mmol/l	14 mmol/ OD600·h (max)	2.0/glucose	[34]
		C. acetobutylicum ATCC 824	Batch (large headspace)	defined	glucose	5-22% headspace		1.79/glucose	[36]
		C. acetobutylicum ATCC 824	Batch, shaken	complex	cellobiose	52.3 mmol/l	3.1 ml/mg protein·day	2.3/hexose	[41]
		C. acetobutylicum ATCC 824	Continuous trickle-bed	defined	glucose	70-79% headspace	8.9 mmol/l·h	0.9/glucose	[37]
		C. acetobutylicum X9	Batch	complex	cellulose	755 ml/l	6.4 mmol/gdcw·h	3.6 mmol/g	[115]
		C. butyricum CGS5	Batch, pH-controlled	defined	sucrose	64% headspace	8.7 mmol/gdcw·h	1.45/hexose[b]	[42]
		C. butyricum EB6	Batch	complex	glucose	948 ml/l	172 ml/l·h		[116]
		C. butyricum W5	Batch, stirred	complex	molasses	9.8 l/l	17.4 mmol/l·h	1.85/hexose	[43]
		C. beijerinckii Fanp3	Batch, shaken	complex	glucose		390 ml/l·h	2.52/glucose	[106]
		C. beijerinckii ATCC 8260	Batch, shaken	defined	glucose		71 ml/l·h	10.3ml/g	[44]
		C. cellobioparum DSM 1351	Batch, shaken	complex	cellobiose	42.3 mmol/l	3.0 ml/mg protein·day	1.9/hexose	[41]
		C. populeti DSM 5832	Batch, shaken	complex	cellobiose	45.3 mmol/l	2.9 ml/mg	1.9/hexose	[41]

							protein·day		
Butyrate - producer s (cont.)	Clostridiales (cont.)	*C. celerecrescens* DSM 5628	Batch, shaken	complex	cellobiose	38.0 mmol/l	2.5 ml/mg protein·day	1.6/hexose	[41]
		Thermoanaerobacterium thermosaccharolyticum W16	Batch	complex	glucose		12.9 mmol/l·h	2.42/glucose	[48]
		Th. thermosaccharolyticum PSU-2	Batch	complex	sucrose		12.1 mmol/l·h	2.53/hexose	[47]
Non-butyrate producer s	Clostridiales	*Clostridium cellulolyticum* ATCC 35319	Batch, shaken	complex	cellobiose	40.6 mmol/l	2.8 ml/mg protein·day	1.8/hexose	[41]
		C. phytofermentans ATCC 700394	Batch, shaken	complex	cellobiose	35.6 mmol/l	2.4 ml/mg protein·day	1.6/hexose	[41]
		C. thermocellum ATCC 27405	Batch	complex	cellulose		6.4 mmol/gdcw·h	1.9/glucose	[52]
		C. thermocellum ATCC 27405	Batch	complex	wood	2.5 l/l		2.32/hexose	[117]
		C. thermocellum JN4	Batch	complex	cellulose	15 mmol/l		0.8/hexose	[107]
		C. thermocellum ATCC 27405	Continuous, stirred, pH-controlled	complex	cellulose		24.8 ml l h (max)	1.65/hexose	[51]
		C. thermocellum ATCC 27405	Batch	complex	cellobiose			0.82/hexose	[2]
		C. thermolacticum DSM 2910	Continuous, stirred, pH-controlled	complex	lactose		5.8 mmol/gdcw·h	1.5/hexose	[3]
		Caldicellulosiruptor saccharolyticus DSM 8903	Continuous, stirred, pH-controlled	complex	glucose		26.5 mmol/gdcw·h	4.0/glucose	[60]
		Ca. saccharolyticus DSM 8903	Batch, stirred, pH-controlled, N_2-sparged	complex	glucose/ xylose	188 mmol/l	17.9 mmol/gdcw·h[a]	3.4/hexose	[55]
		Ca. saccharolyticus DSM 8903	Batch, stirred, pH-controlled, N_2-sparged	complex	glucose	129.5 mmol	10.7 mM l h	2.52/glucose	[58]
		Ca. saccharolyticus DSM 8903	Batch, stirred, pH-controlled, N_2-sparged	complex	sucrose		11.7 mmol/gdcw·h	3.3/hexose	[59]
		Ethanoligenens harbinense B49	Batch, shaken	complex	glucose	2.5 l/l	27.8 mmol/gdcw·h	2.26/glucose	[40]
	Thermotoga	*Thermotoga maritima* MSB8	Batch	complex	glucose			4.0/glucose	[71]
		T. maritima DSM 3109	Batch, shaken	complex	glucose		8.2 mM l h	1.67/glucose	[118]
Non-butyrate producer s (cont.)	Thermotoga (cont.)	*T. neapolitana* DSM 4359	Batch, shaken	complex	glucose		8.7 mM l h	1.84/glucose	[118]
		T. neapolitana DSM 4359	Batch, stirred, pH-controlled, N_2-sparged	complex	glucose/ xylose		14.9 mmol/gdcw·h[a]	3.3/hexose	[55]
		T. neapolitana DSM 4359	Batch, stirred, pH-controlled	complex	glucose		51 ml l h	3.85/glucose	[72]
		T. neapolitana DSM4359	Batch, N_2-sparged	complex	glucose	1.7 l/l		3.24/glucose	[113]
	Archaea	*Pyrococcus furiosus* DSM 3638	Batch, stirred, N_2-sparged	complex	maltose			3.0/hexose	[119]
		P. furiosus DSM 3638	Continuous, stirred, pH-controlled	complex	maltose		100 mmol/gdcw·h	3.18/hexose[b]	[76]
		Thermococcus kodakaraensis KOD1	Continuous, stirred, pH-controlled	complex	starch		14 mmol/gdcw·h	3.33/glucose	[114]

[a] Rate re-calculated for indicated values according to information provided in reference (see text for details); [b] Yield adjusted to hexose equivalent for indicated values on disaccharide substrates.

A continuous culture study of the ATCC824 strain in a trickle-bed reactor found an increased H_2 production rate with increased glucose concentration (from 3.3 mmol/l·h at 1g/l to 8.9 at 10.5g/l) [37]. Maximum yield was found to be 0.9 (Table **2**). Acetate was the main end product under these conditions, therefore it is surprising that yield was nearly half of that reported by Chin *et al.* [34], since acetate is clearly associated with an increased molar yield of H_2. Similarly, a recent study of *C. tyrobutyricum* ATCC 25755 found an increased H_2 yield (reported as g H_2 per g glucose) with a mutated acetate kinase (*ack*) gene designed to skew end product formation towards butyrate [38].

A study of *C. acetobutylicum* X9, a strain closely related to the ATCC strain, was found to ferment microcrystalline cellulose and produce H_2 at a rate of 6.4 mmol/gdcw·h (Table **2**), with slightly more butyrate than acetate produced. Interestingly, the production rate increased nearly tenfold (to 55.9) when X9 was co-cultured with an *Ethanoligenens harbinense* strain not able to metabolize cellulose. H_2 yield (per gram cellulose) also doubled in the co-culture, and end products shifted to primarily acetate and ethanol, with lower amounts of butyrate [39]. Another study of the same *E. harbinense* strain on glucose found a H_2 production rate of 27.8 mmol/gdcw·h (Table **2**) [40].

Although *C. acetobutylicum* was found to ferment cellulose in the above study, another study found no growth of *C. acetobutylicum* on cellulose. However, this organism produced the highest yield of H_2 compared to 5 other mesophilic clostridia growing on cellobiose [41]. Yields for all 6 mesophiles studied were much higher on cellobiose compared to cellulose, with production rates ranging from 2.4 to 3.1 ml H_2 per mg protein per day and final concentrations ranging from 35.6 to 52.3 ml H_2 per l (Table **2**). On cellulose, *C. populeti* and *C. cellulolyticum* had the highest H_2 yield and rate, consistent with producing the most butyrate and acetate respectively, while *C. phytofermentans* had the lowest H_2 yield and also produced the most ethanol [41].

Highest yield of H_2 in batch cultures of *C. butyricum* CGS5 was 2.9 mol H_2/mol sucrose (1.45 mol H_2/mol hexose) with 20g COD/l, with a production rate of 212 ml/gdcw·h (Table **2**), or 8.5 mmol/gdcw·h. A higher production rate of 254 ml/gdcw·h (10.2 mmol/gdcw·h) was observed at 10g COD/l, although molar yield was much lower at 1.85. Butyrate was the most important metabolite (36-60% of total volatile fatty acids), and was highest in cultures with the highest H_2 yield, indicating that H_2 is a product of the butyrate pathway [42].

A study with a different strain of *C. butyricum*, strain W5, producing H_2 from a molasses-based substrate found that H_2 yield (1.85 H_2/glucose) and production rate (17.4 mmol/l·h) were highest at 100g/l (Table **2**). Butyrate was the predominant end product, with increasing proportions of acetate and lactate as molasses concentration increased [43]. Since a value for gdcw/l was provided, the volumetric production rate could be converted to 3.9 mmol/gdcw·h, about half that observed on glucose in the study by Chen *et al.* [42]. It would be interesting to know whether these differences were due to the growth conditions or small differences at the genomic level between the 2 strains of *C. butyricum* tested above.

A study of H_2 production in *C. beijerinckii* ATCC8260 found a volumetric production rate of 71 ml/l·h, with end products dominated by butyrate, ethanol and butanol and almost no acetate (Table **1**) [44]. In contrast, a strain of *C. beijerinckii* isolated from anaerobic sludge had similar end products, although with approximately equal amounts of acetate and butyrate produced (Table **1**), and a much higher production rate (390 ml/l·h) [45]. The molar yield in the second study was 2.52 H_2/glucose, however in the first study the yield was given as 10.3 ml/g glucose, or only 0.08 H_2/glucose, likely due to high solvent (ethanol/butanol) production by this strain under these experimental conditions.

Because of the relationship between temperature and reaction thermodynamics, mesophilic H_2 producers are more sensitive to H_2 partial pressure in the growth medium. This may cause a shift away from H_2 production to an enhanced production of organic reduced end products such as ethanol as the cells grow [1]. Alternately, continual removal of H_2 will stimulate continued production of H_2. As a further example, *Clostridium cellobioparum* increased growth rates and cell yields when H_2 was either periodically purged from the medium, trapped on palladium black or removed from the culture through interspecies H_2 transfer and methane production by a methanogenic Archaeon [46].

H_2 Production by Saccharolytic, Thermophilic and Hyperthermophilic Firmicutes

Saccharolytic and cellulolytic thermophiles, including organisms from the genera *Clostridium, Caldicellulosiruptor, Thermoanaerobacter* and *Thermoanaerobacterium*, are known H_2 producers. A key advantage for several of these is being able to generate H_2 from cellulose and cellulosic waste products rather than more costly refined carbohydrate sources at higher rates than their mesophilic counterparts. As well as H_2 and CO_2, most thermophiles produce a similar range of metabolites, including acetate, lactate, ethanol and/or butyrate depending on species (Table **1**).

In *Thermoanaerobacterium thermosaccharolyticum* PS1 [47], H_2 yields have been shown to be quite different depending on the growth medium, with yields of 0.86/hexose when the cells were grown on sucrose a minimal medium, but up to 2.5/hexose when the cells were supplemented with an organic source of nitrogen, peptone. A shift away from acetate and ethanol production towards an acetate plus butyrate type fermentation was also observed. Since this organism can use peptone as a carbon and energy source, the authors were careful to demonstrate that the peptone alone, at the concentration used could only account for a small proportion of the fermentation products and H_2 produced (<20%). The latter yield and end-product distribution is similar to that observed in the W16 strain of the same species when grown on glucose in a complex medium including yeast extract and tryptone as sources of organic nitrogen [48].

Because many of these organisms naturally produce both H_2 and ethanol as major electron sinks, many authors have chosen to investigate their potential as bioethanol producers, rather than H_2 producers. Nevertheless, insights into hydrogenases and H_2 production can also be found in such articles. In this one, for example, the goal was to knock out H_2 production in order to enhance ethanol production in *Thermoanaerobacterium saccharolyticum* [49].

Another major example is *Clostridium thermocellum* [50]. Nevertheless, in experiments focusing on hydrogen production in *Clostridium thermocellum*, yields of up to 1.6 H_2/hexose have been observed in continuous cultures on cellulose [51], and in batch with a yield of 1.9/glucose rate of 6.4 mmol/gdcw·h [52]. It should be noted that the molar yield of hydrogen dropped to 0.8/hexose when the cells were growing on cellobiose [2]. Since the cells grow at a much faster rate on cellobiose relative to cellulose, it is likely that the flux through the fermentation pathways have an impact on the proportion of carbon and electrons going to different metabolic pathways. Indeed the proportion of formate produced increased significantly when cells were grown of cellobiose [2] relative to growth on cellulose [51,53].

Changes in end-product profiles with growth rate have also been observed in *Clostridium thermolacticum* [3]. They observed yields of 1.5 H_2/lactose (0.75 per hexose) with a rate of 5.7 mmol/gdcw·h in *C. thermolacticum* grown in continuous cultures on lactose. It was noted, however, that at higher dilution rates ethanol steady state concentrations decrease, however, the electrons were passed to lactate rather than H_2, whose yield remained unaffected.

First described in 1994 [54], *Caldicellulosiruptor saccharolyticus* is a hyperthermophile producing H_2 and acetate at high rates, with yields approaching the Thauer limit for these end products from a variety of simple and complex carbohydrate sources (Table **1**) [55]. Metabolism has been found to shift to lactate production at H_2 pressures above 20kPa [56]. The full genome was recently sequenced, and transcription analysis reveals numerous metabolic pathways for degradation of cellulose and hemicellulose. Further, this organism was found to produce H_2 directly from NADH without (thermodynamically unfavourable) transfer to ferredoxin as observed in mesophiles [57].

Reported yields and production rates of H_2 by *C. saccharolyticus* are very consistent. Early studies found 83% of theoretical yield of H_2 on sucrose (3.32 H_2/hexose equivalent) and 63% of theoretical yield of H2 on glucose (2.52 H_2/glucose) (Table **2**) [58,59]. Production rate in the earlier study was reported as 11.7 mmol/gdcw·h, while in the later study it was 10.7 mmol/l·h (Table **2**). Since cell yield was not reported, rate per gdcw could not be derived. Interestingly, the yield was 93% of theoretical maximum on medium containing hydrolyzed paper sludge, but production rate was nearly half that observed on glucose (5.3 mmol/l·h) [58], indicating that yield and rate may be inversely correlated.

In continuous culture experiments using *C. saccharolyticus* DSM8903, H_2 production rate and molar yield at steady state varied with glucose concentration and dilution rate (*i.e.*, the rate at which new media replaced spent media in the culture vessel). The highest molar yield, reaching Thauer's limit of 4 H_2 and 2 acetate per glucose, was observed at the lowest glucose concentration and dilution rate, while the fastest production rate (11.6 mmol/l·h or 26.5 mmol/gdcw·h) was observed at the highest glucose concentration and dilution rate (Table 2) [60]. These observations corroborate an inverse relationship between production rate and yield related to glucose concentration. Another study using batch cultures of the same strain on a mixture of glucose and xylose found a similar yield of H_2/acetate (3.4 and 1.6 respectively per glucose) and a production rate of 12 mmol/l·h. Since biomass in this experiment was 0.67 gdcw/l, production rate in cell yield terms was 17.9 mmol/gdcw·h, similar to previous reports. Interestingly, yields and rates were similar when hydrolyzed plant material was used as substrate [55].

Syntrophic H_2 Production

Associations between syntrophic and methanogenic microorganisms are critical to degradation of organic substrates to methane and CO_2 under anaerobic conditions. As a group, some have even been shown to be responsible for the methanogenic degradation of crude oil [61]. While the complete oxidation of such compounds to CO_2 plus H_2 is thermodynamically unfavourable under standard conditions, syntrophs are capable of such reactions, provided the H_2 partial pressure is maintained at very low levels by H_2-consuming organisms during interspecies H_2 transfer. Propionate is an important intermediary in this pathway and is degraded to acetate and CO_2, *via* succinate, fumarate, malate, oxaloacetate and pyruvate (the methylmalonyl-CoA pathway), by organisms of the *Syntrophobacter* cluster of the δ-Proteobacteria, with protons as the electron acceptor resulting in the liberation of 3 moles of H_2 per mole substrate [62]. Alternatively, bicarbonate can act as electron acceptor resulting in formate production [63]. Although thermodynamically unfavorable ($\Delta G^{o'} = +76$ kJ/mol), this process is facilitated by the presence of H_2- (or formate-) utilizing methanogens such as *Methanospirillum hungatei,* which reduces H_2 partial pressure and decreases ΔG to -10 kJ/mol [64]. Several *Syntrophobacter* organisms can grow in pure culture on propionate with sulfate or fumarate as electron acceptors, or ferment fumarate to succinate and CO_2 in the absence of propionate [64].

Another species of δ-Proteobacteria, *Syntrophus acidotrophicus* SB, degrades benzoate and fatty acids to acetate, H_2 and CO_2 in the presence of H_2-utilizing methanogens [65]. The only substrate that supported pure culture growth of this organism is crotonate, with acetate, butyrate, caproate, CO_2 and H_2 as end products (Table **1**), with a yield of 1.38 H_2 per crotonate (Table **2**) [65]. The genome of this strain was recently sequenced, revealing a severely limited potential to use external electron acceptors such as O_2, sulfate or iron [66]. The organism is dependent on benzoate reduction and H_2 production to re-oxidize required cofactors. The presence of an Rnf-like membrane complex for "reverse electron transport", also observed in *Syntrophobacter fumaroxidans*, catalyzes the reduction of ferredoxin needed for synthesis of pyruvate, a thermodynamically unfavorable reaction [66].

Syntrophomonas wolfei Göttingen strain degrades butyrate and other short-chain fatty acids to acetate or acetate/propionate, CO_2 and H_2 in a similar syntrophic context [67]. Although Gram-negative in appearance, it is most closely related to organisms of order Firmicutes and does not encode LPS [68]. Although no Rnf-like genes are observed in the genome, reverse electron transport is likely accomplished by a NADH:acceptor oxidoreductase as part of a process called "electron bifurcation", in which H_2 production from reduced ferredoxin increases the thermodynamic favorability of H_2 production from NADH, as recently described in *Thermotoga maritima* [69]. Elements of this trimeric hydrogenase are also observed in the other syntrophs discussed above, as well as thermophilic anaerobes such as *C. thermocellum, C. saccharolyticus* and *T. tengcongensis*. Complete pathways for production of PHA also indicate that energy storage plays a critical role in the metabolism of *S. wolfei* [68].

Little direct information is available regarding H_2 yield and production rate in syntrophs, since they mainly function at very low H_2 concentrations where the thermodynamics are favourable thanks to interspecies H_2 transfer in the presence of methanogens, sulphate reducers or other H_2 consumers. An examination of metabolic pathways reveals highly specialized H_2-evolving mechanisms. Therefore, the complete oxidation of organic carbon to CO_2 with release of H_2, from a biological point of view, is feasible and occurs in most anaerobic environment. The constraint is the maximal concentration of H_2 at which the reactions can take place. The development of effective technologies capable of trapping H_2 to maintain low H_2 partial pressure would allow us to use these organisms for H_2 production. For example, syntrophs in pure culture have been shown to grow in the presence of palladium acting as an *in situ* H_2 trap [70].

Fermentative H_2 Production in Thermophiles and Hyperthermophiles from Other Phyla

Production of H_2 by thermophiles in Domain Archaea and in evolutionary ancient lineages of Bacteria is frequently studied, with some organisms producing yields of H_2 approaching the theoretical maximum of 4

H_2/hexose. The phylum Thermotogales represents one of the deepest branches in Domain Bacteria and is considered to be among its most ancient lineages [71]. Several species have been found to produce H_2 under anaerobic, thermophilic conditions. The species *T. maritima* produces acetate, lactate, H_2 and CO_2 from glucose (Table **1**). Although earlier studies found that this organism fermented glucose mainly to lactate, more recent studies have found only very low levels of lactate, with 100% of the theoretical yield of both H_2 and acetate produced [71].

Thermotoga neapolitana has been shown to produce very close to theoretical yields of H_2 on glucose (Table **2**) [72], as does *Coprothermobacter* [73]. However, alanine can also be used as an alternate electron sink when H_2 concentrations become elevated. Alanine, a 3-carbon amino acid, is at the same oxidation level as lactate and serves a similar purpose as an alternate electron sink for electrons from glycolysis. As an end product, because of the presence of the amino group, it does have the advantage of buffering the pH compared to lactate. While alanine production has not been reported as a major alternate end product in most H_2-producing organisms described (see Table **1**), alanine could potentially serve as an alternate electron sink in response to molecular engineering blocking other pathways that compete against H_2 production. Since alanine is required for protein synthesis, genes for alanine production from pyruvate are available in most organisms.

Hyperthermophilic H_2-producing Archaea, such as *Pyrococcus furiosus* and *Thermococcus celer*, circumvent the need for electrons from NADH to generate H_2 by utilizing a glyceraldehyde-3-phosphate ferredoxin oxidoreductase (GAPFOR) rather than the more common NAD^+-dependent glyceraldehyde-3-phosphate dehydrogenase (GAPDH), which is virtually ubiquitous in the domain Bacteria [74,75]. Yields of H_2, approaching the Thauer limit are facilitated by GAPFOR due to increased electronegative electrons, as well as by-passing production of 1-3-bisphosphoglycerate and its ATP-yielding step. Therefore, the maximum ATP made by substrate level phosphorylation per hexose is 2 rather than 4. From a genetic engineering point of view, GAPFOR may show promise for enhancing H_2 production yields if it could be transferred to other organisms.

PATTERNS OF HYDROGEN PRODUCTION

Table **1** shows the diversity of end products associated with H_2 production. Few organisms yield H_2 at levels approaching the Thauer limit. All mesophiles and most thermophiles described to date also use alternate organic electron sinks when grown in pure culture. Interestingly, different fermentation profiles are associated with surprisingly similar specific rates and yields of H_2. Mesophilic enteric organisms producing H_2 solely *via* FHL generate H_2 at rates between 5 and 12 mmol/gdcw·h and yields near 1 per hexose. Butyrate-producing mesophilic Firmicutes typically generate 2.1 to 2.6 mmol H_2/gdcw·h (recalculated from [41] assuming cell dry weight consists of 50% protein) and yields of 1.6 to 2.5 per hexose on soluble substrates. *C. cellulolyticum* and *C. phytofermentans* produce 2.1 and 2.3 mmol H_2/gdcw·h with yields of 1.6 to 1.8 per hexose on soluble substrate. Of note, a rate of 8.7 mmol H_2/gdcw·h and yield of 1.45 H_2/hexose was observed for *C. butyricum* CGS5 [42].

When considering wild-type mesophiles, it would appear that the enteric strategy of producing H_2 *via* FHL may yield less H_2 per hexose, but at a faster rate than what is observed in most mesophilic Firmicutes despite an average two-fold greater yield of H_2. Thermophiles such as *C. thermocellum* produce H_2 at rates of 6.4 mmol/gdcw·h and a yield of 1.9 H_2/hexose when grown on cellobiose [52], while extremely thermophilic Bacteria and Archaea are shown to have production rates of 11 to 28 mmol H_2/gdcw·h and yields of 2.5 to as high as 4 H_2/hexose (Table **2**). In one experiment, *Pyrococcus furiosus* was reported to produce 100 mmol/gdcw·h [76]. A further excellent review of hyperthermophilic H_2 producers has been published recently [77].

When comparing organisms, normalization relative to cell mass is needed for direct comparison of performance. Some organisms are capable of higher rates of H_2 evolution despite lower yields. This is not necessarily a bad thing, since other major end products may have applications in their own right. Clearly, high rates and yields are associated with extreme thermophiles. While some of the above relationships between H_2 production yields and rates can be explained thermodynamically, the reported differences between strains is somewhat more difficult to explain and may be due to differences at the genome level.

A Survey of Genomes of Representative H_2-Producing Organisms

Current understanding of fermentation pathways is based on analysis of individual enzymatic activities in cell extracts and the pathway explained relative to the various end products detected. Decades of such fermentation research have been compiled [1,78] and the utility of this approach has been clearly demonstrated. For example, this approach not only led to the elucidation of the multiple pathways associated with inulin fermentation in *Clostridium thermosuccinogenes*, but also led to the observation that its acetate kinase is inactive at 70°C, resulting in flow of carbon to other end products rather than acetate at that temperature [79]. Other biochemically well characterized hydrogenogenic organisms within the Firmicutes include *Thermoanaerobacter brockii* [80], *Clostridium cellulolyticum* [81] and *Clostridium thermocellum* [2]. A summary of H_2 production pathways relative to competing pathways is presented in Fig. **2**.

Describing the enzymatic steps of fermentation pathways is critical to understand flux of energy, carbon and electrons through the various parts and to lay the groundwork for genetic engineering. Whole-genome sequencing allows for assignment of function to specific genes, and to predict the function of enzymes that may not have previously been recognized as important, for example the PFL of *C. thermocellum* [82]. Whole genome comparisons may also help explain difference in behavior of different but related organisms under similar growth conditions. An example to illustrate this point is the difference in growth on cellobiose of the mesophilic *Clostridium cellulolyticum* and the thermophilic *Clostrodium thermocellum*. The first [83] is inhibited in the presence of high concentrations of cellobiose, and responds by excreting high amounts of pyruvate, while *C. thermocellum* is not affected adversely [53]. It can be speculated that key differences in the genome might be responsible for this difference in response to high concentrations of soluble substrate.

An overview of genes associated with sugar fermentation in sequenced organisms is provided below (Tables **3-5**). An overview of the sequencing centers responsible for each organism's genome considered in this study is provided in Table **3**. All sequence data were retrieved from the US Department of Energy Joint Genome Institute and are freely available online (http://www.jgi.doe.gov/). Gene loci presented here reflect the numbering of the final assembly or available draft (as of 01/01/2010). Automated gene and function predictions were manually checked using RPS-BLAST available through the Conserved Domain Database website (http://www.ncbi.nlm.nih.gov/Structure/cdd/wrpsb.cgi) with an e-value threshold of <1e-10 [84]. Comparative analysis was primarily conducted based on the assignment and classification of Clusters of Orthologous Groups (COGs) [85] and protein families (pfam) using the Integrated Microbial Genomes (IMG) tool set made available through the Joint Genomes Institute (http://img.jgi.doe.gov/cgi-bin/pub/main.cgi) [86]. Predicted polycistronic mRNAs were identified using the Microbes Online Operon Predictions tool (http://www.microbesonline.org/).

Table 3: National Center for Biotechnology Information taxon IDs and corresponding sequencing centers responsible for the generation of the genome sequences data analyzed in this study.

Organism	Taxon ID	Sequencing Center
Acidothermus cellulolyticus 11B	351607	DOE Joint Genome Institute
Bacillus cereus 03BB102	572264	J. Craig Venter Institute
Caldicellulosiruptor saccharolyticus DSM	351627	DOE Joint Genome Institute
Clostridium acetobutylicum ATCC 824	272562	Genome Therapeutics
Clostridium beijerinckii NCIMB 8052	290402	DOE Joint Genome Institute, Univ of Illinois
Clostridium butyricum 5521	447214	J. Craig Venter Institute
Clostridium cellulolyticum H10	394503	DOE Joint Genome Institute
Clostridium kluyveri DSM 555	431943	Goettingen Genomics Laboratory, Max Planck Institute
Clostridium perfringens ATCC 13124	195103	J. Craig Venter Institute
Clostridium phytofermentans ISDg	357809	DOE Joint Genome Institute

Clostridium thermocellum ATCC 27405	203119	DOE Joint Genome Institute, Univ of Rochester
Coprothermobacter proteolyticus DSM 5265	309798	J. Craig Venter Institute
Escherichia coli K12 DH10B	316385	BCM-HGSC, Univ of Wisconsin-Madison
Geobacillus thermodenitrificans NG80-2	420246	Nankai Univ
Klebsiella pneumoniae 342	507522	J. Craig Venter Institute
Thermotoga maritima MSB8	243274	J. Craig Venter Institute
Thermotoga neapolitana DSM 4359	309803	Genotech corp.
Syntrophobacter fumaroxidans MPOB	335543	DOE Joint Genome Institute
Syntrophomonas wolfei wolfei Goettingen	335541	DOE Joint Genome Institute
Syntrophus aciditrophicus SB	56780	Integrated Genomics Inc, UCLA, Univ of Oklahoma
Thermoanaerobacter pseudethanolicus	340099	DOE Joint Genome Institute
Thermoanaerobacter tengcongensis MB4	273068	Beijing Genomics Institute, The Institute of
Thermococcus gammatolerans EJ3	593117	Genoscope, Univ Paris-SUD
Thermococcus kodakaraensis KOD1	69014	Kwansei Gakuin Univ, Kyoto Univ
Thermococcus onnurineus NA1	523850	KORDI
Thermococcus sibiricus MM 739	604354	Center of Bioengineering
Pyrococcus abyssi GE5	272844	Genoscope, IGBMC
Pyrococcus furiosus DSM 3638	186497	Univ of Maryland, Univ of Utah
Pyrococcus horikoshii OT3	70601	NITE, Univ of Tokyo

Table 4: List of relevant enzymes that participate in central catabolism, biofuels production (H_2, ethanol, butanol), or the synthesis of other reduced end-products.

Abbreviation	Common name	EC	Reaction Catalysed
RNF	Rhodobacter-like NADH:Fd oxidoreductase	1.18.1.3	Fdr_{ed} + NAD^+ ↔Fd_{ox} + NADH
FDH	Formate dehydrogenase	1.2.2.1 1.2.1.2	Formate + 2 ferricytochrome b1 ↔CO_2 + 2 ferrocytochrome b1 + 2 H^+ Formate + NAD^+ ↔CO_2 + NADH
ALDH	Acetaldehyde dehydrogenase	1.2.1.10	Acetyl-CoA + NADH ↔Acetaldehyde + CoA + NAD^+
ADH	Alcohol dehydrogenase	1.1.1.1	Aldehyde + NADH ↔Alcohol + NAD^+
ADHE	acetaldehyde/alcohol dehydrogenase	1.1.1.1 1.2.1.10	Acetyl-CoA + NADH ↔Alcohol + NAD^+
GAPDH	Glyceraldehyde-3-phosphate dehydrogenase	1.2.1.12	D-glyceraldehyde-3-P + Phosphate + NAD^+ ↔3-phospho-D-glyceroyl-P + NADH
GAPFOR	Glyceraldehyde-3-phosphate:ferredoxin oxidoreductase	1.2.7.6	D-glyceraldehyde-3-P + H_2O + Fd_{ox} ↔3-phospho-D-glycerate + $2H^+$ + $2Fd_{red}$
ENO	Enolase	4.2.1.11	2-phospho-D-glycerate ↔phosphoenolpyruvate + H_2O
PPK	Pyruvate phosphate kinase	2.7.1.40	ATP + pyruvate ↔ADP phosphoenolpyruvate
PPDK	Pyruvate phosphate dikinase	2.7.9.1	ATP + pyruvate + phosphate ↔AMP + phosphoenolpyruvate + diphosphate
PEPCK	Phosphoenolpyruvate carboxykinase	4.1.1.49 4.1.1.38	ATP + oxaloacetate ↔ADP + phosphoenolpyruvate + CO_2 Diphosphate + oxaloacetate ↔phosphate + phosphoenolpyruvate + CO_2
ME	Malic enzyme	1.1.1.38 1.1.1.39 1.1.1.40	malate + $NAD(P)^+$ ↔pyruvate + CO_2 +NADH
MDH	Malate dehydrogenase	1.1.1.37	malate + NAD^+ ↔oxaloacetate + NADH
PDH	Pyruvate dehydrogenase	E1 1.2.4.1	Pyruvate + [dihydrolipoyllysine-residue acetyltransferase] lipoyllysine ↔[dihydrolipoyllysine-residue acetyltransferase] S-acetyldihydrolipoyllysine +CO_2

		E2 2.3.1.12 E3 1.8.1.4	Acetyl-CoA + enzyme N(6)-(dihydrolipoyl)lysine ↔ CoA + enzyme N(6)-(S-acetyldihydrolipoyl)lysine Protein N(6)-(dihydrolipoyl)lysine + NAD(+) ↔protein N(6)-(lipoyl)lysine + NADH
POR	Pyruvate:ferredoxin oxidoreductase	1.2.7.1	Pyruvate + CoA + 2 Fd_{ox} ↔acetyl-CoA + CO_2 + 2 Fd_{red} + 2 H^+
PFL	Pyruvate formate lyase	2.3.1.54	Acetyl-CoA + formate ↔CoA + pyruvate
PDC	Pyruvate decarboxylase	4.1.1.1	2-oxo acid ↔aldehyde + CO_2
LDH	Lactate dehydrogenase	1.1.1.27	lactate + NAD^+ ↔pyruvate + NADH

Table 5A: Presence of genes in central metabolic pathways for selected H_2-producing organisms.

Organism	RNF	[Ni-Fe] H_2ase[a]				[Fe-Fe] H_2ase[a]				Aldh	ADH	ADHE	GAPDH	GAPFOR	ENO	PPK	PPDK	PEPCK	ME	MDH	PDH	POR	PFL	PDC	LDH
		I	II	III	IV	Mono	Di	Tri	Tetra																
Acidothermus cellulolyticus 11B		•		•							•		•		•	•	•		•	•	•	•			
Bacillus cereus 03BB102											•	•	•		•	•	•		•	•	•		•		•
Geobacillus thermodenitrificans NG80-2										•	•		•		•	•			•	•	•	•			•
Caldicellulosiruptor saccharolyticus DSM 8903					•	•				•	•		•		•	•	•	•	•			•			•
Clostridium acetobutylicum ATCC 824		•				•					•	•	•		•	•	•		•	•		•	•	•	•
C. beijerinckii NCIMB 8052	•	•				•		•			•	•	•		•	•	•		•	•		•	•		•
C. butyricum 5521						•	•			•	•	•	•		•	•	•		•	•		•	•		•
C. cellulolyticum H10		•			•	•	•	•			•	•	•		•	•	•	•	•	•		•	•		•
C. kluyveri DSM 555	•					•			•		•	•	•		•	•	•		•		•	•	•		•
C. perfringens ATCC 13124	•					•					•	•	•		•	•	•					•	•		•
C. phytofermentans ISDg	•				•	•		•		•		•	•		•	•	•		•	•		•	•		•
C. thermocellum ATCC 27405	•				•	•	•	•		•	•	•	•		•		•	•	•	•		•	•		•
Coprothermobacter proteolyticus DSM 5265					•			•					•		•	•	•		•			•	•		
Escherichia coli K-12 MG1655	•	•			•					•	•	•	•		•	•	•		•	•	•	•	•		
Klebsiella pneumoniae 342	•				•					•	•	•	•		•	•	•		•	•	•	•	•		•
Syntrophobacter fumaroxidans MPOB	•	•					•	•			•		•		•	•	•	•	•	•	•	•			
Syntrophomonas wolfei wolfei Goettingen								•			•		•		•	•	•					•	•		
Syntrophus aciditrophicus SB	•			•			•				•		•		•	•	•		•			•			
Thermoanaerobacter tengcongensis MB4		•			•			•			•		•		•	•	•		•			•			
T. pseudethanolicus ATCC 33223	•							•			•	•	•		•	•	•		•		•	•			•
Thermotoga maritima MSB8	•					•		•			•		•		•	•	•		•			•			•
Tt. neapolitana DSM 4359	•					•		•			•		•		•	•	•		•			•			•
Thermococcus gammatolerans EJ3					•						•			•	•	•	•	•	•			•			
Th. kodakaraensis KOD1				•	•						•			•	•	•	•	•	•			•	•		
T. onnurineus NA1				•	•						•			•	•	•	•	•				•			
T. sibiricus MM 739				•	•						•			•	•	•	•	•	•	•		•	•		
Pyrococcus abyssi GE5				•	•						•			•	•	•	•	•	•			•			
P. furiosus DSM 3638				•	•						•			•	•	•	•	•	•			•			
P. horikoshii OT3				•	•						•			•	•	•	•	•	•			•			

[a] Hydrogenase nomenclature based on [120].

Table 5B (Part 1): Identification of locus tags for genes in central metabolic pathways for selected H_2-producing organisms.

				[Ni-Fe] H_2ase[a]			[Fe-Fe] H_2ase[a]						
Organism	**Prefix**	**RNF**	**FDH**	**I**	**III**	**IV**	**Mono**	**Di**	**Tri**	**Tetra**	**Aldh**	**ADH**	**ADHE**
Acidothermus cellulolyticus 11B	Acel_			0731-0726	1015-1019							0138, 0584	
Bacillus cereus 03BB102	BCA_											2306, 2350	4483
Geobacillus thermodenitrificans NG80-2	GNTG_										3152	1754	
Caldicellulosiruptor saccharolyticus DSM 8903	Csac_					1534-1540	1857, 1864				2154	0407, 0554, 0711	
Clostridium acetobutylicum ATCC 824	CA_			0142-0141			0028, 3230					0059	0612
C. beijerinckii NCIMB 8052	Cbei_	2449-2454		2953-2952			0327, 1773, 1901, 4000		3797-3795, 4110-4112			3864	0305
C. butyricum 5521	CBY_						2300, 2676, 3049	2047-2048			3753	3424, 3772	3735
C. cellulolyticum H10	Ccel_			1069-1071		1686-1691, 3371-3366	2467	2300-2301	2234-2232, 2305-2303			0894	3198
C. kluyveri DSM 555	CKL_	1264-1269					0141, 2257			0838-0841		1067, 2405, 2967, 2978, 3000, 3425, 3436	1614
C. perfringens ATCC 13124	CPF_	1318-1321, 1663-1664					0270, 1076, 2655, 2900					0451	2855
C. phytofermentans ISDg	Cphy_	0211-0216				1730-1735	0090, 0092, 2056		0089-0087, 3803-3805		1428		3925
C. thermocellum ATCC 27405	Cthe_	2430-2435				3024-3019	0355	0425-0426, 3004-3003	0340-0342, 0428-0430		2238	0394, 2579	0423
Coprothermobacter proteolyticus DSM 5265	COPRO_					0934-0921			0179-0181				
Escherichia coli K-12 MG1655	b	1627-1632	1474-1476	0972-0974		2481-2489					0351	3589	1241
Klebsiella pneumoniae 342	KPK_	2384-2379	2489-2491			1060-1066					2205	2503, 4882	2109
Syntrophobacter fumaroxidans MPOB	Sfum_	2694-2699		2953-2952				0847-848, 1845-1846	0846-0844			0829	
Syntrophomonas wolfei wolfei Goettingen	Swol_		0800-0803						1019-1017, 1925-1927			1727	
Syntrophus aciditrophicus SB	SYN_	1658-1664	0634-0632		02219-02222			1369-1370				01159	
Thermoanaerobacter tengcongensis MB4	TTE_			1697-1706		0123-0128			0892-0894			0696	
T. pseudethanolicus ATCC 33223	Teth39_	2119-2124							1458-1456			0220, 1597	0206
Thermotoga maritima MSB8	TM_	0244-					0201		1420-			0111,	

		0249							1422, 1424-1426			0920	
Tt. neapolitana DSM 4359	CTN_	0442-437					0485		1069-1068			0580, 1655	
Thermococcus gammatolerans EJ3	TGAM_					0047-0034, 0069-0056, 0251-0241						1572	
Th. kodakaraensis KOD1	TK_				2072-2069	2080-2093						1008, 1569	
Th. onnurineus NA1	TON_				0055-0052, 0537-0534	0281-0266, 1063-1572						0936	
Th. sibiricus MM 739	TSIB_				0282-0285, 1520-1523	1309-1296, 1295-1282						1456	
Pyrococcus abyssi GE5	PAB_				1784-1787	1884-1894						1511	
P. furiosus DSM 3638	PFO_				0891-0894, 1329-1332	1423-1436						0075, 0608	
P. horikoshii OT3	PHO_				1290-1294	1427-1441						0743	

[a] Hydrogenase nomenclature based on [120].

Table 5B (Part 2): Identification of locus tags for genes in central metabolic pathways for selected H_2-producing organisms.

Organism	Prefix	GPFOR	GPDH	ENO	PPK	PPDK	PEPCK	ME	MDH	PDH	POR	PFL	PDC	LDH
Acidothermus cellulolyticus 11B	Acel_		1114	1909	1080	0801		1647	0387	0031-0033, 0394, 0588-0590, 0936	0283-0284, 0700-0703			
Bacillus cereus 03BB102	BCA_		4693, 5250	5246	3417, 4708	3175		0610, 1808, 4713	4702	4074-4077	3870-3871	0528		1989, 5005, 5136
Geobacillus thermodenitrificans NG80-2	GNTG_		2651, 3007	3003	2663			1301, 1385	2658	0922-0925	1155-1156			0487
Caldicellulosiruptor saccharolyticus DSM 8903	Csac_		1953	1950	1831	1955	0274	2059			1458-1461, 2115, 2248-2249			1027
Clostridium acetobutylicum ATCC 824	CA_		0709	0713	0518, 1036	0534		1589, 1596	3552		2229, 2499	0980	0025	0267
C. beijerinckii NCIMB 8052	Cbei_		0597, 2282, 2572	0602	0485	0849, 2063, 2066, 3516, 3859, 3862		4008, 5075	2789, 4072, 4903		1458, 1853, 3642, 4041-4042, 4318	0315, 0705		1014
C. butyricum 5521	CBY_		4068	0318, 4073	0049, 1825	3510		2357, 2744	1691, 2341		0233, 1889, 3642	0465, 0479, 2761, 3421		0742, 2757
C. cellulolyticum H10	Ccel_		2275	2254	2569	2388	0212	0138	0137		0016, 0554-0556, 1164	2224, 2582		2485
C. kluyveri DSM 555	CKL_		1457, 3382	3378	3558	0066, 0920,		2408		2268-2271	2138, 2996-	1107		1381

						2571, 3749					2999, 3337			
C. perfringens ATCC 13124	CPF_		1511	1505	0351, 2404	2268					2318	1356		0098
C. phytofermentans ISDg	Cphy_		2229, 2876	3001	0741, 2900	0651		0409	1117		0603, 3122-3123	1417, 2823		1232
C. thermocellum ATCC 27405	Cthe_		0137	0143		1253, 1308	2874	0344	0345		2390-2393, 2794-2797, 3120	0505		1053
Coprothermobacter proteolyticus DSM 5265	COPRO_		0728	1076	0875	0495		0583			1196-1199	1565		
Escherichia coli K-12 MG1655	b		1779, 4493	2779	1676, 1854	1702		1479t, 2463	3236	0114-0116	1378	0823, 0903, 3951		
Klebsiella pneumoniae 342	KPK_		2662, 2958, 3247	0999	1919, 2188	2159		1340t, 1916, 2508	0477	4619-4621	1717, 3026	2017, 3553, 3628, 3709		2747
Syntrophobacter fumaroxidans MPOB	Sfum_		1468, 2061, 2062	0078	1475, 2959	0630, 0634, 0649, 2063, 2562, 2666, 3264	0901	2777	0460	2645-2648	0299-0301, 2792-2795			
Syntrophomonas wolfei wolfei Goettingen	Swol_		0272	0276	2040	1512					1374-1377	1048		
Syntrophus aciditrophicus SB	SYN_		01810	01703	02952	02966		00517			00157-00157, 02201, 00691-00694			
Thermoanaerobacter tengcongensis MB4	TTE_		1762	1759	1815	0981		2332			0445, 0960			
T. pseudethanolicus ATCC 33223	Teth39_		0781	0735	0684	1358		0337		0785, 0790-0791, 0793	0289, 1842			1997
Thermotoga maritima MSB8	TM_		0688	0877	0208	0272		0542			0015-0018			1867
Tt. neapolitana DSM 4359	CTN_		1898	1698	0477	0413		0126			0263-0264, 0680-0683, 1410-1411			0802
Th. gammatolerans EJ3	TGAM_	0122		0077	1550	1043	1771	0399			0267, 0263-0261			
Th. kodakaraensis KOD1	TK_	2163		2106	0511	1292	1405	1963			1978, 1982-1984	0289		
Th. onnurineus NA1	TON_	1498		1613	0785	0311	0192				1476, 1480-1482			
Th. sibiricus MM 739	TSIB_	1686		0792	0471	0271	0151	0123	1633		1384, 1380-1378	0631		
Pyrococcus abyssi GE5	PAB_	1315		1126	1441	0057	1253	1792			1470, 1473-1476			
P. furiosus DSM 3638	PFO_	0464		0215	1188	0043	0289	1026			0971, 0967-0965-			
P. horikoshii OT3	PHO_	0457		1943	0570	0092	0312	1277			0678, 0682, 0684-0685			

Mining genomic data in a wide variety of hydrogenogenic organisms, as shown in Table **5,** reveals several overarching themes, surprises and potential pitfalls. Such is the purpose of our present exercise. Excellent detailed comparative analyses of specific enzymes have recently become available, including for the various types of hydrogenases [87], as well as in at least two other chapters from this eBook (see Chapters 6 and 9). Here are conceptual observations from our comparison. It is hope that these may be useful for further in-depth comparisons of specific aspects of these various enzymes and enzyme types.

The sequenced hydrogenogenic Firmicutes that do not produce butyrate all possess genes encoding a ferredoxin-dependent membrane-associated NiFe type 4 hydrogenase similar to that found in *E. coli* and to that purified from *Thermoanaerobacter tengcongensis* [88]. This hydrogenase type may play different roles in different organisms, making it difficult to predict functionality from the genome annotation alone. For example, while the Ech type hydrogenase was clearly demonstrated to be constitutively expressed at high levels in *T. tengcongensis* [88], the homologous gene cluster in *Thermoanaerobacterium saccharolyticum* is not expressed under the growth conditions tested, and knock out mutants do not show any alteration in growth or H_2 production [49]. Since most hydrogenogenic microorganisms have multiple putative hydrogenases, it is difficult to predict the function of each specific hydrogenase on the basis of structural homology.

Furthermore, the possibility that a trimeric Fe-hydrogenase may simultaneously accept electrons from both ferredoxin and NADH [69] virtually eliminates the need for an elusive soluble NAD:ferredoxin oxidoreductase whose activity has been described in earlier work [89] but which does not seem to be present in most genomes. What has been found in the genome of many clostridia is a putative Rnf-type membrane-integral NADH-ferredoxin oxidoreductase, which is proposed to be involved in the interconversion of electrons from NADH to ferredoxin or *vice versa* with the assistance of an ion motive force [90]. However, its function has not been directly elucidated in the saccharolytic, hydrogenogenic organisms.

Pyruvate is a key intermediate branchpoint in fermentation (see Fig. **2**), both as a possible substrate for H_2-producing reactions and for competing electron-consuming reactions. Summarizing the comparison of genomes in Table **5**, it appears that the NAD^+-dependent PDH typically associated with aerobic pyruvate metabolism is replaced primarily with ferredoxin-dependent POR in most anaerobic H_2-producing bacteria. The genome annotations also indicate the presence of PFL in facultative anaerobes from phyla Proteobacteria and Firmicutes. The formate dehydrogenase and hydrogenase that make up FHL have not so far been observed in the Firmicutes, and formate is released as an end product in these organisms. The annotation shows that several members of the genus *Clostridium,* both butyrate producers and non-butyrate producers, also contain a putative PFL in their genomes (Tables **5a** and **5b**). This is somewhat surprising since formate is not a reported end product in many of these species (Table **1**). Lack of detection may in some cases be an artifact of the detection method used for quantifying short-chain fatty acids. Indeed, formate production and PFL activity were only recently detected in *C. thermocellum* [82], and may have been overlooked in other clostridia as well.

The acetyl-CoA released from PDH, POR or PFL can either be used to generate ATP with the release of acetate or be reduced to ethanol or butyrate. The pathway for butyrate synthesis is sufficiently similar to synthesis of other fatty acids that it is difficult to distinguish based on genome analysis alone, which is why the pathway is not presented in Table **5**. With respect to ethanol, however, ADHE, a bifunctional enzyme that takes acetyl-CoA through acetaldehyde to ethanol, appears to be diagnostic of an organism's capacity to generate high amounts of ethanol. Many organisms have several other putative alcohol dehydrogenases, which may or may not be associated with ethanol formation. For example, while ethanol formation has not been observed in strains of *Pyrococcus* or *Thermococcus* (Table **1**), putative ADH genes have been found in both genera (Table **5**). These, however, correspond to the secondary alcohol dehydrogenase characterized in *Pyrococcus furiosus* [91]. The actual function of these alcohol dehydrogenases may prevent them from competing directly with hydrogenases for electrons. Conversely, several representatives of *Thermoanaerobacter* and *Clostridium* possess genes corresponding to the NADPH-dependent ethanol-producing AdhA of *Moorella thermoacetica* [92]. An NADPH-dependent alcohol dehydrogenase activity has been observed in *C. thermocellum* [2], however the gene annotation is still not sufficiently refined to distinguish NADPH from NADH dependent alcohol dehydrogenases.

With respect to the synthesis of pyruvate from phosphoenolpyruvate (PEP), the organisms whose sequences have been analysed indicate the presence of the expected pyruvate kinase. However, the absence of an overt sequence for pyruvate kinase has been noted in the genome of *C. thermocellum*. While it is possible that it may use a novel form of the enzyme not captured by current genome annotation techniques, other possibilities should be considered. First, the genome suggests the presence of PEPCK, MDH and ME, which together could also yield pyruvate (Fig. **2**), and may be important in the transfer of electrons from NADH to NADPH [93]. Second, while PPDK is typically associated with gluconeogenesis in many organisms, it has been shown to have a role in glycolysis in fermentative Eukaryotes such as *Giardia lamblia* [94]. A glycolytic role has been proposed in sugar-fermenting *Clostridium* sp. [95], as well as in *Caldicellulosiruptor saccharolyticus* [96], suggesting that there may be a role for PPi-yielding enzymes in some saccharolytic fermentative organisms. These organisms have very active pyrophosphate-dependent phosphofructokinase activity during glycolysis [95,96], suggesting that the cycling of this high energy molecule [1] may play an active role in glycolysis in several anaerobes [97,98].

Finally, a difference that distinguishes the glycolytic pathway in Bacteria from that of hydrogenogens in the hyperthermophilic Archaea, is the replacement of the NADH- yielding glyceraldehyde 3-phosphate dehydrogenase (GAPDH) and ATP-yielding 1,3 bisphophoglycerate kinase with a single glyceraldehyde-3-phosphate ferredoxin oxidoreductase (GAPFOR) [75]. The fact that this reaction does not yield ATP, but only electrons with an E_h more negative than H_2 may help explain why these organisms are capable of generating consistently near 4 H_2 per hexose equivalent.

As an increasing number of genomes are sequenced and compared and as more enzymes are specifically characterized, the distinctive metabolic strategies employed and enzymological contexts under which H_2 evolution is controlled in different organisms will be clarified. This knowledge will allow researchers to develop more effective strategies to modulate competing pathways, and help in the design of molecular engineering strategies leading to enhancement of H_2 evolution.

Table 6: Standard enthalpies of formation (ΔH°_f), standard free energies of formation (ΔG°_f), and free energies of formation as a function of temperature as calculated by the Gibbs-Helmholtz equation.

			Temperature (°C)					
Compound[a,b,c]	ΔH°_f	ΔG°_f	$\Delta G^{37^\circ}_f$	$\Delta G^{45^\circ}_f$	$\Delta G^{60^\circ}_f$	$\Delta G^{70^\circ}_f$	$\Delta G^{80^\circ}_f$	$\Delta G^{100^\circ}_f$
$C_{graphite}$	0	0	0	0	0	0	0	0
O_2	0	0	0	0	0	0	0	0
H_2	0	0	0	0	0	0	0	0
H_2O	-285.8	-237.1	-235.1	-233.8	-231.4	-229.7	-228.1	-224.8
CO_2	-393.5	-394.4	-394.4	-394.5	-394.5	-394.5	-394.6	-394.6
HCO_3^-	-689.9	-586.8	-582.7	-579.9	-574.7	-571.2	-567.8	-560.9
Formate$^-$	-425.4	-350.9	-347.9	-345.9	-342.1	-339.6	-337.1	-332.1
Acetate$^-$	-486.0	-369.3	-364.6	-361.5	-355.6	-351.7	-347.8	-340.0
Lactate$^-$	-686.5	-512.7	-505.7	-501.0	-492.3	-486.4	-480.6	-469.0
Butyrate$^-$	-538.2	-354.2	-346.8	-341.8	-332.6	-326.4	-320.2	-307.9
Ethanol	-277.6	-174.8	-170.7	-167.9	-162.7	-159.3	-155.8	-148.9
Glucose	-1274.4	-910.5	-895.9	-886.1	-867.8	-855.6	-843.4	-819.0

[a]Values for graphite, O_2, H_2, H_2O, CO_2, ethanol, and glucose were obtained from the CRC handbook [121].

[b]Values for organic acids were obtained from [122].

[c]Values for bicarbonate obtained from [123].

THERMODYNAMIC REGULATION OF HYDROGEN PRODUCTION

Most organisms described above use a combination of the reactions for the fermentation of glucose leading to a diversity of end products. Each of these reactions has a Gibb's free energy change ($\Delta G^{o'}$) indicating the amount of added energy required or energy released in the reaction. The $\Delta G^{o'}$ of a given reaction is controlled by 3 factors: the temperature, the pH, and the concentration of substrates and end products in the environment. Chemical reactions of biological relevance are typically reported at 25°C, a pH of 7.0 and concentrations of water soluble substrates and end products each at 1 M, or 1 atm for gases (for example, [1]). If one knows the enthalpy and free energy of formation of the various substrates and compounds (Table **6**), it is possible, through the Gibbs-Helmholtz equation (eq.1), to correct for reactions occurring at different temperatures (Table **7**).

$$\Delta G^{\circ}_{Tact} = \Delta G^{\circ}_{Tref} \text{ x } (T_{act}/T_{ref}) + \Delta H^{\circ}_{Tref} \text{ x } (T_{ref} - T_{act})/T_{ref} \qquad \textbf{(1)}$$

Many of the end products of glucose fermentation are organic acids, which will reduce the culture's pH. Lowering the pH at which a reaction occurs also changes its ΔG, according to the following equation:

$$\Delta G^{\circ} = \Delta G^{\circ} + nRT \ln [H^{+}] \qquad \textbf{(2)}$$

Table 7: Calculated free energy changes per balanced reaction showing possible combinations of end products at different temperatures and pH.

Reaction (ATP produced)	**pH**	**Temperature (°C)**						
		25	**37**	**45**	**60**	**70**	**80**	**100**
Glucose to H_2	7	-5	-28	-43	-72	-91	-110	-148
$C_6H_{12}O_6 + 12H_2O \rightarrow 12H_2 + 6HCO_3^- + 6H^+$	6	29	8	-6	-33	-51	-69	-105
	5	64	44	30	5	-12	-29	-62
Glucose to H_2/acetate (+4ATP):	7	-213	-224	-232	-246	-255	-265	-283
$C_6H_{12}O_6 + 4H_2O \rightarrow 4H_2 + 2CH_3COO^- + 2HCO_3^- +$	6	-190	-201	-208	-220	-229	-238	-255
$4H^+$	5	-168	-177	-183	-195	-203	-211	-226
Glucose to H_2/butyrate (+3ATP):	7	-263	-271	-276	-285	-292	-298	-311
$C_6H_{12}O_6 + 2H_2O \rightarrow 2H_2 + CH_3CH_2CH_2COO^- +$	6	-246	-253	-257	-266	-272	-278	-290
$2HCO_3^- + 3H^+$	5	-229	-235	-239	-247	-252	-258	-268
Glucose to H_2/ethanol/acetate (+3 ATP):	7	-216	-224	-229	-240	-247	-253	-267
$C_6H_{12}O_6 + 3H_2O \rightarrow 2H_2 + CH_3CH_2OH + CH_3COO^-$	6	-199	-206	-211	-221	-227	-233	-246
$+ 2HCO_3^- + 3H^+$	5	-182	-188	-193	-201	-207	-213	-224
Glucose to H_2/ethanol/formate/acetate (+3 ATP):	7	-217	-224	-229	-239	-245	-251	-263
$C_6H_{12}O_6 + 2H_2O \rightarrow H_2 + CH_3CH_2OH + HCOO^- +$	6	-200	-207	-211	-219	-225	-231	-242
$CH_3COO^- + HCO_3^- + 3H^+$	5	-183	-189	-193	-200	-205	-210	-220
Glucose to H_2/acetate/formate (+3 ATP):	7	-216	-225	-232	-244	-252	-260	-276
$C_6H_{12}O_6 + 2H_2O \rightarrow 2H_2 + 2CH_3COO^- + 2HCOO^- +$	6	-193	-201	-207	-218	-225	-233	-247
$4H^+$	5	-170	-178	-183	-193	-199	-205	-218

Glucose to ethanol (+2 ATP):	7	-218	-224	-227	-234	-238	-242	-251
$C_6H_{12}O_6 + 2H_2O \rightarrow 2CH_3CH_2OH + 2HCO_3^- + 2H^+$	6	-207	-212	-215	-221	-225	-229	-237
	5	-196	-200	-203	-208	-212	-215	-222
Glucose to lactate (+2 ATP):	7	-195	-199	-201	-206	-209	-212	-219
$C_6H_{12}O_6 \rightarrow 2CH_3CH(OH)COO^- + 2H^+$	6	-183	-187	-189	-193	-196	-199	-205
	5	-172	-175	-177	-181	-183	-185	-190
Glucose to ethanol/formate/acetate (+3 ATP):	7	-218	-225	-229	-237	-243	-248	-259
$C_6H_{12}O_6 + H_2O \rightarrow CH_3CH_2OH + 2HCOO^- +$	6	-201	-207	-211	-218	-223	-228	-238
$CH_3COO^- + 3H^+$	5	-184	-189	-193	-199	-203	-208	-217
Acetate to H_2	7	104	98	94	87	82	77	68
$CH_3COO^- + 4H_2O \rightarrow 4H_2 + 2HCO_3 + H^+$	6	110	104	101	93	89	84	75
	5	116	110	107	100	95	91	82

When comparing all the reactions from glucose yielding different specific organic products, the $\Delta G^{o'}$ of each reaction is between -200 and -220 kJ/mol, with the exception of butyrate production at -263 kJ/mol. There would appear to be no thermodynamic advantage to produce 2 ethanol+$2CO_2$ relative to $4H_2$+$2CO_2$+2 acetate, and yet, in mid-log phase of a typical batch run with *C. thermocellum* grown on cellobiose, the organism produced a mixture of products (Table **8**) maintaining a ΔG *of* -232 kJ/mol hexose consumed [2]. Indeed, there is little evidence of mesophilic or even moderate thermophilic organisms in pure culture producing $4H_2$ from glucose for any sustained length of time. Examples are mainly available for growth conditions where the H_2 partial pressure is maintained at very low levels by rapid sparging with nitrogen [88], H_2 consumption by a hydrogenotroph [1,65], or removal by chemical means [70].

Table 8: Free energy of reaction under the conditions listed below (mid-exponential *C. thermocellum* in batch culture on cellobiose) is -232 kJ/mol.

	Growth Conditions				**Product Concentration Observed (mM)**					
Rxn	**Time (h)**	**OD_{600}**	**pH**	**Temp. (°C)**	**Glucose equivalents**	**H_2**	**Ethanol**	**Formate**	**Acetate**	**CO_2 + bicarbonate**
5	5.5	0.366	7.0	60	2.86	2.23	2.19	0.87	2.19	3.04

Different branches are associated with different amounts of ATP synthesis. Calculations of ΔG for chemical reactions corresponding to various fermentation pathways do not take into account different amounts of ATP associated with each pathway. While different pathways shown in Table **7** have similar $\Delta G^{o'}$ values (between -200 and -220 kJ/mol), the *in vivo* ΔG would also need to take into account the adenylate energy charge. As end products (especially H_2) build up, organisms shift towards pathways yielding less ATP per mole glucose through substrate level phosphorylation. Because of the relationship between ΔG and temperature, thermophiles and especially hyperthermophiles are more likely to approach yields of 4 H_2 per glucose, throughout growth. The pH value may also play a role in the ΔG value depending on the pathway used. For example, H_2 plus butyrate provides better thermodynamics at lower pH (pH 6, pH 5). This would be consistent with higher butyrate production in organisms capable of fermenting at lower pH, decreasing H_2 yields.

The complete oxidation of glucose to CO_2 with the release of H_2 as sole reduced product has a free energy near zero at room temperature, but of -148 kJ/mol at 100°C at pH 7, and yet hydrogenogenic microorganisms capable of breaking the Thauer limit in pure culture, even hyperthermophiles, have not yet been discovered. This does not appear to be a strictly thermodynamic issue. However, because of the relationship between pH and free energy, such organisms, if discovered, would be expected to be more vulnerable to falling pH values.

In a living cell, the standard equations do not take into account that steady state concentrations of cofactors are themselves not at equilibrium. Cells regulate their thermodynamic efficiency under varying growth conditions in response to shifts in *in vivo* concentration of cofactors such as NAD^+, $NADP^+$, NADH, NADPH, Acetyl-CoA, HS-CoA, AMP, ADP, ATP, and PP_i. Such shifts impact significantly on the flow of carbon and electrons through the various branches of the fermentation pathways available [1]. One way of "breaking the Thauer limit" has been developed to oxidize acetate biologically to CO_2 and electrons, by supplying *Geobacter* with sufficient electrical current to allow the reaction to become thermodynamically favourable in an electrochemically assisted microbial fuel cell [99]. This process is capable of yielding a net 8 to 9 H_2 per glucose. The potential of this technology is reviewed in another chapter (see Chapter 12). Another alternative that is currently being investigated is to develop *in vitro*, cell free, synthetic enzymatic pathways that would not require the added energy to maintain cell homeostasis and biosynthesis, and may thus bypass the Thauer limit [100].

CONCLUSIONS: BRINGING END-PRODUCTS, GENOMICS AND THERMODYNAMICS TOGETHER

By combining genomic and thermodynamic information, kinetic models of metabolism are now being built for specific organisms [101]. For biofuel-producing organisms (*e.g. C. thermocellum* [102]), such models are now being generated and validated or corrected against transcriptomic and metabolic data [103]. In *E. coli*, these models are now also being refined using actual proteomic and metabolomic data [104]. The development of such models is very important since they can be used as predictive tools for suggesting growth conditions that may be optimal for hydrogen production or the effects of specific mutations likely to enhance production of H_2 or other metabolites.

As our pallet of sequenced organism expands and our knowledge of metabolic pathways that generate H_2 or compete against H_2 generation increases, direct comparison of strategies used by different organisms to select alternate branches of their metabolic pathways may lead to an ability to coax selected organisms to produce more H_2 faster, within the thermodynamic constraints imposed by the chemical reactions involved in the living cell. It may also guide improvements through molecular engineering of specific steps, or guide enzyme selection for development of cell-free enzymatic systems. Whatever the case, the starting point for all this is ensuring that researchers are on the same page with respect to how primary data on organism performance is presented in the literature in order to facilitate objective comparisons.

REFERENCES

[1] Thauer RK, Jungermann K, Decker K. Energy conservation in chemotrophic anaerobic bacteria. Bacteriol Rev 1977; 41: 100-80.

[2] Rydzak T, Levin DB, Cicek N, Sparling R. Growth phase-dependant enzyme profile of pyruvate catabolism and end-product formation in *Clostridium thermocellum* ATCC 27405. J Biotechnol 2009; 140: 169-75.

[3] Collet C, Adler N, Schwitzguebel J, Peringer P. Hydrogen production by *Clostridium thermolacticum* during continuous fermentation of lactose. Int J Hydrogen Energy 2004; 29: 1479-85.

[4] Levin D, Pitt L, Love M. Biohydrogen production: prospects and limitations to practical application. Int J Hydrogen Energy 2004; 29: 173-85.

[5] Mnatsakanyan N, Bagramyan K, Trchounian A. Hydrogenase 3 but not hydrogenase 4 is major in hydrogen gas production by *Escherichia coli* formate hydrogenlyase at acidic pH and in the presence of external formate. Cell Biochem Biophys 2004; 41: 357-66.

[6] Penfold DW, Sargent F, Macaskie LE. Inactivation of the *Escherichia coli* K-12 twin-arginine translocation system promotes increased hydrogen production. FEMS Microbiol Lett 2006; 262: 135-7.

[7] Inoue J, Saita K, Kudo T, Ui S, Ohkuma M. Hydrogen production by termite gut protists: characterization of iron hydrogenases of Parabasalian symbionts of the termite *Coptotermes formosanus*. Eukaryot Cell 2007; 6: 1925-32.

[8] Lloyd D, Kristensen B. Metronidazole inhibition of hydrogen production *in vivo* in drug-sensitive and resistant strains of *Trichomonas vaginalis*. J Gen Microbiol 1985; 131: 849-53.

[9] Ordal EJ, Halvorson HO. A Comparison of Hydrogen Production from Sugars and Formic Acid by Normal and Variant Strains of *Escherichia coli*. J Bacteriol 1939; 38: 199-220.

[10] Stephenson M, Stickland LH. Hydrogenlyases: Bacterial enzymes liberating molecular hydrogen. Biochem J 1932; 26: 712-24.

[11] Fan Z, Yuan L, Chatterjee R. Increased hydrogen production by genetic engineering of *Escherichia coli*. PLoS One 2009; 4: e4432.

[12] Wang Q, Ou MS, Kim Y, Ingram LO, Shanmugam KT. Metabolic Flux Control at the Pyruvate Node in an Anaerobic *Escherichia coli* Strain with an Active Pyruvate Dehydrogenase. Appl Environ Microbiol 2010; 76: 2107-14.

[13] Sawers G. The hydrogenases and formate dehydrogenases of *Escherichia coli*. Antonie Van Leeuwenhoek 1994; 66: 57-88.

[14] Maupin JA, Shanmugam KT. Genetic regulation of formate hydrogenlyase of *Escherichia coli*: role of the fhlA gene product as a transcriptional activator for a new regulatory gene, fhlB. J Bacteriol 1990; 172: 4798-806.

[15] Yoshida A, Nishimura T, Kawaguchi H, Inui M, Yukawa H. Enhanced hydrogen production from glucose using ldh- and frd-inactivated *Escherichia coli* strains. Appl Microbiol Biotechnol 2006; 73: 67-72.

[16] Niu K, Zhang X, Tan W, Zhu M. Characteristics of fermentative hydrogen production with *Klebsiella pneumoniae* EXU-15 isolated from anaerobic sewage sludge. Int J Hydrogen Energy 2010; 35: 71-80.

[17] Wu KJ, Saratale GD, Lo YC, *et al.* Simultaneous production of 2,3-butanediol, ethanol and hydrogen with a *Klebsiella* sp. strain isolated from sewage sludge. Bioresource Technol 2008; 99: 7966-70.

[18] Minnan L, Jinli H, Xiaobin W, *et al.* Isolation and characterization of a high H2-producing strain *Klebsiella oxytoca* HP1 from a hot spring. Res Microbiol 2005; 156: 76-81.

[19] Ito T, Nakashimada Y, Kakizono T, Nishio N. High-yield production of hydrogen by *Enterobacter aerogenes* mutants with decreased alpha-acetolactate synthase activity. J Biosci Bioeng 2004 ; 97: 227-32.

[20] Kurokawa T, Tanisho S. Effects of formate on fermentative hydrogen production by *Enterobacter aerogenes*. Mar Biotechnol (NY) 2005; 7: 112-8.

[21] Shin J, Yoon J, Ahn E, *et al.* Fermentative hydrogen production by the newly isolated *Enterobacter asburiae* SNU-1. Int J Hydrogen Energy 2007; 32: 192-9.

[22] Kim S, Seol E, Raj M, Park S, Oh Y, Ryu D. Various hydrogenases and formate-dependent hydrogen produciton in *Citrobacter amalonaticus* Y19. Int J Hydrogen Energy 2008; 33: 1509-15.

[23] Oh YK, Park S, Seol EH, *et al.* Carbon and energy balances of glucose fermentation with hydrogen-producing bacterium *Citrobacter amalonaticus* Y19. J Microbiol Biotechn 2008; 18: 532-8.

[24] Jung GY, Kim, J.R., Park, J., Park, S. Hydrogen production by a new chemoheterotrophic bacterium *Citrobacter* sp. Y19. Int J Hydrogen Energy 2002; 27: 601-10.

[25] Traore AS, Fardeau ML, Hatchikian CE, Le Gall J, Belaich JP. Energetics of Growth of a Defined Mixed Culture of *Desulfovibrio vulgaris* and *Methanosarcina barkeri*: Interspecies Hydrogen Transfer in Batch and Continuous Cultures. Appl Environ Microbiol 1983; 46: 1152-6.

[26] Nakano MM, Dailly YP, Zuber P, Clark DP. Characterization of anaerobic fermentative growth of *Bacillus subtilis*: identification of fermentation end products and genes required for growth. J Bacteriol 1997 ;179 :6749-55.

[27] Tang YJ, Sapra R, Joyner D, *et al.* Analysis of metabolic pathways and fluxes in a newly discovered thermophilic and ethanol-tolerant *Geobacillus* strain. Biotechnol Bioeng 2009; 102: 1377-86.

[28] Ouhib-Jacobs O, Lindley ND, Schmitt P, Clavel T. Fructose and glucose mediates enterotoxin production and anaerobic metabolism of *Bacillus cereus* ATCC14579(T). J Appl Microbiol 2009; 107: 821-9.

[29] Kotay SM, Das D. Microbial hydrogen production with *Bacillus coagulans* IIT-BT S1 isolated from anaerobic sewage sludge. Bioresource Technol 2007; 98: 1183-90.

[30] Kalia VC, Jain SR, Kumar A, Joshi AP. Fermentation of biowaste to H2 by *Bacillus licheniformis*. World J Microb Biot 1994; 10: 224-7.

[31] Veith B, Herzberg C, Steckel S, *et al.* The complete genome sequence of *Bacillus licheniformis* DSM13, an organism with great industrial potential. J Mol Microb Biotech 2004; 7: 204-11.

[32] Budhavaram NK, Fan Z. Production of lactic acid from paper sludge using acid-tolerant, thermophilic *Bacillus coagulans* strains. Bioresource Technol 2009; 100: 5966-72.

[33] Cripps RE, Eley K, Leak DJ, *et al.* Metabolic engineering of *Geobacillus thermoglucosidasius* for high yield ethanol production. Metab Eng 2009; 11: 398-408.

[34] Chin HL, Chen ZS, Chou CP. Fedbatch operation using *Clostridium acetobutylicum* suspension culture as biocatalyst for enhancing hydrogen production. Biotechnol Prog 2003; 19: 383-8.

[35] Kim BH, Bellows P, Datta R, Zeikus JG. Control of Carbon and Electron Flow in *Clostridium acetobutylicum* Fermentations: Utilization of Carbon Monoxide to Inhibit Hydrogen Production and to Enhance Butanol Yields. Appl Environ Microbiol 1984; 48: 764-70.

[36] Oh S, Zuo Y, Zhang H, *et al.* Hydrogen production by *Clostridium acetobutylicum* ATCC 824 and megaplasmid-deficient mutant M5 evaluated using a large headspace volume technique. Int J Hydrogen Energy 2009; 34: 9347-53.

[37] Zhang H, Bruns MA, Logan BE. Biological hydrogen production by *Clostridium acetobutylicum* in an unsaturated flow reactor. Water Res 2006; 40: 728-34.

[38] Liu X, Zhu Y, Yang ST. Construction and characterization of ack deleted mutant of *Clostridium tyrobutyricum* for enhanced butyric acid and hydrogen production. Biotechnol Prog 2006; 22: 1265-75.

[39] Wang A, Ren, N., Shi, Y., Lee, D. Bioaugmented hydrogen production from microcrystalline cellulose using co-culture - *Clostridium acetobutylicum* X9 and *Ethanoligenens harbinense* B49. Int J Hydrogen Energy 2008; 33: 912-7.

[40] Xu L, Ren N, Wang X, Jia Y. Biohydrogen production by *Ethanoligenens harbinense* B49: Nutrient optimization. Int J Hydrogen Energy 2008; 33: 6962-7.

[41] Ren Z, Ward TE, Logan BE, Regan JM. Characterization of the cellulolytic and hydrogen-producing activities of six mesophilic *Clostridium* species. J Appl Microbiol 2007; 103: 2258-66.

[42] Chen WM, Tseng ZJ, Lee KS, Chang JS. Fermentative hydrogen production with *Clostridium butyricum* CGS5 isolated from anaerobic sewage sludge. Int J Hydrogen Energy 2005; 30: 1063-70.

[43] Wang X, Jin B. Process optimization of biological hydrogen production from molasses by a newly isolated *Clostridium butyricum* W5. J Biosci Bioeng 2009; 107: 138-44.

[44] Skonieczny MT, Yargeau V. Biohydrogen production from wastewater by *Clostridium beijerinckii*: Effect of pH and substrate concentration. Int J Hydrogen Energy 2009; 34: 3288-94.

[45] Pan C, Fan Y, Zhao P, Hou H. Fermentative hydrogen production by the newly isolated *Clostridium beijerinckii* Fanp3. Int J Hydrogen Energy 2008; 33: 5383-91.

[46] Chung KT. Inhibitory effects of H2 on growth of *Clostridium cellobioparum*. Appl Environ Microbiol 1976; 31: 342-8.

[47] O-Thong S, Prasertsan P, Karakashev D, Angelidaki I. Thermophilic fermentative hydrogen production by the newly isolated *Thermoanaerobacterium thermosaccharolyticum* PSU-2. Int J Hydrogen Energy 2008; 33: 1204-14.

[48] Ren N, Cao G, Wang A, *et al.* Dark fermentation of xylose and glucose mix using isolated *Thermanaerobacterium thermosaccharolyticum* W16. Int J Hydrogen Energy 2008; 33: 6124-32.

[49] Shaw AJ, Hogsett DA, Lynd LR. Identification of the [FeFe]-hydrogenase responsible for hydrogen generation in *Thermoanaerobacterium saccharolyticum* and demonstration of increased ethanol yield *via* hydrogenase knockout. J Bacteriol 2009; 191: 6457-64.

[50] Demain AL, Newcomb M, Wu JH. Cellulase, clostridia, and ethanol. Microbiol Mol Biol Rev 2005; 69: 124-54.

[51] Magnusson L, Cicek N, Sparling R, Levin D. Continuous hydrogen production during fermentation of alpha-cellulose by the thermophillic bacterium *Clostridium thermocellum*. Biotechnol Bioeng 2009; 102: 759-66.

[52] Islam R, Cicek N, Sparling R, Levin D. Influence of initial cellulose concentration on the carbon flow distribution during batch fermentation by *Clostridium thermocellum* ATCC 27405. Appl Microbiol Biotechnol 2009; 82: 141-8.

[53] Islam R, Cicek N, Sparling R, Levin D. Effect of substrate loading on hydrogen production during anaerobic fermentation by *Clostridium thermocellum* 27405. Appl Microbiol Biotechnol 2006; 72: 576-83.

[54] Rainey FA, Donnison AM, Janssen PH, *et al.* Description of *Caldicellulosiruptor saccharolyticus* gen. nov., sp. nov: an obligately anaerobic, extremely thermophilic, cellulolytic bacterium. FEMS Microbiol Lett 1994; 120: 263-6.

[55] de Vrije T, Bakker RR, Budde MA, *et al.* Efficient hydrogen production from the lignocellulosic energy crop Miscanthus by the extreme thermophilic bacteria *Caldicellulosiruptor saccharolyticus* and *Thermotoga neapolitana*. Biotechnol Biofuels 2009; 2: 12-8.

[56] van Niel EW, Claassen PA, Stams AJ. Substrate and product inhibition of hydrogen production by the extreme thermophile, *Caldicellulosiruptor saccharolyticus*. Biotechnol Bioeng 2003; 81: 255-62.

[57] van de Werken HJ, Verhaart MR, VanFossen AL, *et al.* Hydrogenomics of the extremely thermophilic bacterium *Caldicellulosiruptor saccharolyticus*. Appl Environ Microbiol 2008; 74: 6720-9.

[58] Kadar Z, de Vrije T, van Noorden GE, *et al.* Yields from glucose, xylose, and paper sludge hydrolysate during hydrogen production by the extreme thermophile *Caldicellulosiruptor saccharolyticus*. Appl Biochem Biotechnol 2004; 113-116: 497-508.

[59] van Niel E, Budde M, de Haas G, *et al.* Distinctive properties of high hydrogen producing extreme thermophiles, *Caldicellulosiruptor saccharolyticum* and *Thermotoga elfii*. Int J Hydrogen Energy 2002; 27: 1391-8.

[60] de Vrije T, Mars AE, Budde MA, *et al.* Glycolytic pathway and hydrogen yield studies of the extreme thermophile *Caldicellulosiruptor saccharolyticus*. Appl Microbiol Biotechnol 2007; 74: 1358-67.

[61] Dolfing J, Larter SR, Head IM. Thermodynamic constraints on methanogenic crude oil biodegradation. ISME J 2008; 2: 442-52.

[62] Stams AJ, Van Dijk JB, Dijkema C, Plugge CM. Growth of Syntrophic Propionate-Oxidizing Bacteria with Fumarate in the Absence of Methanogenic Bacteria. Appl Environ Microbiol 1993; 59: 1114-9.

[63] de Bok FA, Luijten ML, Stams AJ. Biochemical evidence for formate transfer in syntrophic propionate-oxidizing cocultures of *Syntrophobacter fumaroxidans* and *Methanospirillum hungatei*. Appl Environ Microbiol 2002; 68: 4247-52.

[64] Harmsen HJ, Van Kuijk BL, Plugge CM, *et al.* *Syntrophobacter fumaroxidans* sp. nov., a syntrophic propionate-degrading sulfate-reducing bacterium. Int J Syst Bacteriol 1998; 48 Pt 4:1 383-7.

[65] Jackson BE, Bhupathiraju VK, Tanner RS, Woese CR, McInerney MJ. *Syntrophus aciditrophicus* sp. nov., a new anaerobic bacterium that degrades fatty acids and benzoate in syntrophic association with hydrogen-using microorganisms. Arch Microbiol 1999; 171: 107-14.

[66] McInerney MJ, Rohlin L, Mouttaki H, *et al.* The genome of *Syntrophus aciditrophicus*: life at the thermodynamic limit of microbial growth. Proc Natl Acad Sci U S A 2007; 104: 7600-5.

[67] McInerney MJ, Bryant MP, Hespell RB, Costerton JW. *Syntrophomonas wolfei* gen. nov. sp. nov., an Anaerobic, Syntrophic, Fatty Acid-Oxidizing Bacterium. Appl Environ Microbiol 1981; 41: 1029-39.

[68] Sieber JR, Sims DR, Han C, *et al.* The genome of *Syntrophomonas wolfei*: new insights into syntrophic metabolism and biohydrogen production. Environ Microbiol 2010; 12: 2289-301.

[69] Schut GJ, Adams MW. The iron-hydrogenase of *Thermotoga maritima* utilizes ferredoxin and NADH synergistically: a new perspective on anaerobic hydrogen production. J Bacteriol 2009; 191: 4451-7.

[70] Mountfort DO, Kaspar HF. Palladium-Mediated Hydrogenation of Unsaturated Hydrocarbons with Hydrogen Gas Released during Anaerobic Cellulose Degradation. Appl Environ Microbiol 1986; 52: 744-50.

[71] Schroder C, Selig M, Schonheit P. Glucose fermentation to acetate, CO2 and H2 in the anaerobic hyperthermophilic eubacterium *Thermotoga maritima*: involvement of the Embden-Meyerhoff pathway. Arch Microbiol 1994; 161: 460-70.

[72] d'Ippolito G, Dipasquale L, Vella F, *et al.* Hydrogen metabolism in the extreme thermophile *Thermotoga neapolitana*. Int J Hydrogen Energy 2010; 35: 2290-5.

[73] Etchebehere C, Muxi L. Thiosulfate reduction and alanine production in glucose fermentation by members of the genus *Coprothermobacter*. Antonie Van Leeuwenhoek 2000; 77: 321-7.

[74] Selig M, Xavier KB, Santos H, Schonheit P. Comparative analysis of Embden-Meyerhof and Entner-Doudoroff glycolytic pathways in hyperthermophilic archaea and the bacterium *Thermotoga*. Arch Microbiol 1997; 167: 217-32.

[75] van der Oost J, Schut G, Kengen SW, *et al.* The ferredoxin-dependent conversion of glyceraldehyde-3-phosphate in the hyperthermophilic archaeon *Pyrococcus furiosus* represents a novel site of glycolytic regulation. J Biol Chem 1998; 273: 28149-54.

[76] Schicho RN, Ma K, Adams MW, Kelly RM. Bioenergetics of sulfur reduction in the hyperthermophilic archaeon *Pyrococcus furiosus*. J Bacteriol 1993; 175: 1823-30.

[77] Verhaart MR, Bielen AA, van der Oost J, Stams A, Kengen SW. Hydrogen production by hyperthermophilic and extremely thermophilic bacteria and archaea: mechanisms for reductant disposal. Environmental Technology 2010; 31: 993-1003.

[78] Gottschalk G. *Bacterial metabolism, 2nd ed.* 1986, Berlin: Springer-Verlag.

[79] Sridhar J, Eiteman MA, Wiegel JW. Elucidation of enzymes in fermentation pathways used by *Clostridium thermosuccinogenes* growing on inulin. Appl Environ Microbiol 2000; 66: 246-51.

[80] Lamed R, Zeikus JG. Glucose fermentation pathway of *Thermoanaerobium brockii*. J Bacteriol 1980; 141: 1251-7.

[81] Desvaux M. *Clostridium cellulolyticum*: model organism of mesophilic cellulolytic clostridia. FEMS Microbiol Rev 2005; 29: 741-64.

[82] Sparling R, Islam R, Cicek N, *et al.* Formate synthesis by *Clostridium thermocellum* during anaerobic fermentation. Can J Microbiol 2006; 52: 681-8.

[83] Guedon E, Desvaux M, Payot S, Petitdemange H. Growth inhibition of *Clostridium cellulolyticum* by an inefficiently regulated carbon flow. Microbiology 1999; 145 (Pt.8): 1831-8.

[84] Altschul SF, Gish W, Miller W, Myers EW, Lipman DJ. Basic local alignment search tool. J Mol Biol 1990; 215: 403-10.

[85] Tatusov RL, Fedorova ND, Jackson JD, *et al.* The COG database: an updated version includes eukaryotes. BMC Bioinformatics 2003; 4: 41.

[86] Markowitz VM, Korzeniewski F, Palaniappan K, *et al.* The integrated microbial genomes (IMG) system. Nucleic Acids Res 2006; 34: D344-8.

[87] Calusinska M, Happe T, Joris B, Wilmotte A. The surprising diversity of clostridial hydrogenases: a comparative genomic perspective. Microbiology+ 2010; 156: 1575-88.

[88] Soboh B, Linder D, Hedderich R. A multisubunit membrane-bound [NiFe] hydrogenase and an NADH-dependent Fe-only hydrogenase in the fermenting bacterium *Thermoanaerobacter tengcongensis*. Microbiology+ 2004; 150: 2451-63.

[89] Lamed R, Zeikus JG. Ethanol production by thermophilic bacteria: relationship between fermentation product yields of and catabolic enzyme activities in *Clostridium thermocellum* and *Thermoanaerobium brockii*. J Bacteriol 1980; 144: 569-78.

[90] Muller V, Imkamp F, Biegel E, Schmidt S, Dilling S. Discovery of a ferredoxin:NAD+-oxidoreductase (Rnf) in *Acetobacterium woodii*: a novel potential coupling site in acetogens. Ann N Y Acad Sci 2008; 1125: 137-46.

[91] Machielsen R, Uria AR, Kengen SW, van der Oost J. Production and characterization of a thermostable alcohol dehydrogenase that belongs to the aldo-keto reductase superfamily. Appl Environ Microbiol 2006; 72: 233-8.

[92] Inokuma K, Nakashimada Y, Akahoshi T, Nishio N. Characterization of enzymes involved in the ethanol production of *Moorella* sp. HUC22-1. Arch Microbiol 2007; 188: 37-45.

[93] Sauer U, Eikmanns BJ. The PEP-pyruvate-oxaloacetate node as the switch point for carbon flux distribution in bacteria. FEMS Microbiol Rev 2005; 29: 765-94.

[94] Feng XM, Cao LJ, Adam RD, Zhang XC, Lu SQ. The catalyzing role of PPDK in *Giardia lamblia*. Biochem Biophys Res Commun 2008; 367: 394-8.

[95] Cruden DL, Durbin WE, Markovetz AJ. Utilization of PP(i) as an Energy Source by a *Clostridium* sp. Appl Environ Microbiol 1983; 46: 1403-8.

[96] Bielen AA, Willquist K, Engman J, *et al.* Pyrophosphate as a central energy carrier in the hydrogen-producing extremely thermophilic *Caldicellulosiruptor saccharolyticus*. FEMS Microbiol Lett 2010; 307: 48-54.

[97] Mertens E. Pyrophosphate-dependent phosphofructokinase, an anaerobic glycolytic enzyme? FEBS Lett 1991; 285: 1-5.

[98] Siebers B, Klenk HP, Hensel R. PPi-dependent phosphofructokinase from *Thermoproteus tenax*, an archaeal descendant of an ancient line in phosphofructokinase evolution. J Bacteriol 1998; 180: 2137-43.

[99] Liu H, Grot S, Logan BE. Electrochemically assisted microbial production of hydrogen from acetate. Environ Sci Technol 2005; 39: 4317-20.

[100] Zhang YH, Sun J, Zhong JJ. Biofuel production by *in vitro* synthetic enzymatic pathway biotransformation. Curr Opin Biotechnol 2010; 21: 663-9.

[101] Jamshidi N, Palsson BO. Formulating genome-scale kinetic models in the post-genome era. Mol Syst Biol 2008; 4: 171.
[102] Roberts SB, Gowen CM, Brooks JP, Fong SS. Genome-scale metabolic analysis of *Clostridium thermocellum* for bioethanol production. BMC Syst Biol 2010; 4: 31.
[103] Gowen CM, Fong SS. Genome-scale metabolic model integrated with RNAseq data to identify metabolic states of *Clostridium thermocellum*. Biotechnol J 2010; 5: 759-67.
[104] Yizhak K, Benyamini T, Liebermeister W, Ruppin E, Shlomi T. Integrating quantitative proteomics and metabolomics with a genome-scale metabolic network model. Bioinformatics 2010; 26: i255-60.
[105] Kim S, Seol E, Raj M, *et al.* Various hydrogenases and formate-dependent hydrogen production in *Citrobacter amalonaticus* Y19. Int J Hydrogen Energy 2008; 33: 1509-15.
[106] Pan C, Fan Y, Zhao P, Hou H. Fermentative hydrogen production by the newly isolated *Clostridium beijerinckii* Fanp3. Int J Hydrogen Energy 2008; 33: 5383-91.
[107] Li F, Hinderberger J, Seedorf H, *et al.* Coupled ferredoxin and crotonyl coenzyme A (CoA) reduction with NADH catalyzed by the butyryl-CoA dehydrogenase/Etf complex from *Clostridium kluyveri*. J Bacteriol 2008; 190: 843-50.
[108] Cao GL, Ren NQ, Wang AJ, *et al.* Acid hydrolysis of corn stover for biohydrogen production using *Thermoanaerobacterium thermosaccharolyticum* W16. Int J Hydrogen Energy 2009; 34: 7182-8.
[109] Guedon E, Payot S, Desvaux M, Petitdemange H. Carbon and electron flow in *Clostridium cellulolyticum* grown in chemostat culture on synthetic medium. J Bacteriol 1999; 181: 3262-9.
[110] Lamed RJ, Lobos JH, Su TM. Effects of Stirring and Hydrogen on Fermentation Products of *Clostridium thermocellum*. Appl Environ Microbiol 1988; 54: 1216-21.
[111] Fardeau ML, Ollivier B, Garcia JL, Patel BK. Transfer of Thermobacteroides leptospartum and Clostridium thermolacticum as Clostridium stercorarium subsp. leptospartum subsp. thermolacticum subsp. nov., comb. nov. and C. stercorarium subsp. thermolacticum subsp. nov., comb. nov. Int J Syst Evol Microbiol 2001; 51: 1127-31.
[112] Xue Y, Xu Y, Liu Y, Ma Y, Zhou P. *Thermoanaerobacter tengcongensis* sp. nov., a novel anaerobic, saccharolytic, thermophilic bacterium isolated from a hot spring in Tengcong, China. Int J Syst Evol Microbiol 2001; 51: 1335-41.
[113] Nguyen TA, Han SJ, Kim JP, Kim MS, Sim SJ. Hydrogen production of the hyperthermophilic eubacterium, *Thermotoga neapolitana* under N2 sparging condition. Bioresource Technol 2010; 101 Suppl 1: S38-41.
[114] Kanai T, Imanaka H, Nakajima A, *et al.* Continuous hydrogen production by the hyperthermophilic archaeon, *Thermococcus kodakaraensis* KOD1. J Biotechnol 2005; 116: 271-82.
[115] Wang A, Ren N, Shi Y, Lee D. Bioaugmented hydrogen production from microcrystalline cellulose using co-culture - Clostridium acetobutylicum X9 and Ethanoligenens harbinense B49. Int J Hydrogen Energy 2008; 33: 912-7.
[116] Chong M, Rahim RA, Shirai Y, Hassan MA. Biohydrogen production by *Clostridium butyricum* EB6 from palm oil effluent. Int J Hydrogen Energy 2009; 34: 764-71.
[117] Levin DB, Islam R, Cicek N, Sparling R. Hydrogen production by *Clostridium thermocellum* 27405 from cellulosic biomass substrates. Int J Hydrogen Energy 2006; 31: 1496-1503.
[118] Nguyen TA, Kim JP, Kim MS, Oh YK, Sim SJ. Optimization of hydrogen production by hyperthermophilic eubacteria, *Thermotoga maritima* and *Thermotoga neapolitana* in batch fermentation. Int J Hydrogen Energy 2008; 33: 1483-8.
[119] Schafer T, Schonheit P. Maltose fermentation to acetate, CO2 and H2 in the anaerobic hyperthermophilic archaeon *Pyrococcus furiosus*: evidence for the operation of a novel sugar fermentation pathway. Arch Microbiol 1992; 158: 188-202.
[120] Vignais PM, Billoud B. Occurrence, classification, and biological function of hydrogenases: an overview. Chem Rev 2007; 107: 4206-72.
[121] *CRC Handbook of Chemistry and Physics, 2009−2010, 90th ed.*, ed. DR Lide and WM Haynes. 2009, Boca Raton, FL: CRC Press.
[122] Shock EL. Organic acids in hydrothermal solutions: standard molal thermodynamic properties of carboxylic acids and estimates of dissociation constants at high temperatures and pressures. Am J Sci 1995; 295: 496-580.
[123] Robie RA, Hemingway BS. *Thermodynamics of minerals and related substances at 298.15 K and 1 Bar (105 Pascals) pressure and higher temperatures*. US Geological Survey Bulletin 2131. 1995, Washington, DC.: United States Government printing office.

CHAPTER 11

Role of Metabolic Engineering in Enhancing Hydrogen Yields

C. Carere and D.B. Levin*

Department of Biosystems Engineering, University of Manitoba, Canada

Abstract: Dark fermentation has the potential to convert carbohydrate-rich waste-streams into hydrogen (H_2). However, current yields of H_2 production during fermentation fall far short of theoretical maximum values. If biohydrogen production *via* dark fermentation is to become practical, yields must be increased. The application of recombinant DNA techniques to direct metabolism towards the production of industrially valuable substrates is an emerging field of study. Metabolic engineering seeks to "improve" cellular function through the modulation of enzymatic, transport, or other regulatory functions of the cell. In contrast to traditional strain improvement approaches involving mutagenesis followed by the screening of colonies for a desired phenotype, metabolic engineering often involves the introduction of heterologous genes or regulatory elements that are often employed to confer novel metabolic configurations.

Keywords: Biohydrogen, microbial metabolism, modeling, metabolic engineering, genetic engineering, genomics, bioinformatics.

INTRODUCTION

Hydrogen is a promising alternative to conventional petrochemical fuels as it possesses greater energy content than oil (142 MJ kg^{-1} for H_2 compared to 42 MJ kg^{-1} for oil) [1,2]. Hydrogen is not chemically linked to carbon (hence oxidation does not contribute to greenhouse gas emissions), and it may be microbially produced from renewable substrates. The microbial production of hydrogen, however, represents an energy loss to the cell, with associated metabolic and regulatory networks illustrative of an evolutionary pressure towards the rational use of energy and the optimization of specific growth rates. While several strategies are being investigated to enhance hydrogen production yields, the enormous potential of metabolic engineering for redirecting electron flux towards hydrogen production within both photosynthetic and non-photosynthetic microorganisms has yet to be fully exploited. The following presents an overview of the disciplines integrated into developing rational metabolic engineering strategies. Where appropriate, examples of their application towards improving hydrogen yields are discussed.

THE MOLECULAR BASIS OF BIOLOGICAL HYDROGEN PRODUCTION

Biologically mediated H_2 synthesis is fundamentally dependent upon the presence of enzymes capable of catalyzing the reduction of protons to dihydrogen (H_2), namely hydrogenase and nitrogenase. Both enzymes are metalloproteins containing iron-sulphur [Fe/S] clusters, while nitogenase enzymes also contain molybdenum-iron [MoFe] clusters. The [MoFe]-nitrogenase, has been extensively characterized and contains, in addition to iron- and sulphur-containing cofactors, a central molybdenum atom in the active site. Alternatively, in some species, molybdenum may be replace by vanadium or iron [3]. The nitrogenase complex, *in vivo*, typically catalyzes the reduction of dinitrogen to ammonia using the electrons from low potential reductants (ferredoxin (fd) or flavodoxin) and ATP according to the reaction:

$$N_2 + 8H^+ + 8e^- + 16ATP \rightarrow 2NH_3 + H_2 + 16ADP + 16P_i$$

The reaction is irreversible and is capable of maintaining catalysis in the direction of H_2 synthesis even under an atmosphere of 100 % H_2 [4]. Turnover rates among nitrogenases are extremely low, which

*Address correspondence to D.B. Levin: Department of Biosystems Engineering, University of Manitoba, Canada; E-mail: levindb@cc.umanitoba.ca

Nuri Azbar and David B. Levin (Eds)

necessitates the large-scale synthesis of both nitrogenase, and associated maturation proteins. Considering the significant energy inputs necessary for nitrogenase catalysis, and the additional enzymatic machinery, energy and time required for their biosynthesis, the use of nitrogenase does not represent a metabolically efficient means of H_2 production [5].

In comparison, hydrogenases are incapable of nitrogen fixation, generally display higher turnover rates and catalysis is not ATP-dependent. Hydrogenases are typically distributed into two pylogenetically distinct classes on the basis of the metal co-factors located at the active site. The [NiFe] and [FeFe] hydrogenases catalyze the reversible oxidation of hydrogen:

$$H_2 \leftrightarrow 2H^+ + 2e^-$$

While all hydrogenases are capable of catalysis in both directions *in vitro*, they are typically committed to either hydrogen uptake or hydrogen synthesis *in vivo*. As a general rule of thumb, [NiFe] hydrogenases are involved in H_2-uptake, and [FeFe] hydrogenases are involved in H_2-evolution, however there are several exceptions to this rule [7]. The NiFe-hydrogenase is a heterodimeric protein consisting of a small (S) and a large (L) subunit. The small subunit contains two [4Fe-4S] and one [3Fe-4S] iron-sulfur clusters, and has been shown to be homologous to soluble ferredoxins and parts of NADH-ubiquinone oxidoreductase (Complex I) of the respiratory chain [7]. The large subunit contains a complex nickel-iron center coordinated to three ligands, one CO and two cyanide (CN^-) molecules. [FeFe] hydrogenases are generally single subunit proteins that contain a unique iron-sulfur cluster in which one of the Fe atoms is also complexed with CO and CN^-. The metallocenters of hydrogenases are oxygen liable and are irreversibly inactivated in the presence of oxygen.

Hydrogenases play a central role in microbial energy metabolism shuttling electrons between H_2 and either NAD(P)H, ferredoxin (fd), flavodoxin, coenzyme F_{420}, or cytochromes, with their function generally dependant on their localization [7]. In fermentative bacteria such as clostridia, hydrogenases are used to dispose of excess electrons through hydrogen evolution and are often cytosolic. Conversely, organisms such as methanogens, sulfate reducers, and photosynthetic bacteria use hydrogen as a source of electrons *via* uptake hydrogenases, which are most often periplasmic or membrane-localized, although may be cytoplasmic in some cases [7].

BIOLOGICAL HYDROGEN SYSTEMS

Microorganisms produce hydrogen *via* two main methods: photosynthetic light-dependent processes such as direct biophotolysis, indirect biophotolysis and photofermentation, as well as light-independent, anaerobic dark fermentations. Photosynthetic hydrogen production occurs in algae (*Chlamydomonas reinhardtii* [8], photosynthetic bacteria (*Rhodovulum sp.* [9]) and cyanobacteria (*Synechocystis* [10]). While the fermentative production of hydrogen has been described within obligate anaerobes (*Clostridium sp.* [11]), hyperthermophilic Archea (*Pyrococcus sp.* [12]), rumen bacteria (*Ruminococcus albus* [13]), methanogens [14], and facultative anaerobes (*Enterobacter* strains [15], *Escherichia coli* [16], and *Citrobacter* species [17]).

In direct biophotolysis, energy recovered from light capture at the photosynthetic apparatus is coupled to water-splitting and the generation of a low potential reductant (ferredoxin), which in turn is used to reduce hydrogenase enzyme and produce hydrogen [5]. This process was first demonstrated in a cell-free chloroplast-ferredoxin-hydrogenase system and has since been demonstrated, *in vivo*, under low-light intensities and very low O_2 partial pressures within green microalgae, including *Chlamydomonas reinhardtii*. The simultaneous photosynthetic production of O_2 and H_2 can only occur under very low concentrations of O_2 (0.1 % O_2 in the liquid phase) as the [FeFe] hydrogenase that participates in proton reduction is extremely oxygen labile. Despite its simplicity and conceptual attractiveness, direct biophotolysis faces several major hurdles in the development of a commercially viable hydrogen-production process: (a) evolved oxygen strongly inhibits the hydrogenase enzyme; (b) the resultant mixture of H_2 and O_2 can be highly volatile; (c) the closed photobioreactors required are likely cost prohibitive; and (d) low solar conversion efficiencies necessitates large land-use requirements [18].

In indirect biophototlysis, light-dependent carbon fixation is separated spatially/temporally from a light-independent fermentative, hydrogen-evolving phase in which the carbon fixed during photosynthesis is utilized. This separation of photosynthetic carbon fixation and oxygen evolution from hydrogen production, in theory, is able to circumvent problems associated with hydrogenase sensitivity to O_2, however, even experimental demonstrations of indirect biophotolysis have proven problematic [5,18].

Photofermentation refers to the process by which photosynthetic bacteria produce hydrogen from organic substrates in a light-dependent process that utilizes nitrogenase to catalyze hydrogen evolution. For example, photofermentation by purple non-sulphur bacteria, such as *Rhodobacter sphaeroides* has been employed to convert small molecular fatty acids, such as acetic acid, into hydrogen [19]. Low solar energy conversion efficiencies and the use of nitrogenase with its inherent high-energy demands, however, represent some of the drawback to this method of hydrogen production.

Compared with photosynthetic processes, the fermentative production of hydrogen generally displays higher rates of production, does not rely on the availability of light, requires less energy, is simpler to perform and is capable of utilizing a variety of carbon sources including organic compounds, biomass obtained during first stage light conversion, low-cost wastes and lignocellulosic substrates [20]. Fermentative hydrogen production, within both facultative and obligate anaerobes, begins with the catabolism of glucose to pyruvate and the generation of reduced electron carriers (NADH in bacteria and ferredoxin in Archea). In facultative anaerobes such as *Escherichia coli*, pyruvate is then converted to acetyl-CoA and formate *via* pyruvate formate lyase (PFL). Hydrogen may next be produced through the oxidation of formate by the formate hydrogen lyase complex (FHL). Among obligate anaerobes the conversion of pyruvate to acetyl-CoA is catalyzed by pyruvate ferredoxin: oxidoreductase (POR), which yields, in addition to acetyl-CoA, both CO_2 and reduced ferredoxin. Hydrogen can then be produced from the reduced ferredoxin *via* the action of hydrogenase [16]. In principle, the fermentative production of hydrogen can yield at most, 4 molecules of H_2 from the complete breakdown of one glucose molecule through a traditional Emden-Meyerhoff Parnas pathway.

$$C_6H_{12}O_6 \rightarrow CH_3COOH + 2CO_2 + 4H_2$$

This maximum yield, also known as the "Thauer limit" [21], assumes common electron carriers (NADH, reduced ferredoxin, and formate) are able to reduce protons to form molecular hydrogen [22]. Observed yields of hydrogen production, however, fall short of the Thauer limit as 1) hydrogenase catalyzed H_2 production is dependent upon both the reduction potential of the intracellular electron carrier utilized (reduced Fd, NADH, formate) and the concentration of accumulated H_2 and 2) other intracellular reactions that utilize NADH as an electron carrier may divert electrons away from H_2 towards the production of other reduced products (lactate, alcohols *etc.*) [22].

METABOLIC ENGINEERING

Metabolic engineering is an emerging multi-disciplinary field that combines the systemic analysis of metabolism with molecular biological tools and techniques to favorably alter cellular properties by designing and implementing rational genetic modifications [23]. The application of metabolic and cellular engineering can typically be divided into five categories: 1) improving production of chemicals natively produced by the host organism, 2) Extending substrate utilization range, 3) the addition of novel catabolic activities for the degradation of toxic compounds, 4) the production of compounds novel to the host organism and 5) the modification of cell properties [24]. These groupings are not mutually exclusive and many examples may fit into multiple categories. For example, work on the gamma proteobacteria, *Klebsiella oxytoca,* has focused concurrently on improving ethanol production and extending the substrate utilization range of the organism [24].

Although metabolic engineering has only recently been described as an academic discipline, the manipulation of cellular pathways for novel phenotype development has in the past been successfully achieved utilizing both selective breeding techniques and *via* mutagenesis [25]. Genetic modification *via* random mutagenesis relies heavily on the usage of chemical mutagens and creative selection techniques as a means of identifying strains with a desirable phenotype. This approach has been applied in the past to fields including the production of solvents, antibiotics and vitamin production [23]. While these approaches may effectively be used to achieve many of the same aims as metabolic engineering, it is important to make the distinction that neither selective breeding nor mutagenesis require a substantive *a priori* knowledge of metabolism at the physiological or genetic level and consequently neither involve the rational genetic modifications that is a defining characteristic of metabolic engineering. Furthermore, both selective breeding and mutagenesis are limited to the manipulation of pre-existing metabolic pathways and do not involve the introduction of heterologous genetic elements.

Metabolic engineering is rooted heavily in the areas of molecular genetics, genetic engineering, biochemistry and bioinformatics. As a consequence, a set of tools have been developed that enable the application of refined engineering strategies to achieve the goals of metabolic engineering. Conventionally these tools are classified as: 1) molecular biology tools, 2) microbiological and analytical tools, and 3) mathematical and computational tools (Fig. **1**). While historically the field has exploited simple organisms, such as bacteria, more complex systems involving complex cells, tissues and organ systems are now utilized [25].

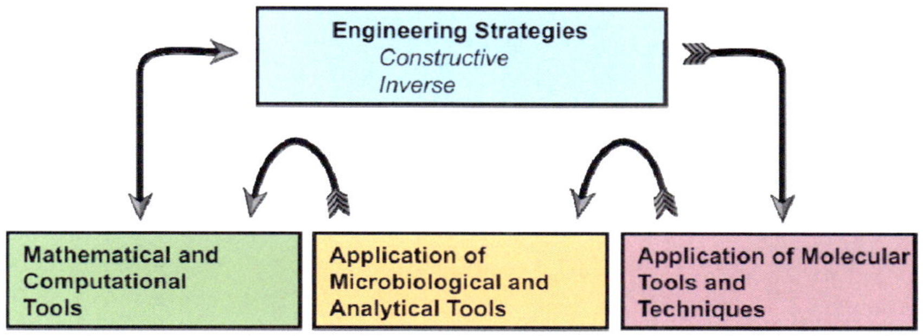

Figure 1: Schematic representation of the tools involved in metabolic and cellular engineering and how they are related to associated engineering principles.

METABOLIC ENGINEERING STRATEGIES

Constructive Metabolic Engineering

Applying the principles of engineering design to metabolic engineering has classically involved extensive physiological studies including knowledge of enzyme kinetics, the systems network, and the intermediate pools involved [25]. Often referred to as "constructive metabolic engineering" the goals of this approach are to first identify flux-limiting steps in metabolism and then to propose genetic manipulations for some predictable benefit [26,27]. This formulation of the objective assumes that several layers about the network pathway are known. Not only are the pathway steps often assumed, but also the identity of the catalysts involved should be known and the identification of a flux-limiting step requires knowledge of reaction kinetics and intermediate metabolite concentrations [28].

This approach was applied to the mesophilic cellulose degrading bacterium, *Clostridium cellulolyticum* in a series of studies that culminated with the introduction of an alternative pathway to ethanol production [29-37] (Fig. **2**). In these studies, Desvaux and Guedon, investigated cellulose catabolism, enzyme kinetics and carbon flux distribution in *C. cellulolyticum* during steady state, batch-fed and batch growth. These studies identified carbon flow through pyruvate:ferredoxin oxidoreductase (POR) as a flux-limiting step in central catabolism and introduced a heterologous pyruvate decarboxylase and acetaldehdye dehydrogenase

(*Zymomonas mobilis*) as a means circumventing this bottleneck and increasing ethanol production. In comparison with that of the wild strain, growth of the recombinant strain led to a 150 % increase in cellulose consumption and a 180 % increase in cell dry weight. While fermentation patterns were altered, the consequence of introducing an alternative route to ethanol only had a modest effect on its production (53 % increase) in comparison to lactate and acetate synthesis (48 % decrease and 93 % increase respectively) [35]. While the engineering of a recombinant strain of *C. cellulolyticum* was successful in that observable changes were recorded with respect to both cell growth and fermentation profiles, it is illustrative that a central limitation of constructive metabolic engineering is the difficulty in anticipating how metabolic flux will respond to genetic manipulation [38].

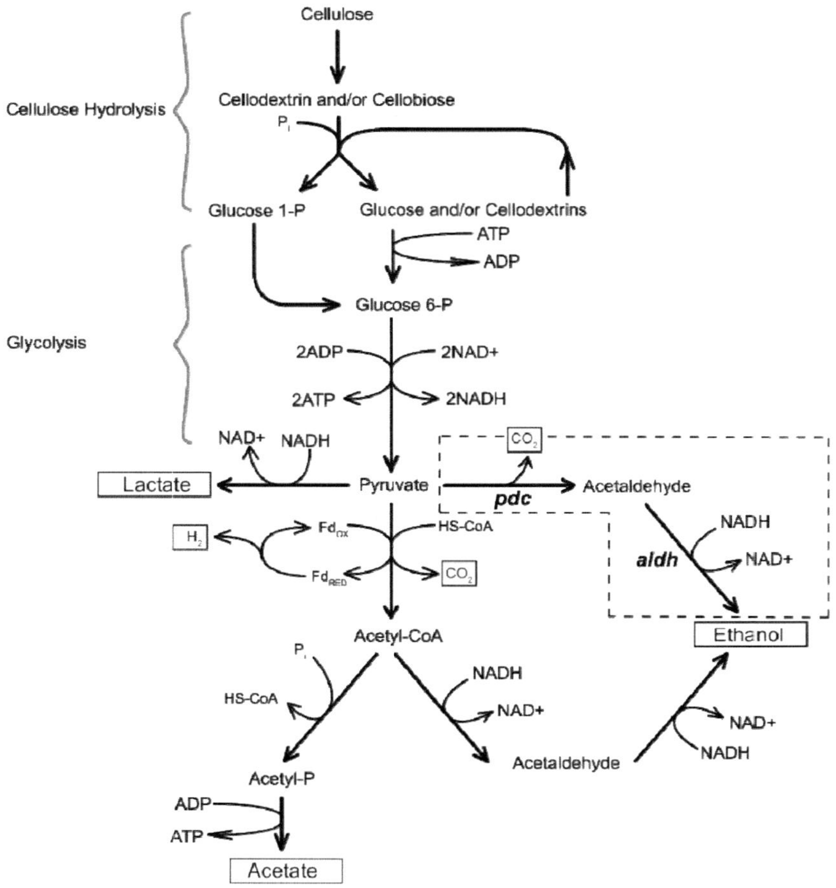

Figure 2: Central metabolism of *Clostridium cellulolyticum*. Heterologous pathways involving the introduction of a pyruvate decarboxylase (*pdc*) and acetaldehyde dehydrogenase (*aldh*) as described by Guedcn *et al.* are outlined [35].

Results from the constructive metabolic engineering approach, as illustrated in the *C. cellulolyticum* example, have been mixed. While some notable successes have been achieved involving the microbial production of amino acids and other fine chemicals, in many instances the metabolic consequences of genetic manipulation based on a constructive strategy differ significantly from the desired effect [35,39]. These unforeseen consequences, or "secondary responses" to the constructive approach to metabolic engineering are fundamentally linked to limitations in the understanding of metabolic networks. Although the accumulated benefit of hindsight may, in the future, enhance the predictive ability of constructive metabolic engineering, it is important to understand that any network contemplated in the constructive design will always represent a sub-network of a more complex global system involving not only whole cell metabolism but often the complex interactions of multi-cellular populations [28].

INVERSE METABOLIC ENGINEERING

In contrast to the classical approach to metabolic engineering, the strategy underlying inverse metabolic engineering is to first identify a desirable phenotype, determine the environmental conditions or genetic basis that confer that phenotype, and finally through genetic modification to alter the phenotype of the selected host for industrial application [27,28] (Fig. **3**). Inverse metabolic engineering was first described with respect to the heterologous expression of *Vitreoscilla* hemoglobin (VHb). *Vitreoscilla*, a Gram-negative obligate aerobic bacterium, produces a soluble, homodimeric hemoglobin-like molecule in response to hypoxic environments [40]. Given that the delivery of oxygen to obligate aerobes is perhaps the defining problem of biochemical engineering, it was hypothesized that cloning and expression of VHb into aerobic microbes of industrial importance would enhance cell productivity [28,40]. Following a demonstration that expression of VHb in *Escherichia coli* enabled higher overall cell densities to be achieved in micro-aerobic cultivations, VHb was subsequently expressed in a variety of organisms including *Pseudomonas* [41], *Streptomyces* [42], and *Saccharomyces* [43] with significant effects on cell growth or productivity.

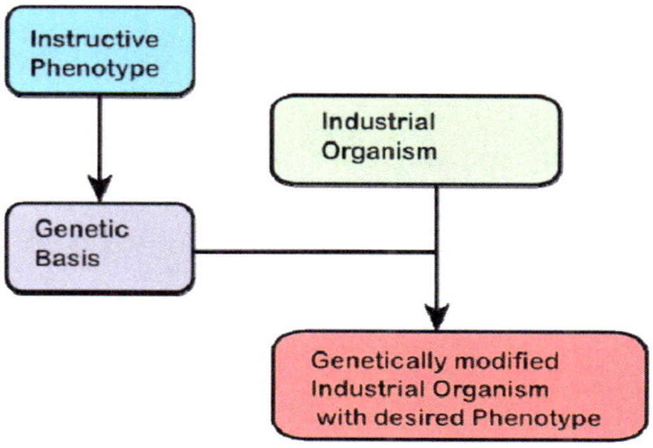

Figure 3: Schematic representation of the inverse metabolic engineering strategy as described by Bailey *et al.* [28].

Currently, the largest obstacle with respect to inverse metabolic engineering is discerning or discovering the genetic basis for a desirable phenotype. While many phenotypes may be multi-factorial in nature, inverse metabolic engineering requires the identification of a discreet set of targets that can yield a predictable phenotype. A common first step in determining the genetic basis of a desirable phenotype is a comparison to a control or parent strain at the transcriptome or proteome level. Finally, if the genetic manipulations proposed on the basis of an inverse metabolic engineering strategy fail to achieve the desired objective, information on the genetic stimulus-phenotype response may further the understanding of the host's physiology and help refine future strategies.

MOLECULAR AND MICROBIOLOGICAL TOOLS OF METABOLIC ENGINEERING

The emergence of metabolic and cellular engineering has largely been precipitated by the development of molecular biology tools enabling the specific manipulation and characterization of genes and the modification of organisms. Nearly all the tools and techniques of genetic manipulation are required for metabolic engineering including: 1) transformation systems for relevant microorganisms (*e.g. Corynebacterium* commonly used in amino acid production, *Pseudomonads* for the degradation of Xenobiotics, or *S. cerevisiae* used for the production of ethanol); 2) Promoters and special vectors used in such transformations (*e.g.* the yeast retroposon Ty3 employed for site specific integration of heterologous genes into the host chromosome); 3) multicistronic expression vectors allow single-step multi-gene metabolic engineering in mammalian cells; and 4) genetic markers for the identification and characterization of metabolic pathways [24,44]. In general the application of molecular tools in the field of metabolic engineering follows the development of a suitable engineering strategy (constructive or inverse) that is based on either a mathematical modeling of the biochemical network or extensive physiological studies.

Microbiological and analytical techniques are often applied both pre- and post- genetic manipulation of an organism as a means of evaluating the effectiveness of a modified metabolic pathway. These tools include: 1) Optimal culturing of the modified microorganism with respect to the suitability of subsequent mathematical modeling. Continuous culture is generally favored over batch or fed-batch operations but genetic stability of the transformed organism may preclude the possibility of continuous maintenance; 2) Optimization of the growth medium with respect to the production of desired metabolites; 3) Mass balance is often used as a means of calculating the yield of desired end products with respect to various substrates; 4) Isotopic labeling allows one to determine how a metabolic pathway contributes to the consumption of a particular substrate; and 5) the employment of non-invasive techniques such as NMR and flow cytometry permits a direct evaluation of microorganisms under conditions similar to those used in industrial processes [44].

The earliest examples of metabolic engineering (1974) included the construction of recombinant *Pseudomonas* strains capable of degrading toxic chemicals including camphor and naphthalene [24,44]. Since then, the utilization of metabolic engineering to enable catabolic activities for the detoxification, degradation and mineralization of toxic chemicals has since been applied to numerous other organisms including *Deinococcus radiodurans* (toluene and trichloroethylene), *Pseudomonads* (biphenyls) and *E. coli* (trichloroethane) [24]. Work in this area is troublesome, however, due to the difficulty in elucidating the genetic basis for desirable phenotype, unfavorable energetics, and generally low substrate concentrations and toxicity effects.

Genome-wide or large-scale methodologies employed in functional genomics such as DNA sequencing, transcription profiling, proteome and metabolite profiling have emerged as vital tools in the development and assessment of metabolic engineering strategies [45]. Until recently, elucidating the genetic basis for an observable phenotype had been achieved through a reductionist approach where changes in cellular phenotype were associated with modifications of a particular gene in a targeted manner. The incorporation of methods from the field of functional genomics, however, permits the application of *inverse strategies* to identify the genes or environmental conditions responsible for a desired phenotype. Genome wide analysis can prove very useful in determining the differences between strains with varying degrees of a desired phenotypic property. These differences may include mutations, presence/absence of genes or differences in regulatory mechanisms governing expression.

An inverse metabolic engineering strategy to identify factors affecting ethanol production and tolerance in *E. coli* was performed *via* transcriptome analysis. Comparing the transcriptome of an ethanol tolerant strain of *E. coli* (obtained by directed evolution) against its' parent strain revealed changes in the expression of several genes affecting ethanol tolerance and production [46,47]. Proteome analysis has been used to evaluate poly(3-hydroxybutyrate) production and accumulation in engineered strains of *E. coli* [48] and ethanol tolerance in *Clostridium thermocellum* [6]. Whole-cell metabolite profiling (metabolomics) and comparative flux analysis, using 13C-labelled substrates, can provide insight into complex metabolic networks and assist in the construction of genome-scale metabolic models. Although these examples may not directly relate to biological hydrogen production, it is not difficult to appreciate the utility of these tools and their potential for application.

MATHEMATICAL MODELING OF METABOLISM

Metabolic networks are highly complex nonlinear reaction systems whose functions are tightly coordinated to meet the physiological demands of living organisms. Many modeling approaches are used to model cellular processes, including kinetic [49,50], stochastic [51,52], cybernetic approaches [53,54], metabolic control analysis, biochemical systems theory, and constraint-based methods. Of these, constraint-based models have been most successful at incorporating diverse "-omic" datasets and building genome-scale models of metabolism [55].

Pre-Genome Modeling of Metabolism

There are few comprehensive metabolic systems for which detailed kinetic information is available and kinetic parameters cannot be estimated from genome sequence alone. In the absence of kinetic information, however, it is still possible to assess the theoretical capabilities of metabolism and examine the feasible metabolic flux distributions under a steady-state assumption imposed on the network by the stoichiometry of metabolic reactions. This process, known as metabolic flux analysis (MFA), differs from detailed kinetic modeling of cellular processes, in that it does not attempt to predict the exact behavior of metabolic networks. Rather, it uses known constraints on the integrated function of multiple enzymes to separate the states that a system can reach from those that it cannot. Within the domain of allowable behavior one can then study the genotype-phenotype relation. Metabolic flux analysis, also referred to as "flux balance analysis", is a matrix-based method used to quantify the flux of unknown intracellular metabolites. MFA is based on the law of mass conservation and the pseudo-steady state hypothesis (PSSH) that purports metabolic intermediates are neither accumulated nor depleted [27]. Following the construction of an appropriate metabolic network map, flux determinations of measurable metabolites are attained [25]. Two techniques commonly used to determine metabolic flux include the application of extracellular metabolite balance models or *via* mass isotopomer analysis that utilizes gas chromatography/mass spectrometry or nuclear magnetic resonance spectroscopy [44]. In order to solve the system of equations produced with MFA, a minimum number of fluxes must be measurable. In the absence of adequate flux information, linear programming can be used to assign potential values if one assumes the system is constrained by an objective function such as the hypothesis that growth rate or yield on a defined substrate with respect to a microbial system should find the most "favorable" permutation [25].

A general formalism to perform metabolic flux analysis was developed by Savinell and Palsson in which the dynamic nature of intracellular metabolism was represented by a system of linear equations [56,57]. Illustrated below (EQ 1), x represents the vector of metabolites per cell, the stoichiometric matrix of dimensions *n* by *m* where *n* and *m* represent the number of metabolites and metabolic fluxes respectively is represented as *A*, *v is* the vector of *m* metabolic fluxes, and *b*, the vector of biosynthetic fluxes.

$$r = \frac{dx}{dt} = A \cdot v - b \qquad \textbf{(1)}$$

Although the predictive capabilities of metabolic flux analysis are limited by the omission of kinetic data, they can provide useful information about the product yield and flux distribution of large metabolic networks [44]. Examples of metabolic engineering based on the results of MFA have appeared for the production of succinic acid, lactic acid, lycopene and L-valine in *E. coli* and ethanol in *S. cerevisiae* [58]. With respect to biological hydrogen production, metabolically engineered *E. coli* strains were developed and characterized by batch experiments and metabolic flux analysis. Deletion of *hycA*, a negative regulator for FHL, resulted in two-fold increase of FHL activity. Deletion of two uptake hydrogenases (*hya* and *hyb*) increased H_2 production yields from 1.20 mol mol^{-1} glucose to 1.48 mol mol^{-1} glucose. Deletion of lactate dehydrogenase (*ldhA*) and fumarate reductase (*frdAB*) further improved hydrogen yields; 1.80 mol mol^{-1} glucose under high H_2 pressure or 2.11 mol mol^{-1} glucose under reduced H_2 pressure. In this instance, MFA revealed that deletion of *ldhA* and *frdAB* directed carbon from glucose through the glycolytic pathway towards H_2 production by FHL and not to the pentose-phosphate pathway [59]. Metabolic flux analysis was also applied to explore the hydrogen producing potential of the chemoheterotrophic bacterium *Citrobacter amalonaticus* Y19. The metabolic pathway was constructed from 81 biochemical reactions. When the specific growth rate of the cells was selected as an objective function, the model described the batch culture reasonably well. When the specific H_2 production rate was selected as an objective function, however, the model reported an achievable maximal H_2 production yield of 8.7 mol H_2 mol^{-1} glucose [60]; far above theoretical yields of fermentative hydrogen production.

Metabolic Control Analysis

In response to some of the shortcomings of MFA, metabolic control analysis (MCA) was developed in the mid 1990's as a means of identifying the kinetic constraints in a biochemical network. Although often believed to determine the rate-limiting step in a reaction network, MCA does not assume a single rate-limiting step exists, but rather seeks to quantify how the control of steady state fluxes is distributed amongst the entire biochemical network [27]. Through manipulations of the reaction network *via* genetic manipulation, enzymatic titration or enzyme inhibition, changes in metabolic flux distribution can subsequently be used to calculate the level of control each enzyme exerts on flux through the pathway. Three commonly used normalized sensitivity measures have been defined as:

Flux Control Coefficient (FCC):

$$C_E^J = \frac{dJ}{dE}\frac{E}{J} \tag{2}$$

Concentration Control Coefficient (CCC):

$$C_E^X = \frac{dX}{dE}\frac{E}{X} \tag{3}$$

Elasticity:

$$\varepsilon_X^{V_A} = \frac{\partial V_A}{\partial X}\frac{X}{v_A} \tag{4}$$

The flux control coefficient (EQ 2) illustrates how changes in the concentration of enzyme E affect flux through that pathway and therefore is analogous to measuring the degree of control exerted by that enzyme on steady state flux. An enzyme with an FCC value close to unity is considered to have a high degree of control over the system network [25]. The concentration control coefficient (EQ 3) is a measure of the control exerted by enzyme E on the steady-state concentration of metabolite X, while elasticity (EQ 4) illustrates how reaction rate responds to changes in metabolite concentration [27].

Two different approaches to MCA have been described. The original “bottom-up” approach screens for control points in the system through the analysis of individual reactions in the metabolic network considered [27]. Conversely, the “top-down” approach divides the global metabolic network into lumped reaction blocks and control coefficients are calculated based on relative flux alone and values of enzyme activities are avoided [61]. The ”top-down approach”, however, is rendered invalid if more than one enzyme catalyzes the same reaction and therefore has limited applications. Furthermore, as the “top-down” approach treats the system as reaction blocks rather than discrete pathways, a small overall FCC value may potentially mask a stringent control point.

As neither MFA nor MCA (top-down approach) explicitly require kinetic information about the enzymes in the biochemical network, the application of these approaches is dramatically simplified with respect to experimental design and execution. As a consequence however, the lack of this data can limit the predictive capacity of these tools. Other mathematical modeling tools have been developed that incorporate elements of flux distribution, stoichiometry and reaction kinetics. Hybrid models integrate limited enzymatic kinetic information with stoichiometric models and “biochemical systems theory”, where metabolic reaction rates are modeled as power-law expressions, have been explored with some success [25,44].

Whole Genome Models of Metabolism

The quantity of genome, metagenome and other high-throughput experimental data aimed at exploring biochemical properties has been steadily increasing in recent years. As of March 26, 2010, more than 6498 genome projects have been initiated, 3986 of those relating to ongoing Bacterial genome projects, 1327 and 202 representing Eukaryotic and Archaeal genomes respectively (http://www.genomesonline.org). Increasingly efficient annotation methods now enable the systematic identification of most enzymes encoded in the genomes of corresponding species. This information, in conjunction with data gathered from other “-omic” technologies, offers a snapshot of the dynamic nature of metabolism and the construction of genome-scale metabolic models for the first time. Traditionally, biochemists have explored metabolism by modularizing a set of reactions into pathways such that *input metabolites* may be converted into *output metabolites*. The modularization of metabolism in this way necessitates an almost arbitrary definition of pathway boundaries as *input* and *output* metabolites are often involved in several reactions. This Pathway-based view of metabolism, although amenable to manual expertise, is of limited value to researchers that hope to formulate hypotheses and evaluate, *in silico*, the potential of genetic modifications. Genome-scale metabolic models, as described by Durot *et al.* [62], bridge the gap between genome-derived biochemical information and metabolic phenotypes in such a way as to both provide a solid interpretative framework for experimental data and enable simple *in silico* experiments with respect to whole-cell metabolism [62]. To date, genome-scale models of metabolism have only been analyzed with the constraints-based modeling philosophy [55].

The level of detail required to build a constraint-based genome-scale model of metabolism is relatively low. In order to account for mass balance and irreversibility constraints, the only information required are precise reaction stoichiometries and directions. The integration of genome annotation into a model is typically referred to as *metabolic network reconstruction*. The process relies heavily on indentifying the specific chemical conversions catalyzed by each enzyme and their corresponding reaction stoichiometry. The Enzyme Commision (EC) numbers classification, which is often integrated into publically available genome annotations, offers an unambiguous system of reaction identification. The broad specificity of some enzymes, such as alcohol dehydrogenase (EC 1.1.1.1), however, may necessitate a survey of the available literature, or enzyme databases such as BRENDA (http://www.brenda-enzymes.org/), ENZYME (http://ca.expasy.org/enzyme/) or UniProt (http://www.uniprot.org/) in order to determine which metabolites are preferentially recognized. The process of metabolic network reconstruction also benefits from metabolic databases, such as MetaCyc [63], KEGG [64] and SEED [65], that are able to group reactions into modules on the basis of their co-occurrence in other organisms and may be integrated with systems to identify candidate genes with so-called *gap-filling* protocols, to resolve inconsistent growth phenotypes and dead-end metabolite inconsistencies. In addition to pathway recognition and stoichiometry, the reversibility and localization of reactions should be incorporated into genome-scale models of metabolism. Reversibility is typically determined using simple thermodynamic considerations based on Gibbs free energy [59] while localization can be inferred directly from predicted amino acid sequences or determined experimentally using fluorescence microscopy [66].

As reported by Durot *et al.* [62], genome-scale models have been developed and experimentally validated for at least 17 bacterial species including: *Clostridium acetobutylicum* [67-69], *E. coli* [70], *Geobacter sulfurreducens* [71], and *Bacillus subtilis* [72]. In general these models have performed well during experimental assessment in predicting both wild-type growth phenotypes and knockout mutant growth phenotypes.

The application of genome-scale metabolic models to metabolic engineering includes: 1) the analysis of network properties to evaluate the maximum theoretical efficiency of a pathway, 2) the prediction and analysis of growth phenotype and flux following perturbations to the network, and 3) a model-based interpretation of the metabolism for engineered strains based on experimental data [62]. Additionally, metabolic modeling can be used to explore the effect of adding new pathways for the production of novel metabolites.

With respect to biological hydrogen production, OptStrain, a computational network aimed at guiding rational pathway modification, has been applied to propose elimination strategies for improving H_2 yields using two different growth substrates in three separate production hosts (glucose: *E. coli* and *C. acetobutylicum*; methanol: *Methylobacterium extorquens*) [73]. The framework identified that no non-native pathways were required for H_2 production in *E. coli* from both reduced Fd (EC 1.12.7.2) and NADH (EC 1.12.1.2). Notably, the maximum theoretical H_2 yield was higher under aerobic conditions, suggesting the sensitivity of hydrogenase to O_2 was not properly taken into account. However, under anaerobic conditions hydrogen was formed at maximum growth, leading to a growth-coupled production mode. Phenotype restriction was explored to determine whether removing functionalities that were in direct or indirect competition with H_2 production could enhance yields in wild-type *E. coli*. It was proposed that deletion of both enolase (EC 4.2.1.11) and glucose-6-phosphate dehydrogenase (EC 1.1.1.49) would promote H_2 formation by directing carbon flux through the 3-phosphoglycerate branch of glycolysis towards serine biosynthesis and eventually formate production. Formate would subsequently by oxidized to

CO_2 and H_2 *via* FHL. The dehydrogenase elimination would prevent the shunting of glucose-6-phosphate into the pentose phosphate pathway. The second strategy, a three-reaction deletion study, involved the removal of ATP synthase (EC 3.6.3.14), α-ketoglutarate dehydrogenase, and acetate kinase (EC 2.7.2.1). The removal of the first reaction enhances proton availability, whereas the other two deletions would ensure that maximum carbon flux was directed toward pyruvate, which would then converted into formate *via* PFL and then H_2 and CO_2 *via* FHL [73].

In *C. acetobutylicum*, OptStrain similarly recognized that no non-native pathways were required for H_2 production. Using a stoichiometric model for *C. acetobutylicum* [74,75], OptKnock suggested the deletion of the acetate-forming and butyrate-transport pathways to effectively maintain a favorable NADH:NAD+ ratio and achieve a theoretical H_2 yields of 3.17 mol mol-1 glucose at the expense of a reduced growth rate [73].

The H_2 producing potential of *Methylobacterium extorquens*, a facultative methylotroph capable of utilizing methanol as a sole carbon source, was investigated for *de novo* pathway addition. OptStrain identified that the introduction of either ferredoxin dependent hydrogenase (EC 1.12.7.2) or N_5, N_{10}-methenyltetrahydromethanopterin hydrogenase (EC 1.12.98.2) would result in H_2 production [73].

Computational frameworks, such as OptStrain, that aim to predict how reshaping whole-genome metabolic networks can result in the production of a desired metabolite, are illustrative of the potential for metabolic engineering to enhance H_2 yields. It is important, however, to note that the validity and relevance of predictions made by these methods are strongly dependent upon the completeness and accuracy of the models being employed. While much of the processes involved in metabolic network reconstruction and application can be automated, manual curation remains a valuable but time-consuming necessity. Furthermore, while all models inherently represent simplifications of an actual network, a purely stoichiometric representation of metabolism can lead to unrealistic predictions of flux by not accounting for kinetic barriers, regulatory interactions (eg., allosteric interference) and other factors.

METABOLIC ENGINEERING TO ENHANCE BIOHYDROGEN YIELDS

There are many examples of metabolic engineering being used to increase H_2 production in fermentative systems. In many organisms, the activity of *uptake hydrogenases*, enzymes that catalyze the oxidation of hydrogen *in vivo*, may contribute to sub-optimal H_2 yields. Down-regulation of the *hupCBA* operon using antisense RNA in *Clostridium saccharoperbutylacetonicum* resulted in a 3.1 fold increase in hydrogen production and a 75.6 % reduction in butanol yields relative to the control strain. The aim of this modification was, in fact, to reduce hydrogen production and shuttle NADH oxidation towards butanol production. Although the antisense RNA was effective at reducing expression of the encoded NAD-dependent hydrogenase, the physiological role of the *hupCBA* gene product was determined to be an uptake hydrogenase. Thus down-regulation of the *hupCBA* operon resulted in a smaller available pool of reduced electron carriers (NADH) poised for oxidation *via* butanol producing pathways [76].

Other strategies to increase fermentative hydrogen production include the over-expression of hydrogen-evolving enzymes, the knockout of metabolic pathways that compete for reducing equivalents and the introduction/over-expression of genes (cellulases, hemicellulases and lignases) to enhance carbohydrate availability to the cell. Overexpression of the natively encoded *hydA* hydrogenase in *Clostridium paraputrificum* M-21 was shown to increase H_2 production 1.7 fold during fermentation of N-acetylglucosamine. The increase in yield within the recombinant strain, 2.4 mol H_2 mol^{-1} N-acetylglucosamine compared to 1.4 mol mol^{-1} N-acetylglucosamine in the wildtype strain, corresponded with enhanced acetic acid production and a dramatic reduction in lactic acid yields [77].

In *Clostridium tyrobutylicum*, inactivation of acetate kinase (*ack*) resulted in a 1.5 fold increase in H_2 production from glucose. *C. tyrobutyricum* produces butyrate, acetate, H_2, and CO_2 as its main fermentation products. To improve butyric acid and hydrogen production, integrational mutagenesis was used to create an engineered mutant with an inactivated *ack* gene. Enzyme activity assays showed that the ACK activity in the recombinant strain decreased by approximately 50 %, while phosphotransacetylase (PTA) and hydrogenase activities increased by approximately 40 %. Compared to the wild type, the recombinant strain grew more slowly at pH 6.0, likely due to the partially impaired, ATP-forming PTA-ACK pathway, produced 23.5 % more butyrate and 50 % more hydrogen [78].

Although theoretical yields of H_2 production are greater with strict anaerobes, facultative anaerobic microorganisms offer several advantages as hydrogen producers including: being less sensitive to oxygen and able to recover hydrogenase activity if exposed to O_2, they display faster growth rates and are often amenable to genetic manipulations [79]. Various members of the order Enterobacteriales have been examined for their hydrogen producing potential. *Enterobacter cloacae* IIT-BT 08 display a high maximum volumetric hydrogen production rate of 447 ml $H_2\ l^{-1}h^{-1}$ [15] and expression of the *E. cloaceae hydA*, a Fe-hydrogenase, as a glutathione S-transferase-HydA fusion protein in *E. coli* BL21 yielded 3.12 mol H_2 mol^{-1} glucose [80]. *Enterobacter aerogenes* strains HU-101, AY-2, and HO-39 have been also investigated. Enhanced H_2 production was achieved by redirecting metabolism away from the formation of alcohol and some organic acids in *Enterobacter cloacae* IIT-BT 08. Double mutants, with defects in both alcohol and organic acid formation pathways, displayed higher yields (3.4 mol H_2 mol^{-1} glucose) compared to the wild type strain (2.1 mol H_2 mol^{-1} glucose) [81]. Although these results were achieved through mutagenesis, they are demonstrative of the potential for metabolic engineering to increase H_2 among these species.

E. coli has been extensively engineered to enhance H_2 yields. Maeda *et al.* [82] altered the regulation of FHL by inactivating the repressor encoded by *hycA* and by overexpressing the activator encoded by *fhlA*. Hydrogen uptake activity was removed by deleting *hyaB* and *hybC*. Glucose metabolism was directed towards formate production and oxidation in a series of mutations targeting formate dehydrogenase (*fdnG*, *fdoG*), nitrate reductase (*narG*), pyruvate oxidase (*poxB*) and dehydrogenase (*aceE*) and formate transport (*focA*, *focB*). Succinate and lactate producing pathways, pathways competing for reducing equivalents, were inactivated by deleting *frdC* and *ldhA*, respectively. The best of the metabolically engineered strains, BW25113, increased hydrogen production 4.6 fold from glucose and increased the hydrogen yield by two (0.65 to 1.3 mol H_2 mol^{-1} glucose) [82].

Among phototrophic bacteria, sustained hydrogenase-mediated H_2 production has been reported in mutant strains of the oxygenic cyanobacterium *Synechocystis* PCC 6803 that lack a NADPH-dehydrogenase. This mutant strain (M55) produced only low-levels of O_2 in light and displayed a poor capacity to fix CO_2. Sustained production of H_2 from NADPH, whose reduction potential is nearly 100 mV higher than that of the reduction H^+/H_2, is only thermodynamically possible in conditions of a very high level reduced NADPH. Although the accumulation of NADPH in the mutant strain redirected electron flux towards a bidirectional NiFe-hydrogenase, encoded by the *hox* genes, photosystem II independent H_2 production was limited by the production of reductant. The bidirectional NiFe-hydrogenase of *Thiocapsa roseopersicnia*, in mutants expressing only the *hox*-type enzyme, demonstrated the enzyme can indeed function in H_2 production, however due to its' reversibility, it cannot sustain continuous production in the same way as nitrogenase [4].

Non-autotrophic mutant strains of the photosynthetic bacterium *Rhodobacter capsulatus* have displayed increased stoichiometry of H_2 production following the introduction of a point mutation in the global regulator, Integration Host Factor (IHF). The affect of the mutated protein was a reduction in hydrogenase biosynthesis to 25 % compared to the wildtype strain. It was proposed that the increase in H_2 production was linked to better substrate conversion; hence the nitrogenase-mediated reduction of protons to H_2 (an energetically costly reaction) could maintain its activity over a longer period of time [4].

CONCLUDING REMARKS

The field of metabolic and cellular engineering is now more than 35 years old (dating from a 1974 patent by Chakrabarty *et al.* [see 83]) and although much has been accomplished in this time, the technology remains far from mature [24]. The successful application of metabolic engineering to the production of primary and secondary metabolites, and the alteration and improvement of cellular properties is contingent upon an understanding of the complex and dynamic nature of metabolic pathways. Metabolic and cellular engineering provides a much-needed framework for analyzing measurements at the transcript- and proteomic levels and in the future, these data sets; in combination with emerging flux analysis technologies will provide the best promises for the systematic study and elucidation of metabolic networks at the genome-scale. Regarding the biological production of hydrogen, metabolic engineering is a promising approach that has been focused on redirecting and optimizing the flow of reducing equivalents towards H_2-producing enzymes. The application of more mature engineering strategies and the predictive utility of genome-scale metabolic models offers the potential to further enhance yields, perhaps *via* less intuitive alterations, in the future.

REFERENCES

[1] Demirbas A. Fuel characteristics of olive husk and walnut, hazelnut, sunflower, and almond shells. Energ Source 2002; 24(3): 215-21.

[2] Islam MN, Islam MN, Beg MRA, Islam MR. Pyrolytic oil from fixed bed pyrolysis of municipal solid waste and its characterization. Renew Energ. 2005; 30(3): 413-20.

[3] Kim J, Rees D. Structural models for the metal centers in the nitrogenase molybdenum-iron protein. Science 1992; 257(5077): 1677-82.

[4] Vignais PM, Magnin J-P, Willison JC. Increasing biohydrogen production by metabolic engineering. Int J Hydrogen Energy 2006; 31(11): 1478-83.

[5] Hallenbeck PC, Benemann, J.R. Biological hydrogen production; fundamentals and limiting processes. Int J Hydrogen Energy 2002; 27: 1185-93.

[6] Williams T, Combs J, Lynn B, Strobel H. Proteomic profile changes in membranes of ethanol-tolerant *Clostridium thermocellum*. Appl Microbiol Biotechnol 2007; 74(2): 422-32.

[7] Vignais PM, Billoud, B., Meyer, J. Classification and phylogeny of hydrogenases. FEMS Microbiol Rev 2001; 25: 455-501.

[8] Greenbaum E. Energetic Efficiency of hydrogen photoevolution by algal water splitting. Biophys J 1988; 54(2): 365-8.

[9] Matsunaga T, Hatano T, Yamada A, Matsumoto M. Microaerobic hydrogen production by photosynthetic bacteria in a double-phase photobioreactor. Biotechnol Bioeng 2000; 68(6): 647-51.

[10] Dutta D, De D, Chaudhuri S, Bhattacharya SK. Hydrogen production by Cyanobacteria. Microb Cell Fact 2005; 4: 36-47.

[11] Levin DB, Carere CR, Cicek N, Sparling R. Challenges for biohydrogen production *via* direct lignocellulose fermentation. Int J Hydrogen Energy 2009; 34(17): 7390-403.

[12] Schicho RN, Ma K, Adams MWW, Kelly RM. Bioenergetics of Sulfur Reduction in the Hyperthermophilic Archaeon *Pyrococcus furiosus*. J Bacteriol 1993;175(6): 1823-30.

[13] Nandi R, Sengupta S. Microbial production of hydrogen: An overview. Crit Rev Microbiol 1998; 24(1): 61-84.

[14] Valentine DL, Blanton DC, Reeburgh WS. Hydrogen production by methanogens under low-hydrogen conditions. Arch Microbiol 2000; 174(6): 415-21.

[15] Kumar N, Das D. Continuous hydrogen production by immobilized *Enterobacter cloacae* IIT-BT 08 using lignocellulosic materials as solid matrices. Enzyme Microb Tech 2001; 29(4-5): 280-7.

[16] Vardar-Schara Gl, Maeda T, Wood TK. Metabolically engineered bacteria for producing hydrogen *via* fermentation. Microb Biotechnol 2008; 1(2): 107-125.

[17] Vatsala TM. Hydrogen production from (cane-molasses) stillage by *Citrobacter freundii* and its use in improving methanogenesis. Int J Hydrogen Energy 1992; 17(12): 923-7.

[18] Huesemann M, Hausmann T, Carter B, Gerschler J, Benemann J. Hydrogen generation through indirect biophotolysis in batch cultures of the non-heterocystous nitrogen-fixing cyanobacterium *Plectonema boryanum*. Appl Biochem Biotechnol 162(1): 208-20.

[19] Nath K, Kumar A, Das D. Hydrogen production by *Rhodobacter sphaeroides* strain O.U.001 using spent media of *Enterobacter cloacae* strain DM11. Appl Microbiol Biotechnol 2005; 68(4): 533-41.

[20] Levin DB, Pitt L, Love M. Biohydrogen production: prospects and limitations to practical application. Int J Hydrogen Energy 2004; 29(2): 173-85.

[21] Thauer RK, Jungermann K, Decker K. Energy conservation in chemotrophic anaerobic bacteria. Bacteriol Rev 1977; 41(1): 100-80.

[22] Chou CJ, Jenney FE, Jr., Adams MW, Kelly RM. Hydrogenesis in hyperthermophilic microorganisms: implications for biofuels. Metab Eng. 2008;10(6): 394-404.

[23] Koffas M, Roberge C, Lee K, Stephanopoulos G. Metabolic engineering. Annu Rev Biomed Eng 1999; 1: 535-57.

[24] Cameron DC, Tong IT. Cellular and metabolic engineering. An overview. Appl Biochem Biotechnol. 1993; 38(1-2): 105-40.

[25] Yarmush ML, Banta S. Metabolic engineering: advances in modeling and intervention in health and disease. Annu Rev Biomed Eng 2003; 5: 349-81.

[26] Bailey JE, Sburlati A, Hatzimanikatis V, Lee K, Renner WA, Tsai PS. Inverse metabolic engineering: A strategy for directed genetic engineering of useful phenotypes. Biotechnol Bioeng 1996; 52(1): 109-21.

[27] Yang Y-T, Bennett G, N., San KY. Genetic and metabolic engineering. Electron J Biotechn 1998; 1(3): 134-41.

[28] Bailey JE, Sburlati A, Hatzimanikatis V, Lee K, Renner WA, Tsai PS. Inverse metabolic engineering: a strategy for directed genetic engineering of useful phenotypes. Biotechnol Bioeng 2002; 79(5): 568-79.

[29] Desvaux M, Guedon E, Petitdemange H. Cellulose catabolism by *Clostridium cellulolyticum* growing in batch culture on defined medium. Appl Environ Microbiol 2000; 66(6): 2461-2470.

[30] Desvaux M, Guedon E, Petitdemange H. Kinetics and metabolism of cellulose degradation at high substrate concentrations in steady-state continuous cultures of *Clostridium cellulolyticum* on a chemically defined medium. Appl Environ Microbiol 2001; 67(9): 3837-45.

[31] Desvaux M, Guedon E, Petitdemange H. Metabolic flux in cellulose batch and cellulose-fed continuous cultures of *Clostridium cellulolyticum* in response to acidic environment. Microbiology 2001; 147(6): 1461-71.

[32] Desvaux M, Guedon E, Petitdemange H. Carbon flux distribution and kinetics of cellulose fermentation in steady-state continuous cultures of *Clostridium cellulolyticum* on a chemically defined medium. J Bacteriol 2001; 183(1): 119-30.

[33] Guedon E, Desvaux M, Payot S, Petitdemange H. Growth inhibition of *Clostridium cellulolyticum* by an inefficiently regulated carbon flow. Microbiology+ 1999; 145 (8): 1831-8.

[34] Guedon E, Desvaux M, Petitdemange H. Kinetic analysis of *Clostridium cellulolyticum* carbohydrate metabolism: importance of glucose 1-phosphate and glucose 6-phosphate branch points for distribution of carbon fluxes inside and outside cells as revealed by steady-state continuous culture. J Bacteriol 2000; 182(7): 2010-7.

[35] Guedon E, Desvaux M, Petitdemange H. Improvement of cellulolytic properties of *Clostridium cellulolyticum* by metabolic engineering. Appl Environ Microbiol 2002; 68(1): 53-8.

[36] Guedon E, Payot S, Desvaux M, Petitdemange H. Carbon and electron flow in *Clostridium cellulolyticum* grown in chemostat culture on synthetic medium. J Bacteriol 1999;181(10): 3262-9.

[37] Guedon E, Payot S, Desvaux M, Petitdemange H. Relationships between cellobiose catabolism, enzyme levels, and metabolic intermediates in *Clostridium cellulolyticum* grown in a synthetic medium. Biotechnol Bioeng 2000; 67(3): 327-35.

[38] Bailey JE. Toward a science of metabolic engineering. Science 1991; 252(5013): 1668-75.

[39] Deng MD, Severson DK, Grund AD, *et al.* Metabolic engineering of *Escherichia coli* for industrial production of glucosamine and N-acetylglucosamine. Metab Eng 2005; 7(3): 201-14.

[40] Khosla C, Bailey JE. Heterologous expression of a bacterial haemoglobin improves the growth properties of recombinant *Escherichia coli*. Nature 1988; 331(6157): 633-5.

[41] Liu SC, Webster DA, Stark BC. Cloning and expression of the *Vitreoscilla* hemoglobin gene in pseudomonads: effects on cell growth. Appl Microbiol Biotechnol 1995; 44(3): 419-24.

[42] Magnolo SK, Leenutaphong DL, DeModena JA, *et al.* Actinorhodin production by *Streptomyces coelicolor* and growth of *Streptomyces lividans* are improved by the expression of a bacterial hemoglobin. Nat Biotechnol 1991; 9(5): 473-6.

[43] Chen W, Hughes DE, Bailey JE. Intracellular expression of *Vitreoscilla* hemoglobin alters the aerobic metabolism of *Saccharomyces cerevisiae*. Biotechnol Prog 1994; 10(3): 308-13.

[44] Cortassa S, Aon MA, Iglesias AA, D. L. An Introduction to Metabolic and Cellular Engineering. New Jersey: Word Scientific; 2002.

[45] Bro C, Nielsen J. Impact of 'ome' analyses on inverse metabolic engineering. Metab Eng 2004; 6(3): 204-11.

[46] Gonzalez R, Tao H, Purvis JE, York SW, Shanmugam KT, Ingram LO. Gene array-based identification of changes that contribute to ethanol tolerance in ethanologenic *Escherichia coli*: comparison of KO11 (parent) to LY01 (resistant mutant). Biotechnol Prog 2003; 19(2): 612-23.

[47] Yomano LP, York SW, Ingram LO. Isolation and characterization of ethanol-tolerant mutants of *Escherichia coli* KO11 for fuel ethanol production. J Ind Microbiol Biotechnol 1998; 20(2): 132-8.

[48] Han MJ, Yoon SS, Lee SY. Proteome analysis of metabolically engineered *Escherichia coli* producing Poly(3-hydroxybutyrate). J Bacteriol. 2001; 183(1): 301-8.

[49] Loew LM, Schaff JC. The Virtual Cell: a software environment for computational cell biology. Trends Biotechnol 2001; 19(10): 401-6.

[50] Tomita M, Hashimoto K, Takahashi K, *et al.* E-CELL: software environment for whole-cell simulation. Bioinformatics 1999; 15(1): 72-84.

[51] Arkin A, Ross J, McAdams HH. Stochastic kinetic analysis of developmental pathway bifurcation in phage lambda-infected *Escherichia coli* cells. Genetics 1998; 149(4): 1633-48.

[52] McAdams HH, Arkin A. Simulation of prokaryotic genetic circuits. Annu Rev Biophys Biomol Struct 1998; 27:199-224.

[53] Guardia MJ, Gambhir A, Europa AF, Ramkrishna D, Hu W-S. Cybernetic Modeling and Regulation of Metabolic Pathways in Multiple Steady States of Hybridoma Cells. Biotechnol Prog 2000; 16(5): 847-53.

[54] Varner J, Ramkrishna D. Mathematical models of metabolic pathways. Curr Opin Biotechnol 1999; 10(2): 146-50.

[55] Price ND, Papin JA, Schilling CH, Palsson BO. Genome-scale microbial *in silico* models: the constraints-based approach. Trends Biotechnol 2003; 21(4): 162-9.

[56] Savinell JM, Palsson BO. Optimal selection of metabolic fluxes for *in vivo* measurement. I Development of mathematical methods. J Theor Biol 1992; 155(2): 201-14.

[57] Savinell JM, Palsson BO. Optimal selection of metabolic fluxes for *in vivo* measurement. II. Application to *Escherichia coli* and hybridoma cell metabolism. J Theor Biol 1992; 155(2): 215-42.

[58] Kim HU, Kim TY, Lee SY. Metabolic flux analysis and metabolic engineering of microorganisms. Mol Biosyst 2008; 4(2): 113-120.

[59] Kim S, Seol E, Oh Y-K, Wang GY, Park S. Hydrogen production and metabolic flux analysis of metabolically engineered *Escherichia coli* strains. Int J Hydrogen Energy 2009; 34: 7417-27.

[60] Oh Y-K, Kim H-J, Park S, Kim M-S, Ryu DDY. Metabolic-flux analysis of hydrogen production pathway in *Citrobacter amalonaticus* Y19. Int J Hydrogen Energy 2008; 33(5): 1471-82.

[61] Brown GC, Hafner RP, Brand MD. A 'top-down' approach to the determination of control coefficients in metabolic control theory. Eur J Biochem. 1990 Mar 10; 188(2): 321-5.

[62] Durot M, Bourguignon PY, Schachter V. Genome-scale models of bacterial metabolism: reconstruction and applications. FEMS Microbiol Rev 2009; 33(1): 164-90.

[63] Caspi R, Foerster H, Fulcher CA, *et al.* MetaCyc: a multiorganism database of metabolic pathways and enzymes. Nucleic Acids Res 2006; 34(Database issue): D511-6.

[64] Aoki-Kinoshita KF, Kanehisa M. Gene annotation and pathway mapping in KEGG. Methods Mol Biol 2007; 396: 71-91.

[65] Overbeek R, Begley T, Butler RM, *et al.* The subsystems approach to genome annotation and its use in the project to annotate 1000 genomes. Nucleic Acids Res 2005; 33(17): 5691-702.

[66] Meyer P, Dworkin J. Applications of fluorescence microscopy to single bacterial cells. Res Microbiol 2007; 158(3): 187-94.

[67] Lee J, Yun H, Feist AM, Palsson BO, Lee SY. Genome-scale reconstruction and *in silico* analysis of the *Clostridium acetobutylicum* ATCC 824 metabolic network. Appl Microbiol Biotechnol 2008; 80(5): 849-62.

[68] Senger RS, Papoutsakis ET. Genome-scale model for *Clostridium acetobutylicum*: Part I. Metabolic network resolution and analysis. Biotechnol Bioeng 2008; 101(5): 1036-52.

[69] Senger RS, Papoutsakis ET. Genome-scale model for *Clostridium acetobutylicum*: Part II. Development of specific proton flux states and numerically determined sub-systems. Biotechnol Bioeng 2008; 101(5): 1053-71.

[70] Feist AM, Henry CS, Reed JL, *et al.* A genome-scale metabolic reconstruction for *Escherichia coli* K-12 MG1655 that accounts for 1260 ORFs and thermodynamic information. Mol Syst Biol 2007; 3: 121.

[71] Mahadevan R, Bond DR, Butler JE, *et al.* Characterization of metabolism in the Fe(III)-reducing organism *Geobacter sulfurreducens* by constraint-based modeling. Appl Environ Microbiol 2006; 72(2): 1558-68.

[72] Oh YK, Palsson BO, Park SM, Schilling CH, Mahadevan R. Genome-scale reconstruction of metabolic network in *Bacillus subtilis* based on high-throughput phenotyping and gene essentiality data. J Biol Chem 2007; 282(39): 28791-9.

[73] Pharkya P, Burgard AP, Maranas CD. OptStrain: a computational framework for redesign of microbial production systems. Genome Res 2004;14(11): 2367-76.

[74] Desai RP, Nielsen LK, Papoutsakis ET. Stoichiometric modeling of *Clostridium acetobutylicum* fermentations with non-linear constraints. J Biotechnol 1999; 71(1-3): 191-205.

[75] Papoutsakis ET. Equations and calculations for fermentations of butyric acid bacteria. Biotechnol Bioeng 1984; 26(2): 174-87.

[76] Nakayama S, Kosaka T, Hirakawa H, Matsuura K, Yoshino S, Furukawa K. Metabolic engineering for solvent productivity by downregulation of the hydrogenase gene cluster hupCBA in *Clostridium saccharoperbutylacetonicum* strain N1-4. Appl Microbiol Biotechnol 2008; 78(3): 483-93.

[77] Morimoto K, Kimura T, Sakka K, Ohmiya K. Overexpression of a hydrogenase gene in *Clostridium paraputrificum* to enhance hydrogen gas production. FEMS Microbiol Lett 2005; 246(2): 229-34.

[78] Liu X, Zhu Y, Yang ST. Construction and characterization of ack deleted mutant of *Clostridium tyrobutyricum* for enhanced butyric acid and hydrogen production. Biotechnol Prog 2006; 22(5): 1265-75.

[79] Shin J-H, Hyun Yoon J, Eun Kyoung A, Kim M-S, Jun Sim S, Park TH. Fermentative hydrogen production by the newly isolated *Enterobacter asburiae* SNU-1. Int J Hydrogen Energy 2007; 32(2): 192-9.

[80] Mishra J, Khurana S, Kumar N, Ghosh AK, Das D. Molecular cloning, characterization, and overexpression of a novel [Fe]-hydrogenase isolated from a high rate of hydrogen producing *Enterobacter cloacae* IIT-BT 08. Biochem Biophys Res Commun 2004; 324(2): 679-85.

[81] Kumar N, Ghosh A, Das D. Redirection of biochemical pathways for the enhancement of H_2 production by *Enterobacter cloacae*. Biotechnol Lett 2001; 23(7): 537-41.

[82] Maeda T, Sanchez-Torres V, Wood TK. Metabolic engineering to enhance bacterial hydrogen production. Microb Biotechnol 2008; 1(1): 30-9.

[83] Palombi L. The Patenting of biological materials In the context of the agreement on trade-related aspects of intellectual property rights New South Wales: The University of New South Wales; 2004.

CHAPTER 12

Hydrogen Production by Microbial Electrohydrogenesis

Nathan Wrana and David B. Levin*

Department of Biosystems Engineering, University of Manitoba, Canada

Abstract: Microbial electrohydrogenesis is a novel process capable of generating high hydrogen (H_2) yields from organic waste streams and fermentation end products. A small voltage is applied to a Microbial Electrolysis Cell (MEC) to force the microbial oxidation of organic material at the anode and drive the chemical reduction of protons at the cathode. In this chapter, the microbe-electrode relationship is studied and three mechanisms for electron transfer to an electrode surface are explored. The thermodynamics of electrode reactions and cell potentials is described to calculate the systems theoretical limitations. Deviations from ideal behavior are characterized by quantifying the energy losses associated with overpotentials at the electrode-solution interface and ohmic resistances inherent in electronic and ionic conduction. Finally, strategies are discussed to overcome the technical challenges electrohydrogenesis currently faces. The purpose of this chapter is to introduce the fundamental principles of microbial electrolysis, to identify the biological and electrochemical challenges currently under investigation, and to discuss the future role microbial electrohydrogenesis will play towards sustainable H_2 production.

Keywords: Biohydrogen, microbial electrohydrogenesis, microbial electrolysis cells, thermodynamics, over potential, ohmic resistance, *Geobacter sulfereducens.*

INTRODUCTION

Current Status of the Hydrogen Production Economy

A hydrogen (H_2) economy is well established, including the technology to produce it. Global H_2 consumption is estimated at 50 million metric tons at an overall growth rate between 5 and 10% [1,2]. The majority of H_2 produced is consumed as a reactant in hydrogenation processes, in either chemical manufacturing or petroleum industries [3]. According to Fig. **1a**, approximately 50% of all H_2 produced is used in the production of ammonia fertilizer. Petroleum refining or hydrocracking processes account for roughly 37% while; petroleum production process account for 8% (*e.g.* methanol synthesis). Furthermore, H_2 is used as an oxygen scavenger in the metals industry (during heat treatment processes), as a jet fuel, a coolant, and in the food industry (oil and fat hydrogenation processes). The demand for H_2 makes it a valuable chemical. With a higher average selling price ($6 kg^{-1}) than electricity ($0.1 kWh) and natural gas ($0.43 kg^{-1}), H_2 production is economically attractive [1].

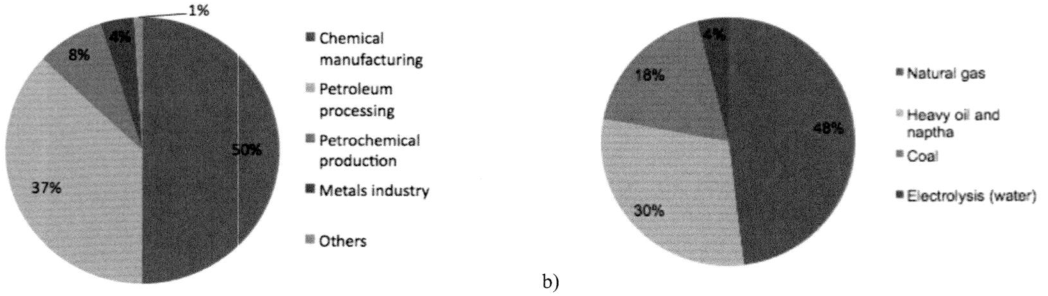

a) b)

Figure 1: An overview of the global H_2 production economy. a) Major industrial applications for H_2 production (adapted from Ramachandran and Menon [3]), and b) major energy sources for H_2 production (adapted from Logan [1]).

***Address correspondence to David B. Levin:** Department of Biosystems Engineering, University of Manitoba, Canada; E-mail: levindb@cc.umanitoba.ca

Unfortunately, 96% of global H_2 production is derived from fossil sources (Fig. **1b**) [1,2]. Nearly half of all H_2 is produced from natural gas streams, in particular steam methane reforming (SMR). Heavy oil, naphtha, and coal make up an additional 48% of the market, while only 4% of H_2 production is derived from renewables *via* the electrolysis of water. The dependency on fossil fuels places H_2 production in a vulnerable position, where pricing is influenced by a potentially unreliable and financially volatile oil market. With an industrializing and motorizing population base approaching 9 billion, energy consumption will only increase as crude oil reserves continue to shrink. Estimated oil demand is expected to double in the pre-industrialized countries of China and India. According to the International Energy Agency (IEA), by the year 2025, global oil demand will rise by 52%, making an already stretched fossil fuel market in higher demand and even shorter supply. Clearly, there is strong incentive to reduce our dependency on hydrocarbon reforming and to develop renewable H_2 producing strategies.

MICROBIAL ELECTROHYDROGENESIS

Principles of Operation

A novel process known as microbial electrohydrogenesis is capable of generating high H_2 yields using organic waste as an energy source. Microbial electrolysis cells (MECs) are bio-electrochemical reactors in which anode-respiring, electrochemically active bacteria oxidize soluble organic material at the anode, generating carbon dioxide (CO_2), electrons, and protons. Electrons are shuttled from the anode to the cathode through an external electric circuit while protons diffuse through an electrolyte to the cathode. Although not required, a membrane can be used to separate the electrode reactions as well as facilitate proton transfer. Finally, electrons and protons are chemically reduced at the cathode-electrolyte interface, to form H_2. Thus, H_2 is produced chemically in a MEC. The bacteria serve as a microbial catalyst, reducing total energy consumption while enabling H_2 generation from biomass-based alternatives.

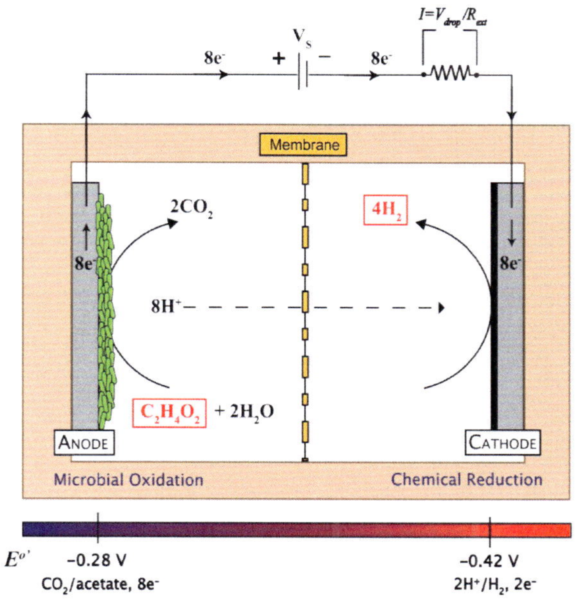

Figure 2: Principles of operation of a microbial electrolysis cell (MEC). Microbial oxidation of organic material at the anode, *e.g.* acetate oxidation, liberates electrons (e^-) and protons (H^+), which undergo chemical reduction at the cathode to form H_2. The reduction potential at the anode is higher than at the cathode and thus a small voltage must be supplied to drive this non-spontaneous reaction. Because current production is very small, the voltage drop (V_{drop}) across an external resistor is measured and the current is calculated according to Ohm's law.

Unfortunately, thermodynamic evaluation confirms H_2 generation by microbial electrohydrogenesis is not a spontaneous reaction. The standard reduction potential, $E^{o\prime}$, for H_2 production is lower than the reduction potential at the anode. In practice, this potential difference is even greater due to ohmic resistance through the system and overpotentials at the electrode-electrolyte interface. Consequently, energy from an external power source must be supplied to the cell in order to drive electrons from the anode to the cathode.

Current Status and Comparison with other H_2 Producing Technologies

The number of studies investigating H_2 generation *via* microbial electrohydrogenesis is rapidly increasing. High H_2 yields and recovery efficiencies can be achieved using energy sources that were once considered waste, such as fermentation end products and wastewater streams. Several bench-scale reactors have been summarized in Table **1** to demonstrate the technology's rapid progression from proof-of-concept design to its current status. The performance of each reactor was compared in terms of H_2 yield, volumetric H_2 production rate, and coulombic efficiency[1]. For the purpose of consistency, all papers selected in this evaluation were inoculated with mixed cultures and used acetate as the sole carbon source.

It should be noted that acetate is typically used as a carbon source because it is a preferred substrate by electrochemically active bacteria (see section on electricigens). Additionally, the removal of acetate has potential industrial applications, which include the treatment of wastewater, landfill lecheate, and fermentation waste. It can be shown that the production of acetate is the highest H_2-yielding pathway by dark fermentation (4 mol-H_2 mol^{-1}-glucose). Coupling microbial electrohydrogenesis with dark fermentation could theoretically yield an additional 8 mol-H_2 mol^{-1}-glucose [4]. In practice, however, acetate generally does not exist alone and is typically found mixed with other compounds (organic acids, fatty acids, alcohols, biogas). Studies have tested other substrates, including butyric acid, lactic acid, propionic acid, and valeric acid [5] as well as diluted carbon sources found in wastewaters [6,7]. Results showed that overall H_2 recovery efficiencies were much lower when wastewater was used as a substrate (10–28%) *versus* acetate-only MECs.

Even though over 95% of current H_2 producing technologies depend on fossil sources, many non-reforming approaches also exist which convert alternative resources such as biomass and water into H_2 rich streams [8]. A brief description of the most popular H_2-from-biomass and H_2-from-water technologies is presented in Table **2**. The major advantages and disadvantages of each process are described, and H_2 production rates and efficiencies are evaluated with respect to microbial electrohydrogenesis. All technologies are compared against the industry standard, H_2 production using SMR.

Table 1: Overview of Various MEC Architectures and System Performances Reported in Literature. All studies used mixed consortiums as inoculums and acetate as the sole substrate. Overall H_2 production ($Y_{H2,th}$), volumetric H_2 production rate, and coulombic efficiency (C_E), measured at specific applied voltages (E_{ap}), were compared to reactor volume and electrode and membrane material and size.

Ref.	Operation	Applied voltage E_{ap} (V)	Overall H_2 yield $Y_{H2,th}$ (%)	Vol. H_2 production rate (m^3-H_2 m^{-3} d^{-1})	Coulombic efficiency C_E (%)	Total liquid vol. (L)	Membrane Material	Membrane Area (cm^2)	Anode Material	Anode Area (cm^2)	Cathode Material	Cathode Area (cm^2)	Electrode spacing (cm)
[9] [Bottle]	B	0.25	54^a	0.005	60^a	0.4	CEM (Nafion 117)	3.5	C-cloth (plain)	12	C-paper (0.5 mg Pt/cm^2)	12	15
[10]	B	0.5	53	0.02	92	6.6	CEM (Nafion 117)	256	G-felt (disc)	452 $(400)^b$	Ti-mesh (5 mg Pt/cm^2)	452 $(400)^b$	Direct contact w/ CEM
[11]	B	1.0	23	0.31 0.33	23	3.3	AEM (Fumasep FAB) CEM (Nafion 117)	256	G-felt (disc)	400	MEA Ti-mesh (5 mg Pt/cm^2)	400	Direct contact w/ membrane
[5]	B	0.6	88^c	1.1	88^a	$.04^c$	AEM (AMI-7001)	n/a	G-granules	528	C-cloth (0.5 mg Pt/cm^2)	1	4
[12]	B	0.6^c	88^c	3.12	92^c	0.028	No membrane		G-brush (NH_3)	2,200	C-cloth (0.5 mg Pt/cm^2)	7	n/a
[13][†]	C	1.0	98	6.3	98	0.1	No membrane (J-cloth)		C-felt	50	GDE (E-tek) (0.5 mg Pt/cm^2)	n/a	0.03

n/a not available; $Y_{H2,th}$ is calculated based on the number of moles H_2 produced *versus* theoretical amount of H_2 that can be extracted form substrate (*i.e.* 4 mol-H_2 mol^{-1}-acetate); *B* batch, *C* continuous, *CEM* cation exchange membrane, *AEM* anion exchange membrane, *C* carbon, *G* graphite, *Ti* titanium, *MEA* membrane electrode assembly, *GDE* gas diffusion electrode; [†] equivalent acetate loading rate of 4 g L^{-1} d^{-1}; [a] referenced from Logan [14]; [b] surface area of electrode submerged in electrolyte; [c] referenced from Logan *et al.* [15]

[1] Coulombic efficiency is defined as the fraction of electrons recovered as current *versus* the total number of electrons available in the substrate

From Table **2**, it becomes apparent that even though microbial electrohydrogenesis is competitive with many H_2-from-biomass processes, it is still a technology in its infancy. To be considered a mature process, H_2 production rates and volumetric throughput must be comparable with SMR, biomass gasification, and alkaline electrolyzers. Target objectives have been proposed [16,17]. As an example, for MECs to be competitive with high performance wastewater treatment systems[2], volumetric production rates must be greater than 10 m^3-H_2 m^{-3} d^{-1}, operating at 90% H_2 efficiency in a reactor scaled-up to 10^4 – 10^6 L in size. Despite bench-scale reactors operating near some of these goals, the major hurdle facing electrohydrogenesis is scale-up: the ability to maintain this performance in reactor volume's six orders of magnitude greater, and doing so at a competitive price.

The purpose of this chapter is twofold: i) to explain the fundamental principles of microbial electrohydrogenesis in an attempt to identify major bottlenecks within this process and ii) to discuss strategies to overcome these technical limitations that currently impede MEC technology. Investigation into the microbial communities that develop at the anode and the various mechanisms of electron transfer responsible for cell growth are examined. The thermodynamics of electrode reactions and electrode potentials are derived to determine theoretical operating conditions. Finally, energy losses associated with internal resistance and current flow are examined to explain deviations from ideal behaviour. First, however, the history of MEC technology and its evolution from bio-electrochemical systems (BES) is explored.

History of the MEC

The first connection between electricity and biology was made in 1791 by Luigi Galvani, who discovered that severed frog's legs contract every time the muscle and nerve endings are connected to a static electricity generator [18,19]. Around the same time, the concept of the fuel cell was developed by W.R. Grove who successfully produced an electric current and water by combining hydrogen gas with oxygen [20]. Surprisingly, the first correlation between a half cell and microorganisms was not made until 1911 when M.C. Potter observed small amounts of electricity could be derived from microbial communities, introducing the possibility that bacteria were capable of generating an electric current [18,21]. This discovery led to the BES in which anodic and/or cathodic reactions are biologically catalyzed by microorganisms. When electricity is produced in a BES, the term microbial fuel cell (MFC) is used. However, the term MEC is employed when electricity is consumed to drive electrochemical reactions [17]. In the context of this paper, BESs apply specifically to microbial catalysis and not isolated enzymatic catalysis when assisting redox reactions in electrochemical cells [18,22,23].

Table 2: Comparison of H_2 Production Technologies. Hydrogen producing technologies are evaluated based on maturity, capacity, and efficiency. Only processes relevant to electrohydrogenesis and renewables (biomass and water sources) were selected. Steam methane reforming was included in this analysis because it is the largest net producer of H_2 (adapted from [8,24]).

Technology	Description	Advantages	Disadvantages	Maturity	H_2 production	Efficiency	Ref.
Microbial Electrohydrogenesis	electrochemical oxidation of organic material (microbial catalysis) and reduction of H^+ and e^- to from H_2 (chemical, biological catalysis)	organic waste materials can be used as energy sources	requires additional energy input (~0.6 V)	L	6.3 m^3-H_2 m^{-3} d^{-1}	81%[a]	[13]
		near complete substrate utilization (>90%)	methanogenic competition in mixed cultures (H_2 consumption for CH_4 production)				
		very high H_2 recovery efficiency (>93%)	high internal energy losses				
		low operating temperatures and	requires anodic anaerobic conditions				

[2] Target based on wastewater systems from BIOPAQ (Netherlands): high rate aerobic systems (up to 5 kg-COD m^{-3} day^{-1}) and high rate anaerobic systems (>10 kg-COD m^{-3} day^{-1})

		pressures	technology still in its infancy (more research needed in catalyst development, membrane design, scale-up testing)				
Hydrocarbon reforming							
Steam reforming (methane)	steam at high temperatures reacts with CH_4 to produce CO and H_2. WGS reaction further converts CO into H_2 and CO_2	extensive industrial use	highest CO and CO_2 emissions vs. other fossil fuel H_2 producing technologies	C	1.5 million $Nm^3\ d^{-1}$	70-85%[b]	[25]
		no O_2 required	operating near theoretical limits – little room for improvement				
		High H_2/CO_2 ratio (3:1)					
Hydrogen from biomass							
Biomass gasification	chemical conversion of biomass at high temperature and controlled O_2 (endothermic, partial oxidation reaction) into a producer gas or syngas (H_2, CH_4, CO, CO_2, N_2, NH_3).	extensive industrial use	significant accumulation of tar	C	40,482 Nm^3-$H_2\ d^{-1}$[†]	35-50%[b]	[26,27]
		fuel flexibility	low thermal efficiency and energy density				
		simplicity	H_2 separation from syngas required				
		syngas can be used as feed for steam reforming and WGS reaction to increase H_2 yields	commercial systems requires large quantities of biomass – high transportation costs				
Photolysis	**Direct** - photosynthetic production of H_2 from H_2O by algae and bacteria	water is primary feed for direct photolysis	low light utilization efficiency (5-15%)	L	0.04 m^3-$H_2\ m^{-3}\ d^{-1}$	0.5%[c]	[8,24]
		ability to metabolically engineer designer strains	co-production of O_2 inhibits H_2 production				
	Indirect - production of H_2 from carbohydrates produced by photosynthesis (photoheterotrophs)	use sunlight as energy source			0.19 m^3-$H_2\ m^{-3}\ d^{-1}$		
		simple nutritional requirements					
Dark fermentation	fermentation of carbohydrate rich substrates by anaerobic bacteria (sometimes algae). Typical fermentation end-products include mixed gas (H_2, CO_2, CH_4, H_2S), organic acids, alcohol (ethanol, butanol)	ability to metabolize soluble and insoluble biomass	separation step required to produce high quality H_2	L	4.4 – 65.1 m^3-$H_2\ m^{-3}\ d^{-1}$	60-80%[d]	[8,24]
		ability to metabolically engineer strains for enhanced H_2 production	H_2 competition with reduced fermentation end-products				
		low temperature demand	strongly depends on environmental growth conditions				
			process inhibited by O_2				
Photo fermentation	light energy and reduced compounds (organic acids) are catalyzed by a nitrogenase to form H_2 under N_2-deficient conditions	ability to use reduced organic compounds as a carbon source	H_2 consumption by uptake hydrogenases	L	0.09 m^3-$H_2\ m^{-3}\ d^{-1}$	0.1%[e]	[8,24]
		can be grown using a variety of processes (batch, continuous, porous glass, activated glass, polyurethane foam)	slow nitrogenase enzyme activity				
		process not inhibited by O_2	not efficient solar collectors (current collection efficiency, 1.9%)				

Biological water-gas-shift (WGS)	microbial oxidation of CO to CO_2, following the WGS reaction	low temperature and pressure reaction (enzymatic catalysis)	requires a CO source	L	51.7 m^3-H_2 m^{-3} d^{-1}	n/a	[8,24]
		thermodynamics favor high conversion of CO to CO_2 and H_2	requires darkness (photoheterotrophic)				
Hydrogen from water							
Alkaline electrolyzer	H_2O is introduced at the cathode and is separated into H_2 and OH^-. The OH^- travels through the alkaline electrolyte (30 wt% KOH or NaOH), forming O_2 at the anode.	mature technology	expensive Pt catalysts required	C	11,640 Nm^3-H_2 d^{-1}††	50-60%^{e}	[28,29]
		low operating temperature	electrolyte must be replenished				
		modest current density (100-300 mA cm^{-1})	H_2 must be separated from electrolyte				
PEM electrolyzer	similar to electrohydrogenesis. H_2O is split into H^+ and O_2 at the anode. H^+ combine at the cathode to produce H_2	low operating temperature	stringent regulation of fuel quality (CO and H_2S poisoning)	N	24 Nm^3-H_2 d^{-1}†††	55-70%^{e}	[30]
		high current density (>1600 mA cm^{-1})					
		no H_2 separation unit required					

n/a = not available

C = Commercial; N = Near term; L = Long term

Nm^3 = cubic meter of gas volume at normal (*i.e.* atmospheric) pressure conditions

† MTCI (Manufacturing and Technology Conversion International) gasifier; type: BFB (bubbling fluidized bed); feedstock: pulp sludge

†† Norsk Hydro Electrolyser – Atmospheric Type No.5040.

††† HOGEN RE PEM electrolyzer by Proton EnergySystems

[a] Energy-conversion efficiency (ECE) includes applied voltage but not energy in substrate (see [31]). Does not include H_2 purification.

[b] Thermal efficiency based on higher heating values

[c] Does not include H_2 purification

[d] Theoretical maximum, 4 mol-H_2 mol^{-1} glucose

[e] lower heating value of generated H_2 divided by the electrical energy to the electrolysis cell

Electricity production has been and remains the principle driving force behind the development of BES technology. In 1931, B. Cohen conducted the first bacteria-electrode interaction studies by examining overall potentiometric intensities of chemical reactions during bacterial growth [32,33]. He demonstrated bacterial cultures are electrical half-cells that can perform work. With the introduction of artificial electron mediators (potassium ferricyanide or benzoquinone), Cohen was able to build the first bacterial battery, producing a voltage of 35 V when six cells were connected in electrical series. During the late 1950s and throughout the1960s, interest in converting organic material into electrical energy surfaced. Initiated by the US space program, MFCs were seen as potential waste disposal units that could generate power during space missions [26,34]. Despite an improved understanding of the electrical connections between electrodes and microorganisms as well as the introduction of potentially new applications, it was determined that current could not be produced at a consistent rate or in quantities large enough to be a viable source of electrical energy [30,33-36].

A revived interest in electricity production from microbes came in the 1990's. Apart from the search for new, environmentally responsible sources of energy, studies showed increases in power densities using exogenous chemical mediators to deliver electrons to acceptors outside of the cell [37]. However, the breakthrough came with the discovery of the electricigen, bacteria capable of direct extracellular electron transfer to fuel cell anodes through the complete oxidation of organic compounds [36,38]. It has been known for almost 100 years that electrode-reducing microorganisms can transfer electrons *via* direct contact with the anode or through the synthesis of chemical mediators, both exogenous and endogenous.

However, it was observed for the first time that organic compounds were completely oxidized to CO_2 with all available electrons converted to current [39].

ELECTROGENIC BACTERIA

Taxonomic profiles of electrode-reducing microbial communities from numerous MFC systems have been reported [25,40-43]. The bacterial communities that develop show great diversity and typically depend on the enrichment conditions used to colonize the electrode surface [44]. However, amongst this group of electrode-respiring microorganisms, bacteria from the phylum Proteobacteria dominate anode communities. According to an eight system comparative study conducted by Aelterman *et al.* [43], 64% of the anode population belonged to the class of α-, β-, γ-, or δ-Proteobacteria, the most studied of these belonging to the families of *Shewanella* and *Geobacteraceae* [38]. The complete genome of *Shewanella oneidensis* was sequenced in 2002 [45] and subsequently that of *Geobacter sulfurreducens* in 2003 [46]; both organisms will serve as excellent models to elucidate the mechanisms of electron transfer between microorganism and electrode.

Mechanisms for Microbe-Electrode Electron Transfer

Understanding the principles of electron transfer between a microorganism and an electrode are essential in optimizing the current generated by any BES. Closer investigation into these mechanisms could not only influence material selection used to improve the electrical connection between the bacteria and the electrode but also manipulate the surface design of the electrode to facilitate electron delivery [36]. Although the exact mechanisms for electron transfer are not completely understood, three methods have been proposed (Fig. 3): i) long-range electron transfer *via* electron shuttles, ii) direct electron transfer *via* outer-surface *c*-type cytochromes, and iii) long-range electron transfer *via* conductive pili or “microbial nanowires”.

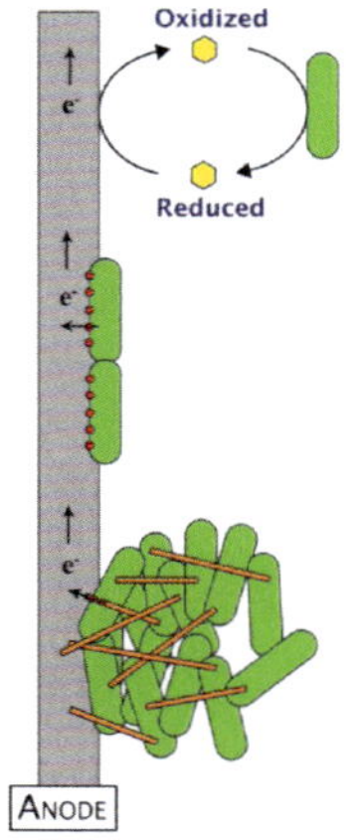

Figure 3: Proposed mechanisms for electron transfer to the anode of a MEC. i) Long-range electron transfer *via* electron shuttles (yellow hexagon), ii) direct electron transfer *via* outer-surface *c*-type cytochromes (red circles), and iii) long-range electron transfer *via* conductive pili or “microbial nanowires” (orange rods).

Electron Shuttles

Certain microorganisms produce soluble exogenous mediators that shuttle electrons from cells to insoluble compounds *via* diffusion. This phenomenon was first identified in mutant species of *Shewanella putrifaciens* [47] and later demonstrated in MFCs inoculated with *S. oneidensis* MR-1 [48]. Results showed that as many as half of the *S. oneidensis* cells were planktonic, suggesting substrate oxidation and concomitant current generation were coupled to long-range electron shuttles. Various techniques used to detect redox-active molecules eventually identified riboflavin secretion as the mechanism for extracellular electron transfer in *Shewanella* sp. [44,49,50]. This mechanism for electron transfer has been observed in other bacteria, such as *Geothrix fermentans* [51]. Unfortunately, despite the advantages of long-distance interaction with an electrode, electron shuttles are energetically taxing and may not be the most desirable system for the bacterium [36,52].

Direct Contact *via* Outer Membrane c-Type Cytochromes

In contrast to synthesized shuttles that indirectly transfer electrons to an electrode, direct contact between the cell and anode is also possible through outer membrane *c*-type cytochromes. In the same study conducted by Lanthier *et al.* [48], the other half of *S. oneidensis* cells that were not planktonic were attached to the surface of the anode, suggesting multiple strategies for electron transfer exist. In order to differentiate between direct and indirect mineral reduction pathways, further investigation was required. Nanoporous glass beads deposited with Fe(III) (hydr)oxide were used to measure iron reduction by *S. oneidensis* MR-1 both indirectly and as a biofilm [53]. Results from strains with mutations in *cyma* and *omcB* identified a potential reduction process in which cytoplasmic *c*-type cytochrome CymA is required for both direct and indirect mineral reduction whereas outer membrane *c*-type cytochrome OmcB is not necessary for indirect iron reduction [53]. It has been proposed that CymA functions as the terminal electron acceptor in the electron transport chain of *S. oneidensis* MR-1 whereas OmcB acts as a direct electrical contact between the microbe and the electrode surface [38].

Gene deletion studies in *G. sulfurreducens* further highlight the importance outer membrane *c*-type cytochromes have as direct electrical contacts between microbe and electrode. Two genes that encode outer membrane cytochromes, *omcS* and *omcE*, were targeted based on high transcript levels during growth on an electrode as the sole electron acceptor [54]. Deletion of *omcS* significantly reduced current production, which was restored after re-expression of the gene *in trans* on a plasmid. However, current inhibition was only temporary following the deletion of *omcE*. Over time, *G. sulfurreducens* found alternative routes to transfer electrons to the anode surface [36]. Results from this study outline the significance of outer membrane *c*-type cytochromes in direct electron transfer to an electrode. However, some organisms capable of reducing Fe(III) oxides lack *c*-cytochromes [55]. In addition, electron transport proteins attached directly to the outer membrane of the cell cannot explain how thick biofilms up to 75 μm [56] develop along anode surfaces as reported in many MFC studies [57-59]. Further investigation using *G. sulfurreducens* led to the discovery of electrically conductive pili known as "nanowires" [60].

Electron Transfer *via* Microbial Nanowires

According to Reguera [61], the nanowires of *G. sulfurreducens* are protein filaments composed of the repeated single unit PilA. They are classified as type IV pili but differ from convention due to their size (4-5 nm in diameter and 20 μm in length) and function. Unlike other pili that aid in cell motility or adhesion to solid surfaces, nanowires are electrically conductive conduits that enable microbial cell-cell communication. They are responsible for maximizing biofilm health by coordinating a cooperative electronic community, aggregating and interconnecting cells into a network capable of effectively distributing and dissipating electrons. Nanowires are responsible for high current and power production in MFCs, enabling active participation from cells located not only on the surface of the electrode but also at the outer boundaries of the biofilm.

A study by Reguera *et al.* [60] investigated the relationship between the nanowires of *G. sulfurreducens* and soluble and insoluble electron acceptors. Deletion of the *pilA* gene inhibited the production of pili and the reduction of insoluble Fe(III) oxides. The mutant was still able to reduce the soluble electron acceptors fumarate and Fe(III) citrate. Introducing a copy of the *pilA* gene *in trans* restored pili production and the ability to reduce Fe(III) oxides. Further studies revealed that PilA may have a structural role in biofilm formation [62]. However, it was observed that *pilA*-deficient mutants of *G. sulfurreducens* still formed biofilms along the surface of a graphite anode that was not connected in electrical series to a cathode, using fumarate as the terminal electron acceptor [36]. These findings suggest that although nanowires may not be necessary for biofilm growth, they are required for high-current production in BESs.

BIO-ELECTROCHEMICAL PRINCIPLES AND CHALLENGES

Introduction

As previously mentioned, an electrolysis cell is an electrochemical system composed of two electrodes connected in electrical series and separated by an ionic conductor, or electrolyte. Unlike a galvanic cell that produces energy from spontaneous chemical reactions, electrolysis requires an input of energy to drive chemical changes that would otherwise not occur. When a current is added to the cell, charge is transferred by two separate means: electrons are carried externally by electrical conduction and ions flow internally through the electrolyte by ionic conduction. The interface where these two chemical phases meet is particularly important because this is the site of charge transfer between electrode and electrolyte: the reactions that occur here determine the operation and performance of the electrolysis cell. Thus, understanding the principles of electrode reactions with respect to microbial electrolysis is required in order to calculate the theoretical limitations of electrohydrogenesis and explain deviations from ideal behavior.

The energy losses measured within a MEC determine the amount of potential that must be supplied to the reactor to generate H_2. However, total resistance within an electrochemical system is complicated and depends on the interrelation between three phenomena – thermodynamics, kinetics, and transport. The thermodynamic properties of the electrode and electrolyte determine the driving force for chemical reactions occurring at its interface. Kinetic parameters determine the rate at which these reactions will occur, while the transport of reactants to the interface and the continuous flow of charge through the cell is governed by principles of mass and charge transfer. Additionally, BESs must consider any biological losses associated with microbial activity, which determine the extent of electron generation and transfer to the anode.

It will be emphasized here the importance for sufficient withdrawal of spent electrolyte. This statement is typically not mentioned in research papers but its significance cannot be overstressed. Operating any MEC in batch mode will inevitably lead to total system failure. Numerous complications may arise, however the most prominent include the accumulation of undesired products (catalyst fouling, sluggish kinetics), the consumption of buffer (large fluctuations in solution pH), and the utilization of substrate (carbon limiting conditions for microorganisms). Adequate removal of contaminants and replacement of nutrients and salts is necessary for the continued operation of any MEC.

In this section, the thermodynamics of electrode reactions in terms of the electrochemical potential will be reviewed. The rates of these reactions in terms of current densities, j (A m^{-2}), will be discussed, introducing several models typically used to determine the kinetic parameters for microbial systems. Finally, mechanisms for the transport of charge and mass will be examined with an emphasis on the technical challenges that MEC technology currently faces. This section is intended to be an introduction to electrochemistry and its application to electrohydrogenesis. For a closer analysis on electrochemical principles and methods, many excellent textbooks are available [63-65].

Thermodynamics in Terms of Electrochemical Potential Fundamentals

The chemical potential, μ, is a thermodynamic quantity used to describe the energy level of a chemical species. It is a function of temperature, pressure, and concentration. Analogous to a ball rolling down a hill due to gravitational potential, a species will always move from a higher chemical potential to a lower chemical potential. When the species is charged, the electrical state must also be considered. In the presence of an electric field, the contribution of the electrostatic potential, ϕ, is very strong and is evaluated separately from the internal chemical potential components of that species. Thus, the electrochemical potential, $\bar{\mu}_i$, is the sum of the contributions from both the chemical potential and electrostatic potential of species i:

$$\bar{\mu}_i = \mu_i + z_i F\phi \tag{1}$$

where z is the charge of species i and F is Faraday's constant, 96,485 C mol^{-1} [3].

The electrode-electrolyte interface of an electrochemical cell is a multi-component, open system in which mass (*i.e.* the electron) of each species is free to move between the liquid, α, and solid, β, phases. According to Gibbs, once a state of equilibrium is achieved, the electrochemical potential of each species will be the same in both phases:

---

[3] It is easily shown that if the species is uncharged, $z_i = 0$, the electrochemical potential equals the chemical potential, $\bar{\mu}_i = \mu_i$.

$$\bar{\mu}_i^{\alpha} = \bar{\mu}_i^{\beta} \quad \mathbf{(2)}$$

By introducing the phase equilibrium constraint (Eq. 2) to a multi-component, open system in phase equilibrium, the electrochemical potential can now be related in terms of its natural variables temperature, T, pressure, P, and mole number, n. This is accomplished using the fundamental thermodynamic differential equation Gibbs energy, G:

$$dG = -SdT + VdP + \sum_i \bar{\mu}_i dn_i \quad \mathbf{(3)}$$

where S is the entropy of the system. What makes Eq. 3 so important is that once a thermodynamic function is described in terms of its natural variables, all thermodynamic properties of that system can be determined [66]. Assuming isothermal and isobaric reaction conditions typical of MECs, the Gibbs equation varies only with composition. Upon integration, the energy change for a chemical reaction is given by a change in Gibbs free energy for each half-reaction:

$$\Delta G = \left(\sum_i \bar{\mu}_i n_i\right)_{Products} - \left(\sum_i \bar{\mu}_i n_i\right)_{Reactants} \quad \mathbf{(4)}$$

Gibbs free energy is a convenient measure for describing the spontaneity of a process and can be viewed as the tendency for a reaction to reach equilibrium. If ΔG is negative, the electrochemical potential for the reactants is at a higher energy level and electrons will spontaneously flow to products. If ΔG is positive, the electrochemical potential is lower for reactants and additional energy must be supplied to drive electrons to products. When ΔG is zero, the system is said to be in equilibrium.

It may be more convenient to measure the electromotive force, E_{emf}, of a MEC rather than calculate the Gibbs free energy. An electromotive force is a measure of the work required to produce an electrical potential difference between two electrodes:

$$W = E_{emf} Q = E_{emf} nF \quad \mathbf{(5)}$$

where Q is the charge and n is the molar number of electrons transferred in the reaction. The E_{emf} of a MEC is the minimum amount of energy required to drive electrons from the anode to the cathode but can only be determined after chemical equilibrium of each half-cell is achieved. Allowing current to flow through an electrode gives rise to polarization effects, or departures from the equilibrium potential. When the rate of forward reactions equals the rate of backwards reactions, no net current is flowing allowing the true difference in equilibrium potentials of each electrode half-reaction to be compared. This is also termed the open-circuit potential (OCP) because the voltage is measured between two open terminal electrodes. Under equilibrium conditions, the amount of reversible work to establish a potential difference is equal to the Gibbs free energy:

$$E_{emf} = -\frac{\Delta G}{nF} \quad \mathbf{(6)}$$

The evolution of H_2 is a thermodynamically unfavorable reaction. For typical MECs, the cell E_{emf} is negative making the ΔG positive. These calculations provide the minimum requirements to drive H_2 production but more energy is required due to various potential losses throughout the system (described in detail below).

Electrode Potentials

Any electrochemical cell can be viewed as a controlled redox reaction, where oxidation and reduction are spatially separated into half-reactions located at the anode and cathode respectively. The minimum cell potential to drive electrohydrogenesis, expressed as an electromotive force, can be determined by evaluating each electrode separately:

$$E_{emf} = E_c - E_a \tag{7}$$

The Nernst equation is used to determine the equilibrium potential for each half-reaction:

$$E_{electrode} = E^o_{electrode} - \frac{RT}{nF}\ln(\Pi) \tag{8}$$

where R is the universal gas constant, 8.31447 J mol^{-1} K^{-1}, and T is the absolute temperature. According to IUPAC convention, all reactions are reported as a reduction potential: $oxidized + e^- \rightarrow reduced$. Therefore, the reaction quotient, Π, is the ratio of products divided by reactants expressed as activities. To simplify this treatment, dilute solution conditions are assumed and the Nernst equation can be expressed directly in terms of concentrations (i.e. activity coefficients are assumed to be unity). The reaction quotient becomes:

$$\Pi = \frac{[reduced]^p}{[oxidized]^r} \tag{9}$$

where r and p are the stoichiometric coefficients for reactants and products respectively.

The standard potential, $E^o_{electrode}$, is the equilibrium reduction potential under standard temperature and pressure (STP) conditions (298.15 K, 1 M for all species, 1 bar). All standard potentials are reported relative to a stable, well-defined reference electrode, which is typically the standard hydrogen electrode (SHE) or normal hydrogen electrode (NHE) (assigned a value of 0.000 V under STP conditions). Other standard reference electrodes are used, which include the saturated calomel electrode (SCE) and saturated Ag/AgCl electrode with potentials of +0.242 V and +0.197 V *versus* NHE respectively [63].

Typically when calculating the thermodynamics of biological reactions, neutral conditions are assumed [67]. Microbial electrohydrogenesis is no exception and the NHE is adjusted to a pH of 7.0. However, caution is advised and thermodynamic calculations should be adjusted according to actual operating pH values. The corrected NHE potential for the reduction of H_2 can be calculated using Eq. 8:

$$2H^+ + 2e^- \rightarrow H_2 \tag{10}$$

$$E^{o/}_{(H_2/H^+)} = E^o - \frac{RT}{nF}\ln\frac{H_2}{[H^+]^2}$$

$$E^{o/}_{(H_2/H^+)} = 0 + \frac{(8.31 \cdot 298.15)}{(2 \cdot 9.65 \cdot 10^4)}\ln\frac{1}{(10^{-7})^2} = -0.414\,V\ \textbf{(vs NHE)}$$

Depending on the substrate consumed at the anode by the bacteria and the operating conditions of the MEC, the theoretical total cell potential will vary. Standard and actual potentials of the typical electrode reactions found in MECs are summarized in Table **3**. For clarity purposes, all anode reactions are written as oxidation potentials to indicate electron loss. Additionally, many excellent review papers and textbooks compare the standard potentials of the reactions typically found in BESs [14, 64, 65, 67-69].

Total Energy Losses

As mentioned above, the driving force required for an electrochemical reaction to occur is the difference in potential energy between the anode and the cathode. This was quantified as a voltage, E_{emf}, or OCP, and is the minimum amount of energy required to drive an electrolytic process. However, this value represents a theoretical limit based on thermodynamic relationships whereas in practice, voltage requirements are considerably higher. The additional amount of energy required to generate H_2 in a MEC can be treated as a series of resistances. It is the sum of the overpotentials at the anode, η_a, and the cathode, η_c, as well as ohmic losses, IR_Ω, within the system:

$$E_{cell} = E_{emf} + (\sum \eta_a + \sum \eta_c + I \sum R_\Omega) \quad \textbf{(11)}$$

When a current runs through an electrochemical cell, a departure from equilibrium conditions is observed. This phenomenon is known as polarization. The magnitude of this potential drop is caused by a resistance to the passage of current and is appropriately termed overpotential, η. The extent of polarization at any one of the two electrodes depends on the current density and the nature of the reaction but is independent of the other electrode and the processes occurring there [65]. Therefore, each electrode can be studied individually when treating this phenomenon. A more thorough investigation into the variables that influence overpotentials and their effect on the reaction rates at the electrode-solution interface is discussed below.

It should be noted that the contribution of overpotentials and ohmic resistances to total energy loss is represented by polarization curves. By plotting the potential as a function of current density, deviations from OCP conditions, and thus overall MEC performance, can be observed [25, 63, 64].

Table 3: Reduction half-reactions and reduction potentials for common reactions in MEC studies.

	Half-reaction	**E^o (V *vs.* NHE)**	**E (V *vs.* NHE)**	**Refs.**
Anode reactions				
Acetate†	$8C_2H_3O_2^- + 3H_2O \rightarrow 8CO_2 + 8HCO_3^- + 8H^+ + 8e^-$	0.130	-0.284	[14,70]
Propionate†	$C_3H_5O_2^- + 5H_2O \rightarrow 2CO_2 + HCO_3^- + 14H^+ + 14e^-$	0.128	-0.286	[70]
Butyrate†	$20C_4H_7O_2^- + 7H_2O \rightarrow 3CO_2 + 20HCO_3^- + 20H^+ + 20e^-$	0.127	-0.287	[70]
Glucose	$C_6H_{12}O_6 + 12H_2O \rightarrow 6HCO_3^- + 30H^+ + 24e^-$	0.104	-0.289	[68]
Ethanol†	$C_2H_6O + 3H_2O \rightarrow 2CO_2 + 12H^+ + 12e^-$	0.090	-0.324	[70]
Cathode reactions				
H_2	$2H^+ + 2e^- \rightarrow H_2$	0.000	-0.414	[14,68]
Methane	$HCO_3^- + 9H^+ + 8e^- \rightarrow CH_4 + 3H_2O$	0.227	-0.248	[68]
H_2O_2	$O_2 + 2H^+ + 2e^- \rightarrow H_2O_2$	0.695	0.328	[14]
N_2	$2NO_3^- + 12H^+ + 10e^- \rightarrow N_2 + 6H_2O$	1.246	0.734	[68]
H_2O	$O_2 + 4H^+ + 4e^- \rightarrow 2H_2O$	1.229	0.805	[14,68]

The third column represents reduction potentials under STP conditions (T = 298.15 K, P = 1 bar, [ ] = 1 M). The fourth column represents actual experimental conditions:

$T = 298.15\ K$

$[H^+] = 10^{-7}\ M$

$[H_2O] = 1\ M$

$[HCO_3^-], [NO_3^-], glucose = 0.05\ M$

$[H_2O_2] = 0.5\ M$

$pCO_2 = 1\ bar$

$pO_2 = 0.2\ bar$

All other product concentrations = 1 M

† $[HCO_3^-] = 1\ M$

Ohmic Losses

Ohmic voltage losses are determined by resistance to electron flow through electrical conductors (*i.e.* electrodes and external circuitry) and resistance to ion flow through ionic conductors (*i.e.* electrolyte and proton exchange membrane) [25,69]. To compare different MEC architectures, resistances should be calculated using current density (A m^{-2}) rather than current. Additionally, current density should be expressed as a function of the projected anode surface area to normalize against biofilm growth [71].

Sleutels *et al.* [71] conducted ion transport resistance studies in a MEC using AEM and CEM membranes. Resistance to the transport of ions through the AEM configuration was much lower both at the beginning of the experiment and end of the experiment (12 and 8 mΩ m^2 respectively) compared to the CEM configuration (48 and 128 mΩ m^2). In addition to transport losses across the membrane, the total resistance based on electrode and electrolyte conductivity was estimated [71,72]. At the end of the experiment, the total resistance using the AEM configuration was significantly lower than the CEM, at 192 and 435 mΩ m^2 respectively. As a result, using an AEM produced higher rates of H_2 (2.1 m^3-H_2 m^{-3} d^{-1} at a current density of 5.3 A m^{-2}) compared to the CEM configuration (0.4 m^3-H_2 m^{-3} d^{-1} at a current density of 2.3 A m^{-2}) [71]. These findings are consistent with other published work comparing AEM and CEM configurations [5,10,11].

In addition to transport losses, potential losses associated with pH gradients across the membrane have also been observed [10,71]. Because membranes carry a charge, the concentration of ions will be different inside the membrane than outside in solution. This ion gradient can be expressed as a Donnan equilibrium [73]. Measured potential losses due to changes in ion concentration between CEM and AEM configurations is significant, at 0.16 and 0.10 V (vs. NHE) respectively [71]. A closer examination into the design of membranes is given in subsequent sections (see Future Outlook for MEC Technology).

Besides membrane selection, reducing electrode spacing, increasing electrolyte conductivity, and selecting electrode materials with low resistivity are options necessary to manage voltage losses and increase system performance.

Overpotentials at the Cathode

The reaction at the cathode in a MEC is known as the hydrogen evolution reaction (HER) and is driven by the reduction of protons into H_2 (Eq. 10). The HER has been heavily studied under extreme conditions of low pH or high alkalinity [74], but with the introduction of microorganisms at the anode (and possibly the cathode), neutral operating conditions (pH 7, mesophilic temperatures) is required. Two overpotentials account for the majority of energy lost at the cathode: charge-transfer losses, η_{ct}, and mass-transfer losses, η_{mt}:

$$\sum \eta_C = \eta_{ct} + \eta_{mt} \tag{12}$$

A third overpotential may also exist, termed a reaction overpotential, η_{rxn}. These losses are associated with chemical reactions preceding the transfer of electrons, which include protonation or dimerization effects [63]. With respect to microbial electrolysis, the energy losses associated with η_{rxn} are very small when compared with η_{ct} and η_{mt} and are therefore ignored [25,69,74].

Charge-Transfer (Activation) Overpotential

Slow kinetics related to the sluggish transfer of electrons at the electrode surface results in charge-transfer or activation losses. Factors that determine the nature of this reaction include electrode material, electrode surface area, nature of the electron transfer reaction, and operating temperature [74]. Assuming that the reaction is reversible, current can be related to the activation overpotential using the Butler-Volmer equation [63,64]. The general form of this equation is given by:

$$j = -j_o \left[\exp\left(\frac{\alpha_a F}{RT} \eta_{ct} \right) - \exp\left(\frac{\alpha_c F}{RT} \eta_{ct} \right) \right] \tag{13}$$

where j_o is the exchange current density (A m^{-2}), analogous to the rate constant in chemical kinetics, and α is the apparent transfer coefficient, relating potential to reaction direction (*i.e.* determines whether a reaction is an anodic or cathodic current[4]). For H_2 evolution and thus a cathodic current, the first term in Eq. 13 is negligible and can be removed. Assuming no concentration polarization [25], large activation overpotentials are linearly correlated to the logarithmic value of the reaction current density, and a Tafel plot can be created:

$$\eta_{ct} = \frac{RT}{\alpha_C F} \ln\left(\frac{j}{j_o}\right) \tag{14}$$

Tafel plots show that activation polarization rates decrease as current densities increase, meaning associated potential losses are most prominent under low current conditions, between 0 and 1 mA cm^{-2} [75]. This result has serious implications since MECs operate under low current conditions. One strategy to minimize the impact of activation losses is to increase the specific surface area of the cathode in the order of 10^6 m^2 m^{-3} [76]. Increasing the surface area decreases the current density and thus the activation losses. Other strategies aim to improve catalyst reaction kinetics by testing new catalysts, increasing operating temperatures, or increasing catalyst loading to effectively boost the number of reactive sites [14,25,69].

Microbial electrolysis cells have typically used platinum (Pt) to overcome the slow reaction kinetics of the HER on carbon-based [5,6,9,12,77] and titanium (Ti)-based [10,11] cathodes. The highest H_2 volumetric production rate in a MEC was achieved using a 0.5 mg cm^{-2} Pt catalyst loaded on a gas diffusion electrode (6.3 m^3-H_2 m^{-3} d^{-1} at an applied voltage of 1.0 V) [13]. Low overpotentials for Pt catalyzed HERs (-0.05 V at 15 A m^{-2}) have also been reported under optimized mass transport conditions [74]. However, Pt is very costly and not a practical solution for MEC scale-up.

Stainless steel (SS), nickel (Ni), and Ni alloys have been used to catalyze the HER in water electrolysis [78]. They have shown good performance but at high alkalinity and high temperatures (T = 450-700°C) [79]. The potential of using these metals as a replacement for platinum has been investigated in several MECs. Stainless steel brushes produced H_2 at a rate of 1.7 m^3-H_2 m^{-3} d^{-1} at an applied voltage of 0.6 V [80]. Increasing the surface area of the cathode decreased activation overpotentials by 10.5% (η_{ct} with 50% bristle loading *versus* η_{ct} with brush core only). At an applied voltage of 0.9 V, metal sheets composed of SS alloy A286 produced H_2 at twice the rate than Ni 625 (1.5 *versus* 0.8 m^3-H_2 m^{-3} d^{-1} respectively) [81]. The use of Ni 210 catalyst powder rather than a solid Ni plate increased H_2 production rates to 1.3 m^3-H_2 m^{-3} d^{-1}, despite operating at a lower applied voltage of 0.6 V [79]. The use of SS and Ni to replace platinum catalyzed cathodes shows promise, however long-term corrosion testing and its impact on current density and H_2 evolution is required.

The use of a microbial biofilm to produced H_2 at the cathode has been successfully reported [82,83]. To achieve a functional biocathode, the group exploited the principal that hydrogenases are reversible. A three-step process was exercised which: i) developed a healthy, acetate-fed bioanode, ii) adapted the bioanode for H_2 oxidation by feeding the chamber with H_2 only, and iii) reversed the polarity of the cell to naturally select for an H_2-producing, electrochemically active community at the cathode. Initial experiments produced H_2 at a rate of 0.63 m^3-H_2 m^{-3} d^{-1} (current density of 1.2 A m^{-2} and cathode potential of -0.7 V) [82]. Jeremiasse *et al.* [83] demonstrated for the first time that an MEC with both a bioanode and biocathode can produce H_2. Two MECs were operated at an applied voltage of 0.5 V. Both MECs had cathode potential of -0.7 V, however the current density of 1.9 A m^{-2} for MEC 1 was less than the 3.3 A m^{-2} generated in MEC 2. The overpotential at the cathode for both experiments, based on a theoretical potential of -0.42 V for the HER at pH 7 was -0.28 V. For biocathodes to be a viable option, the overpotential must be reduced to -0.1V and the current density increased to 10 A m^{-2} [24,83].

---

[4] An anodic current results in the flow of electrons from species in solution to the electrode where electrons flow from the electrode back into solution in a cathodic current.

The number of available reaction sites and total surface area of an electrode heavily influence the rate of a reaction. However, the rate of mass transport to that surface will also play a significant role in the overall reaction rate. The different modes of mass transport must be discussed.

Mass-Transfer (Concentration) Overpotential

Poor transfer of material in and out of the electrode-electrolyte interface is important in predicting current flow because the rate of the reaction will depend on the concentration of reactants and products. Poor mass-transfer develops chemical and potential gradients between the bulk solution and the interface, creating overpotentials at the electrode. Three modes of mass transport can influence an electrolysis reaction: i) migration (electrical potential gradient), ii) convection (bulk fluid movement, and iii) diffusion (chemical potential gradient).

Migration. Electrolysis places a positive charge on the anode and a negative charge on the cathode, creating an electric field and a driving force for ions in solution. Assuming no concentration effects (discussed below), anions and cations will move in opposite directions creating a build-up of charge at the electrode interface. Migration is essentially an electrostatic effect where any charged species near the interface will be attracted or repelled by electrostatic forces. However, due to the complexity of real solutions, such as diffuse layer interactions and solvation effects, migratory flux in MECs has never been reported.

Convection. Material can be transported to the interface by the bulk movement of the solution. Convection can be caused naturally by the formation of density gradients or thermal differences or may be forced by mechanical stirring or pressure gradients. Natural convection in electrolysis is discouraged because of its unpredictability. Although mixing a solution cannot cause a current, it can have a large impact on the concentration profile of ions and may be an effective means of bringing reactants to or removing products from an electrode surface.

To enhance the mass transport of substrate, protons, and buffer at the anode, a forced flow was introduced both through and perpendicular to a porous carbon felt anode [84]. At an applied voltage of 1 V, a high current density of 16.4 A m^{-2} (1 mm thick electrode) was achieved at a H_2 production rate of 5.6 m^3-H_2 m^{-3} d^{-1}. However, these results represent a combination of strategies that also included a higher buffer concentration. The contribution of forced flow to improve current density and H_2 production was not reported.

Diffusion (concentration). During current flow, the surface concentrations of the substances involved in the reaction change relative to the bulk concentrations in solution. Concentration polarization is observed when the supply or elimination of reactants and products is limited. Migration of ions to the electrode, diffusion gradients resulting from improper mixing, or unbalanced ratios of oxidized to reduced species at the electrode surface are all contributing factors that will result in a potential shift away from the electrode's equilibrium position [25,69]. Concentration polarization is the most significant mass-transport process in MEC cathode reactions [25,68,74].

Jeremiasse *et al.* [74] investigated the use of biocompatible buffers to lower the concentration polarization for the HER at Pt-cathodes (0.5 mg cm^{-2} Pt/Ir mixed metal oxide coated Ti). It was shown that the overpotential at the cathode was strongly dependent on the pH. At 50 mM and 15 A m^{-2}, the use of phosphate and ammonia led to the lowest overpotentials of -0.05 V at a pH of 6.2 and 9.0 respectively. Under the same experimental conditions, Tris, Hepes, and carbonate each had overpotentials of -0.07 (pH 7.8), -0.08 (pH 7.2), and -0.09 V (pH 9.3) respectively. Increasing the buffer concentration, increasing the linear flow speed, and decreasing the current density further reduced the concentration overpotential to a minimum, under conditions where pH approached the buffer dissociation constant (pKa). Higher current densities were reported by other groups by increasing the buffering capacity of the electrolyte [84]. A higher phosphate concentration in the anode chamber increased H^+ away from the biofilm and lowered the ionic resistance of the electrolyte.

Sleutels *et al.* [71] developed a method to compare the performance of different systems based on resistance rather than potential. Because the cell voltage of an MEC remains constant, expressing the sum of potential losses as partial resistances normalizes energy losses with respect to current density. The cathode potential loss with a CEM configuration (0.15 V at a current density of 2.3 A m^{-2}) was smaller than the AEM configuration (0.17 V at a current density of 5.3 A m^{-2}). However, when current density is taken into account, the CEM cathode resistance is much large than the AEM configuration, at 65 and 32 mΩ m^2 respectively [71].

Overpotentials at the Anode – A Kinetics Perspective

The reaction at the anode is catalyzed microbially. A carbon source is oxidized by anode-respiring bacteria (ARB), which in turn produce electrons (transferred to the electrode), protons (migrate and diffuse to the cathode), CO_2, and biomass. Clearly, the same potential losses that exist at the cathode are also present at the anode. However, in addition to charge-transfer and mass-transfer overpotentials, the anode is the site for metabolic losses associated with biofilm growth, η_{ARB}:

$$\sum \eta_a = \eta_{ct} + \eta_{mt} + \eta_{ARB} \quad (15)$$

The reported OCP for acetate oxidation in an MEC (-0.200 V *vs.* NHE) is close to the theoretical value (-0.284 *vs.* NHE) [11]. Typically, overpotentials are much smaller at the anode than at the cathode, ranging from 0.12 to 0.15 V (*vs.* NHE) [11,71]. However, these values represent total energy losses. To quantify specific overpotentials within an ARB community, electron flux from the electron donor (substrate) to the electrode surface must be fully understood.

Lee *et al.* [85] have proposed a three-stage conceptual model that relates the potential of the substrate to the potential of the anode: i) the substrate is oxidized and cellular electron carriers (*e.g.* NAD^+ to NADH) are reduced, ii) electrons travel down the potential gradient to outer-membrane, terminal cellular electron acceptors, and iii) electrons are passed to the anode *via* an extracellular electron-transfer mechanism. In order to achieve high yields of H_2, the ARB community must be able to produce high current densities at the lowest possible potential (*i.e.* as close to the potential of the substrate) [85]. Steps i) and ii) represent the intracellular potential drop. This is the only process where ARB can gain energy because it is assumed that potential losses related to extracellular activity, step iii), are mainly dissipated as heat given electron transfer occurs outside of the cell [44,86]. Therefore, in order to maximize cell growth, and thus current densities, it is necessary to keep extracellular potential losses to a minimum.

The Monod equation is typically used to model bacterial growth kinetics, particularly rates of substrate utilization (*i.e.* substrate oxidation and cellular electron carrier reduction) [87,88]. According to Torres *et al.* [70], substrate flux within a biofilm can be expressed under substrate limiting conditions as:

$$j = j_{\max} \frac{S}{K_s + S} \quad (16)$$

where j_{max} is the maximum biofilm current density (A m^{-2}) (j_{max} is a function of the maximum rate of substrate utilization, biofilm density, biofilm thickness, and fraction of electrons used for cell synthesis; see [70]), S is the substrate concentration (g m^{-3}), and K_s is the half-saturation coefficient (g m^{-3}) which corresponds to the concentration at which j_{max} is one-half of its maximum.

Monod-type growth was observed and kinetic parameters were determined for *G. sulfurreducens* using acetate (5.5 mM) as the limiting substrate and fumarate (30 mM) and Fe(III) (60 mM) as soluble electron acceptors [89]. High rates of acetate consumption coupled with low K_s values were observed, making *G. sulfurreducens* an ideal candidate for MECs due to high substrate oxidation rates at low substrate concentrations [90].

The Monod equation can be used only when the electron acceptor concentration is known. Unfortunately, the terminal electron acceptor for ARB is a solid anode lacking a defined concentration [91]. In order to describe the biofilm's response to changes in anode potentials, the Nernst-Monod equation was developed [92]:

$$j = j_{\max}\left[\frac{1}{1+\exp\left(-\frac{F}{RT}\eta_{ARB}\right)}\right] \quad (17)$$

where $\eta_{ARB} = E_a - E_{KA}$. E_{KA} is an electrochemical parameter, which is a measure of potential when $j = 1/2 j_{max}$. E_{KA} is analogous to K_s and is indicative of an ARBs desire to transfer electrons to the anode - the more negative E_{KA} is, the greater the affinity to transfer electrons [90]. Combining Eq. 16 with Eq. 17 introduces both the biological and electrochemical factors required to describe the current density of an ARB community as a function of the potential at the anode [85]:

$$j = j_{\max}\frac{S}{K_S + S}\left\{\frac{1}{1+\exp\left[-\frac{F}{RT}(E_a - E_{KA})\right]}\right\} \quad (18)$$

Under substrate limiting conditions, biological parameters such as K_S and substrate utilization rates play an important role and must be determined experimentally. Under substrate excess conditions ($S >> K_s$), the rate of substrate utilization will approach unity (*i.e.* $S/K_s+S = 1$) and intracellular ARB kinetics will be a function of the anode potential only. It should be noted that although anode potential has the greatest impact on current flux through the ARB [90], biofilm kinetics should also consider substrate gradients that develop within the thick biofilms, up to 50 μm thick [58].

The effects of anode potential on the kinetic parameters of the Nernst-Monod equation were evaluated for an ARB community [90]. A very negative E_{KA} of -0.448 V (vs Ag/AgCl) was reported, indicating fast biofilm growth and high current flux to the anode close to saturation levels (where j_{max} = 9.25 A m^2). At a poised anode potential of -0.400 V (vs Ag/AgCl), η_{ARB} was calculated to be 0.048 V. Thus having E_{KA} values lower than the potential of the anode is the most effective strategy for ARBs to achieve high current densities in MECs.

FUTURE OUTLOOK FOR MEC TECHNOLOGY

For microbial electrohydrogenesis to become an economically competitive technology, high H_2 production rates must be achieved at low applied voltages. This can only be accomplished if current flux to the anode is maximized while total energy loss is minimized. Several strategies are discussed below.

Positive Energy Balance

For microbial electrohydrogenesis to have a net positive energy balance, the summation of all energy losses associated with all aspects of the system must be lower than the energy stored in the production of H_2. A number of studies have characterized total energy losses within MECs. In order to have a better understanding for the contribution of overpotentials at the anode and cathode, ohmic resistance, and theoretical requirements to the total losses in the system, two studies were compared (Fig. **4**).

By examining the figures adapted from Rozendal *et al.* [10] (Fig. **4a**) and Lee and Rittmann [31] (Fig. **4b**), two themes immerge: i) the use of a membrane increases ohmic resistance by more than 50%, and ii) the highest energy losses are overpotentials at the electrodes, particularly at the cathode. Additionally, despite operating at different applied voltages, the overall contribution of each energy loss due to overpotentials at the electrodes was relatively similar. It was later determined that the operating voltage for any MEC must be 0.6 V or lower in order to achieve any energy benefit from the process, assuming a cathodic energy recovery of at least 80% [31].

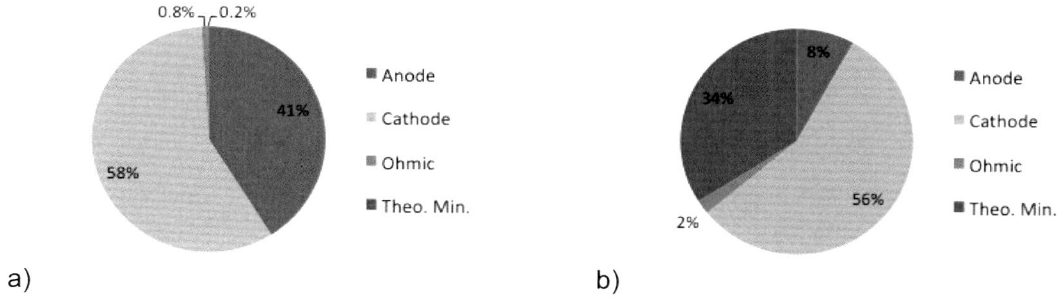

Figure 4: Total energy losses within a MEC. The contribution of overpotentials at the anode and cathode, ohmic losses, and theoretical requirements were compared. a) Rozendal *et al.* [10] used a dual chamber MEC at an applied voltage of 0.5V. b) Lee and Rittmann [31] used a single chamber MEC at an applied voltage of 1.1 V. Both reactors were operated in batch mode and used acetate as the sole carbon source (figures adapted from [10] and [31] respectively).

Membrane Design

The decision to include a membrane into the design of a MEC has serious implications. The most obvious advantage of a membrane is that it acts as a barrier, separating the anode chamber from the cathode chamber. This produces a clean H_2 stream, preventing mixing with CO_2 and other metabolic bio gases (CH_4, H_2S) that may be produced at the anode. Additionally, greater coulombic efficiencies and higher cathodic recoveries can be achieved since proton transfer is facilitated through the electrolyte [10,68,71].

Unfortunately, improved reactor performance comes at a cost. Not only will resistance within the membrane increase total ohmic energy losses by as much as 50% [31,93], but a significant pH gradient will also develop across the membrane. Another way to describe proton consumption at the cathode is the formation of hydroxyl ions (OH^-)[5]. According to Rozendal *et al.* [11], as [OH^-] increases with time, an uneven distribution of charge is created. As the membrane attempts to maintain electroneutrality, ionic species are transported between the two chambers. Unfortunately, under neutral conditions (pH 7), typical MEC membranes predominantly transport ionic species other than OH^-. A concentration gradient develops, inducing the diffusion of charged species in the opposite direction. This pH gradient will continue to build until the migration of cations and anions in one direction equals the diffusion of ionic species in the opposite direction. The result of this phenomenon is severe, with experiments showing cathode pH's greater than 12. In terms of energy loss, every additional increment in pH unit results in an ohmic loss of approximately 0.059 V, requiring an additional energy input of 0.13 kWh m^{-3}-H_2 [16]. This explanation is general and applies to all semi-permeable membranes. Actual pH increases in the cathode chamber will depend on the type of membrane being used. For example, the cathode pH is lower for MECs using AEMs rather than CEMs based on their different affinities for hydroxyl ions [16]. This statement is in agreement with experimental results [11].

One way to lower pH gradients and energy losses is to remove the membrane from the design. By doing so, single-chambered MECs have achieved the highest volumetric H_2 production rates – 3.12 m^3-H_2 m^{-3} d^{-1} [12] and 6.3 m^3-H_2 m^{-3} d^{-1} [13]. Additionally, membrane-less operation simplifies the MEC design, lowering construction costs. Unfortunately, several major drawbacks do exist. First, H_2 is no longer a pure product. As mentioned above, additional purification steps are required to remove metabolic gases being produced at the anode by electrochemically active bacteria (mainly CO_2). However, the greatest challenge for membrane-less operation is the microbial consumption of H_2 *via* methanogens, homoacetogens, and H_2 scavengers able to convert H_2 back into current (H_2 recycling) [85].

A number of studies have reported CH_4 production by hydrogenotrophic and acetoclastic methanogens, not only at the anode but also at the cathode [7,91,93-95]. One study by Clauwaert and Verstraete [93] reported that CH_4 was the main energetic product (78% of total gas) despite continuous operation and slightly acidic

---

[5] based on the equilibrium of H^+ and OH^- through the water dissociation constant ($K_w=[H^+][OH^-]\approx 10^{-14}$)

conditions (pH 6.1–6.2). Another study by Lee and Rittmann [91] not only observed CH_4 production (6.9% of total gas) but also determined H_2 recycling accounted for 62–76% of the total current generation. Controlling methanogenesis and preventing H_2 recycling present significant challenges for the practical application of membrane-less MECs.

Reducing Electrode Overpotentials

Designer consortiums – anode. Greater current densities can be achieved by packaging electricigens as designer consortiums. Designing experiments that force bacteria to produce current is a promising approach since no previous evolutionary selection pressure on microorganisms has existed [36]. This strategy was successful in selecting a variant of *G. sulfurreducens* with an enhanced capacity for current production [96]. *Geobacter sulfurreducens* strain KN400 was recovered from a biofilm after 5 months of growth in a MFC. Although further functional analysis is required, phenotypic changes to the outer surface of the cell were observed, resulting in a greater abundance of conductive pili with a lower internal resistance and greater tendency to form biofilms.

Table 4: Comparison of Cathode Catalysts used in MECs. Advantages and disadvantages are discussed to qualitatively compare commonly used catalysts.

Material	Advantages	Disadvantages	Ref.
Platinum (Pt)	low overpotentials for HER (0.05 V)	very costly – not feasible for scale-up: Pt = \$630 m^{-2}	see Table 1
	high volumetric H_2 production rates (6.3 m^3-H_2 m^{-3} d^{-1})		
Stainless Steel (SS 305) & Nickel Alloys (Ni 625)	have been used to catalyze HER in water electrolysis	long term corrosion / performance testing required	[81]
	bench scale performance similar to Pt		
	economical: SS = \$63 m^{-2}; Ni = \$370 m^{-2}		
Biological	no metal catalyst required	high overpotentials (0.28 V)	[83]
	robust biofilm	further testing required	

Improved cathode – catalysts. A qualitative summary of the most commonly used cathode catalysts is presented in Table **4** (for a quantitative comparison, see section on charge-transfer (activation) overpotentials). Replacing Pt with inexpensive cathode materials or even the use of biocathodes to catalyze the HER is necessary if MEC scale-up is to be economically feasible. Pilot plant testing has begun at the Napa Wine Company in Oakville, CA, USA, using SS cathodes [97]. However, the electrode reaction kinetics, although promising, must be improved and the long-term mechanical stability of the electrodes requires further testing [81].

Despite the many challenges facing electrohydrogenesis this is a technology experiencing rapid growth and development. Expensive Pt catalysts are being replaced with SS and Ni alloys. Showing even more promise is the development of microbial biocathodes has successfully catalyzed H_2 production as well. Electron flux at the anode-electrolyte interface is improving not only by engineering anodes from more suitable materials with higher surface areas electrode surface area, but by selecting bacterial strains with enhanced electrochemical activity. By implementing smarter reactor designs to reduce energy losses and testing these MECs at pilot plant scales the potential to commercialize this technology has never been greater. With the world now realizing sustainable H_2 can be achieved from waste streams, the future for microbial electrohydrogenesis and MEC technology looks very promising indeed.

REFERENCES

[1] Logan BE. Extracting hydrogen and electricity from renewable resources. Environ Sci Technol 2004; 38(9):160-7.

[2] Lattin WC, Utgikar VP. Transition to hydrogen economy in the United States: a 2006 status report. Int J Hydrogen Energy 2007; 32: 3230-7.
[3] Ramachandran R, Menon RK. An overview of industrial uses of hydrogen. Int J Hydrogen Energy 1998; 23(7): 593-8.
[4] Lu L, Ren N, Xing D, Logan BE. Hydrogen production with effluent from an ethanol-H2 coproducing fermentation reactor using a single-chamber microbial electrolysis cell. Biosens Bioelectron 2009; 24(10): 3055-60.
[5] Cheng S, Logan BE. Sustainable and efficient biohydrogen production *via* electrohydrogenesis. Proc Natl Acad Sci USA 2007; 104(47): 18871-3.
[6] Ditzig J, Liu H, Logan BE. Production of hydrogen from domestic wastewater using a bioelectrochemically assisted microbial reactor (BEAMR). Int J Hydrogen Energy 2007; 32: 2296-304.
[7] Wagner RC, Regan JM, Oh SE, Zuo Y, Logan BE. Hydrogen and methane production from swine wastewater using microbial electrolysis cells. Water Res 2009; 43(5): 1480-8.
[8] Holladay JD, Hu J, King DL, Wang Y. An overview of hydrogen producing technologies. Catal Today 2009; 139: 244-60.
[9] Liu H, Grot S, Logan BE. Electrochemically assisted microbial production of hydrogen from acetate. Environ Sci Technol 2005; 39(11): 4317-20.
[10] Rozendal RA, Hamelers HV, Buisman CJ. Effects of membrane cation transport on pH and microbial fuel cell performance. Environ Sci Technol 2006; 40(17): 5206-11.
[11] Rozendal RA, Hamelers HV, Molenkamp RJ, Buisman CJ. Performance of single chamber biocatalyzed electrolysis with different types of ion exchange membranes. Water Res 2007; 41(9): 1984-94.
[12] Call D, Logan BE. Hydrogen production in a single chamber microbial electrolysis cell lacking a membrane. Environ Sci Technol 2008; 42(9): 3401-6.
[13] Tartakovsky B, Manuel MF, Wang H, Guiot SR. High rate memrane-less microbial electrolysis cell for continuous hydrogen production. Int J Hydrogen Energy 2009; 34: 672-7.
[14] Logan BE. Microbial fuel cells. New Jersey: John Wiley & Sons, Inc.; 2008.
[15] Logan BE, Call D, Cheng S, Hamelers HV, Sleutels TH, Jeremiasse AW, *et al.* Microbial electrolysis cells for high yield hydrogen gas production from organic matter. Environ Sci Technol 2008; 42(23): 8630-40.
[16] Rozendal RA. Hydrogen production through biocatalyzed electrolysis: Wageningen University; 2007.
[17] Clauwaert P, Aelterman P, Pham TH, De Schamphelaire L, Carballa M, Rabaey K, *et al.* Minimizing losses in bio-electrochemical systems: The road to applications. Appl Microbiol Biotechnol 2008; 79(6): 901-13.
[18] Bullen RA, Arnot TC, Lakeman JB, Walsh FC. Biofuel cells and their development. Biosens Bioelectron 2006; 21(11): 2015-45.
[19] Kipnis N. Luigi Galvani and the debate on animal electricity. Ann Sci 1987; 44: 107-42.
[20] Grove WR. Philosophy Magazine Series. 1839; 3(14): 127.
[21] Potter MC. Electrical effects accompanying the decomposition of organic compounds. P R Soc Lond B-Conta 1911; 84(571): 260-76.
[22] He Z, Angenent LT. Application of bacterial biocathodes in microbial fuel cells. Electroanal 2006; 18(19-20): 2009-15.
[23] Rabaey K, Rodriguez J, Blackall LL, *et al.* Microbial ecology meets electrochemistry: electricity-driven and driving communities. ISME J 2007; 1(1): 9-18.
[24] Levin DB, Pitt L, Love M. Biohydrogen production: prospects and limitations to practical application. Int J Hydrogen Energy 2004; 29: 173-85.
[25] Spath PL, Mann MK. Life cycle assessment of hydrogen production *via* natural gas steam reforming. Golden, Colorado: National Renewable Energy Laboratory, 2001 Contract No.: NREL/TP-570-27637.
[26] Jameel H, Keshwani DR, Carter SF, Treasure TH. Thermochemical conversion of biomass to power and fuels. In: Cheng J, Ed. Biomass to Renewable Energy Processes. Boca Raton, FL, USA: CRC Press - Taylor & Francis Group; 2010. pp. 437-89.
[27] Ciferno JP, Marano JJ. Benchmarking biomass gasification technologies for fuels, chemical, and hydrogen production. Pittsburgh, PA: National Energy Technoogy Laboratory, US Department of Energy, 2002.
[28] Sigurvinsson J, Werkoff F. On the cost of the hydrogen produced by alkaline electrolysis. Int Hydr Energy Cong Exhibit (IHEC); 2005; Istanbul, Turkey.
[29] Ivy J. Summary of electrolytic hydrogen production. Golden, Colorado: National Renewable Energy Laboratory, 2004. Contract No.: NREL/MP-560-36734.
[30] Barbir F. PEM electrolysis for production of hydrogen from renewable energy sources. Sol Energy 2005; 78: 661-9.
[31] Lee H-S, Rittmann B. Characterization of energy losses in an upflow single-chamber microbial electrolysis cell. Int J Hydrogen Energy 2010; 35: 920-7.

[32] Berk RS, Canfield JH. Bioelectrochemical energy conversion. Appl Microbiol. 1964; 12:10-2.

[33] Cohen B. The bacterial culture as an electrical half-cell. J Bacteriol 1931; 21(1): 18-9.

[34] Shukla AK, Suresh P, Berchmans S, Rajendran A. Biological fuel cells and their applications. Curr Sci 2004; 87(4): 455 -68.

[35] Lewis K. Symposium on bioelectrochemistry of microorganisms. IV. Biochemical fuel cells. Bacteriol Rev 1966; 30(1): 101-13.

[36] Lovley DR, Nevin KP. Electricity production with electricigens. In: Wall JD, Harwood CS, Demain A, Eds. Bioenergy. Washington, D.C.: ASM Press; 2008. pp. 295-306.

[37] Allen RM, Bennetto HP. Microbial fuel-cells: electricity production from carbohydrates. Appl Biochem Biotechnol 1993; 39/40(1): 27-40.

[38] Debabov VG. Electricity from microorganisms. Mikrobiologiia 2008; 77(2): 149-57.

[39] Bond DR, Holmes DE, Tender LM, Lovley DR. Electrode-reducing microorganisms that harvest energy from marine sediments. Science 2002; 295(5554): 483-5.

[40] Jung S, Regan JM. Comparison of anode bacterial communities and performance in microbial fuel cells with different electron donors. Appl Microbiol Biotechnol 2007; 77(2): 393-402.

[41] Chae KJ, Choi MJ, Lee JW, Kim KY, Kim IS. Effect of different substrates on the performance, bacterial diversity, and bacterial viability in microbial fuel cells. Bioresource Technol 2009; 100(14): 3518-25.

[42] Logan BE, Regan JM. Electricity-producing bacterial communities in microbial fuel cells. Trends Microbiol 2006; 14(12): 512-8.

[43] Aelterman P, Rabaey K, Schamphelaire LD, Clauwaert P, Boon N, Verstraete W. Microbial fuel cells as an engineered ecosystem. In: Wall JD, Harwood CS, Demain AL, editors. Bioenergy. Washington, D.C.: ASM Press; 2008. p. 307-22.

[44] Lovley DR. The microbe electric: conversion of organic matter to electricity. Curr Opin Biotech. 2008 Dec; 19(6): 564-71.

[45] Heidelberg JF, Paulsen IT, Nelson KE, *et al.* Genome sequence of the dissimilatory metal ion-reducing bacterium *Shewanella oneidensis*. Nat Biotechnol 2002; 20(11): 1118-23.

[46] Methe BA, Nelson KE, Eisen JA, *et al.* Genome of *Geobacter sulfurreducens*: metal reduction in subsurface environments. Science 2003; 302(5652): 1967-9.

[47] Newman DK, Kolter R. A role for excreted quinones in extracellular electron transfer. Nature. 2000; 405(6782): 94-7.

[48] Lanthier M, Gregory KB, Lovley DR. Growth with high planktonic biomass in *Shewanella oneidensis* fuel cells. FEMS Microbiol Lett 2008; 278(1): 29-35.

[49] Marsili E, Baron DB, Shikhare ID, Coursolle D, Gralnick JA, Bond DR. Shewanella secretes flavins that mediate extracellular electron transfer. Proc Natl Acad Sci USA 2008;105(10): 3968-73.

[50] von Canstein H, Ogawa J, Shimizu S, Lloyd JR. Secretion of flavins by Shewanella species and their role in extracellular electron transfer. Appl Environ Microbiol 2008; 74(3): 615-23.

[51] Bond DR, Lovley DR. Evidence for involvement of an electron shuttle in electricity generation by *Geothrix fermentans*. Appl Environ Microbiol 2005; 71(4): 2186-9.

[52] Mahadevan R, Bond DR, Butler JE, *et al.* Characterization of metabolism in the Fe(III)-reducing organism *Geobacter sulfurreducens* by constraint-based modeling. Appl Environ Microbiol 2006; 72(2): 1558-68.

[53] Lies DP, Hernandez ME, Kappler A, Mielke RE, Gralnick JA, Newman DK. Shewanella oneidensis MR-1 uses overlapping pathways for iron reduction at a distance and by direct contact under conditions relevant for Biofilms. Appl Environ Microbiol 2005; 71(8): 4414-26.

[54] Holmes DE, Chaudhuri SK, Nevin KP, *et al.* Microarray and genetic analysis of electron transfer to electrodes in *Geobacter sulfurreducens*. Environ Microbiol 2006; 8(10): 1805-15.

[55] Lovley DR, Phillips EJ, Lonergan DJ, Widman PK. Fe(III) and S0 reduction by *Pelobacter carbinolicus*. Appl Environ Microbiol 1995; 61(6): 2132-38.

[56] Lovley DR. Extracellular electron transfer: wires, capacitors, iron lungs, and more. Geobiology 2008; 6(3): 225-31.

[57] Bond DR, Lovley DR. Electricity production by *Geobacter sulfurreducens* attached to electrodes. Appl Environ Microbiol 2003; 69(3): 1548-55.

[58] Reguera G, Nevin KP, Nicoll JS, Covalla SF, Woodard TL, Lovley DR. Biofilm and nanowire production leads to increased current in *Geobacter sulfurreducens* fuel cells. Appl Environ Microbiol 2006; 72(11): 7345-8.

[59] Nevin KP, Richter H, Covalla SF, *et al.* Power output and coulombic efficiencies from biofilms of *Geobacter sulfurreducens* comparable to mixed community microbial fuel cells. Environ Microbiol 2008; 10(10): 2505-14.

[60] Reguera G, McCarthy KD, Mehta T, Nicoll JS, Tuominen MT, Lovley DR. Extracellular electron transfer *via* microbial nanowires. Nature 2005; 435(7045): 1098-101.

[61] Reguera G. Are microbial conversations being lost in translation? Microbe 2009: 506-12.

[62] Reguera G, Pollina RB, Nicoll JS, Lovley DR. Possible nonconductive role of *Geobacter sulfurreducens* pilus nanowires in biofilm formation. J Bacteriol 2007; 189(5): 2125-7.

[63] Bard AJ, Faulkner LR. Electrochemical methods: Fundamentals and applications. 2nd ed. New York: Wiley; 2001.

[64] Newman JS, Thomas-Alyea KE. Electrochemical systems. 3rd ed. Hoboken, N.J.: Wiley-Interscience; 2004.

[65] Bagotsky VS. Fundamentals of Electrochemistry. 2 ed. New Jersey: John Wiley & Sons, Inc.; 2005.

[66] Alberty RA. Use of legendre transforms in chemical thermodynamics. Pure Appl Chem 2001; 73(8): 1349-80.

[67] Thauer RK, Jungermann K, Decker K. Energy conservation in chemotrophic anaerobic bacteria. Bacteriol Rev 1977; 41(1): 100-80.

[68] Hamelers HV, Ter Heijne A, Sleutels TH, Jeremiasse AW, Strik DP, Buisman CJ. New applications and performance of bioelectrochemical systems. Appl Microbiol Biotechnol 2010; 85(6): 1673-85.

[69] Logan BE, Hamelers B, Rozendal R, *et al.* Microbial fuel cells: methodology and technology. Environ Sci Technol 2006; 40(17): 5181-92

[70] Torres CI, Marcus AK, Rittmann BE. Kinetics of consumption of fermentation products by anode-respiring bacteria. Appl Microbiol Biotechnol 2007; 77(3): 689-97.

[71] Sleutels THJA, Hamelers HVM, Rozendal RA, Buisman CJN. Ion transport resistance in microbial electrolysis cells with anion and cation exchange membranes Int J Hydrogen Energy 2009; 34(9): 3612-20.

[72] Ter Heijne A, Hamelers HV, De Wilde V, Rozendal RA, Buisman CJ. A bipolar membrane combined with ferric iron reduction as an efficient cathode system in microbial fuel cells. Environ Sci Technol 2006; 40(17): 5200-5.

[73] Prausnitz JM, Lichtenthaler RN, de Azevedo EG. Molecular thermodynamics of fluid-phase equilibria. 3rd ed: Prentice Hall; 1998.

[74] Jeremiasse AW, Hamelers HV, Kleijn JM, Buisman CJ. Use of biocompatible buffers to reduce the concentration overpotential for hydrogen evolution. Environ Sci Technol. 2009; 43(17): 6882-7.

[75] Rabaey K, Lissenns G, Verstraete W. Microbial fuel cells: performances and perspectives. In: Lens P, Westermann P, Haberbauer M, Moreno A, Eds. Biofuels for fuel cells: Renewable energy from biomass fermentation. London: IWA Publishing; 2005. pp. 377-99.

[76] Freguia S, Rabaey K, Yuan Z, Keller J. Sequential anode-cathode configuration improves cathodic oxygen reduction and effluent quality of microbial fuel cells. Water Res 2008; 42(6-7): 1387-96.

[77] Hu H, Fan Y, Liu H. Hydrogen production using single-chamber membrane-free microbial electrolysis cells. Water Res. 2008; 42(15): 4172-8.

[78] Couper AM, Pletcher D, Walsh FC. Electrode materials for electrosynthesis. Chem Rev 1990; 90(5): 837-65.

[79] Selembo PA, Merrill MD, Logan BE. Hydrogen production with nickel powder cathode catalysts in microbial electrolysis cells. Int J Hydrogen Energy 2010; 35: 428-37.

[80] Call DF, Merrill MD, Logan BE. High surface area stainless steel brushes as cathodes in microbial electrolysis cells. Environ Sci Technol. 2009; 43(6): 2179-83.

[81] Selembo PA, Merrill MD, Logan BE. The use of stainless steel and nickel alloys as low-cost cathodes in microbial electrolysis cells. J Power Sources 2009; 190(2): 271-8.

[82] Rozendal RA, Jeremiasse AW, Hamelers HV, Buisman CJ. Hydrogen production with a microbial biocathode. Environ Sci Technol 2008; 42(2): 629-34.

[83] Jeremiasse AW, Hamelers HV, Buisman CJ. Microbial electrolysis cell with a microbial biocathode. Bioelectrochemistry 2009; 78: 39-43.

[84] Sleutels THJA, Lodder R, Hamelers HVM, Busiman CJN. Improved performance of porous bio-anodes in microbial electrolysis cells by enhanced mass and charge transport. Int J Hydrogen Energy 2009; 34: 9655-61.

[85] Lee HS, Vermaas WF, Rittmann BE. Biological hydrogen production: prospectives and challenges. Trends Biotechnol 2010; 28: 262-71.

[86] Torres CI, Marcus AK, Lee HS, Parameswaran P, Krajmalnik-Brown R, Rittmann BE. A kinetic perspective on extracellular electron transfer by anode-respiring bacteria. FEMS Microbiol Rev 2010; 34(1): 3-17.

[87] Rittmann BE, McCarty PL. Environmental biotechnology: Principles and applications. Boston: McGraw-Hill; 2001.

[88] Monod J. Resherches sur la croissance des cultures bacteriennes. Paris: Hermann; 1942.

[89] Esteve-Nunez A, Rothermich M, Sharma M, Lovley D. Growth of *Geobacter sulfurreducens* under nutrient-limiting conditions in continuous culture. Environ Microbiol 2005; 7(5): 641-8.

[90] Lee HS, Torres CI, Rittmann BE. Effects of substrate diffusion and anode potential on kinetic parameters for anode-respiring bacteria. Environ Sci Technol 2009; 43(19): 7571-7.

[91] Lee HS, Rittmann BE. Significance of biological hydrogen oxidation in a continuous single-chamber microbial electrolysis cell. Environ Sci Technol 2010; 44(3): 948-54.

[92] Kato Marcus A, Torres CI, Rittmann BE. Conduction-based modeling of the biofilm anode of a microbial fuel cell. Biotechnol Bioeng 2007; 98(6): 1171-82.

[93] Clauwaert P, Verstraete W. Methanogenesis in membraneless microbial electrolysis cells. Appl Microbiol Biotechnol 2009; 82(5): 829-36.

[94] Lee HS, Torres CI, Parameswaran P, Rittmann BE. Fate of H2 in an upflow single-chamber microbial electrolysis cell using a metal-catalyst-free cathode. Environ Sci Technol 2009; 43(20): 7971-6.

[95] Cheng S, Xing D, Call DF, Logan BE. Direct biological conversion of electrical current into methane by electromethanogenesis. Environ Sci Technol 2009; 43(10): 3953-8.

[96] Yi H, Nevin KP, Kim BC, *et al.* Selection of a variant of *Geobacter sulfurreducens* with enhanced capacity for current production in microbial fuel cells. Biosens Bioelectron 2009; 24(12): 3498-503.

[97] Logan BE. Scaling up microbial fuel cells and other bioelectrochemical systems. Appl Microbiol Biotechnol 2010; 85(6): 1665-71.

[98] Lovely DR. Future shock from the microbe electric. Microb Biotechnol 2009; 2: 139-141.

CHAPTER 13

Use of Immobilized Cell Systems in Biohydrogen Production

Nuri Azbar[1*] and Ilgi K. Kapdan[2]

[1]Bioegineering Department, Ege University and [2]Environmental Engineering Department, Dokuz Eylul University, Turkey

Abstract: Cell and enzyme immobilization is a widely employed technique in many industrial applications. Over the recent years, researchers have been also focusing on benefiting from the merits of immobilization techniques in biohydrogen studies. In this chapter, various immobilization techniques and their application results in terms of enhancing biohydrogen production *via* different processes, namely anaerobic dark and light fermentation, are presented. For this purpose, firstly, principles of various immobilization techniques and protocols are discussed and secondly the literature reports on the results of use of immobilization techniques are reviewed in regards to their biohydrogen productions.

Keywords: Immobilization, bioreactor, biohydrogen, biofilm systems, immobilization materials, carrier material, fixed bed, fluidized bed, expended bed, UASB.

INTRODUCTION

As an alternative to common hydrogen production methods (mainly thermo chemical ones), biological hydrogen fermentation seems to be favorable, because hydrogen can be yielded at a high rate with a concomitant reduction in environmental organic pollutants.

Immobilization in biotechnology is defined as "the technique used for the physical or chemical fixation of cells, organelles, enzymes, or other proteins (*e.g.* monoclonal antibodies) onto a solid support, into a solid matrix or retained by a membrane, in order to increase their stability and make possible their repeated or continued use". In another way, immobilization can be defined as "the confinement or localization of viable microbial cells to a certain defined region of space in such a way as to exhibit hydrodynamic characteristics which differ from those of the surrounding environment" [1].

Suspended culture systems employing continuous stirred tank reactors (CSTR) are more commonly used for continuous hydrogen production. In such a conventional system, hydrogen-producing bacteria are well suspended in the mixed liquor and suffer less from the mass transfer resistance; however, biomass retention is substantially influenced by the culture hydraulic retention time (HRT), and washout of biomass may occur at shorter HRTs. Thus hydrogen production rates are restricted considerably by low CSTR biomass retention and low hydraulic loading rates. On the other hand, immobilized-cell systems have become popular alternatives to suspended-cell systems in continuous operations because they are capable of maintaining higher biomass concentrations and could operate at higher dilution rates without biomass washout. Some other advantages of immobilized cells over free cells are improved biological stability, improved biomass holdup, improved product yields, improved reactor choice, improved downstream processing, improved product stability, improved reaction selectivity, advantages due to cell proximity and advantageous partition effects. The immobilization of viable microbial cells for fermentation purposes has been practiced within the fermentation industry since the introduction of the "quick" process for vinegar production by Schetzenbach in 1823. The major objective of the advancement of the biohydrogen production processes is to improve the hydrogen yield to make the process economically viable. In this regard, immobilization seems to be a promising approach since significant increase in volumetric hydrogen production could be possible in immobilized bioreactors [2]. The immobilized-cell system is also gifted with a feature of creating a local anaerobic environment, which is well suited to oxygen-sensitive

***Address correspondence to Nuri Azbar:** Department of Bioegineering, Ege University, Turkey; E-mail: nuriazbar@gmail.com fermentative H_2 production. Continuous and stable hydrogen production was achieved by immobilizing the combined system of *Phormidium valderianum*, *Halobacterium halobium* and *Escherichia coli* in a PVA-alginate film. The intermittent supply of nitrogen was found to be essential to retain cellular metabolic activities, which in turn showed prolonged production of hydrogen [3].

The numerous techniques for achieving immobilization can be categorized according to the physical process involved, namely; attachment, entrapment, containment and aggregation [2]. From this point of view, biomass immobilization can be achieved through forming granules (self immobilization), biofilm forming either by attachment on inert support materials or entrapment in porous materials, and containment *via* gel-entrapped bioparticles or membranes. In this regard, potential reactor configurations employed are CSTR, fixed- or packed-bed reactor, fluidized bed reactor, upflow anaerobic sludge blanket (UASB) reactor, gas lift reactor and trickling biofilter. From the viewpoint of reactor configuration, those continuous hydrogen-producing reactors can be divided into two types: tank-shaped reactor and column-shaped reactor.

Classification of cell immobilized cell systems according to the physical localization and the nature of micro environment is given in Fig. **1**.

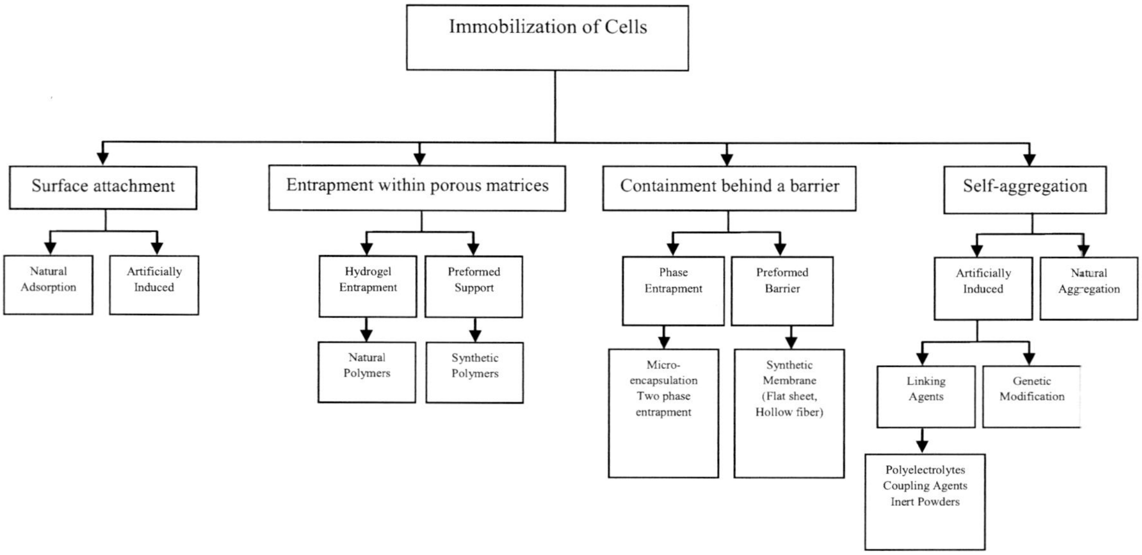

Figure 1: Classification of cell immobilized cell systems according to the physical localization and the nature of micro environment [4].

Most cells are naturally immobilized one way or another, so immobilization provides support for cells. The first point to take into account in selection of support material should be its intended use and application. Some of the points to consider while making a decision are summarized in Table **1**.

SURFACE ATTACHMENT OF CELLS

All forms of immobilization in which cells are in some way bound to the surface of a solid support come into the category of attachment or biofilm forming. Natural adhesion of cells to surfaces seem to be most accepted mechanism [6].

The adsorption of cells to an organic or inorganic support material is governed by Van der Waals and ionic interactions. Immobilization of cells can be achieved naturally or induced artificially by using linking agents. Metal oxides or covalent bonding agents such as glutaraldehyde or amminosilane can act as linking agent in artificial immobilization.

Table 1: Fundamental Considerations in Selecting a Support and Method of Immobilization [5].

Property	Point for considerations
Physical	Strength, noncompression of particles, available surface area, shape/form (beads/sheets/fibers), degree of porosity, pore volume, permeability, density, space for increased biomass, flow rate, and pressure drop.
Chemical	Hydrophilicity (water binding by the support), inertness toward cell, available functional groups for modification, and regeneration/reuse of support
Stability	Storage, residual enzyme activity, cell productivity, maintenance of cell viability, and mechanical stability of support material
Safety	Toxicity of component reagents, health and safety for process workers and end product users, specification of immobilized preparation for food, pharmaceutical, and medical applications
Economic	Availability and cost of support, chemicals, special equipment, reagents, technical skill required, environmental impact, industrial-scale chemical preparation, feasibility for scale-up, continuous processing, effective working life, reusable support.
Reaction	Flow rate, cell loading and catalytic productivity, reaction kinetics, side reactions, multiple cell systems, batch, CSTR, FPBR, FBR and so on; diffusion limitations on mass transfer of cofactors, substrates, and products.

Attachment is one of the oldest and simplest immobilization techniques. Sand and other active porous materials such as coke, wood, etc were widely used [7]. The support particles (< 1mm in diameter) provide a very large surface area for attachment per unit volume especially when used for fluidized bed bioreactors. Cells adhere to the solid surfaces and establish an active biofilm layer as they grow. The biofilm thickness may be as little as a monolayer of cells or may be as much as several millimeters in the case of wastewater treatment organisms [8].

This type of immobilization is performed directly in the bioreactor by introducing both inoculums and carrier to the medium. The carrier is washed to remove no adhering cells before commencing fermentation. Thus, the procedures are simple and inexpensive and the conditions gentle on the cells. The disadvantage lies in the vulnerability of the weakly bound biofilm to detachment from the support.

In regards to the particulate material in the influent, while self immobilizing bacteria requires no or minimum particulate materials in influent, for anaerobic digestion, biofilm reactors, either in upflow or in downflow mode, have proved successful in dealing with particulate feedstocks. It is likely that the preferred reactor for particulate substrates will be the biofilm reactor. For lower volumetric hydrogen production rates, it would be the CSTR system.

The type of immobilization material plays a significant role in production of hydrogen gas. It could be the key factor in determination of operating conditions, long term stability of the system, retrieving produced hydrogen gas, production yield and rate. The desired characteristics of the immobilization materials are large surface area for holding high biomass, high void fraction for easy retrieving the produced gas, better liquid to microorganism contact to decrease the diffusion limitation. Surface adsorption of hydrogen producing sludge has been examined by using different support materials such as activated carbon [9-11], polyurethane-activated carbon [12], granular activated carbon [10,12-14], glass beads [15], a mesoporous pure silica SB15 [11], expended clay [10,16,17], loofah sponge [10], inert stone chips [18,19], polyurethane foam [20] and porous ceramic fittings [21]. Mohan *et al.* [11] compared hydrogen production by SB15 activated carbon and expanded clay. It was pointed out that SB-15 was superior to PAC and GAC in terms of production yield and rate. Activated carbon and expanded clay provided high biomass yields and hydrogen production volume whereas loofah sponge was not efficient in biomass immobilization and hydrogen generation [10]. Hydrogen production potentials of expended clay and polystyrene were compared in another study and no significant difference was observed [22]. Basile *et al.* [23] investigated the effect of surface charge of four different hydrogels as surface immobilization matrix in biohydrogen production. Cationic hydrogels improved cell attachment resulting in faster substrate consumption and hydrogen production. Fritsch *et al.* [21] stated that the hydrogen supersaturation in the liquid phase could be a limiting factor in the hydrogen generation. Porous ceramic fittings with high specific surface was found very effective in enhancing the bubble formation and mass transfer of hydrogen out of liquid phase. As a result, the dissolved hydrogen concentration in the liquid phase was decreased. Some organic materials as pine wood shavings [13], rice straw, bagasse and coir [24] served as microbial support material to hydrogen producing biomass. Pine wood shavings resulted in extremely higher hydrogen production than the other support particles as perlit and GAC. The wood shavings were also consumed as carbon sources in the absence of any other external carbon source. The study on immobilization of *Enterobacter cloacae* IIT-BT 08 on rice straw, bagasse and coir in packed bed reactors indicated that coir was the best carrier compared to rice straw and bagasse in terms of cell retention (0.44 g dry cell/g dry carrier), packing density (100 g/L reactor volume), cell loading (44 g dry cell/L reactor volume), and hydrogen production rate (1.39 L/h L). This is largely attributed to a higher surface area provided by coir because of its fibrous and corrugated properties. These results indicated that packing materials of higher surface area are preferred. However there is no conclusion about utilization of these lignocellulosic materials as carbon sources by the immobilized culture [24].

Washout of support materials might be an intrinsic drawback of biofilm processes. In a recent study, Zhang *et al.* [14] investigated hydrogen production by granular sludge and biofilm sludge growing on GAC in two fluidized bed reactors at a pH of 5.5 and an OLR of 40 g-glucose/L h. A close performance in hydrogen production was observed while two types of immobilized cultures were tested at different HRTs and influent substrate concentrations. However, biofilm sludge was washed out substantially and reactor biomass was replaced by granular sludge after 50 days of operation, albeit the hydrogen production was not influenced during the transition. As washout of support carriers is not a concern of granular sludge systems, this poses to be an advantage of the systems. Thus, support material has important effects on biomass immobilization and consequently hydrogen production in biofilm reactors.

Kim *et al.* [25] investigated hydrogen production by biofilm of hydrogen-producing bacteria formed on the surface of hydrophilic poly vinyl alcohol (PVA) and polymer-induced granular sludge in a stirring reactor. The reactor system using granular sludge produced much more biogas for a longer time than the reactor system with microorganisms attached to the PVA (ca. 300 mL/(L reactor h) *versus* 50 mL/(L reactor h). This was presumably attributed to the facts that granular sludge had a higher density than PVA attached with biofilm. Biogas accumulated in the PVA pores would result in PVA flotation, which caused a poor contact between microorganisms immobilized inside the PVA and substrate. Moreover, more microorganisms were harbored by granular sludge than PVA with biofilm (approximately 7000 mg/L reactor versus 4000 mg/L reactor). These results indicate the granular sludge reactor appears to be a better choice for high-rate fermentative hydrogen production.

Rapid buildup of hydrogen producing bacteria on biofilm would result in a system upset because of the mass transfer limitation. Working on a trickling biofilter, for example, Oh *et al.* [26] noted that microbial growth of hydrogen-producing bacteria was too excessive under a mesophilic condition, which caused system upset just after 1 week of operation.

Finally, a large amount of carrier materials is normally required to support microorganism growth in biofilm processes. On the one hand, supporting materials occupy a considerable space in the reactor, which reduces the effective volume and hence reactor performance and efficiency. Beside, supporting materials need to be replaced periodically due to wear and tear. The cost of material replace could be a major economic consideration of maintenance.

Entrapment- within Porous Matrices

A variety of macro and micro porous materials (brick, ceramics, sintered glass, activated carbon, reticulated foam, stainless steel wire, etc) could be used for the entrapment of cells. The effectiveness of the immobilization varies with the cell type and support material type. Unlike attachment type biofilm forming, cells which have been entrapped in a preformed porous supports are protected from the shear field outside the particles but, like attached cells, they are not confined to the particle by any carrier. The main advantage of this technique is that the immobilized biomass hold-up can be controlled such that cells growing beyond the boundaries of the particle are removed by abrasion, either by the flow field surrounding the particle or by particle-particle contact [1].

One of the other most popular form of cell immobilization in use in research laboratories is the entrapment of cells within porous structures which are formed *in situ* around the cells (**hydrogel entrapment**). The cells, in the form of slurry or paste, are generally mixed with a compound which is then gelled to form a porous matrix under conditions sufficiently mild so as not to affect the viability of the cells [27]. The majority of techniques involving in situ entrapment for fermentation purposes make use of polysaccharide gels. Of these, which include K-carrageenan, agar and alginates, gellan gum, hyaluronic acid, gelatin, calcium alginates gel and LentiKats® (entrapment in hydrogel particles) are the most popular ones. Inferior performance of gel-immobilized sludge processes is mainly attributed to the low mass transfer efficiency. Furthermore, stability and durability of gel-entrapped bioparticles are generally questionable. Cell growth within the particles can occur, though if the cell concentration exceeds approximately 30% v/v the gel will lose its integrity. Calcium alginate gels are unstable in the presence of calcium chelators, such as phosphate, and gas evolution within the gels can also be a cause of particle disruption. In addition to that, immobilization matrix is in most cases highly biodegradable.

The study on hydrogen production by cell entrapment in alginate beads indicated that the bead diameter and the biomass density in the beads are the crucial factors. Increasing bead diameter up to 6 mm provided more anaerobic conditions which favored hydrogen production efficiency. Larger bead diameters limited the availability of substrate and the exchange of product. The accumulated end products inside the beads inhibited the hydrogen production. High cell density in gel beads enhanced hydrogen gas production but gel stability in long term operation was a problem to be considered [28]. The cells were entrapped in flatbed agar gel sheets in order to solve gas retrieving problem associated with beads [29]. However, it was stated that the sheets may not be suitable for high cell densities more that 8 mg dry cells/ml agar because of the difficulty in the preparation and fragility of the flatbeds. Hence, gel immobilization techniques may not be the technology of choice for fermentative hydrogen production.

Biofilm attachment on solid or porous supports has an advantage in improvement of hydrogen production rate compared to gel-entrapped bioparticles. On the other hand, in order to overcome aforementioned problems of these natural polymers, polyvinyl alcohol (PVA) based matrices are suggested. PVA is a synthetic polymer that can be used to form hydrogels by the conventional freeze–thawing method and leads to particles with excellent mechanical properties. In contrast to biopolymers PVA hydrogels are hardly biodegradable and show an excellent mechanical stability. A large elongation at break (350–450%) and even in long-term fermentations, no significant abrasion could be observed. A new method to produce lens-shaped hydrogels based on this material at very gentle conditions, room temperature, and short time has been developed [30]. A LentiKat combines the advantages of small and large particles: minimized diffusion limitation caused by the thin structure is linked to easy separation because of the large diameter of the biocatalyst. It can be retained easily in the bioreactor (*e.g.*, by a sieve). Mechanical stability of particles in extended operation period is one of the parameter taken into account. Mechanical stability and hydrogen production activity of entrapped cells in polymethyl methacrylate (PMMA) and polyethylene –octene –elastomer (POE) and sodium alginate beads indicated that POE –based entrapped sludge gave the highest yield and rate of production, provided better mechanical strength by maintaining 90% production activity in relatively long term operation [10]. PMMA supplemented with collagen and activated carbon support material to modify the physical properties of immobilized cells. The batch hydrogen production by this support material provided 232 ml H_2/L production rate and 2.25 mol H_2/mol sucrose [31]. A novel immobilization matrix was developed by Wu *et al.* includes entrapment of H_2 producing sludge activated carbon slurry in ethylene-vinyl acetate (EVA) [9]. Silicone gel was used instead of EVA in another study [32]. The immobilized discs in both studies were intact, mechanically stable with good settling efficiency for prolonged batch operation.

The mechanistic explanation in improvement of hydrogen production resulting from entrapment within specific gels was drawn by Zhu *et al.* [33]. The authors investigated hydrogen production by *R. sphaeroides* RV cells entrapped in chitosan agar with ammonium (which is an inhibitor/repressor of nitrogenase) as the nitrogen source. The results showed that the positively charged gel resisted the diffusion of ammonium and acted as a pH buffer to some extent, thereby enhancing hydrogen production in the presence of ammonia. This effect is particularly useful for wastewater containing significant ammonium.

Containment

When cell separation from the effluent is required or when some high-molecular-weight or specific product needs to be separated from the effluent, these systems are highly useful. Cells may be immobilized by containment behind a barrier, either preformed or formed in situ. The barrier may be as simple as interface between to immiscible liquids in which case the cell suspension is emulsified with an organic solvent and re-suspended as droplets in an aqueous phase or may involve semi permeable membranes used for microfiltration and ultrafiltration [34,35].

The barrier can be either preformed (hollow fiber systems and flat membrane reactors) or formed around the cells to be immobilized (microcapsules and two-phase entrapment). The synthetic membranes are usually polymeric microfiltration or ultrafiltration membranes, although other types of membranes have been used such as ceramic, silicone rubber or ion exchange membranes. Not only pore size and the structure control the mass transfer through the membrane but the hdyrophobicity/hydrophilicity and charge is also important. Transport can be governed by diffusion and/or by flow induced by application of a pressure difference over the membrane.

Entrapment behind preformed membranes represents a gentle immobilization method since no chemical agents or harsh conditions are employed. Cells are often immobilized by filtration of a cell suspension followed by some growth in the seeded reactor. Two-phase systems can also be used in applications where substrates or products are partitioned separately. The recycling of the cell-containing phase is also possible with these systems.

Efforts to maximize the efficiency of immobilized system have been directed to reducing the transfer limitation of liquid nutrients. A patented technique describes the containment of cells in the space between two layers of microfiltration hollow fiber membranes with a radial width of less than 0.2 mm [36]. The nutrients are fed through the two fibers by diffusion. The width of the space is a key factor controlling the supply of substrate. Increasing the transfer of nutrients to the cells by convection is a solution in overcoming mass transfer limitation, and has been proved successful [37,38]. Alternatively, a higher cell density not limited by mass transfer can also be attained through the use of a cell recycle bioreactor [39,40].

Micro-encapsulation is another method of containment type cell immobilization. This technique involves the formation of a spherical gel mold containing cells, on which is deposited a polymeric membrane. Capsule diameters from 20 μm to 2 mm are possible. Micro-encapsulation can provide simultaneous product separation and cell cultivation, resulting in concentration of high-molecular weight metabolic products within the capsule. The immobilized cells are totally separated from the culture medium, which results in an easier and cost effective downstream processing. This is not possible with conventional gel immobilization growing cell systems, where cell release from the matrix is observed [41].

Self-Aggregation

Also called "Cell to cell attachment" technique is most usually achieved by significantly increasing the effective size or density of the cells by aggregation. Thus flocculated cells in the form of large aggregates can be considered to be immobilized if the flocs can be separated from the bulk liquid by course screen or rapid sedimentation. Self-aggregation can be natural or artificially induced by cross linking agents. Self flocculation of biomass provides certain advantageous over entrapment and surface adsorption methods. It is support particle free, cost effective and more applicable to the large scale hydrogen production systems. The granules formed by self flocculation are stable along the long term operation. The only drawback of granular sludge systems is long time requirement for granule formation.

Anaerobic sludge granulation is a widely used self-immobilization method in anaerobic digestion. In an up flow anaerobic sludge bed (UASB) reactor, sludge agglutinates into granules, which results in an increase in biomass concentration and a reduction in sludge washout. The granules allow the organic loading rate of the UASB reactor to far exceed the typical loading rates applied in conventional activated sludge processes. Granules also have many advantages over other systems, which contribute to the success of the UASB design. First, the granules have faster settling velocity, which explains the reduced sludge washout. Second, the granules provide a protective structure for microorganisms in a harsh environment, which ensures stable operation even if environmental shock occurs [42]. Third, in an anaerobic digestion system, granules are formed naturally and also have a porous structure, which is ideal for mass transfer of the nutrients required by the microorganisms and for the biogas being produced. And finally, it has been found that hydrogen-producing biomass can develop into granules with high bioactivity [43].

Understanding the granulation process is highly relevant for improving reactor design and performance. To achieve the purpose, the mechanisms for granulation should be understood fully. Due to absence of initial cores for cell attachment, microbial granulation is preceded with a self adhesion or aggregation of microbial cells [44]. This process thus can be defined in terms of energy involved in the interaction of cell to cell and is governed by the surface physicochemical characteristics of microbial cells. In a thermodynamic sense, microbial sludge stability is governed by a charge balance among several repulsive forces which include electrostatic, salvation (hydration) and steric forces, and attractive forces including Van der Walls, short range hydrogen bonds, and electrostatic forces [45]. Some physicochemical models therefore have been proposed, for example, secondary minimum adhesion model [46,47], extracellular polymers (ECPs) bonding model [48], inert nuclei model [49] and ion-bonding model [44]. The granulation of microbial cells is a complicated process, in which biological, microbiological and hydrodynamic factors are also involved other than physicochemical forces. For example, different models and hypotheses based on aforementioned factors have been proposed for anaerobic granulation, including structural models, proton translocation-dehydration theory, cellular automaton model and cell-to-cell communication model [41,43,44]. The formation and mechanisms of conventional granulation of anaerobic sludge in UASB reactor have been well documented [46-48].

Hydrogen-producing sludge has been shown to agglutinate into granules after 60 days of operation in continuous-flow tank reactors. Furthermore, granular sludge has been reported to be visible in the UASB systems after 120 days of reactor operation [50]. Rapid and efficient granular sludge formation has also been found in carrier-induced granular sludge bed (CIGSB) bioreactors with the addition of support carriers, especially activated carbon [46]. Overall, without the carrier inducement, direct granulation of hydrogen-producing bacteria is a time-consuming process, and there are many unknown factors that need to be investigated during the process. Long startup period of a several months is needed for complete development of granules [43,51]. For example, during the startup of a UASB hydrogen-producing reactor, Mu and Yu [51] found that small granules were formed with diameters of 400–500 μm at the reactor bottom after 140 days of operation which developed rapidly to 2.0 mm after 60 more days. Although the UASB reactor had reached steady state conditions in terms of hydrogen production and substrate degradation after 5 months from the startup, mature and stable granular sludge were only completely developed and accumulated beyond 8 months of operation.

One possible source for hydrogen producing granules is a conventional UASB reactor treating wastewaters. There are thousands of UASBs running worldwide, and once every 2 to 3 months, part of the methanogenic granules inside each of these reactors must be disposed of to maintain the reactor's efficiency. In previous research, chloroform treatment of granules was found to effectively eliminate methane production and to convert the culture into hydrogen production.

Commercial methane-producing granule-based UASB reactors are undesirable when operating on substrates with high suspended solids contents [52]. The same rationale might be applicable to hydrogen-producing granular sludge reactors.

Granular sludge seems to demonstrate an advantage over the biofilm sludge. First, microbial kinetic constants of hydrogen producing bacteria revealed that the specific growth rate and biomass growth yield are very high compared to methanogens. The maximum specific growth rate and biomass growth yield were reported to be 0.17–0.5 h^{-1} and 0.08–0.33 g VSS/g COD, both of which indicate that hydrogen-producing bacteria would increase rapidly if a higher organic respectively, loading was employed. The OLR for immobilized sludge hydrogen production was reported as high as 80 g glucose/L h [32].

Literature on initiating granulation of a hydrogen-producing culture and the formation mechanisms involved is somewhat limited. There are only few studies reporting rapid formation and accumulation of granular sludge. Zhang *et al.* [53] developed a rapid approach of acid incubation to initiating the formation of hydrogen-producing granules in a CSTR. In the study, hydrogen-producing granules were formed rapidly within 114 h as the seeded microbial culture was subjected to a 24-h period of acid incubation at a pH of 2.0. Formation of granular sludge facilitated biomass retention of up to 32 g VSS/L in 3–5 days which enhanced hydrogen production. It has been highlighted by the authors that changing the culture pH results in improvement in surface physicochemical properties of the culture favoring microbial granulation [54].It was interesting to note that a superior hydrogen production performance was demonstrated by the granular sludge induced by acid incubation and support carriers compared to the self flocculated granules. Lee *et al.* [46] demonstrated that packing of a small quantity of carrier matrices at the bottom of the upflow reactor significantly stimulated sludge granulation that can be accomplished within 80–290 h in a novel carrier-induced granular sludge bed (CIGSB) bioreactor. The column reactors were initially packed with cylindrical activated carbon, spherical activated carbon, sand, or filter sponge at a bed height of 4–8 cm and bed porosities of 90–99%. Granulation of the seed sludge took place in all the carrier-packed reactors as the HRTs were shortened to 4–8 h, dependent on carrier type. The role of carriers in granular sludge formation is still unclear, but it is presumed that the biofilm fraction detached from the carrier surface may serve as starting nuclei to trigger the massive growth of granular sludge. In a recent study conducted by Wu *et al.* [32] it was found that formation of self-flocculated granular sludge occurred during operation at a short HRT in a CSTR, in which gel-entrapped bioparticles were initially loaded. It was concluded that a similar mechanism with gel-entrapped bioparticles working as the inducing carriers concluded the granulation of hydrogen-producing bacteria. By adding cationic polymer (cationic polyacrylamide) and anionic polymer (silica sol), Kim *et al.* [25] reported a rapid granulation of hydrogen-producing culture which was accomplished within 5 minutes. The detailed procedures are described as follows: sludge itself has a negative charge of -26 mV. To make the sludge neutral, high molecular weight cationic polymer (cationic polyacrylamide, molecular weight 15,000,000) of 0.7% (w/w) of dry sewage digester sludge was added to the sludge and stirred at 200 rpm for 2 min. Because residual cations may cause a detrimental effect on the microorganisms, anionic organic materials (anionic silica sol) of 0.7% (w/w) of dry sewage digester sludge were added and stirred at 200 rpm for 2 min. The total time required for the granulation was therefore about 5 min. When granular sludge was operated in a stirred reactor, the granular shape was maintained, its size ranging from 1.0 to 3.0 mm, and the maximum concentration of the granular sludge was found to be granule size ranged between 1.0 to 3.0 approximately 7 g/L. O-Thong *et al.* [54] used heat treated methanogenic granular sludge as support material for immobilization of thermophilic anaerobic organism, *Thermoanaerobacterium thermosaccharolyticum.* A thick and high density biomasses around the granules were observed and the biomass concentration reported as in the range of 12-16 g VSS/L [54]. The hydrogen production rates of spontaneous granular sludge in UASB reactors and CSTRs were reported in a range of 0.16–0.54 L/L h [27,43,51] In contrast, a much higher hydrogen production rate ranging from 3.20 to 15.03 L/(L h) was reported in the systems with acid or carrier induced granular sludge [28,32,46,53]. This was likely attributed to the difference in microbial composition of hydrogen-producing cultures. *C.*

pasteurianum has been regarded as the bacterial species with the best hydrogen production performance of granular sludge. Wu *et al.* [32] pointed out that the *C. pasteurianum* is likely the "superstar" hydrogen-producing species in the mixed bacterial consortium in their study. In fact, the bacterial hydrogen production activity, in terms of high specific hydrogen production rate, of 19.60 mmol H_2/(g VSS·h) was reported in the study, which was much higher than the values reported in other immobilized-cell systems. Although such a "superstar" species, *C. pasteurianum,* was found to be the only microbial strain in the present study, the highest specific hydrogen production rate achieved by suspended sludge was estimated to be 15.62 mmol H_2/(g VSS·h).

On the other hand, carriers have been successfully used in various systems for wastewater treatment in both lab-scale and full-scale systems as the solid support for microbial attachment [55,56]. Zheng *et al.* [57] showed that biohydrogen could efficiently be produced in glucose-fed biofilm reactors filled with plastic carriers and operated at 70°C.

Immobilized Cell Bioreactor Configurations

Bioprocesses for hydrogen gas production by immobilized systems mainly include fixed bed, fluidized bed, expended bed and granular sludge systems (Fig. **2a-f**). As an alternative approach to immobilized systems, continuously stirred tank reactor (CSTR) containing immobilized anaerobic sludge was used for hydrogen production [9]. Anaerobic hydrogen gas production potentials of these bioprocesses by batch, sequential batch, semi-continuous and continuous operations have been investigated. Cumulative hydrogen gas volume, yield and rate of hydrogen gas formation significantly varies with environmental conditions, microbial culture, substrate or carbon sources, bioprocess type, immobilization matrix, organic loading (OLR), hydraulic retention time (HRT) and operation mode.

Immobilized Cell-Continuous Stirred Tank Reactor (CSTR)

The CSTR provides complete mixing by the use of impellers mounted on a rotating shaft (Fig. **2a-b**). The liquid components of a CSTR reactor are assumed to be homogeneous in composition, which has equal concentrations in both effluent and influent of the reactor. With immobilized cells, high fluid velocities and high stirring rates needed to achieve a constant supply of substrate and product removal. Efforts to obtain well mixed conditions may result in abrasion damage to the immobilized catalyst because of high shear forces at the impeller surface. Lacroix *et al.* [58] reported that 4 to 5% of the beads containing immobilized cells in carrageenan and locust bean gum in a CSTR was destroyed by shear, even if the medium was supplemented with KC1.Nevertheless, the CSTR offers the best mixing characteristics and mass transfer. The substrate concentrations in a CSTR are typically lower than the packed bed and fluidized bed reactor, resulting in lower average reaction rates. However, a lower substrate concentration may be advantageous for inhibited cell culture.

Hydrogen gas production by self flocculating sludge in CSTR was studied by Zhang *et al.* [2]. It was reported that liquid stirring significantly influenced biomass retention of granular sludge in the system [2]. A severe washout of granular sludge can occur when the culture was mixed at a stirring rate of 280 rpm, and lowering the stirring rate would favor biomass retention. However, granular sludge could not be mixed thoroughly, and a static granular sludge layer was formed at stirring rates below 160 rpm. At a stirring rate of 160 rpm, a consistent biomass layer of granular sludge was formed which accounted for about 47% of total reactor working volume (6 L) and corresponded to an overall biomass concentration of 16.0 g VSS/L. Improvement on reactor configuration and proper control may increase the reactor performance.

A specific immobilization matrix of POE was added into CSTR system for simultaneous hydrogen gas and ethanol production from sucrose [8]. The maximum yield and rate were obtained as 1.18 mol H2/mol hexose and 37 mmol/h/L at HRT =4 h and 20 g COD/L. For shorter HRTs, ethanol production enhanced with decreased hydrogen production. It was concluded that bacteria population of mixed culture was dominated by hydrogen producer when operating at HRT= 8h -2h. However, the population shifts to ethanol producers when HRT shortened from 1 h to 0.5 h. These rates and yields obtained in CSTR system are higher than ones obtained in fixed bed reactor with the same immobilization material and substrate [59]. The hydrogen production potentials of CSTR and ICSTR were compared by Wu *et al.* [60]. Addition of immobilized sludge substantially improved hydrogen production and biomass holding capacity with marked decrease in HRT. Table 2 depicts the rate and yields of hydrogen gas production in CSTR and ICSTR systems.

Table 2: Rates and Yields of Hydrogen Gas Production from different Carbon Sources in Anaerobic Continuous Stirred Tank Reactor Containing Immobilized Culture.

Biofilm Process	Culture	Substrate Type	Concentration of Substrate	Support particle Type	HRT	H_2 production rate	H_2 production yield	Ref
ICSTR	Mixed culture	Sucrose	20 g COD/L	POE	4 h	37 mmol/h L	1.18 mol/mol hexose	[9]
ICSTR	Mixed culture	Glucose	20 g COD/L	POE	4 h	7.14 mmol/h L	0.13 mol/mol hexose	[9]
ICSAB	Mixed culture	Glucose	20 g COD/L	Silicone gel	4 h	0.6 L/h L	0.87 mol/mol glucose	[55]
CSTR	Mixed culture	Glucose	20 g COD/L	None	10 h	0.6 L/h L	1.63 mol/mol glucose	[55]

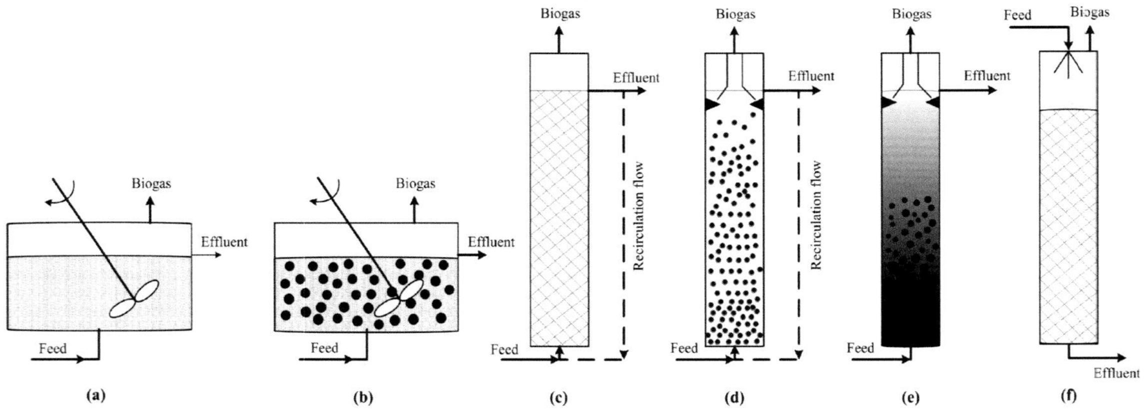

Figure 2: Schematic diagram of hydrogen-producing reactors: (a) conventional CSTR containing suspended sludge, (b) immobilized-cell CSTR, (c) fixed/packed-bed reactor, (d) fluidized bed reactor, (e) UASB reactor, and (f) trickling biofilter [59].

Fixed/Packed Bed Reactor (FPBR)

In a FPBR, the biocatalyst particles are packed into a column through which the substrate solution passes (Fig. **2c**). If the fluid velocity profile is perfectly adjusted, the FPBR operates as a plug-flow reactor that is an ideal behavior. Productivity of the PBR for a given biocatalyst depends on the type of immobilization, whether by entrapment or surface attachment. Higher cell loadings are often achieved by entrapment, resulting in improved productivity. FPBRs have the advantage of simplicity of operation. Anaerobic conditions are easily maintained and may be improved by sparging the substrate solution with nitrogen (should be avoided for photofermentation) or argon (preferred one). High rates of reaction may be achieved throughout the length of the PBR. Lower regions receiving highly concentrated substrate medium exhibited an almost complete substrate conversion [59] in contrast to higher levels in the column. The rate of external mass transfer of substrate and product in a PBR is typically less than that of a CSTR; thus, reaction rate may be limited by mass transfer caused by external diffusion limitation, especially at low substrate concentrations for some immobilization materials (e.g polyacrylamide). Godia *et al.* [61], on the other hand, showed external transfer was not the rate-limiting factor when the gel beads were used.

Gas accumulation during the process, may induce dead space in the reactor. Resultant channeling of fluids may affect substrate conversion. Gas build-up may also result in back-mixing and deviation from ideal plug flow behavior. Tapered column design, in which the cross-sectional area increased with height, could be used in order to provide a better gas removal and reduced pressure drop. Bed compaction also depends on the type of support particle. When attachment on an inert support by self-immobilization or by covalent binding is used, the compaction of the particles is less of a problem because of the stronger resistance to compression of the support. Sieve plate baffles could be incorporated in the reactor in order to reduce compaction problems.

The main parameters that affect the hydrogen production potential of fixed bed system are the characteristics of immobilization material, organic loading rate or hydraulic retention time. Fixed bed system has been widely used in biohydrogen production. Chang *et al.* [10] investigated the effects of immobilization materials and hydraulic retention time on hydrogen production from sucrose (20 g COD/L). It was observed that biomass holding capacity of activated carbon (5.53 g VSS/L) is considerably higher than expended clay (1.85 g VSS/L). The optimum HRT for maximum production rate (1.32 L/h L) was determined as HRT=1 h with butyric acid (HBr) as the main end product. Lower HRTs resulted in cessation of hydrogen production with elevated percentage of CO_2 in the gas phase because of inactivation of hydrogen producers and substrate inhibition effect [9]. Similar result in terms of the optimum HRT was observed when lignocellulosic materials as rice straw, bagasse and coir were used as immobilization surface. The hydrogen production rate increased with increasing dilution rate to D= 0.9 h^{-1} (HRT = 1 h) for all matrixes. The maximum rate was obtained as 62 mmol/h L with coir. Recirculation of effluent increased the production rate to 75.6 mmol/h L with 9.4 h^{-1} dilution rate [13]. However, Jo *et al.* [20] reported that yield and rate significantly decrease when a packed bed system with *C. tyroboytricum* immobilized on polyurethane foam was operated at HRT=1. The accumulation of lactate occurred and a sharp decreases in pH to 4.6 at HRT=1 h was observed due to no reduction of protons to hydrogen gas [17]. Wu *et al.* [59] studied on POE matrix in a packed bed reactor. The maximum production rates for sucrose, glucose and fructose with initial substrate concentrations equal to 20 g COD/L were obtained at HRT=4 h as 10.3 mmol/h L, 16.1 mmol/h L, 7.2 mmol/h L, respectively. The production rates are lower compared to the rate obtained for the same substrate concentration by Chang *et al.* [10] probably because of simultaneous ethanol and hydrogen production [9]. When entrapped cells in PMMA with supplementary materials were used, wash out of suspended biomass and unstable operation of system occurred at HRT=2 h. The optimal HRT was HRT= 6 h with 1.80 L/h L rate and 2 mol/mol sucrose production yield [30]. It was stated that PMMA is a strong matrix and displayed a very stable performance for 6 months in continuous operation. Hydrogen production in fixed bed system containing porous polyurethane- activated carbon support particle with immobilized *C. tyrobuytricum* resulted in the longest HRT (16.7 h) reported so far in fixed film systems for the maximum production yield of 326 ml H_2/g glucose [11]. Long HRT requirement is probably because of high organic loading rate applied with 50 g/L initial glucose concentration. Fristch *et al.* [21] pointed out the importance of hydrogen supersaturation in the liquid phase for the microbial behaviour in biohydrogen production. Therefore, squared shaped ceramic fitting with large surface area was suggested to enhance gas retrieve. The improvement in the production yield and rate was 30% and 35 %, respectively, with ceramic fittings in comparison to the production in standard column [20]. Table **3** summarizes the rates and yield of hydrogen gas production from different carbon sources in fixed bed systems.

The reported hydrogen content in fixed film systems is in the range of 27%- 50% [9,20,30,31,59]. There is no common trend in H_2 content with variation in HRT, substrate concentration or substrate type. The H_2 content was around 39% for sucrose but it was lower for fructose and glucose at the same HRT [59]. Increasing sucrose concentration did not effect the gas composition [20,30,31]. A decrease in the hydrogen content from 37% to 30 % was observed when HRT was reduced from 2 h to 0.5 h [9]. It reached to peak value of 50% when HRT shortened from 24 h to 2 h with a sudden decrease to 40% at HRT=1 h [19]. However, almost constant H_2 content around 38% was reported for HRTs between 4h and 8h [30].

The optimum HRT varies depending on the support particle and bioprocess used in the production of hydrogen. Wu *et al.* [31] concluded that more mass transfer limitation occur in entrapped cells than surface attachment, self flocculation or in suspended cultures resulting in the difference in the effect of HRT on H_2 production. The most common conclusion about the effect of HRT on hydrogen production in fixed film systems is that the yield and rate of production respond reversely to the changes in HRT. The rate increases but the yield decreases when HRT lowered. The possible reasons that have been reported in the literature for this result are the inhibition of production because of over accumulation of hydrogen gas in the liquid phase [20,59] and increasing in the substrate concentration or organic loading rates as HRT decreased [9,13]. The other explanations are changes in the soluble product profile from HBr to HPr [59], shift in the microbial population from hydrogen producers to non-hydrogen producers when HRT decreased [6,7,57]. Leita *et al.* [62] observed that a gradual decrease from HRT= 2 h to HRT=0.5 h resulted in decreasing in butyric acid production without effecting concentrations of HAc and HPr. Similarly, HBr was the main end product above HRT=2 h and then HBr replaced by lactate at HRT=1 h [19]. Most of the continuous hydrogen production studies in fixed bed systems were carried out by using pure carbohydrate sources such as glucose and sucrose. Some studies reported that the substrate inhibition concentrations of these carbon sources were 2 % or above for glucose [13] and 20 g COD/L of sucrose [8,20,30].

Alkalinity was found to be a significant parameter affecting the production of hydrogen gas and organic acids. The average hydrogen production yield was 2.48 mol H_2/mol of glucose without $NaHCO_3$ and 2.15 mol H_2/mol of glucose with 1000 mg/L $NaHCO_3$ [62]. The adverse effect of $NaHCO_3$ could be due to the generation of CO_2 which is considered as inhibitory to H_2 production at high concentrations [63] or shift in the pH value from 4 up to 7.5 which corresponds to significant reduction in HBr concentration. Low yields of formations in some studies can be related with the high pH values as 6.7 [19,20].

Fluidized Bed Reactor (FBR)

FBRs provide conditions that are intermediate to those of the CSTR and PBR (Fig. **1d**). Mixing is better than in the PBR and lower levels of shear are encountered compared with the CSTR. A FBR consists in a column in which the biocatalyst particles are maintained in motion by a continuous flow of the substrate solution. The pressure drop of the fluid flow supports the weight of the bed. The FBR offers higher productivity than the CSTR because liquid approximates plug flow similar to the PBR. However, the FBR is more advantageous for fermentation with substrate inhibition than the PBR because of the mixing caused by fluid flow. In contrast with the PBR, the FBR facilitates solid-fluid mixing, gas removal, and minimizes pressure drop. FBRs can expand to accommodate growing biomass, so they are less sensitive to plugging. Compared with the CSTR, the FBR has the advantage of reduced chance of contamination because of the absence of a drive shaft seal. To achieve good fluidization characteristics, the density difference between the biocatalyst particles and the substrate solution should be as high as possible. Hydrated hydrocolloid gels have a similar density to the substrate solution. Thus, fluid velocity is often reduced to avoid expanding the bed, resulting in wash-out conditions. Changes in bead density are observed with time because of cell growth and gas formation inside the beads. The loss of immobilized cells by wash-out from the top of the column can be avoided by a mesh protection of the outlet or the provision of overflow weirs or expanded upper regions of the column where flow rate is reduced [64].

EVA was utilized to immobilize H_2-producing sludge for continuous production of H_2 in a three-phase fluidized bed bioreactor (FBR). It was stated that a proper packing ratio of the immobilized cells is a requisite design parameter for a successful three-phase bioH_2-producing bioreactor. Using a packing ratio of 10% resulted in a 23–30% higher hydrogen production rate over that for 5% packing ratio. It was concluded that more biomass loading enhanced the H_2 production rate. However, 15% packing ratio caused a decrease in both the H_2 production rate and yield. This result was evaluated as inefficient mass transfer arising from the improper immobilized cell packing ratio. The yield of production reached to maximum level of 4.26 mol/mol sucrose at HRT=6 h and at 10% packing ratio. Severe washout of suspended biomass occurred when the continuous bioreactor was operated at a short HRT of 2 h. [65]. Silicone –gel was successfully used in hydrogen production in batch operation [32]. However, in fluidized bed system, biogas formation around the beads forced them to float and the immobilized cells eventually accumulated towards the liquid surface on the top of the bioreactor during long-term operations, resulting in inefficient and unstable hydrogen fermentation. Therefore a modified draft tube fluidized bed (DTFBR) reactor was developed [66]. Although, the particles moving towards the top of the reactor were forced to move down into the draft tube by the bulk/local circulating liquid flow, a small portion of particles still floated on the top of the DTFBR reactor. A good bed motion and complete mixing in AFBR was achieved by applying high flowrate (Q= 128 L/h). There was low coalescence of rising biogas bubbles and biogas easily detached immobilized particles [17].

The yield and rate of H_2 formations respond to the changes in operation conditions differently in FBBR systems. Some studies concluded that increasing HRT or substrate concentration could lead to increasing in volumetric production rate but decreasing in the yield of formation. For instance, the maximum production rate (2.27 L/h L) occurred at 40 g COD/L and HRT= 2.2h whereas, the yield was maximum (4.98 mol/mol sucrose) at HRT= 8.9 at the same substrate concentration [66]. On the other hand, a significant increase in both HPR and yield was reported when HRT was shortened [14,17,22]. The rate and yield reached to maximum values of 2.22 L/h L and 1.16±0. mol/mol glucose, respectively, when HRT was shortened from 4 h to 0.5 h for expended clay support particle [14]. Similarly, it was observed that both yield and rate of formation increase with decreasing HRT from 8 h to 1 h which corresponds to increase in OLR from 15.7 to 116.6 kg COD/m d. The maximum values of yield and rate were 2.41 mol H_2 mol/glucose and 0.97 L/h L at short HRT of 1 h, respectively, with expended clay [17]. The findings of Barros *et al.* [22] and Shida *et al.* [16] were almost parallel with the results of Amorim *et al.* [17] for the same type of immobilization material except that optimal HRT that maximizes the yield was 2 h. These results can be interpreted as the effect of organic loading rate. A direct relationship between production rate and yield with respect to changes in organic loading rate was reported by Amorim *et al.* [17]. The observation of Zhang *et al.* [14] was linear correlation between the rate of formation and organic loading rate. However, there are also conclusions that no universal trend of dependence between yield and OLR [66]. In the light of these studies, short HRTs are more productive in terms of H_2 generation and the optimal HRT that maximizes the yield and rate has been selected as between 0.5 h-2 h in AFBR. Barros *et al.* [22] concluded that the minimum HRT needed to maintain specific hydrogen production from glucose conversion is related to maintaining adequate concentrations of H_2-producing microorganisms in the system and contamination of the system by non-H_2-producing organisms, which compete for substrate, must be prevented. The microbial analyses carried out in some efficient hydrogen producing AFBR systems revealed that the surface and interior of the support particles were covered by rod-like bacteria and endospores which were attributed to high biomass retention and presence of *Clostridum* sp. in the system [13,17,66]. Applications of intermittent thermal treatment when H_2 producing activity declined during long term continuous operation, were a solution to restore the hydrogen production [66]. The investigation about the relationship between instable H_2 production and microbial community structure in FBBR at dynamic state indicated that high hydrogen gas production occurs with the rapid growth of *Clostridium butyricum* but increasing organic loading rate enhances the growth of *E coli* rather than *Clostridium butyricum* [67]. A shift from acetate-butyrate to acetate-propionate production due to decrease in *C. butyricum* but enrichment of potential hydrogen consumers and propionate producers results in decline in the rate. The latter ones retained in the system regardless of changes in HRT because of their good adhesion ability on the support particle. It was suggested that the instability of H_2 fermentation in biofilm reactors is due to enrichment and efficient adhesion of H_2 consumers on the carrier. Therefore, Koskinen *et al.* [67] suggests that, biofilm reactors may not favour mesophilic H_2 fermentation. However, Shida *et al.* [16] was able to obtain efficient and stable H_2 production for over 178 days with 2 g/L glucose concentration at 30 °C operating temperature and at different HRTs. The system stability for long term operation was attributed to utilization of expended clay as support particle and the applied sludge pretreatment strategy which consisted of rapid heating and ice cooling to 25 °C [32]. The operation pH is another factor to favor the hydrogen production. The maximum rate and yield formations were obtained in effluent pH value less 4 in FBBR system [14, 16, 22]. But there are also reports about the maximum production in pH= 6.8. It is clearly seen that almost the same H_2 production yields can be achieved for pH range between 4-6.8 [22,66]. The summary of rates and yield of hydrogen gas production in anaerobic fluidized bed systems were given in Table **4**.

The H_2 content produced in fluidized bed system varies around 45% in general [16,65,66]. The highest hydrogen composition of 61.2% was achieved at HRT = 4 h and then a decrease to 57% was observed when HRT was shortened [14]. There are also results about relatively stable H_2 content in the range of 40–46% regardless of changes in HRT [65, 66]. In addition, the effect of substrate concentration on hydrogen content was reported as insignificant [14, 65, 66]. The substrate consumption was over 95% for wide range of HRT (4 h to 9 h) [65] and for initial substrate concentrations up to 20 g COD/L sucrose. Similarly, glucose conversion reached over 93% for the HRT between 8 and 1 h [16]. On the other hand, Zhang *et al.* [14] verified that glucose conversion decreased from 99.47% to 71.44% when HRT was decreased from 4 h to 0.5 h.

The major soluble metabolites during H_2 fermentation in FBBR were butyric acid (HBu) and acetic acid (HAc), accounting for 52–61% and 19–24% of total SMP, respectively. Moreover, the HBu/HAc (B/A) ratio is considered as a performance indicator for dark H_2 fermentation. This ratio ranged between 2.2 and 3.0 in FBBR [65] but slightly increased to B/A=2.9–4.3 with regard to variation in HRT and substrate concentration. Zhang *et al.* [14] concluded that the composition of dissolved metabolites can be associated with the variation in HRT. Metabolite concentration decreased slightly as the HRT was shortened, with the exception of butyrate, which initially increased (HRT from 4 to 1 h), but decreased with further decreases of HRT from 1 to 0.5 h. Beside to other metabolic products, hexanoate formation about 5÷10% was observed. The effect of HRT on soluble material composition was further investigated by Amorim *et al.* [17]. The results indicated that the predominant SMP was butyric acid (HBu) up to a HRT of 6 h, ranging from 37.7% to 44.9%. Acetic acid was produced in higher rates, ranging from 36.1% to 53.3% of the SMP when HRT=4 h. Acetate contribution increased as HRT decreased, despite a slight deviation in butyric acid percentage during HRT of 8 h and 6 h.

Table 3: Rates and Yields of Hydrogen Gas Production from Different Carbon Sources in Anaerobic Fixed Bed Systems.

Biofilm Process	Culture	Substrate Type	Concentration of Substrate	Support particle Type	HRT	H_2 production rate	H_2 production yield	Ref
FB	*Enterobacter cloacae* IIT-BT	Glucose	NA	Baggase	1 h	52 mmol/h L	NA	[24]
FB	*Enterobacter cloacae* IIT-BT	Glucose	NA	Rice straw	1 h	44mmol/h L	NA	[24]
FB	*Enterobacter cloacae* IIT-BT	Glucose	NA	Coir	1 h	62 mmol/h L	NA	[24]
FB	*C. tyrobuytricum* JM1	Glucose	5 g/L	Polyurethane foam	2 h	7.2 L/h L	223 ml/g hexose	[20]
FB	Mixed culture	Glucose	2 g/L	Expended Clay	2 h	NA	2.48 mol/mol glucose	[62]
FB	Mixed culture	Glucose	20 g COD/L	POE	4 h	16.1 mmol/h L	0.7 mol/mol hexose	[59]
FB	Mixed culture	Glucose	50 g/L	Porous polyurethane	8 h	22.9 L/h L	326 ml/g glucose	[12]
FB	Mixed culture	Sucrose	17.8 g/L	Activated carbon	1 h	1.32 L/h L	NA	[10]
FB	Mixed culture	Sucrose	17.8 g/L	Expended Clay	2 h	415 ml/h L	NA	[10]
FB	*C. butryricum* CG2	Sucrose	20 g COD/L	Porous ceramic	5.5 h	305 ml/h L	0.92 mol/mol hexose	[21]
FB	Mixed culture	Sucrose	20 g COD/L	PMMA	6 h	1. 80 L/h L	2 mol/mol sucrose	[31]
FB	Mixed culture	Sucrose	20 g COD/L	POE	4 h	10.3 mmol/h L	0.44 mol/mol hexose	[59]
FB	Mixed culture	Fructose	20 g COD/L	POE	4 h	7.2 mmol/h L	0.58 mol/mol hexose	[59]

POE;polyethtlene octene elastomer, MCR; monolith ceramic, PMMA; polymethyl methacrylate, NA: Not available

Table 4: Rates and Yields of Hydrogen Gas Production from Different Carbon Sources in Anaerobic Fluidized Bed Reactor (AFBR) and Draft Tube Fluidized Bed Reactor (DTFBR).

Biofilm Process	Culture	Substrate Type	Concentration of Substrate	Support particle Type	HRT	H_2 production rate	H_2 production yield	Ref
AFBR	Mixed culture	Glucose	4 g/L	Expended Clay	2 h	1.21 L/h L	2.59 mol/mol glucose	[22]
AFBR	Mixed culture	Glucose	4 g/L	Polsystrene	2 h	0.75 L/h L	1.90 mol/mol glucose	[22]
AFBR	Mixed culture	Glucose	2 g/L	Expended Clay	1 h	1.15 L/h L	2.10 mol/mol glucose	[17]
AFBR	Mixed culture	Glucose	10 g/L	GAC	0.25 h	7.49 L/h L	1.71 mol/mol glucose	[14]
AFBR	Mixed culture	Glucose	20 g COD/L	POE	0.55 cm/s*	37 mmol/h L	1.04 mol/mol hexose	[59]
AFBR	Mixed culture	Glucose	10 g/L	GAC	0.5 h	2.22 L/h L	1.16 mol/nol hexose	[15]
AFBR	Mixed culture	Glucose	2 g/L	Expended Clay	2 h	0.60 L/h L	2.49 mol/mol glucose	[18]
AFBR	Mixed culture	Sucrose	20 g COD/L	POE	0.55 cm/s*	45 mmol/h L	0.64 mol/mol hexose	[59]
AFBR	Mixed culture	Sucrose	20 g COD/L	EVA	6 h	0.89 L/h L	4.26 mol/mol sucrose	[65]
DTFBR	Mixed culture	Sucrose	40 g COD/L	Silicone gel	8.9 h	0.42 L/h L	4.98 mol/mol sucrose	[66]
AFBR	Mixed culture	Fructose	20 g COD/L	POE	0.91cm/s*	33 mmol/h L	0.56 mol/mol hexose	[59]

*; upflow velocity, EVA; ethylene-vinyl acetate, GAC; Granular activated carbon.

Upflow Anaerobic Sludge Bed Reactor (UASB)

As sketched in Fig. **1e**, the upflow anaerobic sludge blanket (UASB) basically is composed of self aggregated granular type biomass. Developed in Netherlands in late seventies (1976-1980) by Prof. Gatze Lettinga-Wageningen University, UASB reactor was originally used for treating wastewater from sugar refining, breweries and beverage industry, distilleries and fermentation industry, food industry, pulp and paper industry. UASB reactor is increasingly gaining popularity for high strength organic wastewater treatment because of its high biomass concentration and rich microbial diversity [48].

Originally, the UASB reactor is a methanogenic (methane-producing) digester composed of microorganisms that naturally form granules (pellets) of 0.5-2 mm diameter that have a high sedimentation velocity and thus resist wash-out from the system even at high hydraulic loads. This technology was evolved from the anaerobic "clarigestor" in South Africa. A similar but variant technology to UASB is the expanded granular sludge bed (EGSB) digester. UASB uses an anaerobic process whilst forming a blanket of granular sludge which suspends in the tank. Wastewater flows upwards through the blanket and is processed (degraded) by the anaerobic microorganisms. The upward flow combined with the settling action of gravity suspends the blanket with the aid of flocculants. The blanket begins to reach maturity at around 3 months. Small sludge granules begin to form whose surface area is covered in aggregations of bacteria. In the absence of any support matrix, the flow conditions creates a selective environment in which only those microorganisms, capable of attaching to each other, survive and proliferate. Eventually the aggregates form into dense compact biofilms referred to as "granules". The formation and mechanisms of conventional granulation of anaerobic sludge in a UASB reactor have been well documented [43].

One possible source for hydrogen producing granules is a conventional UASB reactor treating wastewaters. There are thousands of UASBs running worldwide, and once every 2 to 3 months, part of the methanogenic granules inside each of these reactors must be disposed of to maintain the reactor's efficiency. Various types of pretreatment methods including, thermal, acid and base, and application of inhibiting chemicals were reported to enrich hydrogen producers and eliminate methanogenic groups. For example, chloroform treatment of granules was found to be effectively eliminates methane production and to divert the culture into hydrogen production [68].

UASB systems have been widely applied to hydrogen gas production with some modifications to enhance granulations. Chang *et al.* [50] determined the optimal HRT as 8 h for maximum HPR (47. 2 L/d) and HPY (1.5 mol/mol sucrose) with 42 % hydrogen gas content from 20 g COD/L of sucrose in a conventional UASB system. Sucrose consumption was enhanced with increasing HRT but the consumed substrate was converted to biomass. Shorter HRTs resulted in lower production and washout of organisms [50]. A significant improvement in the production yield at lower HRTs was achieved in carrier induced granular sludge bed systems (CIGSB). Hydrogen production yield varied between 2.38 mol/mol sucrose and 3.88 mol/mol sucrose for all HRTs (0.5 h-4 h) examined indicating that hydrogen producing bacterial populations retained in the systems even at low HRTs [69]. Wu *et al.* [60] achieved almost the same yield of 1.54 mol/mol glucose at HRT= 4 h with activated carbon addition to enhance granulation. The performance of CIGSB system at HRTs less than 1 h was further increased by supplementation of Ca^{2+} ion to improve the granule formation. The size of the granule sludge enlarged from 2 mm to 3.5 mm. Four fold increase in the rate accompanied by a 2 fold increase in the yield with 3.5 fold increase in the biomass concentration at HRT= 0.5 h were achieved in comparison to the operation without Ca^{2+} [70]. Pure thermophilic culture of *Th. thermosaccharolyticum* showed almost the same production yield of 3.09 mol/mol sucrose [54]. When sucrose conversion efficiency (95%) and production rate (144 mmol/l/h) were taken into account, the optimal HRT was 1 h which corresponds to OLR= 46 mmol sucrose/L h. Yield of production and sucrose conversion increased to 3.52 mol/mol sucrose and 99%, respectively, when HRT extended to 24 h, but rate substantially reduced to 8.6 mmol H_2/h L. One of the adverse effects of low HRT was accumulation of granules on the liquid surface due to the force of gas formed around the granules. The optimal HRT for the maximum production yield of 1.68 mol/mol glucose extended up to 48 h when the substrate was starch under thermophilic conditions (50°C). However the rate was maximum at HRT= 6 h corresponding to 127 kg COD/m^3/d organic loading rate [71]. Hydrogen production potentials of AGSB reactors and CIGSB reactors were summarized in Table **5**.

Lee *et al.* [70] observed that there were "dead zones" in granular sludge bed due to poor mixing. Liquid reflux strategy was employed to improve the mixing conditions. Upflow liquid velocity up to 1.09 m/h provided a slight improvement in the reactor performance. But a sharp decrease in hydrogen production occurred for higher upflow velocities between 1.53-2.29 m/h as a result of biomass loss and dilution in substrate concentration. On the contrary, Gavala *et al.* [72] operated a UASB system with higher upflow liquid velocities between 4.5-6 m/h at HRTs=12- 6 h. No differences in hydrogen production and the soluble metabolite profile were obtained. Gas reflux was used as an alternative to liquid reflux [70]. Yield and rate of production were slightly enhanced at 1 m/h reflux rate. The observations indicated that high reflux rates may negatively affect the settling ability of granules and severe reduction in biomass concentration and sludge diameter due to shear stress and upflow driving force.

The composition of end products vary depending on the optimal HRT determined in the process although HAc and HBr are the main soluble metabolites in most of the hydrogen production studies in UASB systems. Chang *et al.* [50] reported that the liquid products at optimal HRT =8 h were 51% butyrate and 31% acetate. This result was evaluated as the dominant bacterial culture was *Clostridium*. HBr and HAc constituted about 77% of total soluble products at HRTs= 4 h and 0.5 h. But, ethanol and HPr formations were observed when HRT was shortened at 45°C operation temperature [69]. Similar result was obtained by Akutsu *et al.* for thermophilic condition [71]. HAc and HBr were the main end product with small fraction of HPr, lactic, caproic, formic acid and ethanol at HRT= 48 h indicating that butyric acid fermentation was the main pathway in hydrogen formation. However, the pattern devoted to lactic and formic acid formation for HRTs less than 24 h. The alkaline or neutral pH conditions and increased organic loading rate were considered as the reasons for lactic and formic acid production. On the other hand decreasing HRT from 24 h to 0.75 h provided 2.5 fold increasing in HAc and HBr formation with *Th. thermosaccharolyticum* without affecting ethanol, lactate and propionic acid concentrations [54]. The predominant soluble metabolite was HAc followed by HBr, EtOH and HPr in CIGSB when supplemented with Ca^{2+} ion at HRT=1 h [70].

High utilization rates over 90% were obtained at long HRTs between 12-48 h in UASB [54,71]. Some studies indicated that it is possible to have more than 95% conversion for HRTs= 1-4 h [60,69]. Shorter HRTs as 0.5 h resulted in decrease in substrate utilization efficiency less than 70% [60,69,71]. On the contrary, Chang *et al.* [50] did not observe a substantial difference in sucrose removal when HRT was decreased from 24 h to 4 h. Low substrate utilization at short HRTs was caused by operation at high organic loading rates which results in exceeding the maximum capacity of H_2 producers for complete consumption of influent carbon substrate [69].

pH in the effluent of UASB reactors were in the range of 5- 6.6 in some studies[60,54]. Akutsu *et al.* [71] reported as pH=5 is the optimal pH value under thermophilic conditions and alkaline or neutral pH values were unfavorable because of shift in the fermentation pattern to formic and lactic acid formation. A decrease in pH from 5.33 to 4.36 was observed as organic loading rate increased in expended granular sludge bed system [73]. During start-up period of UASB system with operation at HRT=24 h, pH value varied around pH=8 and then it remained almost constant at around pH=7 for HRTs=20-4 h [50].

Granule formation in UASB system requires a long start-up period. The maximum average granule diameter of 0.43 mm was obtained after 173 days operation which corresponded to maximum hydrogen production rate and yield [66]. However carrier induced granular sludge bed system provided granule formation within 60 h of operation at HRT=4 h after short period of start-up [69]. The components of extra cellular polymeric substances (EPS) play a crucial role in self-flocculation and the major ones in EPS were found as protein and carbohydrates [74]. The ratio of these polymeric substances could be a factor in granule formation [75]. *Clostridum* is the well known and commonly used hydrogen producing organism. The other well-know hydrogen producing organism is *Enterobacter aerogenes* and the self-flocculation ability of this organism was investigated by Yokoi *et al.* [76]. A novel non-spore forming hydrogen producing organisms *Ethanoligenes harbinense*, YUAN 3, was first isolated by Xing *et al.* from anaerobic hydrogen producing sludge [77]. It was reported that YUAN 3 has excellent self flocculation and hydrogen production capabilities [78]. The culture produced EPS at exponential growth phase and autoaggregation with average 5 mm granule diameter occurred at stationary phase. The protein and carbohydrate were the main components of the EPS with small fraction of DNA. The concentrations of protein and carbohydrate in EPS content of YUAN 3 were 21 mg/g cell dry mass and 16.9 mg/g cell dry mass, respectively, which were considerably higher compared to EPS contents of the other *Ethanoligenes harbinense* strains studied [74].

Operation temperature also affects the concentration and composition of EPS in sludge. The concentration of EPSs decreased with increasing temperature [47]. Lee *et al.* [69] indicated that the biomass concentration in the UASB decreased at high temperatures (40-45°C) without affecting hydrogen production adversely. This result was explained as denaturation of some essential enzymes and proteins associated with cell growth and hydrogen production. However, when heat treated methanogenic granules were used as carrier material for immobilization of H_2 producing organisms under thermophilic operation (60°C), granules with high mechanical strength and thick biomass were obtained. Extensions of new surface layers of EPS on the granules were observed [54]. The same immobilization and granule formation approaches were applied at 55°C [71]. The heat treated methanogenic sludge granules acted as nuclei for the attachment of H_2 producing organisms. The granule size distribution was in the range of 0.5 mm-4 mm. Carrier induced granulation follows the same principle in granule formation. Activated carbon serves as inert nuclei and then enhances the biofilm formation [69,75]. Another novel protocol to shorten granule formation period was developed by Zhang *et al.* [75]. Acid incubation (pH=2) of sludge for 24 h in the column reactor and operating the system at low HRTs (2 h) significantly reduced the time for granulation (92 h) by nearly three fold compared to time required without acid incubation (346 h). Acid incubation promoted the nuclei formation. Moreover extra cellular polymer (ECP) content of sludge increased and a shift from polysaccharides to protein in the ECP was observed.

Table 5: Rates and Yields of Hydrogen Gas Production from Different Carbon Sources in Anaerobic Granular Sludge Bed Reactor Systems.

Biofilm Process	Culture	Substrate Type	Concentration of Substrate	Support Particle	HRT	H_2 production rate	H_2 production yield	Ref
UASB	Mixed culture	Glucose	10 g/L		0.5 h	4.93 L/h L	1.83 mol/mol glucose	[75]
UASB	Mixed culture	Glucose	10 g/L		2 h	19 mmol/h L	0.7 mol/mol glucose	[72]
UASB	Mixed culture	Sucrose	20 g COD/L		8 h	270 mmol/d L	1.5 mol/mol sucrose	[50]
UASB	Mixed culture	Sucrose	20 g COD/L		0.5 h	7.3 L/h L	3.03 mol/mol sucrose	[44]
UASB	Mixed culture	Starch	15 g/L		48 h	1.90 L/h L	1.68 mol/mol glucose	[71]
AGSB	Mixed culture	Glucose	20 g COD/L	AC	4 h	0.97 L/h L	1.54 mol/mol glucose	[60]
AGSB	Mixed culture	Glucose	10 g/L	AC	0.25 h	6.5 L/h L	1.71 mol/mol glucose	[13]
AGSB	*Th. thermosaccharolyticum*	Sucrose	20 g COD/L		18 h	12.1 mmol/h L	3.1 mol/mol sucrose	[54]
CIGSB	Mixed culture	Sucrose	20 g COD/L	AC	1 h	4.75 L/h L	3.88 mol/mol glucose	[69]
CIGSB	Mixed culture	Sucrose	20 g COD/L	AC	2 h	1.38 L/h L	2.38 mol/mol glucose	[70]
CIGSB	*Th. thermosaccharolyticum*	Sucrose	20 g COD/L	heat treated sludge	1 h	144 mmol/h L	3.09 mol/mol sucrose	[54]
EGSB	Mixed culture	Molasses	10 g COD/L	AC	2 h	0.71 L/h L	3.47 mol/mol glucose	[72]

Limited studies focused on determination of kinetic coefficients in UASB system. The yield, coefficient ($Y_{X/S}$ g biomass/g substrate) which is cell mass produced per consumed substrate and k_d endogenous decay coefficient (h^{-1}) were determined by Chang *et al.* as Yg= 0.1 g VSS/g COD and kd= 0.1 day $^{-1}$ [50]. Specific growth rate, μ, and saturation constant, Ks, were calculated as 1 h^{-1} and 15.3 mg/L for *Th. thermosaccharolyticum.* [54].

Trickling Biofilter Reactor (TBR)

This configuration is very similar to FPBR except for the feeding direction (Fig. **1f**). A trickling biofilter consists of various filling materials such as gravel, slag, polyurethane foam, sphagnum peat moss, or plastic media over which sewage or other wastewater flows downward and causes a layer or film of microbial slime to grow, covering the bed of media. The process mechanism, or how the removal of waste from the water happens, involves both absorption and adsorption of organic compounds within the sewage or other wastewater by the layer of microbial slime. The slime layer continues to build until it eventually sloughs off, breaking off longer growth into the treated effluent as a sludge that requires subsequent removal and disposal. Other filters utilizing higher-density media such as sand, foam and peat moss do not produce a sludge that must be removed, but require backwashing.

TBR is expected to offer many advantages to fermentative H_2 production, such as high cell density, reduced gas hold-up, and easy control of pH and temperature. On the other hand, rapid buildup of hydrogen producing bacteria on biofilm would result in a system upset because of the mass transfer limitation. Working on a trickling biofilter, for example, Oh *et al.* [26] noted that microbial growth of hydrogen-producing bacteria was too excessive under a mesophilic condition, which caused system upset just after 1 week of operation.

Gas Lift Reactor (GLR)

GLR is based on the use of gas to circulate the contents of a bioreactor through an external tube or internally, using a draft tube (no rotating parts are involved). Gas lift reactors contain two concentric tubes, a riser (an inner tube) and a downcomer (an outer tube). In these reactors, mixing is achieved by circulating essentially gas at the bottom of the reactor. As a result of force applied by the gas (at the bottom of the inner tube), the liquid in the inner tube moves up which then overflows (the inner tube) downward thus creating eddies to mix the liquid. In some of the gas lift reactors downcomer is replaced with an external loop to circulate fermentation broth. Such bioreactors are of simple construction and operation with low power consumption. Therefore, this bioreactor configuration is attractive for large scale applications and have been adopted by industry in a variety of processes using both free cells and cell aggregates, *e.g.* for cell mass production [1].

Sequencing Batch Reactor

Sequencing batch reactors (SBR) are operated discontinuous mode with fill, react, settle and draw periods in sequence. The react period could contain anaerobic, aerobic and anoxic phases depending on the aim for utilization of the system such as wastewater treatment nutrient removal or product formation. SBR processes have certain advantages over continuous processes, including a high degree of process flexibility, no requirement for a separate clarifier and high biomass holding capacity. It is practical for wastewaters with low flowrate in terms of controlling environmental conditions, operation and shifting from one phase to another when necessary. Suspended culture SBR processes are the most common ones for both hydrogen gas production and wastewater treatment [79,80]. Limited number of studies considered biofilm based SBR systems for hydrogen gas production. Mohan *et al.* [18] assessed molecular hydrogen (H_2) production with simultaneous wastewater treatment in biofilm configured periodic discontinuous/sequencing batch reactor using chemical wastewater as substrate. Anaerobic mixed culture was subjected to sequentially repeated heat-shock (100 °C; 2 h) and acid (pH—3.0; 24 h) treatment. Periodic discontinuous batch operation provided high flexibility in selectively enriching the specific group of microflora which resulted in rapid startup and stabilization. The sequencing batch systems could also provide high cell retention capacity of biomass due to batch mode operation. This property of SBR process enhances the persistence of augmented strain in the systems which could positively affect the hydrogen production yield and rate [19]. The H_2 generation efficiency was found to be dependent on the applied organic loading rate (OLR) in biofilm SBR system. Increase in substrate loading rate resulted in a sharp drop in H_2 production and yield [81].

Membrane Processes

Membrane reactors can be configured as flat sheet or hollow fibre modules. Hollow fibre modules provide a higher surface to volume ratio without the need for membrane support. However, the geometry of flat sheet modules is simpler, providing an accurate regulation of the distances between the membranes. Additionally, these modules can be easily disassembled, providing an easy access to modules compartments and options for membrane cleaning and replacement. In membrane bioreactors, three types of cell immobilization are possible i) cell immobilization on the membrane where a biofilm is formed ii) cell immobilization within the membrane iii) cell immobilization in a cell compartment separated by the membrane. Drawbacks of the use of membrane bioreactors can be a reduced mass transfer rate depending on the membrane properties, available surface area and external flow conditions [82].

Table 6: Rates and yields of hydrogen gas production from different carbon sources in anaerobic membrane bioreactor systems.

Culture	Substrate Type	Concentration of Substrate	Membrane Type	HRT	H_2 production rate	H_2 production yield	Ref
Mixed culture	Glucose	20 g COD/L	hollow-fiber cross-flow microfiltration	4 h	1.02 L/h L	1.72mol/mol glucose	[84]
Mixed culture	Glucose	8.2 g/L	Plate flame	9 h, SRT= 90 d	2.5 L/d L	0.86 mol/mol glucose	[85]
Mixed culture	Glucose	16.0 g/l	Plate flame	9h, SRT=2 d	5.8 L/d L	1.19 mol/mol glucose	[86]
Mixed culture	Glucose	16.0 g/l	Flat flame, microfiltration	9 h	NA	1.6 mol/mol glucose	[88]
Mixed culture	Glucose	87.5 g/COD L	hollow-fiber ultrafiltration	8 h	0.20 mol/L d	1.77mol/mol glucose	[83]
Mixed culture	Sucrose	20 g COD/L	hollow-fiber cross-flow microfiltration	2 h	1.67 L/h L	1.51mol/mol glucose	[84]
Mixed culture	Fructose	20 g COD/L	hollow-fiber cross-flow microfiltration	2	1.87 L/h L	1.55 mol/mol glucose	[84]

Shen *et al.* [83] investigated the effect of organic loading on hydrogen production potentials of CSTR and MBR reactors. MBR contained a hollow-fiber ultrafiltration membrane module. The hydrogen production yields of two reactors were not substantially different for the examined OLRs between 4.0 g COD/L d - 30 g COD/L d. More distinctive hydrogen production potentials were observed between CSTR and MBR by the variation in HRT. A microfiltration membrane module was connected with a conventional CSTR process for dark H_2 fermentation for different carbon sources as glucose, sucrose and fructose. The H_2 production ceased (0.15 L/h L) at HRT=4 h and washout of biomass (0.77 g VSS/l) occurred when the system was operated at CSTR mode without membrane. Integration of membrane into the system provided a substantial increase in the production rate (1.02 L/h L) and biomass concentration (5.7 g VSS/L). Even further improvement in the rate of production was observed at shorter HRTs (1 h). The MBR system was able to be stably operated for long term without suffering membrane fouling problem. In addition to the enhancement in HPR, the HY for all the carbon substrates examined was also higher when MBR operation was applied [84]. Similarly, H_2 production rate in MBR's was 2.43–2.56 l H_2/d L, which was about 2.6 times higher than that (0.95–0.97 L H_2/d L^1) of the CSTR, reflecting the MBR's higher H_2 productivity. No CH_4 was detected in the biogas at 3300 mg-DOC/L substrate concentration although cell retention (90 days) was high enough to sustain the growth of methanogens. A shift from butyrate to lactate and propionate occurred as biomass concentration increased in MBR. This result was explained as change in the food-to-microorganism (F/M) ratio for a continuous MBR operation might influence the anaerobic fermentation pathway away from H_2 production towards lactic acid or propionic acid production [85]. At high initial substrate concentration (6400 mg DOC/L), the system was efficiently operated until SRT=12 days and then membrane fouling and gradual decrease in the flux observed when SRT was increased to 90 days. Lactate was the dominant end product and it reached to maximum concentrations at the highest SRT. The decrease in H_2 yield with the extension of SRT was caused by increase of inert biomass on membrane and devote of metabolic pathway to lactate generation [86]. The extracellular polymeric substances (EPS) are considered another factor in efficient operation of MBR for hydrogen gas production. EPS concentration increases in time, a shift from low to high EPS molecular weights over time during hydrogen fermentation occurs as SRT is extended. These substances are retained by adsorption onto the polymeric membrane surface. EPS deposition on the membrane surface or the interior of the membrane pores results in severe membrane fouling [87]. The effect of $FeSO_4$ concentration as an iron source on the efficiency of continuous H_2 production in a submerged MBR system was investigated by Lee *et al.* [88]. Addition of 10.9 mg $FeSO_4$/L enhanced the H_2 production rate (41.6 L/d), yield (1.32 mol/mol glucose) and formation of HBr and HAc rather than hydrogen consuming end products. A higher hydrogenase activity of 1.08 mgM.B reduced/min was measured in the whole cells at 10.9 mg $FeSO_4$/L, which was about six fold higher than that observed in the control experiment (at 2.7 mg $FeSO_4$/L). Table **6** indicates the rates and yields of hydrogen gas formation in anaerobic membrane bioreactors.

CONCLUSIONS

An intensive research work has been carried out on the advancement of hydrogen production. Among these alternative approaches, such as the development of genetically modified microorganism, use of metabolic engineering, biochemical assisted bioreactor design, development of two-stage processes, use of immobilization techniques for the improvement of the reactor designs and use of different solid matrices for the immobilization of whole cells seem to be promising. Many advantages have been claimed over the years for the use of immobilized cells such as robustness against cell washout, more stable operational conditions, higher cell inventory in bioreactor *etc.* Since the review papers on microbial cell immobilization were published in the late 1970s, there has been considerable growth in research activity related to the application of immobilized cells to fermentation processes including biohydrogen production. Apart from the advantages of using immobilization techniques, it is also important to mention potential disadvantages of this system. Among those, mass transfer problem could actually be solved by judicious selection of bioreactor operations and the use of particles with improved physic-chemical properties. Another problem with the use of immobilization systems for biohydrogen purpose is the lack of expertise in large scale immobilized cell processes. However, the large scale expertise on immobilization in fermentation industry could be used to create a virtuous circle for the use cell immobilization in biohydrogen applications.

REFERENCES

[1] Webb C, Dervakos GA. Studies in viable cell immobilization. R.G. Landes Company, USA: Academic Press 1996.

[2] Zhang ZP, Show KY, Tay JH, Liang DT, Lee DJ. Enhanced continuous biohydrogen production by immobilized anaerobic microflora. Energ Fuel 2008; 22: 87–92.

[3] Bagai R, Madamwar D. Prolonged evolution of photohydrogen by intermittent supply of nitrogen using a combined system of *Phormidium valderiannum*, halo-bacterium halobium and *Escherichia coli*. Int J Hydrogen Energy 1998; 23: 545-50.

[4] Nedovic V,Willaert R. Fundamentals of cell immobilization biotechnology, USA: Kluwer Academic Publisher 1996.

[5] Bickerstaff GF. Immobilization of enzymes and cells. USA: Humana Press, 1997.

[6] Berkeley RCW, Lynch JM, Melling J, Rutter PR, Vincent B. Microbial Adhesion to surfaces. Chichester. Ellis Horwood, 1980.

[7] Vega JL, Clausen EC, Gaddy JL. Biofilm reactors for ethanol production. Enzyme Microb Tech 1998; 10: 390-402.

[8] Cooper PF, Atkinson B. Biological fluidized Bed Treatment of Water and Wastewater. Chichester:ellis Horwood 1981.

[9] Wu SY, Lin CN, Chang JS, Chang JS. Biohydrogen production with anaerobic sludge immobilized by ethylene-vinyl acetate copolymer. Int J Hydrogen Energy 2005; 30(13-14): 1375-81.

[10] Chang JS, Lee KS, Lin PJ. Biohydrogen production with fixed-bed bioreactors. Int J Hydrogen Energy 2002; 27(11-12): 1167-74.

[11] Mohan SV, Mohanakrishna G, Reddy SS, Raju BD, Rao KSR, Sarma PN. Self-immobilization of acidogenic mixed consortia on mesoporous material (SBA-15) and activated carbon to enhance fermentative hydrogen production. Int J Hydrogen Energy 2008; 33(21): 6133-42.

[12] Mitchell RJ, Kim JS, Jeon BS, Sang BI. Continuous hydrogen and butyric acid fermentation by immobilized *Clostridium tyrobutyricum* ATCC 25755: Effects of the glucose concentration and hydraulic retention time. Bioresource Technol 2009; 100(21): 5352-5.

[13] Ivanova G, Rákhely G, Kovács KL. Hydrogen production from biopolymers by *Caldicellulosiruptor saccharolyticus* and stabilization of the system by immobilization. Int J Hydrogen Energy 2008; 33(23): 6953-61.

[14] Zhang ZP, Show KY, Tay JH, Liang DT, Lee DJ. Biohydrogen production with anaerobic fluidized bed reactors- A comparison biofilm-based and granular-based systems. Int J Hydrogen Energy 2007; 33(5): 1559-64.

[15] Zhang ZP, Tay JH, Show KY *et al.* Biohydrogen production in a granular activated carbon anaerobic fluidized bed reactor. Int J Hydrogen Energy 2007; 32(2): 185-91.

[16] Shida GM, Barros AR, Reis CM *et al.* Long-term stability of hydrogen and organic acids production in an anaerobic fluidized-bed reactor using heat treated anaerobic sludge inoculum. Int J Hydrogen Energy 2009; 34(9): 3679-88.

[17] Amorim ELC, Barros AR, Damianovic MHRZ, Silva EL. Anaerobic fluidized bed reactor with expanded clay as support for hydrogen production through dark fermentation of glucose. Int J Hydrogen Energy 2009; 34(2): 783-90.

[18] Mohan SY, Bhaskar V, Sarma PN. Biohydrogen production from chemical wastewater treatment in biofilm configured reactor operated in periodic discontinuous batch mode by selectively enriched anaerobic mixed consortia. Water Res 2007; 41(12): 2652-64.

[19] Mohan SV, Mohanakrishna G, Raghavulu SV, Sarma PN. Enhancing biohydrogen production from chemical wastewater treatment in anaerobic sequencing batch biofilm reactor (AnSBBR) by bioaugmenting with selectively enriched kanamycin resistant anaerobic mixed consortia. Int J Hydrogen Energy 2007; 32(15): 3284-92.

[20] Jo JH, Lee DS, Park D, Park JM. Biological hydrogen production by immobilized cells of *Clostridium tyrobutyricum* JM1 isolated from a food waste treatment process. Bioresource Technol 2008; 99(14): 6666-72.

[21] Fritsch M, Hartmeier W, Chang JS. Enhancing hydrogen production of *Clostridium butyricum* using a column reactor with square-structured ceramic fittings. Int J Hydrogen Energy 2008; 33(22): 6549-57.

[22] Barros AR, Amorim ELC, Reis CM, Shida GM, Silva EL. Biohydrogen production in anaerobic fluidized bed reactors: Effect of support material and hydraulic retention time. Int J Hydrogen Energy 2010; 35(8): 3379-88.

[23] Basile MA, Dipasquale L, Gambacorta A. The effect of the surface charge of hydrogel supports on thermophilic biohydrogen production. Bioresource Technol 2010; 101(12): 4386-94.

[24] Kumar N, Das D. Continuous hydrogen production by immobilized *Enterobacter cloacae* IIT-BT 08 using lignocellulosic materials as solid matrices. Enzyme Microb Tech 2001; 29(4-5): 280-7.

[25] Kim JO, KimYH, Ryu JY, Song BK, KimIH, Yeom SH. Immobilization methods for continuous hydrogen gas production biofilm formation versus granulation. Process Biochem 2005; 40: 1331–7.

[26] Oh YK, Kim SH, Kim MS, Park S. Thermophilic biohydrogen production from glucose with trickling biofilter. Biotechnol Bioeng 2004; (88): 690–8.

[27] Kierstan MPJ, Coughan MP. Immobilization of cells and enzymes by entrapment. In: Woodward J, ed. Immobilized Cells and enzymes-A Practical Approach. Oxford: IRL Press, 1985; pp.39-48.

[28] Ishikawa M, Shohei Y, Yuzuru T *et al.* Development of a compact high-density microbial hydrogen reactor for portable bio-fuel cell system. Int J Hydrogen Energy 2006; 31(11): 1484-9.

[29] Ishikawa M, Yamamura S, Ikeda R *et al.* Development of a compact stacked flatbed reactor with immobilized high-density bacteria for hydrogen production. Int J Hydrogen Energy 2008; 33(5): 1593-7.

[30] Jekel M, Buhr A, Willke T, Vorlop KD. Immobilization of biocatalysts in LentiKats®. Chem Eng Technol 1998; 21(3): 275–8.

[31] Wu KJ, Chang JS. Batch and continuous fermentative production of hydrogen with anaerobic sludge entrapped in a composite polymeric matrix. Process Biochem 2007; 42(2): 279-84.

[32] Wu SY, Hung CH, Lin CN, Chen HW, Lee AS, Chang JS. Fermentative hydrogen production and bacterial community structure in high-rate anaerobic bioreactors containing silicone-immobilized and self-flocculated sludge. Biotechnol Bioeng 2006; 93(5): 934-46.

[33] Zhu H, Wakayama T, Suzuki T, Asada Y, Miyake J. Entrapment of *Rhodobacter sphaeroides* RV in cationic polymer/agar gels for hydrogen production in the presence of NH_4^+. J Biosci Bioeng 1999; 88(5): 507–12.

[34] Rupp RG. Use of cellular microencapsulation of yeasts for use in the tower fermenter. Biotechnol Lett 1985; 4: 621–26.

[35] Chang HN. Membrane bioreactors: engineering aspects. Biotechnol Adv 1987; 5: 129–45.

[36] Robinson JR. Dual hollow fiber bioreactor for cell culture. US Patent 5015585, 1991.

[37] Efthymiou GS, Shuler ML. Elimination of diffusional limitation in a membrane entrapped cell reactor by pressure cycling. Biotechnol Prog 1987; 3: 259–64.

[38] Jeong YS, Vieth WR, Matsuura T. Study of transport phenomena in an immobilized yeast membrane bioreactor. Ind Eng Chem Res 1989; 28: 231–6.

[39] Chang HN, Yoo IK, Kim BS. High density cell culture by membrane-based cell recycle. Biotechnol Adv 1994; 12: 467–87.

[40] Shi YF, Jin FX, Wu YY. Microfiltration membrane bioreactor in stirred backflush operation for biotransformation using intact cells. Process Biochem 1997; 32: 387–90.

[41] El-Mansi EMT, Bryce CFA, Demain AL, Allman AR. Fermentation Microbiology and Biotechnology. Taylor & Francis Group, FL.CRC Press 2007.

[42] Hu B, Chen S. Pretreatment of methanogenic granules for immobilized hydrogen fermentation. Int J Hydrogen Energy 2007; 32: 3266-73.

[43] Fang HHP, Liu H, Zhang T. Characterization of a hydrogen producing granular sludge Biotechnol Bioeng 2002; (78): 44–52.

[44] Tay JH, Tay STL, Yu L, Yeow SK, Ivanov V. Biogranulation Technologies for Wastewater Treatment, Waste Management Series 6. Elsevier 2006.

[45] Liu Y, Xu HL, Yang SF. Tay JH. Mechanisms and models for anaerobic granulation in upflow anaerobic sludge blanket reactor. Water Res 2003; 661-73.

[46] Lee KS, Wu JF, Lo YS, Lo YC, Lin PJ, Chang JS. Anaerobic hydrogen production with an efficient carrier-induced granular sludge bed bioreactor. Biotechnol Bioeng 2004; 87: 648-57.

[47] Liu LL, Wang ZP, Yao J, Sun XJ, Cai WM. Investigation on the formation and kinetics of glucose-fed aerobic granular sludge. Enzyme Microb Tech 2005; 36: 712-6.

[48] Hulshoff LW, de Castro Lopes SI, Lettinga G, Lens PNL. Anaerobic sludge granulation. Water Res 2004; 1376-89.

[49] Lettinga G, van Velsen AFM, Hobma SW, de Zeeuw, W, Klapwijk A. Use of the upflow sludge blanket (USB) reactor concept for biological waste water treatment especially for anaerobic treatment. Biotechnol Bioeng 1980; 22: 699–734.

[50] Chang FY, Lin CY. Biohydrogen production using an up-flow anaerobic sludge blanket reactor. Int J Hydrogen Energy 2004; 29: 33-9.

[51] Yu HQ, Mu Y. Biological hydrogen gas production in a UASB reactor with granules, II Reactor performance in 3 year operation. Biotechnol Bioeng 2006; 94: 988–95.

[52] Hawkes FR, Hussy I, Kyazze G, Dinsdale R, Hawkes DL. Continuous dark fermentative hydrogen production by mesophilic microflora: Principles and progress Int J Hydrogen Energy 2007; 32: 172–84.

[53] Zhang ZP, Show KY, Tay JH, Liang DT, Lee DJ, Jiang WJ. Rapid formation of hydrogen-producing granules in an anaerobic continuous stirred tank reactor induced by acid incubation. Biotechnol Bioeng 2007; 96(6): 1040-50.

[54] O-Thong S, Prasertsan P, Karakashev D, Angelidaki I. High-rate continuous hydrogen production by *Thermoanaerobacterium thermosaccharolyticum* PSU-2 immobilized on heat-pretreated methanogenic granules. Int J Hydrogen Energy 2008; 33(22): 6498-6508.

[55] Rodgers M, Zhan XM. Biological nitrogen removal using a vertically moving biofilm system. Bioresource Technol 2004; 93(3): 313–9.

[56] Von Munch E, Barr K, Watts S, Keller J. Suspended carrier technology allows upgrading high-rate activated sludge plants for nitrogen removal *via* process intensification. Water Sci Technol 2000; 41(4–5): 5–12.

[57] Zheng H, Zeng RJ, Angelidaki I. Biohydrogen production from glucose in upflow biofilm reactors with plastic carriers under extreme thermophilic conditions (70°C). Biotechnol Bioeng 2008; 100(5): 1034-8.

[58] Lacroix C, Paquin C, Arnaud JP. Batch fermentation with entrapped growing cells of *Lactobacillus casei.* Appl Microbiol Biotechnol 1990; 32: 403–8.

[59] Wu KJ, Chang CF, Chang JS. Simultaneous production of biohydrogen and bioethanol with fluidized-bed and packed-bed bioreactors containing immobilized anaerobic sludge. Process Biochem 2007; 42(7): 1165-71.

[60] Wu SY, Hung CH, Lin CY *et al.* HRT-dependent hydrogen production and bacterial community structure of mixed anaerobic microflora in suspended, granular and immobilized sludge systems using glucose as the carbon substrate. Int J Hydrogen Energy 2008; 33(5): 1542-9.

[61] Godia F, Casas C, Sola C. Mathematical modelization of a packed-bed reactor performance with immobilized yeast for ethanol fermentation. Biotechnol Bioeng 1987; (30): 836.

[62] Leite JAC, Fernandes BS, Pozzi E, Barboza M, Zaiat M. Application of an anaerobic packed-bed bioreactor for the production of hydrogen and organic acids. Int J Hydrogen Energy 2008; 33(2): 579-86.

[63] Park W, Hyun SH, Oh SE, Logan BE, Kim IS. Removal of headspace CO_2 increases biological hydrogen production. Environ Sci Technol 2005; 39: 4416-20.

[64] Groboillot DK, Poncelet BD, Neufeld RJ. Immobilization of cells for application in the food industry. Crit Rev Biotechnol 1994; 14(2): 75-107.

[65] Lin CN, Wu SY, Chang JS. Biohydrogen production in a three-phase fluidized bed bioreactor using sewage sludge immobilized by ethylene–vinyl acetate copolymer. Bioresource Technol 2009; 100(13): 3298-3301.
[66] Lin CN, Wu SY, Chang JS. Fermentative hydrogen production with a draft tube fluidized bed reactor containing silicone-gel-immobilized anaerobic sludge. Int J Hydrogen Energy 2006; 31(15): 2200-10.
[67] Koskinen PEP, Kaksonen AH, Puhakka LA. The relationship between instability of H_2 production and compositions of bacterial communities within a dark fermentation fluidized-bed bioreactor. Biotechnol Bioeng 2007; 97(4): 742-58.
[68] Hu B, Chen S. Biological hydrogen production using chloroform-treated methanogenic granules. Appl Biochem Biotech 2008; 148: 83–95.
[69] Lee KS, Lin PJ, Chang JS. Temperature effects on biohydrogen production in a granular sludge bed induced by activated carbon carriers. Int J Hydrogen Energy 2006; 31(4): 465-72.
[70] Lee KS, Lo YS, Lo YC, Lin PJ, Chang JS. Operation strategies for biohydrogen production with a high-rate anaerobic granular sludge bed bioreactor. Enzyme Microb Tech 2004; 35(6-7): 605-12.
[71] Akutsu Y, Lee DY, Chi YZ, Li YY, Harada H, Yu HQ. Thermophilic fermentative hydrogen production from starch-wastewater with bio-granules. Int J Hydrogen Energy 2009; 34(12): 5061-71.
[72] Gavala HN, Skiadas IV, Ahring BK. Biological hydrogen production in suspended and attached growth anaerobic reactor systems. Int J Hydrogen Energy 2006; 31(9): 1164-75.
[73] Guo WQ, Ren NQ, Wang XJ *et al.* Biohydrogen production from ethanol-type fermentation of molasses in an expanded granular sludge bed (EGSB) reactor. Int J Hydrogen Energy 2008; 33(19): 4981-8.
[74] Ren N, Xie T, Xing D. Composition of extracellular polymeric substances influences the autoaggregation capability of hydrogen-producing bacterium Ethanoligenens harbinense. Bioresource Technol 2009; 100(21): 5109-13.
[75] Zhang ZP, Show KY, Tay JH, Liang DT, Lee DJ, Su A. The role of acid incubation in rapid immobilization of hydrogen-producing culture in anaerobic upflow column reactors. Int J Hydrogen Energy 2008; 33(19): 5151-60.
[76] Yokoi H, Aratake T, Hirose J, Hayashi S,Takasaki Y. Simultaneous production of hydrogen and bioflocculant by *Enterobacter* sp. BY-29, World J Microbiol Biotech 2001; 17: 609–13.
[77] Xing D, Ren N, Li Q, Lin M, Wang A, Zhao L. *Ethanoligenens harbinense* gen. nov., sp. nov., isolated from molasses waste water. Int J Syst Evol Micr 2006; 56: 755–60.
[78] Xing D, Ren N, Wang A, Li Q, Feng Y, Ma F. Continuous hydrogen production of auto-aggregative *Ethanoligenens harbinense* YUAN-3 under non-sterile condition. Int J Hydrogen Energy 2008; 33: 1489–95.
[79] Chen SD, Lo YC, Lee KS, Huang TI, Chang JS. Sequencing batch reactor enhances bacterial hydrolysis of starch promoting continuous bio-hydrogen production from starch feedstock. Int J Hydrogen Energy 2009; 34(20): 8549-57.
[80] Arooj MF, Han SK, Kim SH, Kim DH, Shin HS. Effect of HRT on ASBR converting starch into biological hydrogen. Int J Hydrogen Energy 2008; 33(22): 6509-14.
[81] Bhaskar YV, Mohan SV, Sarma PN. Effect of substrate loading rate of chemical wastewater on fermentative biohydrogen production in biofilm configured sequencing batch reactor. Bioresource Technol 2008; 99(15): 6941-8.
[82] Nedovic V, Willaert R. Fundamentals of cell immobilization biotechnology. USA: Kluwer Academic Publisher 2004.
[83] Shen L, Bagley DM, Liss SN. Effect of organic loading rate on fermentative hydrogen production from continuous stirred tank and membrane bioreactors. Int J Hydrogen Energy 2009; 34(9): 3689-96.
[84] Lee KS, Lin PJ, Fangchiang K, Chang JS. Continuous hydrogen production by anaerobic mixed microflora using a hollow-fiber microfiltration membrane bioreactor. Int J Hydrogen Energy 2007; 32(8): 950-7.
[85] Lee DY, Li YY, Noike T. Continuous H_2 production by anaerobic mixed microflora in membrane bioreactor. Bioresource Technol 2009; 100(2): 690-5.
[86] Lee DY, Li YY, Noike T. Influence of solids retention time on continuous H_2 production using membrane bioreactor. Int J Hydrogen Energy 2010; 35(1): 52-60.
[87] Lee DY, Li YY, Noike T, Cha GC. Behavior of extracellular polymers and bio-fouling during hydrogen fermentation with a membrane bioreactor. J Membrane Sci 2008; 322(1): 13-8.
[88] Lee DY, Li YY, Oh YK, Kim MS, Noike T. Effect of iron concentration on continuous H_2 production using membrane bioreactor. Int J Hydrogen Energy 2009; 34(3): 1244-52.

CHAPTER 14

Prospects for Biohydrogen Production

Nuri Azbar[1]* and David B. Levin[2]

[1]Department of Bioengineering, Ege University and [2]Department of Biosystems Engineering, University of Manitoba

Abstract: Hydrogen is seen as a "fuel of the future" that will replace the petroleum-based economy. It is a versatile energy carrier with the potential for extensive use as a transportation fuel, in power generation, and in many other applications. Hydrogen is currently produced from fossil sources (steam reforming of methane), but technologies utilizing renewable sources are urgently needed for sustainability. Biological hydrogen (biohydrogen) production is one of the challenging areas of technology development for sustainability. There are a wide range of biohydrogen technologies, including direct biophotolysis, indirect biophotolysis, photo-fermentations, and dark-fermentation. The current scientific results are promising, but substantial improvements in biohydrogen production through research advances (*i.e.* improvement in efficiency through genetically engineered microorganism, development of bioreactors *etc.*) are needed. In this study, attempts have been made to highlight not only the advantages, but also the bottlenecks that limit biohydrogen production.

Keywords: Biohydrogen, dark fermentation, direct biophotolysis, indirect biophotolysis, photofermentation, bioengineering.

INTRODUCTION

Mankind is now facing major challenges due to climate change associated with increasing concentrations of atmospheric CO_2 generated by extensive combustion of fossil fuels. In this regard, a substantial amount of research has been dedicated to find environmentally benign and sustainable alternatives to carbon-based fuels. Hydrogen is viewed as a very promising energy carrier, since it does not generate carbon emissions when burned and could be produced in a sustainable manner. It is known that more than 500 billion m^3 of hydrogen are produced each year equivalent to 6.5 EJ of energy corresponding to the 1.5 % of world energy consumption [1]. The vast majority (99%) of this hydrogen generated by steam reformation of natural gas (methane), which is fossil-based, and all of the CO_2 generated is released to atmosphere. The transportation sector is responsible for 1/3 of green house gas (GHG) emissions from fossil fuel combustion, and hydrogen could play a significant role in reducing GHG emissions by displacing petroleum-based transportation fuels [1]. Vehicles can be powered with hydrogen fuel cells, which are three times more efficient than a gasoline powered engine. About 40 million tonnes of hydrogen per year would be required to fuel about 100 million fuel cell-powered cars after full market penetration.

Biological hydrogen production (biohydrogen) is a very inexpensive and very simple prospective method for renewable hydrogen production, in contrast to electro-chemical and thermo-chemical processes, which are both energy and pressure intensive. Interest in biohydrogen started 20 years ago in parallel to concerns regarding the atmospheric pollution by fossil fuels and climate change debates. As a result, biohydrogen option started to take its place in the agenda of Governments, particularly in Germany, the US and Japan. In this chapter, attempts have been made to highlight not only the advantages but also the bottlenecks of biohydrogen production towards improvements of production and process efficiency under the light of available scientific reports in literature.

THE PROSPECTS OF BIOHYDROGEN

Dark Fermentation

Dark fermentation is a special type of anaerobic digestion comprising only biopolymer hydrolysis and

Address correspondence to Nuri Azbar: Department of Bioegineering, Ege University, Turkey; E-mail: nuriazbar@gmail.com

Nuri Azbar and David B. Levin (Eds)

acidogenesis. It leads to production of hydrogen, carbon dioxide, and some simple organic compounds such as volatile fatty acids (VFAs) and alcohols. The advantages of this technology are that it can produce hydrogen all day long without any constraints of light, a variety of carbon sources including waste organic materials can be used as substrate, and it is anaerobic process so there is no O_2 limitation. Dark fermentation produces valuable metabolites, such as volatile fatty acids (acetic, butyric, lactic acids) and alcohols (ethanol, butanol). The main problem with anaerobic bacteria is that they are incapable of further breaking down the acids into more H_2. The key obstacle appears to be constraints imposed by the metabolic pathways involved. Depending upon the metabolites produced, maximum hydrogen yields vary between 2 moles H_2/mole glucose from facultative anaerobic pathway and 4 moles H_2/mole glucose from strict anaerobic pathway [2]. If the metabolite is propionic acid then consumption of H_2 occurs instead of production [3]. Thus, the highest theoretical yields of H_2 are associated with acetate as the fermentation end-product.

In practice, however, high H_2 yields are associated with a mixture of acetate and butyrate fermentation products, and low H_2 yields are associated with propionate and reduced end-products (alcohols, lactic acid). Clostridium pasteurianum, C. butyricum, and C. beijerinkii are high H_2 producers, while C. propionicum is a poor H_2 producer [4]. Hydrogen production by these bacteria is highly dependent on the process conditions such as pH, hydraulic retention time (HRT), and gas partial pressure, which affect metabolic balance. Thus, fermentation end-products produced by a bacterium depend on the environmental conditions in which it grows. Reduced fermentation end-products like ethanol, butanol, and lactate, contain hydrogen that has not been liberated as gas. To maximize the yield of H_2, the metabolism of the bacterium must be directed away from alcohols (ethanol, butanol) and reduced acids (lactate) towards volatile fatty acids (VFA). C. pasteurianum is a classic H_2 and VFA producer, but its metabolism can be directed away from H_2 production and towards solvent production by high glucose concentrations (12.5% w/v), by CO (which inhibits Fe-hydrogenase), and by limiting Fe concentrations [5]. The partial pressure of H_2 (pH_2) is an extremely important factor for continuous H_2 synthesis. Hydrogen synthesis pathways are sensitive to H_2 concentrations and are subject to end-product inhibition. As H_2 concentrations increase, H_2 synthesis decreases and metabolic pathways shift to production of more reduced substrates such as lactate, ethanol, acetone, butanol, or alanine. As the temperature increases, however, H_2 production is less affected negatively, which suggests the use of thermophilic bioreactors for better H_2 production [6]. Gas sparging has been found to be useful to reduce hydrogen partial pressure for enhancement of hydrogen yield. Thereby 50% increase in H_2 yield was achieved [7].

In practice, H2 yields are below the values mentioned above, and various efforts have been made to increase yields. Several approaches have been taken to maximize yields with existing pathways. For example, Increasing H_2 yields by elimination of reactions competing for reductants either pyruvate or NADH is one possibility. Second option is the elimination of the activity of uptake hydrogenases [8]. Thereby, 10 to 35% increase in hydrogen production was achieved. Another approached proposed is the metabolic engineering of new pathways to overcome thermodynamic/metabolic barriers. A direct artificial pyruvate to hydrogen pathway was constructed using a variation of the approach but with generation of reduced ferredoxin from pyruvate by overexpression of YdbK, an endogenous "cryptic" PFOR (pyruvate:ferredoxin oxidoreductase). High yields (1.8 mole H_2/mole glucose) were achieved with TPP (thiamine pyrophosphate) supplementation [9]. Hallenbeck and Benemann [10] proposed another approach using reverse electron flow to reduce ferredoxin with NADH thus generating enough reducing power to drive hydrogen evolution by hydrogenase, which would require energy input to overcome the thermodynamic barrier. It was suggested that ferredoxin reduction could be achieved by a small amount of respiration generating an electrochemical gradient. Systems that function in this manner in nature are known, although little understood. One such system is the membrane bound *rnf* complex. This has been shown to be present in a number of bacteria where it is thought to function to conserve energy under anaerobic conditions by generating an ion gradient when reduced ferredoxin is used to reduce NAD [11]. Reversed electron flow was reported for some anaerobic bacteria such as *Acidaminococcus fermentans* and *Fusobacterium nucleatum*. In these organisms, an electrochemical Na^+ gradient has been proposed to reverse the action of membrane-bound NADH:ferredoxin oxidoreductases thereby reducing ferredoxin with NADH. Reduced ferredoxin in turn drives hydrogen production. Research in this direction might make possible to adapt this system to driving biohydrogen production by dark fermentation past the present metabolic barrier [11]. Metabolic modifications have been shown to increase hydrogen production.

However, further increases are necessary to compete with conventional hydrogen production methods and this might be possible through the application of various metabolic engineering strategies.

Direct Biophotolysis

In direct biophotolysis, solar energy is used to convert water to oxygen and hydrogen. Sunlight is converted into useful, stored chemical energy *via* photosynthesis. Hydrogen production by the green algae *Chlamydomonas reinhardtii* has been studied extensively in laboratory experiments [12]. The hydrogenase enzyme responsible for evolution of molecular hydrogen is highly sensitive to oxygen and the overall efficiency is too low for practical application [11]. Therefore, photosynthetic production of H_2 and O_2 must be temporally and/or spatially separated. In a two-phase process, CO_2 is first fixed into H_2-rich substrates during normal photosynthesis (Phase 1), followed by light-mediated generation of molecular H_2 when the microalgae are incubated under anaerobic conditions (Phase 2). Phase 2 of the two-stage process can be achieved by incubating the microalgae in medium that does not contain sulfur-containing nutrients [12]. Hydrogen production by direct photolysis using green algae is currently limited by three parameters [13]: (i) solar conversion efficiency of the photosynthetic apparatus; (ii) H_2 synthesis processes (*i.e.* the need to separate the processes of H_2O oxidation from H_2 synthesis); and (iii) bioreactor design and cost. A number of approaches to improve H_2 production by green algae are currently under investigation. These include genetic engineering of light gathering antennae [14], optimization of light input into photobioreactors [15], and improvements to the two-phase H_2 production systems used with green algae [16,17].

Indirect Biophotolysis

In indirect biophotolysis, oxygen evolution and hydrogen evolution are temporally and spatially separated. *Cyanobacteria* are ideal organisms for use in indirect biophotolysis [10,18,19]. During photofermentation, photosynthetic bacteria produce hydrogen through the action of their nitrogenase system. It has been considered that this process is ineffective and it would be simpler and more efficient to extract the hydrogen from organic substrates using a dark fermentation process [10]. Rates of H_2 production by non-nitrogen-fixing cyanobacteria range from 0.02 μmol H_2/mg chl a/h (*Synechococcus PCC 6307*) to 0.40 μmol H_2/mg chl a/h (*Aphanocapsa montana*) [20]. These rates are very low compared with those of heterocystous cyanobacteria, which range from 0.17 μmol H_2/mg chl a/h (*Nostoc linckia IAM M-14*) to 4.2 μmol H_2/mg chl a/h (*Anabaena variabilis IAM M-58*) [10].

Because of the higher rates of H_2 production by Anabaena species and strains, these have been subject to intense study for the past several years. Molecular biology seems to be able provide substantial improvement in production. Mutant strains of A. variabilis have demonstrated significantly higher rates of H_2 production compared with wild-type strains. A. variabilis PK84, for example, produced H_2 at a rate of 6.91 nmol/μg of protein/h (in 350 ml cultures). When A. variabilis PK84 was cultured under conditions of nitrogen starvation, the rate of H_2 synthesis was 12.6 mol/μg of protein/h (in 350 ml cultures). The concentration of total protein in the culture was 28200 μg/ml of culture [21]. Assuming no change as the culture volume is scaled up, a 1 l culture would contain 28,200 μg protein and would produce 355 nmol $H_2/(l \times h)$ or approximately 0.355 mmol $H_2/(l \times h)$.

Hydrogen production *via* indirect photolysis using *cyanobacteria* can be improved by screening for wild-type strains possessing highly active hydrogen evolving enzymes (nitrogenases and/or hydrogenases), in combination with high heterocyst formation [19]. Genetic modification of strains to eliminate uptake hydrogenases and increase levels of bidirectional hydrogenase activity may yield significant increases in H_2 production. For example, a mutant strain of Anabaena (AMC 414), in which the large subunit of the uptake hydrogenase (hupL) was inactivated by a deletion event [22], produced H_2 at a rate that was more than twice that of the parent wild-type strain, *Anabaena* PCC 7120 [23]. Finally, optimization of cultivation conditions such as light intensity, pH, temperature, and nutrient content, as well as maintaining low partial pressures of H_2 and CO_2 will contribute to increased H_2 production.

Photofermentation

Production of hydrogen by photosynthetic bacteria takes place under illumination and in the presence of an inert anaerobic atmosphere (such as argon or helium) from the breakdown of organic substrates such as malate and lactate, preferably. Many of the parameters that limit H_2 production by green algae and cyanobacteria also apply to photoheterotrophic bacteria used in photo-fermentation systems. On the other hand, lack of oxygen evolving activity in photofermentation, which otherwise causes oxygen inactivation problem, could be considered as advantage in comparison to photolysis. Efficiency of phototrophic bacteria is determined by photochemical efficiency, which is 3-10% and photofermentative limitations such as C/N ratios, maximal activity of nitrogenase and minimal activity of hydrogenase. Efforts to improve hydrogen production includes the use of co-cultures, novel photobioreactor designs and the use of specific waste streams as substrates for photo-fermentation. The main challenge for practical application of photobiological hydrogen production is the required scaling up of the system. Constraints related to the need of large surface/volume ratio and associated high costs have to be overcome [1].

Two-Stage Systems

"Two-stage" or "hybrid" biohydrogen systems integrate H_2 production by dark fermentation with hydrogen production by photo-fermentation. Volatile fatty acids synthesized by anaerobic bacteria during dark fermentation reactions are converted to hydrogen by photo-heterotrophic bacteria in the second step, in a photobioreactor. Oh *et al.* [24,25] suggested a two-stage system for hydrogen production using the photo-heterotrophic bacterium *Rhodopseudomonas palustris* P4. In the first step, glucose was fermented in dark conditions to H_2 and acetate. In the second step, acetate was utilized *via* photo-fermentation by *R. palustris* P4. The conversion yield of acetate to H_2 was estimated to be 2.4–2.8 mol H_2/mol acetate, indicating that the overall yield from glucose to H_2 by a two-step process (dark- and photo-fermentation) was increased two-fold compared to that by a dark-only process. However, the low H_2 volumetric production rate in the second step remains a problem for the economic feasibility of the whole process.

Process Configurations and Use of Immobilization Systems

Immobilization of microbes seems to be promising approach for biohydrogen. The specific H_2 production rates achieved with immobilized biomass reactors are not substantially higher than the rates obtained in CSTRs. Also, H_2 yields are similar for both systems. However, owing to the higher biomass concentration, the volumetric H_2 production rate (L hydrogen/day/L reactor) can be increased by up to three times, since theoretically, biomass retention can allow growth of slow-growing bacteria. Immobilization using self-flocculating bacteria or a support medium are also possible [26].

In general, rates of hydrogen production by photo-heterotrophic bacteria are higher when the cells are immobilized in or on a solid matrix, than when the cells are free-living. Continuous cultures of *Rhodopseudomonas capsulata* and *Rhodobacter spheroides* were reported to produce H_2 at rates that range from 40 to 50 ml H_2/l of culture/h, 80 ml to 100 ml H_2/l of culture/h [27,28]. Continuous cultures of *Rhodospirillum rubrum* were reported to produce H_2 at a rate of 180 ml H_2/l of culture/h [29]. Cultures of *Rb. spheroides* immobilized on porous glass, on the other hand, were reported to produce H_2 at a rate of 1.3 l H_2/l of immobilized culture/h [30]. Rates of H_2 production by *Rb. spheroides* GL1 immobilized on activated glass were 3.6–4.0 l H_2/l/h [31-33]. If the system of culturing *Rb. spheroides* on porous glass could be scaled-up without compromising the rate of H_2 synthesis, this would result in rates of 3.6–4.0 l H_2/l of immobilized culture/h.

Engineering Aspects of Using Biohydrogen

Levin *et al.* [19] compared hydrogen production rates of various biohydrogen systems by first standardizing the units of hydrogen production and then by calculating the size of biohydrogen systems that would be required to power proton exchange membrane (PEM) fuel cells of various sizes [32]. In order to assess the potential application of the various biohydrogen systems, Levin *et al.* [19] calculated the size of bioreactors that would be required to supply sufficient H_2 to proton exchange membrane fuel cells (PEMFC) to generate enough electricity to meet the energy demands of a typical house located in the Pacific Northwest of North America (British Columbia, Canada). The choice of a PEMFC is based on the idea that biohydrogen systems might best be used as a means of delivering small, distributed power systems to communities. If a biohydrogen system can deliver enough H_2 to power a PEMFC for 24 h, on a continuous basis, and the fuel cell system can produce enough electricity to supply the electrical demand year round, then the biohydrogen system could have a truly useful, and potentially commercial, application. In Levin's work, upon having described several biohydrogen systems and their rates of H_2 production, they also described several fuel cell systems, their fuel requirements, and their limitations. Following this, they have calculated the size of bioreactors required to supply sufficient H_2 to power PEMFCs (up to 5 kW power or up to 43,800 kWh yearly energy use).

H_2 production rates (mmol $H_2/(l \times h)$) for several biohydrogen systems were recalculated and reported in standardized units by Levin *et al.* [19]. This report reveals the wide range of H_2 synthesis by different biohydrogen systems. Light-dependent biohydrogen systems (direct photolysis, indirect photolysis, and photo-fermentation) all have rates of H_2 synthesis well below 1 mmol $H_2/(l\times h)$. Dark-fermentation systems, all produce H_2 at rates that are well above 1 mmol $H_2/(l\times h)$. The rates of H_2 synthesis by an undefined consortium of thermophilic *Clostridium* and by the extreme thermophilic *Caldicellulosiruptor saccharolyticus* are very similar (8.2 and 8.4 mmol $H_2/(l \times h)$, respectively). A pure strain of mesophilic *Clostridium* demonstrated very good rates of H_2 synthesis with xylose as a substrate (21.0 mmol $H_2/(l\times h)$, and two dark-fermentation systems that utilized undefined consortia of mesophilic bacteria had impressively higher rates of H_2 synthesis (64.5 and 121 mmol $H_2/(l \times h)$, respectively).

Levin *et al.* [19] concluded that photosynthesis-based systems do not produce H_2 at rates that are sufficient to meet the goal of providing enough H_2 to power even a 1 kW PEMFC on a continuous basis. This does not mean that these systems should be abandoned. There may be applications other than our hypothetical objective to which they may be more suited. Moreover, continued research will no doubt result in significant improvements in their respective technologies, and thus in the rates of H_2 production. Thermophilic and extreme thermophilic biohydrogen systems would require bioreactors in the range of approximately 2.9–14.6 m^3 to provide sufficient H_2 to power PEMFCs of 1.5–5.0 kW, and a bioreactor of approximately 5.7 m^3 would be required to power the 5.0 kW fuel cell using the pure culture of mesophilic *Clostridium* sp. The size of bioreactors required for these systems are very large, and thus these systems may be considered impractical for hypothetical application at this time. Some dark-fermentation systems, however, appear promising. Bioreactors of reasonable size would be sufficient to power the 5.0 kW fuel cell using undefined consortia of mesophilic bacteria, enriched for *Clostridium* species. A bioreactor of approximately 0.5 m^3 would provide enough H_2 to power a 2.5 kW PEMFC, while a bioreactor of approximately 1.0 m^3 would provide sufficient H_2 to power a 5.0 kW PEMFC.

While dark-fermentation systems may have practical applications, there are a number of technical challenges that must be considered and overcome before these systems can be used to produce H_2 to power a PEMFC. The most significant of these problems is whether the systems can be scaled up to volumes large enough to generate the required flow rate (22.1 mol H_2/h for the 5.0 kW fuel cell). The working volume of the bioreactor described that produced 121 mmol $H_2/(l\times h)$ was 3 L, and used sucrose (at 20 g/l of culture) as the carbon source for bacterial growth [33]. The biogas produced was 25–35% H_2 and 65–75% CO_2. Further research is required to determine if the rate of H_2 production will remain at high levels if these systems are scaled up to much larger volumes (183 l or more), and if carbon sources other than pure sucrose can be used.

Another major challenge is purification and storage of biohydrogen. Gas produced by dark fermentation systems is a mixture of H_2 and CO_2 (often with H_2 in lower concentration than CO_2), with varying amounts of water and ethanol vapor, at near ambient pressure and temperature. PEM FCs require H_2 with less than 0.1% contaminants, and this technologies to separate the of H_2 from the contaminating gases are required if H_2 generated by dark fermentation is to be useful [32].

BIOHYDROGEN: LOOKING FORWARD

The use of H_2 as a transportation fuel has great potential to displace fossil fuel consumption and drastically reduce greenhouse gas emissions generated by hydrocarbon combustion. There are, however, many technical and economic barriers to the implementation of a Hydrogen Economy [34]. Biohydrogen production offers potential to generate renewable H_2 from inexpensive "waste" feedstocks, but there are many scientific and engineering challenges that must be met by current and future biohydrogen researchers if these technologies are to be technically feasible and economically viable.

First, the rates and yields of H_2 production *via* fermentation are low and must be increased significantly if this technology is to become a viable method for generating usable H_2. The factors that limit hydrogen yield include thermodynamic limitations, product inhibition, presence of branched catabolic pathways, media composition, and the nature of the substrate [35]. A much more comprehensive understanding of the relationships between gene and gene product expression, end-product synthesis patterns, and the factors that regulate carbon and electron balance, within the context of the bioreactor conditions must be achieved if we are to improve molar yields of H_2 during cellulose fermentation. Strategies to increase yields of H_2 production include manipulation of carbon and electron flow *via* end-product inhibition (metabolic shift), metabolic engineering at the genetic level, synergistic co-cultures, and bioprocess engineering and bioreactor designs that maintain a neutral pH during fermentation and ensure rapid removal of H_2 and CO_2 from the aqueous phase [36].

Second, biohydrogen technologies must be integrated with H_2 purification and storage technologies [32,36]. Dark fermentation, the most promising biohydrogen processes (in the near term), produces H_2 and CO_2 in nearly equal molar amounts at near ambient temperatures and pressures. If H_2 generated by dark fermentation (or any biohydrogen system) is to be used as a source of usable energy, it must be purified, concentrated, and stored for future use. Biohydrogen researchers must work with scientists and engineers with specialized knowledge of H_2 purification and storage.

Third, biohydrogen systems must be scaled-up to the pilot plant level to demonstrate that large scale biohydrogen production is possible. The development pilot plant systems will present significant challenges with respect to H_2 production, purification and storage, but these need to be addressed if biohydrogen is to move from being a bench level scientific curiosity to a real alternative source of useable energy.

Finally, for biohydrogen to become an economically viable source of usable energy, a market into which the H_2 would be sold must be identified. The key market for fuel cells has always assumed to be the automotive industry. The great expectation that H_2 fuel cell powered vehicles will displace gasoline and diesel powered vehicles has not materialized for a variety of reasons, but primarily because fuel cell technology has not yet matured and the infrastructure required for hydrogen storage, transportation, and refueling has been slow to develop. Add to this the current economic contraction of the North American automotive industry. Thus, it is likely that the automotive industry will not be the first to fully adopt H_2 technologies. It is now more likely that H_2 fuel cell development will come through niche applications, such as portable power, backup power for telecommunications, and fuel-cell powered fork-lifts, that will slowly build as H_2 storage methods improve, and that many methods of H_2 storage will be required to satisfy a broad range of fuel cells applications. This is analogous to battery technology that has advanced from simple alkaline batteries to a multitude of energy storage systems, each with its own applications, but none supplanting alkaline batteries completely.

CONCLUDING REMARKS

Biohydrogen technologies are still in their infancy. Existing technologies offer potential for practical application, but if biohydrogen systems are to become commercially competitive they must be able to synthesize H_2 at rates that are sufficient to power fuel cells of sufficient size to do practical work. Further research and development aimed at increasing rates of synthesis and final yields of H_2 are essential. Optimization of bioreactor designs, rapid removal and purification of gases, and genetic modification of enzyme pathways that compete with hydrogen producing enzyme systems offer exciting prospects for biohydrogen systems. Even a 10-fold increase in the rate of H_2 synthesis by some dark-fermentation systems would reduce bioreactor size dramatically. This would greatly facilitate overcoming the engineering challenges of scale up, and create new opportunities for practical applications.

REFERENCES

[1] Nath K, Das D. Biohydrogen production as a potential energy resource: Present state-of-art. J Sci Indust Res 2004; 63: 729-38.

[2] Hallenbeck PC. Fundamentals of the fermentative production of hydrogen. Water Sci Technol 2005; 52: 21-9.

[3] Shin H-S, Youn J-H, Kim S-H. Hydrogen production from food waste in anaerobic mesophilic and thermophilic acidogenesis. Int J Hydrogen Energy 2004; 29: 1355-63.

[4] Hawkes FR, Dinsdale R, Hawkes DL, Hussy I. Sustainable fermentative biohydrogen: challenges for process optimization. Int J Hydrogen Energy 2002; 27: 1339-47.

[5] Dabrock B, Bahl H, Gottschalk G. Parameters affecting solvent production by *Clostridium pasteurianum*. Appl Environ Microbiol 1992; 58: 1233-9.

[6] Adams MWW. The metabolism of hydrogen by extremely thermophilic sulphur-dependent bacteria. FEMS Microbiol Rev 1990; 75: 219–38.

[7] Mizuno O, Dinsdale R, Hawkes FR, Hawkes DL, Noike L. Enhancement of hydrogen production from glucose by nitrogen gas sparging. Bioresource Technol 2000; 73: 59-65.

[8] Penfold DW, Forster CF, Macaskie LE. Increased hydrogen production by *Escherichia coli* strain HD701 in comparison with the wild-type parent strain MC4100. Enzyme Microb Tech 2003; 33: 185-9.

[9] Akhtar MK, Jones PR. Construction of a synthetic YdbK-dependent pyruvate: H_2 pathway in *Escherichia coli* BL21 (DE3). Metab Eng 2009; 11: 139-47.

[10] Hallenbeck PC, Benemann JR. Biological hydrogen production; fundamentals and limiting processes. Int J Hydrogen Energy 2002; 1185-93.

[11] Boiangiu C, Jayamani E, Brügel D, *et al.* Sodium ion pumps and hydrogen production in glutamate fermenting anaerobic bacteria. J Mol Microbiol Biotechnol 2005; 10: 105-19.

[12] Kosourov S, Patrusheva E, Ghirardi ML, Seibert M, Tsygankov AA. Comparison of hydrogen photoproduction by sulfur-deprived *Chlamydomonas reinhardtii* under different growth conditions. J Biotechnol 2007; 128: 776-87.

[13] Melis T. Green alga production: process, challenges, and prospects. Int J Hydrogen Energy 2002; 27: 1217-28.

[14] Polle JEW, Kanagari S, Masuda JE, Melis T. Truncated chlorophyll antenna size of the photosystems-a practical method to improve microalgal productivity and hydrogen production in mass culture. Int J Hydrogen Energy 2002; 27: 1257-64.

[15] Gordon J. Tailoring optical systems to optimized photobioreactors. Int J Hydrogen Energy 2002; 1175-84.

[16] Laurinavichene TV, Tolstygina IV, Galiulina RR, Ghirardi ML, Seibert M, Tsygankov AA. Dilution methods to deprive *Chlamydomonas reinhardtii* sulphur for subsequent hydrogen photoproduction. Int J Hydrogen Energy 2002; 27: 1245-50.

[17] Tsygankov AA, Kosourov S, Seibert M, Ghirardi ML. Hydrogen photoproduction under continuous illumination by sulphur-deprived, synchronous *Chlamydomonas reinhardtii* cultures. Int J Hydrogen Energy 2002; 27: 1139-44.

[18] Ghirardi ML, Zhang L, Lee JW, *et. al.* Microalge: a green source of renewable H_2. Trends Biotechnol 2000; 18: 506-11.

[19] Levin DB, Pitt L, Love M. Biohydrogen production: prospects and limitations to practical application. Int J Hydrogen Energy 2004; 29: 173-85.

[20] Ashutosh P, Archana P, Priya S, Anjana P. Using reverse micelles as microreactor for hydrogen production by coupled systems of *Nostoc/R. palustris* and *Anabaena/R. palustris*. World J Microbiol Biotechnol 2007; 23: 269-74.

[21] Howarth DC, Codd GA. The uptake and production of molecular hydrogen by unicellular cyanobacteria. J Gen Microbiol 1985; 131: 1561-9.

[22] Carrasco CD, Buettner JA, Golden JW. Programmed DNA rearrangement of a cyanobacterial hupL gene in heterocysts. Proc Natl Acad Sci USA 1995; 92: 791-5.

[23] Lindblad P, Christensson K, Lindberg P, Federov A, Pinto F, Tsygankov AA. Photoproduction of H_2 by wildtype *Anabaena* PCC 1720 and a hydrogen uptake deficient mutant: from laboratory to outdoor culture. Int J Hydrogen Energy 2002; 27: 1271-81.

[24] Masukawa H, Nakamura K, Mochimaru M, Sakurai H. Photobiohydrogen production and nitrogenase activity in some heterocystous cyanobacteria. BioHydrogen 2001; 2: 63-6.

[25] Sveshnikov DA, Sveshnikov NV, Rao KK, Hall DO. Hydrogen metabolism of mutant forms of *Anabaena variabilis* in continuous cultures and under nutritional stress. FEMS Microbiol Lett 1997; 147: 297-301.

[26] Bartacek, J, Zabranska, J, Lens, P.N.L. Developments and constraints in fermentative hydrogen production. Biofuel Bioprod Bior 2007; 1: 201-14.

[27] Oh Y-K, Seol E-H, Lee EY, Park S. Fermentative hydrogen production by a new chemoheterotrophic bacterium *Rhodopseudomonas palustris* P4. Int J Hydrogen Energy 2002; 27: 1373-9.

[28] Oh Y-K, Seol E-H, Kim MSM-S, Park S. Photoproduction of hydrogen from acetate by a chemoheterotrophic bacterium *Rhodopseudomonas palustris* P4. Int J Hydrogen Energy 2004; 29: 1115-21.

[29] Zurrer H, Bachofen R. Aspects of growth and hydrogen production of the photosynthetic bacterium *Rhodospirillum rubrum* in continuous culture. Biomass 1982; 2: 165-74.

[30] Tsygankov AA, Fedorov AS, Talipova IV, *et al.* Application of immobilized phototrophic bacteria for simultaneous wastewater treatment and hydrogen photoproduction. Appl Biochem Microbiol 1998; 34: 1–5 (in Russian).

[31] Tsygankov AA, Fedorov AS, Talipova IV, Laurinavichene TV, Miyake J, Gogotov IN. Use of immobilized phototrophic microorganisms for wastewater treatment and simultaneous production of hydrogen. Appl Biochem Microbiol (in Russian) 1998; 34: 362-6.

[32] Levin DB, Chahine R. Challenges for renewable hydrogen production from biomass. Int J Hydrogen Energy 2010; 35: 4962-9.

[33] Chang J-S, Lee K-S, Lin P-J. Biohydrogen production with fixed-bed bioreactors. Int J Hydrogen Energy 2002; 27: 1167–74.

[34] Edwards PP, Kuznetsov VL, David WIF, Brandon NP. Hydrogen and fuel cells: Towards a sustainable energy future. Energ Policy 2008; 36(12): 4356-62.

[35] Levin DB, Carere C, Ramachandran U, Rydzak T, Saunders J. Fermentative biofuels: prospects of practical application. In: Muradov N, Veziroglu TE, Eds. Carbon-neutral fuels and energy carriers. Boca Raton, Florida, USA: Taylor & Francis Group Publishers 2011; pp. 601-34.

[36] Levin, D.B., C.R. Carere, N. Cicek, R. Sparling. Challenges for biohydrogen production *via* direct lignocellulose fermentation. Int J Hydrogen Energy 2009; 34: 7390-403.

INDEX

A

B

C

D

E

F

G

H

I

J

K

L

M

N

O

P

R

S

T

U

V

W

Y

Made in United States
Orlando, FL
04 November 2026

90200453R00167